COMPUTATIONAL INTELLIGENCE PARADIGMS

THEORY AND APPLICATIONS USING MATLAB®

COMPUTATIONAL
INTELLIGENCE
PARADIGMS

THEORY AND
APPLICATIONS
USING MATLAB®

S. SUMATHI
SUREKHA P.

CRC Press
Taylor & Francis Group
Boca Raton London New York

CRC Press is an imprint of the
Taylor & Francis Group, an **informa** business

A CHAPMAN & HALL BOOK

Contents

Preface

Computational intelligence is a well-established paradigm, where new theories with a sound biological understanding have been evolving. Defining computational intelligence is not an easy task. In a nutshell, which becomes quite apparent in light of the current research pursuits, the area is heterogeneous with a combination of such technologies as neural networks, fuzzy systems, rough set, evolutionary computation, swarm intelligence, probabilistic reasoning, multi-agent systems, etc. Just like people, neural networks learn from experience, not from programming.

Neural networks are good at pattern recognition, generalization, and trend prediction. They are fast, tolerant of imperfect data, and do not need formulas or rules. Fuzzy logic in the narrow sense is a promising new chapter of formal logic whose basic ideas were formulated by Lotfi Zadeh. The aim of this theory is to formalize the "approximate reasoning" we use in everyday life, the object of investigation being the human aptitude to manage vague properties. This work is intended to help, provide basic information, and serve as a first step for individuals who are stranded in the mind-boggling universe of evolutionary computation (EC). Over the past years, global optimization algorithms imitating certain principles of nature have proved their usefulness in various domains of applications.

About This Book

The aim of this book is to furnish some theoretical concepts and to sketch a general framework for computational intelligence paradigms such as artificial neural networks, fuzzy systems, evolutionary computation, genetic algorithms, genetic programming, and swarm intelligence. The book includes a large number of intelligent computing methodologies and algorithms employed in computational intelligence research. The book also offers a set of solved programming examples related to computational intelligence paradigms using MATLAB software. Additionally, such examples can be repeated under the same conditions, using different data sets. Researchers, academicians, and students in computational intelligence can use this book to verify their ideas related to evolution dynamics, self-organization, natural and artificial morphogenesis, emergent collective behaviors, swarm intelligence, evolutionary strategies, genetic programming, and evolution of social behaviors.

Salient Features

The salient features of this book include

- Detailed description on computational intelligence (CI) paradigms

- Worked out examples on neural networks, fuzzy systems, hybrid neuro fuzzy systems, evolutionary computation, genetic algorithms, genetic programming, and swarm intelligence using MATLAB software

- MATLAB toolboxes and their functions for neural networks, fuzzy logic, genetic algorithms, genetic programming, evolutionary algorithms, and swarm optimization

- Research projects, commercial emerging software packages, abbreviations and glossary of terms related to CI

Organization of the Book

Chapter 1 describes the paradigms of computational intelligence (CI), problem classes of CI, and an introductory explanation of neural networks, fuzzy systems, evolutionary computing, and swarm intelligence.

Chapter 2 provides an understanding of the basic neural networks, the historical overview of neural networks, the components of a neural network, and implementation of an electronic neural network.

Chapter 3 outlines some of the most common artificial neural networks based on their major class of application. The categories are not meant to be exclusive, they are merely meant to separate out some of the confusion over network architectures and their best matches to specific applications.

Chapter 4 discusses the major class of neural network based on applications such as data classification, data association, and data conceptualization. Implementations of these networks using MATLAB Neural Network Toolbox are also described.

Chapter 5 presents a set of common neural network real world applications. A few MATLAB simulated examples such as Coin detection, Pattern Recall, Pattern Classification, and Simulink models using different Neural Network architectures are illustrated in this chapter.

Chapter 6 discusses the basic fuzzy sets, operations on fuzzy sets, relations between fuzzy sets, composition, and fuzzy arithmetic. A few MATLAB programs are also illustrated on topics such as membership functions, fuzzy operations, fuzzy arithmetic, relations, and composition.

Chapter 7 focuses on fuzzy rules such as Mamdani fuzzy rules and Takagi-Sugeno fuzzy rules, expert system modeling, fuzzy controllers, and implementation of fuzzy controllers in MATLAB.

Chapter 8 illustrates some applications of fuzzy systems such as Fuzzy Washing Machine, Fuzzy Control System, and approximation of sinusoidal functions in MATLAB.

Chapter 9 concentrates on the different types of fused neuro-fuzzy systems such as FALCON, ANFIS, GARIC, NEFCON, FINEST, FUN, EFuNN, and SONFIN. A detailed description of ANFIS including its architecture and learning algorithm are discussed. The implementation detail of hybrid neuro-fuzzy model is also delineated. An explanation on Classification and Regression trees with its computational issues, computational details, computational formulas, advantages, and examples is given in this chapter. The data clustering algorithms such as hard c-means, Fuzzy c-means, and subtractive clustering are also described.

In Chapter 10, MATLAB illustrations are given on ANFIS, Classification and Regression trees, Fuzzy c-means clustering algorithms, Fuzzy ART Map, and Simulink models on Takagi-Sugeno inference systems.

Chapter 11 depicts a brief history of evolutionary computation (EC). This chapter enlightens the paradigms of EC such as Evolutionary Strategies and Evolutionary Programming. Genetic Algorithms and Genetic Programming will be discussed elaborately in the next chapter. This chapter also describes the advantages and disadvantages of EC.

Solved MATLAB programs are given in Chapter 12 to illustrate the implementation of evolutionary computation in problems such as optimization, proportional-derivative controller, multiobjective optimization, and minimization of functions.

Chapter 13 furnishes a detailed description of genetic algorithm, its operators and parameters are discussed. Further, the schema theorem and technical background along with the different types of GA are also elaborated in detail. Finally MATLAB codes are given for applications such as maximization of a given function, traveling sales man problem, and economic dispatch problem using genetic algorithm.

In Chapter 14, a brief history of genetic programming is discussed. To get an idea about programming a basic introduction to Lisp Programming Language is dealt. The basic operations of GP are discussed along with illustrations and MATLAB routines. The steps of GP are also explained along with a flow chart.

Chapter 15 gives the basic definition of swarms, followed by a description on Swarm Robots. The biological models, characterizations of stability, and overview of stability analysis of Swarms are also elaborated in this chapter. The chapter deals with the taxonomy, properties, studies, and applications of swarm intelligence. The variants of SI such as Particle Swarm Optimization (PSO) and Ant Colony Algorithms for Optimization Problems are discussed. A few applications of Particle Swarm Optimization such as Job Scheduling on Computational Grids and Data Mining and a few applications of Ant Colony Optimization such as Traveling Salesman Problem (TSP), Quadratic Assignment Problem (QAP) and Data Mining and their implementation in MATLAB are explained in this chapter.

About the Authors

Dr. S. Sumathi, born on January 31, 1968, completed a B.E. degree in Electronics and Communication Engineering and a Masters degree in Applied Electronics at Government College of Technology, Coimbatore, Tamil Nadu. She earned a Ph.D. degree in the area of Data Mining and is currently working as assistant professor in the Department of Electrical and Electronics Engineering, PSG College of Technology, Coimbatore with teaching and research experience of 16 years.

Dr. Sumathi received the prestigious Gold Medal from the Institution of Engineers Journal Computer Engineering Division for the research paper titled, "Development of New Soft Computing Models for Data Mining" 2002 to 2003 and also best project award for UG Technical Report titled, "Self Organized Neural Network Schemes: As a Data mining Tool," 1999. She received the Dr. R. Sundramoorthy award for Outstanding Academic of PSG College of Technology in the year 2006. The author has guided a project which received Best M.Tech Thesis award from the Indian Society for Technical Education, New Delhi.

In appreciation for publishing various technical articles Dr. Sumathi has received national and international journal publication awards - 2000 to 2003. She prepared manuals for the Electronics and Instrumentation Lab and Electrical and Electronics Lab of the EEE Department, PSG College of Technology, Coimbatore and also organized the second National Conference on Intelligent and Efficient Electrical Systems in the year 2005 and conducted short term courses on "Neuro Fuzzy System Principles and Data Mining Applications," November 2001 and 2003. Dr. Sumathi has published several research articles in national and international journals/conferences and guided many UG and PG projects. She has also published books on, *Introduction to Neural Networks with MATLAB", Introduction to Fuzzy Systems with MATLAB, Introduction to Data Mining and its Applications, LabVIEW Based Advanced Instrumentation Systems, and Evolutionary Intelligence: An Introduction to Theory and Applications Using MATLAB.* She has reviewed papers in national/international journals and conferences. Her research interests include neural networks, fuzzy systems and genetic algorithms, pattern recognition and classification, data warehousing and data mining, operating systems and parallel computing.

Ms. Surekha. P born on May 1, 1980 has completed her B.E. degree in

Electrical and Electronics Engineering in PARK College of Engineering and Technology, Coimbatore, Tamil Nadu, and Masters degree in Control Systems at PSG College of Technology, Coimbatore, Tamil Nadu. Her current research work includes computational intelligence. She was a rank holder in both B.E. and M.E. degree programs. She received an alumni award for best performance in curricular and co-curricular activities during her Master's degree program. She has presented papers in national conferences and Journals. She has also published books such as, *LabVIEW Based Advanced Instrumentation Systems* and *Evolutionary Intelligence: An Introduction to Theory and Applications Using MATLAB.* She is presently working as a lecturer in Adhiyamaan College of Engineering, Hosur, Tamil Nadu. Her research areas include Robotics, Virtual Instrumentation, Mobile Communication, and Computational Intelligence.

Acknowledgment

The authors are always thankful to the Almighty for perseverance and achievements.

The authors owe their gratitude to Mr. G. Rangaswamy, Managing Trustee, PSG Institutions, and Dr. R. Rudramoorthy, Principal, PSG College of Technology, Coimbatore, for their whole-hearted cooperation and great encouragement given in this successful endeavor.

Dr. S. Sumathi owes much to her daughter S. Priyanka, who has helped a lot in monopolizing her time on book work and substantially realized the responsibility. She feels happy and proud for the steel frame support rendered by her husband, Mr. Sai Vadivel. Dr. S. Sumathi would like to extend wholehearted thanks to her parents who have reduced the family commitments and given their constant support. She is greatly thankful to her brother, Mr. M. S. Karthikeyan, who has always been "Stimulator" for her progress. She is pertinent in thanking parents-in-law for their great moral support.

Ms. Surekha. P. would like to thank her husband, Mr. A. Srinivasan, for supporting her throughout the development of this book in all the efforts taken. She extends her sincere thanks to her parents, Mr. S. Paneerselvam, Mrs. P. Rajeswari, and brother, Mr. Siva Paneerselvam, who shouldered a lot of extra responsibilities during the months this was being written. They did this with the long term vision, depth of character, and positive outlook that are truly befitting of their name.

The authors wish to thank all their friends and colleagues who have been with them in all their endeavors with their excellent, unforgettable help and assistance in the successful execution of the work.

Chapter 1

Computational Intelligence

1.1 Introduction

Computational Intelligence (CI) is a successor of artificial intelligence. CI relies on heuristic algorithms such as in fuzzy systems, neural networks, and evolutionary computation. In addition, computational intelligence also embraces techniques that use Swarm intelligence, Fractals and Chaos Theory, Artificial immune systems, Wavelets, etc. Computational intelligence is a combination of learning, adaptation, and evolution used to intelligent and innovative applications. Computational intelligence research does not reject statistical methods, but often gives a complementary view of the implementation of these methods. Computational intelligence is closely associated with soft computing a combination of artificial neural networks, fuzzy logic and genetic algorithms, connectionist systems such as artificial intelligence, and cybernetics.

CI experts mainly consider the biological inspirations from nature for implementations, but even if biology is extended to include all psychological and evolutionary inspirations then CI includes only the neural, fuzzy, and evolutionary algorithms. The Bayesian foundations of learning, probabilistic and possibilistic reasoning, Markovian chains, belief networks, and graphical theory have no biological connections. Therefore genetic algorithms is the only solution to solve optimization problems. CI studies problems for which there are no effective algorithms, either because it is not possible to formulate them or because they are complex and thus not effective in real life applications. Thus the broad definition is given by: computational intelligence is a branch of computer science studying problems for which there are no effective computational algorithms. Biological organisms solve such problems every day: extracting meaning from perception, understanding language, solving ill-defined computational vision problems thanks to evolutionary adaptation of the brain to the environment, surviving in a hostile environment. However, such problems may be solved in different ways. Defining computational intelligence by the problems that the field studies there is no need to restrict the types of methods used for solution.

The exploration of CI is concerned with subordinate cognitive functions: perceptual experience, object identification, signal analysis, breakthrough of structures in data, simple associations, and control. Solution for this type of problems can be obtained using supervised and unsupervised ascertaining, not only neural, fuzzy, and evolutionary overtures but also probabilistic and statistical overtures, such as Bayesian electronic networks or kernel methods. These methods are used to solve the same type of problems in various fields such as pattern recognition, signal processing, classification and regression, data mining. Higher-level cognitive functions are required to solve non-algorithmizable problems involving organized thinking, logical thinking, complex delegacy of knowledge, episodic memory, projecting, realizing of symbolic knowledge.

These jobs are at present puzzled out by AI community using techniques based on search, symbolic cognition representation, logical thinking with frame-based expert systems, machine learning in symbolic domains, logics, and lingual methods. Although the jobs belong to the class of non-algorithmic problems, there is a slight overlap between jobs solved using low and high-level cognitive functions. From this aspect AI is a part of CI concentrating on problems that require higher cognition and at present are more comfortable to solve using symbolic knowledge representation. It is possible that other CI methods will also find applications to these problems in the future. The main overlap areas between low and high-level cognitive functions are in sequence learning, reinforcement learning, machine learning, and distributed multi-agent systems. All tasks that require rule based reasoning based on perceptions, such as robotics and automation, machine-driven automobile, etc., require methods for solving both low and high-level cognitive problems and thus are a raw meeting ground for AI experts with the rest of CI community.

"Symbol manipulation is the source for all intelligence" - this idea was proposed by Newell and Simon and they declared that the theory of intelligence was about physical symbols rather than symbolic variables. Symbolic representation of physical are based on multi-dimensional approach patterns comprising states of the brain. Representative models of brain processes do not offer precise approximation of any problem that is described by continuous variables rather than symbolic variables. Estimations to brain processes should be done at a proper level to obtain similar functions. Symbolic dynamics may provide useful information on dynamical systems, and may be useful in modeling transition between low to high level processes. The division between low, and high level cognitive functions is only a rough approximation to the processes in the brain. Incarnated cognition has been intensively studied in the last decade, and developmental estimates depicting how higher processes emerge for the lower ones have been adopted by robotics.

In philology it is admitted that real meaning of linguistic terms comes from body-based metaphors and is equivalently true in mathematics also. New CI methods that go beyond pattern identification and help to solve AI problems may eventually be developed, starting from distributed knowledge representation, graphical methods, and activations networks. The dynamics of such models will probably allow for reasonable symbolic approximations. It is instructive to think about the spectrum of CI problems and various approximations needed to solve them. Neural network models are inspired by brain processes and structures at almost the lowest level, while symbolic AI models by processes at the highest level. The brain has a very specific modular and hierarchical structure, it is not a huge neural network. Perceptron model of a neuron has only one internal parameter, the firing threshold, and a few synaptic weights that determine neuron-neuron interactions. Individual neurons credibly act upon brain information processing in an undistinguished manner.

The basic processors used for neural network modelling include bigger neural structures, such as microcircuits or neural cell gatherings. These structures have more complex domestic states and more complex interactions between elements. An electronic network from networks, hiding the complexness of its processors in a hierarchical way, with different aborning properties at each level, will get increasingly more inner knowledge and additional complex interactions with other such systems. At the highest-level models of whole brains with a countless number of potential internal states and very complex interactions may be obtained. Computational intelligence is certainly more than just the study of the design of intelligent agents; it includes also study of all non-algoritmizable processes that humans (and sometimes animals) can solve with various degrees of competence. CI should not be dealt as a bag of tricks without deeper basis. Challenges from good numerical approaches in various applications should be solicited, and knowledge and search-based methods should complement the core CI techniques in problems involving intelligence. Goldberg and Harik view computational intelligence more as a way of thinking about problems, calling for a "broader view of the scope of the discipline."

They have examined the restrictions to build up computational design, finding the exemplars of human behaviors to be most useful. Although this is surely worthy delineating the problems that CI wants to solve and welcoming all methods that can be used in such solutions, independent of their inspirations, is even more important.

Arguably, CI comprises of those paradigms in AI that relate to some kind of biological or naturally occurring system. General consensus sug-

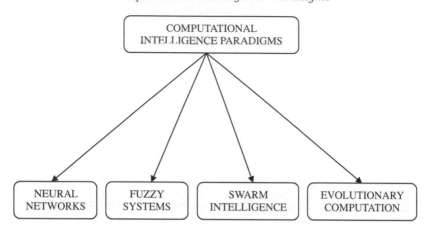

FIGURE 1.1: Computational Intelligence Paradigms

gests that these paradigms are neural networks, evolutionary computing, swarm intelligence, and fuzzy systems. Neural networks are based on their biological counterparts in the human nervous system. Similarly, evolutionary computing draws heavily on the principles of Darwinian evolution observed in nature. Swarm intelligence, in turn, is modeled on the social behavior of insects and the choreography of birds flocking. Finally, human reasoning using imprecise, or fuzzy, linguistic terms is approximated by fuzzy systems. This chapter describes these paradigms of CI briefly. Following the discussion, other paradigms of CI such as Granular Computing, Chaos Theory, and Artificial Immune Systems are also dealt. Hybrid approaches of CI and the challenges that CI faces are elaborated in this chapter.

Figure 1.1 shows these four primary branches of CI and illustrates that hybrids between the various paradigms are possible. More precisely, CI is described as the study of adaptive mechanisms to enable or facilitate intelligent behavior in complex and changing environments. There are also other AI approaches, that satisfy both this definition as well as the requirement of modeling some naturally occurring phenomenon, that do not fall neatly into one of the paradigms mentioned thus far. A more pragmatic approach might be to specify the classes of problems that are of interest without being too concerned about whether or not the solutions to these problems satisfy any constraints implied by a particular definition for CI.

The following section identifies and describes four primary problem classes for CI techniques. A compendious overview of the main concepts behind each of the widely recognized CI paradigms is presented in this chapter.

1.2 Primary Classes of Problems for CI Techniques

Optimization, defined in the following section is undoubtedly the most important class of problem in CI research, since virtually any other class of problem can be re-framed as an optimization problem. This transformation, particularly in a software context, may lead to a loss of information inherent to the intrinsic form of the problem. The major classes of problems in CI are grouped into five categories as Control Problems, Optimization Problems, Classification Problems, Regression Problems, and NP Complete Problems. In the following sections Optimization and NP Complete Problems are discussed.

1.2.1 Optimization

An optimization problem can be represented in the following way

Given: a function $f : A$ R from some set A to the real numbers

Sought: an element x_0 in A such that $f(x_0) \leq f(x)$ for all x in A ("minimization") or such that $f(x_0) \geq f(x)$ for all x in A ("maximization").

This conceptualization is addressed as an optimization problem or a numerical programming problem. Several real life and theoretical problems perhaps may be modeled in this comprehensive framework. Problems phrased using this technique in the fields of physical science and technology may refer to the technique as energy minimization, the function f as comprising the energy of the system being simulated. Typically, A is some subset of the Euclidean space Rn, often specified by a set of constraints, equalities or inequalities that the members of A have to satisfy. The domain A of f is called the search space, while the elements of A are called candidate solutions or feasible solutions. The function f is called an objective function, or cost function. A feasible solution that minimizes (or maximizes, if that is the goal) the objective function is called an optimal solution.

Generally, when the feasible region or the objective function of the problem does not present convexity, there may be several local minima and maxima, where a local minimum x* is defined as a point for which there exists some $\delta > 0$ so that for all x such that

$$|| x - x* || \leq \delta$$

the expression

$$f(x*) \leq f(x)$$

holds; that is to say, on some region around x* all of the function values are greater than or equal to the value at that point. Local maxima are also outlined similarly. An expectant routine of algorithms are nominated for solving non-convex problems - including the majority of commercially accessible solvers that are not capable of making a distinction between local optimum results and stringent optimal solutions, and will treat the former as actual solutions to the original problem. The branch of applied mathematics and numerical analysis that is concerned with the development of deterministic algorithms that are capable of guaranteeing convergence in finite time to the actual optimal solution of a non-convex problem is called global optimization.

1.2.2   NP-Complete Problems

Non-Deterministic Polynomial Time (NP) problems are one of the most common underlying complexity assorts In computational theory. Intuitively, NP contains all determination problems for which the solutions have mere proofs of the fact that the answer is indeed 'yes'. More precisely, these validations have to be confirmable in polynomial time by a deterministic Turing machine. In an equivalent elegant definition, NP is the set of determination problems solvable in polynomial time by a non-deterministic Turing machine. The complexity class P is contained in NP, but NP contains many important problems, called NP-complete problems, for which no polynomial-time algorithms are known. The most important open question in complexity theory, the P = NP problem, asks whether such algorithms actually exist for NP-complete problems. It is widely believed that this is not the case. Several biological computer science problems are covered by the class NP. In specific, the decision versions of many interesting search problems and optimization problems are contained in NP.

Verifier-Based Definition

In order to explain the verifier-based definition of NP, let us consider the subset sum problem: Assume that we are given some integers, such as $-7, -3, -2, 5, 8$, and we wish to know whether some of these integers sum up to zero. In this example, the answer is "yes", since the subset of integers $-3, -2, 5$ corresponds to the sum $(-3) + (-2) + 5 = 0$. The task of deciding whether such a subset with sum zero exists is called the subset sum problem.

As the number of integers that we feed into the algorithm becomes larger, the time needed to compute the answer to the subset sum problem grows exponentially, and in fact the subset sum problem is NP-complete. However, notice that, if we are given a particular subset (often called a certificate,) we can easily check or verify whether the subset sum is zero, by just summing up the integers of the subset. So if the sum is indeed zero, that particular subset is the proof or witness for the fact that the answer is "yes". An algorithm that verifies whether a given subset has sum zero is called verifier. A problem is said to be in NP if there exists a verifier for the problem that executes in polynomial time. In case of the subset sum problem, the verifier needs only polynomial time, for which reason the subset sum problem is in NP.

The verifier-based definition of NP does not require an easy-to-verify certificate for the "no"-answers. The class of problems with such certificates for the "no"-answers is called co-NP. In fact, it is an open question whether all problems in NP also have certificates for the "no"-answers and thus are in co-NP.

Machine Definition

Equivalent to the verifier-based definition is the following characterization: NP is the set of decision problems solvable in polynomial time by a non-deterministic Turing machine.

A list of examples that are classified as NP complete are

- Integer factorization problem

- Graph isomorphism problem

- Traveling salesman problem

- Boolean satisfiability problem

1.3 Neural Networks

The term neural network refers to a network or circuit of biological neurons. Nowadays, the modernization of the term neural network is referred to as artificial neural network. The networks are composed of artificial neurons or nodes. Biological neural networks are made up of real biological neurons that are connected or functionally-related in the peripheral nervous system or the central nervous system. In the field of neuroscience, they are often identified as groups of neurons that perform a specific physiological function in laboratory analysis. Artificial neural networks are made up of interconnected artificial neurons (programming constructs that mimic the properties of biological neurons). Artificial neural networks may either be used to gain an understanding of biological neural networks, or for solving artificial intelligence problems without necessarily creating a model of a real biological system.

1.3.1   Feed Forward Neural Networks

An artificial neural network in which the connections between the units do not form a directed cycle is referred to as a feedforward neural network. The feedforward neural network is the first and arguably simplest type of artificial neural network organized. In this network, the information moves in only one direction, forward, from the input nodes, through the hidden nodes (if any) and to the output nodes. There are no cycles or loops in the network.

Figure 1.2 illustrates a fully connected three layer network. The layers consist of neurons, which compute a function of their inputs and pass the result to the neurons in the next layer. In this manner, the input signal is fed forward from one layer to the succeeding layer through the

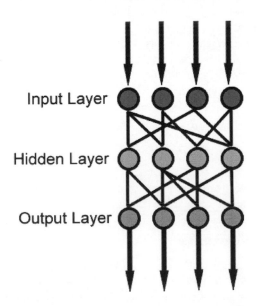

FIGURE 1.2:   Three Layer Feed Forward Neural Network

network. The output of a given neuron is characterized by a nonlinear activation function, a weighted combination of the incoming signals, and a threshold value. The threshold can be replaced by augmenting the weight vector to include the input from a constant bias unit. By varying the weight values of the links, the overall function, which the network realizes, can be altered.

There are other types of Neural networks based on both supervised and unsupervised learning which are discussed in detail with their algorithms in Chapter 2 of this book.

1.4 Fuzzy Systems

Fuzzy sets originated in the year 1965 and this concept was proposed by Lofti A.Zadeh. Since then it has grown and is found in several application areas. According to Zadeh, The notion of a fuzzy set provides a convenient point of departure for the construction of a conceptual frame work which parallels in many respects of the framework used in the case of ordinary sets, but is more general than the latter and potentially, may prove to have a much wider scope of applicability, specifically in the fields of pattern classification and information processing." Fuzzy logics are multi-valued logics that form a suitable basis for logical systems reasoning under uncertainty or vagueness that allows intermediate values to be defined between conventional evaluations like true/false, yes/no, high/low, etc. These evaluations can be formulated mathematically and processed by computers, in order to apply a more human-like way of thinking in the programming of computers. Fuzzy logic provides an inference morphology that enables approximate human reasoning capabilities to be applied to knowledge-based systems. The theory of fuzzy logic provides a mathematical strength to capture the uncertainties associated with human cognitive processes, such as thinking and reasoning. Fuzzy systems are suitable for uncertain or approximate reasoning, especially for the system with a mathematical model that is difficult to derive. Fuzzy logic allows decision making with estimated values under incomplete or uncertain information.

1.4.1 Fuzzy Sets

The concept of a fuzzy set is an extension of the concept of a crisp set. Similar to a crisp set a universe set U is defined by its membership function from U to [0,1]. Consider U to be a non-empty set also known

as the universal set or universe of discourse or domain. A fuzzy set on
U is defined as

$$\mu_A(x) : U \rightarrow [0, 1] \qquad (1.1)$$

Here μ_A is known as the membership function, and $\mu_A(x)$ is known
as the membership grade of x. Membership function is the degree of
truth or degree of compatibility. The membership function is the crucial
component of a fuzzy set. Therefore all the operations on fuzzy sets are
defined based on their membership functions. Table 1.1 defines fuzzy set
theoretic operators that are analogues for their traditional set counterparts.

TABLE 1.1:   Fuzzy Set Theoretic Operators

Operator	Definition
Intersection	The membership function $\mu_{\overline{C}}(x)$ of the inter-section $\overline{C} = \overline{A} \cap \overline{B}$ is defined by $\mu_{\overline{C}}(x) = \min\{\mu_{\overline{A}}(x), \mu_{\overline{B}}(x)\}, x \in X$
Union	The membership function $\mu_{\overline{C}}(x)$ of the union $\overline{C} = \overline{A} \cup \overline{B}$ is defined by $\mu_{\overline{C}}(x) = \max\{\mu_{\overline{A}}(x), \mu_{\overline{B}}(x)\}, x \in X$
Complement	The membership function $\mu_{\overline{C}}(x)$ of the comple-ment of a normalized fuzzy set $\overline{A}$ is defined by $\mu_{\overline{C}}(x) = 1 - \mu_{\overline{A}}(x), x \in X$

Fuzzy relations play a vital role in fuzzy logic and its applications.
Fuzzy relations are fuzzy subsets of X × Y defined as the mapping from
X $\longrightarrow$ Y. If X ,Y $\subseteq$ R are universal sets then the fuzzy relation is defined
as $\overline{R}\{((x,y), \mu_{\overline{R}}(x,y)|(x,y)\epsilon X \times Y\}$. These fuzzy relations in different
product spaces can be combined with each other through the operation
composition. There are two basic forms of composition, namely the min-
max and the max-product composition.

Min-Max Composition: Let $\overline{R}_1 \epsilon X \times Y$ and $\overline{R}_2 \epsilon Y \times Z$ denote two
fuzzy relations, the min-max composition on these relations is defined
as $\overline{R}_1 \circ \overline{R}_2 = max\{min\{\mu_{\overline{R}_1}(x,y), \mu \overline{R}_2(y,z)\}\} x \epsilon X, y \epsilon Y, z \epsilon Z$ Max Product Composition: Let $\overline{R}_1 \epsilon X \times Y$ and $\overline{R}_2 \epsilon Y \times Z$ denote two fuzzy re-
lations, the max product composition on these relations is defined as
$\overline{R}_1 \circ \overline{R}_2 = max\{min\{\mu_{\overline{R}_1}(x,y) * \mu \overline{R}_2(y,z)\}\} x \epsilon X, y \epsilon Y, z \epsilon Z$ The membership of a fuzzy set is described by this membership function $\mu_A(x)$ of
A, which associates to each element $x_o \epsilon$ X a grade of membership $\mu_A(x_o)$
The notation for the membership function $\mu_A(x)$ of a fuzzy set A A :

X $\longrightarrow$ [0,1]. Each fuzzy set is completely and uniquely defined by one particular membership function. Consequently symbols of membership functions are also used as labels of the associated fuzzy sets.

1.4.2 Fuzzy Controllers

Fuzzy logic controllers are based on the combination of Fuzzy set theory and fuzzy logic. Systems are controlled by fuzzy logic controllers based on rules instead of equations. This collection of rules is known as the rule base usually in the form of IF-THEN-ELSE statements. Here the IF part is known as the antecedent and the THEN part is the consequent. The antecedents are connected with simple Boolean functions like AND,OR, NOT etc., Figure 1.3 outlines a simple architecture for a fuzzy logic controller.

The outputs from a system are converted into a suitable form by the fuzzification block. Once all the rules have been defined based on the application, the control process starts with the computation of the rule consequences. The computation of the rule consequences takes place within the computational unit. Finally, the fuzzy set is defuzzified into one crisp control action using the defuzzification module. The decision parameters of the fuzzy logic controller are as follows:

Input Unit**:** Factors to be considered are the number of input signals and scaling of the input signal.

Fuzzification Unit: The number of membership functions, type of membership functions are to be considered.

Rule Base**:** The total number of rules, type of rule structure (Mamdani or Takagi), rule weights etc. are to be considered.

Defuzzification Unit**:** Type of defuzzification procedure is to be considered.

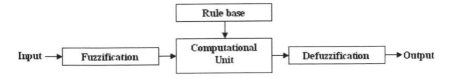

FIGURE 1.3: Fuzzy Controller Architecture

Some of the important characteristics of fuzzy logic controllers are

- - User defined rules, which can be changed according to the application. - Robust - Can be applied to control non linear systems also - Simple design - Overall cost and complexity is low

- User defined rules, which can be changed according to the application.

- Robust

- Can be applied to control non linear systems also

- Simple design

- Overall cost and complexity is low

1.5   Evolutionary Computing

Algorithms formed on the basis of evolution of biological life in the natural world to solve several computational issues is termed as evolutionary computation. Evolutionary Computing (EC) is strongly based on the principles of natural evolution. A population of individuals is initialized within the search space of an optimization problem so that P (t) = $\{x_i(t)$ S $|1 \leq i \leq \mu \}$. The search space S may be the genotype or phenotype depending on the particular evolutionary approach being utilized. The fitness function f, which is the function being optimised, is used to evaluate the goodness individuals so that F (t) = $\{f(x_i(t))$ ϵ R $|1 \leq i \leq \mu\}$. Obviously, the fitness function will also need to incorporate the necessary phenotype mapping if the genotype space is being searched. Searching involves performing recombination of individuals to form offspring, random mutations and selection of the following generation until a solution emerges in the population. The parameters, p_r, p_m, and p_s are the probabilities of applying the recombination, mutation, and selection operators respectively. Recombination involves mixing the characteristics of two or more parents to form offspring in the hope that the best qualities of the parents are preserved. Mutations, in turn, introduce variation into the population thereby widening the search. In general, the recombination and mutation operators may be identity transforms so that it is possible for individuals to survive into the following generation unperturbed. Finally, the new or modified individuals are re-evaluated before the selection operator is used to pare the population back down to a size of . The selection operator provides evolutionary pressure such that the most fit individuals survive into the next generation. While selection is largely based on the fitness of individuals, it is probabilistic to prevent premature convergence of the population.

Three methodologies that have emerged in the last few decades such as: "evolutionary programming (EP)" (L.J. Fogel, A.J. Owens, M.J. Walsh Fogel, 1966), "evolution strategies (ES)" (I. Rechenberg and H.P. Schwefel Rechenberg, 1973), and "genetic algorithms (GA)" (Holland, 1975) are discussed in this section. Although similar at the highest level, each of these varieties implements an evolutionary algorithm in a different manner. The differences touch upon almost all aspects of evolutionary algorithms, including the choices of representation for the individual structures, types of selection mechanism used, forms of genetic operators, and measures of performance which are explained in Chapter 5 of this book.

1.5.1   Genetic Algorithms

Genetic Algorithms (GAs) are adaptive heuristic search algorithms introduced on the evolutionary themes of natural selection. The fundamental concept of the GA design is to model processes in a natural system that is required for evolution, specifically those that follow the principles posed by Charles Darwin to find the survival of the fittest. GAs constitute an intelligent development of a random search within a defined search space to solve a problem. GAs were first pioneered by John Holland in the 1960s, and have been widely studied, experimented, and applied in numerous engineering disciplines. Not only does GA provide alternative methods to solving problems, it consistently outperforms other traditional methods in most of the problem's link. Many of the real world problems involved finding optimal parameters, which could prove difficult for traditional methods but ideal for GAs. In fact, there are many ways to view genetic algorithms other than an optimization tool. Perhaps most users concern GAs as a problem solver, but this is a restrictive view.

GAs are used as problem solvers, as a challenging technical puzzle, as a basis for competent machine learning, as a computational model of innovation and creativity, as a computational model of other innovating systems, and for guiding philosophy.

GAs were introduced as a computational analogy of adaptive systems. They are modeled loosely on the principles of the evolution through natural selection, employing a population of individuals that undergo selection in the presence of variation-inducing operators such as mutation and recombination (crossover). A fitness function is used to evaluate individuals, and reproductive success varies with fitness.

The Algorithm

1. Randomly generate an initial population

2. Compute and save the fitness for each individual in the current population

3. Define selection probabilities for each individual in so that it is proportional to the fitness

4. Generate the next population by probabilistically selecting individuals from current population to produce offspring via genetic operators

5. Repeat step 2 until satisfying solution is obtained

The paradigm of GAs described above is usually the one applied to solving most of the problems presented to GAs. Though it might not find the best solution more often than not, it would come up with a partially optimal solution.

1.5.2   Genetic Programming

In Genetic Programming programs are evolved to solve pre-described problems from both of these domains. The term evolution refers to an artificial process gleaned from the natural evolution of living organisms. This process has been abstracted and stripped off of most of its intricate details. It has been transferred to the world of algorithms where it can serve the purpose of approximating solutions to given or even changing problems (machine learning) or for inducing precise solutions in the form of grammatically correct (language) structures (automatic programming).

It has been realized that the representation of programs, or generally structures, has a strong influence on the behavior and efficiency of the resulting algorithm. As a consequence, many different approaches toward choosing representations have been adopted in Genetic Programming. The principles have been applied even to other problem domains such as design of electronic circuits or art and musical composition.

Genetic Programming is also part of the growing set of Evolutionary Algorithms which apply the search principles of natural evolution in a variety of different problem domains, notably parameter optimization. Evolutionary Algorithms, and Genetic Programming in particular, follow Darwin's principle of differential natural selection. This principle states that the following preconditions must be fulfilled for evolution to occur via (natural) selection

- There are entities called individuals which form a population. These entities can reproduce or can be reproduced.

- There is heredity in reproduction, that is to say that individuals produce similar offspring.

- In the course of reproduction there is variety which affects the likelihood of survival and therefore of reproducibility of individuals.

- There are finite resources which cause the individuals to compete. Due to over reproduction of individuals not all can survive the struggle for existence. Differential natural selections will exert a continuous pressure toward improved individuals.

In the long run, genetic programming and its kin will revolutionize program development. Present methods are not mature enough for deployment as automatic programming systems. Nevertheless, GP has al ready made inroads into automatic programming and will continue to do so in the foreseeable future. Likewise, the application of evolution in machine-learning problems is one of the potentials that is to be exploited over the coming decade.

1.5.3 Evolutionary Programming

Evolutionary programming is one of the four major evolutionary algorithm paradigms. It was first used by Lawrence J. Fogel in 1960 in order to use simulated evolution as a learning process aiming to generate artificial intelligence. Fogel used finite state machines as predictors and evolved them. Currently evolutionary programming is a wide evolutionary computing dialect with no fixed structure or representation, in contrast with some of the other dialects. It is becoming harder to distinguish from evolutionary strategies. Some of its original variants are quite similar to the later genetic programming, except that the program structure is fixed and its numerical parameters are allowed to evolve. Its main variation operator is mutation, members of the population are viewed as part of a specific species rather than members of the same species, therefore each parent generates an offspring, using a survivor selection.

1.5.4 Evolutionary Strategies

Evolution Strategies (ESs) are in many ways very similar to Genetic Algorithms (GAs). As their name implies, ESs too simulate natural evolution. The differences between GAs and ESs arise primarily because the original applications for which the algorithms were developed are different. While GAs were designed to solve discrete or integer optimization problems, ESs were applied first to continuous parameter optimization problems associated with laboratory experiments.

ESs differ from traditional optimization algorithms in some important aspects like:

- They search between populations of solutions, rather than from individual to individual.

- They use only objective function information, not derivatives.

- They use probabilistic transition rules.

The basic structure of an ES is very similar to that of a basic GA. One minor change from the standard optimization routine flow diagram is the use of the word 'population' rather than 'solution'. A more major difference is that the usual operation of generating a new solution has been replaced by three separate activities — population selection, recombination, and mutation. It is in the implementation of these operations that the differences between ESs and GAs lie.

1.6   Swarm Intelligence

The collections of birds and animals, such as flocks, herds and schools, move in a way that appears to be orchestrated. A flock of birds moves like a well-choreographed dance troupe. They veer to the left in unison, then suddenly they may all dart to the right and swoop down toward the ground. How can they coordinate their actions so well? In 1987, Reynolds created a "boid" model, which is a distributed behavioral model, to simulate on a computer the motion of a flock of birds. Each boid is implemented as an independent actor that navigates according to its own perception of the dynamic environment. A boid must observe the following rules. First, the "avoidance rule" says that a boid must move away from boids that are too close, so as to reduce the chance of in-air collisions. Second, the "copy rule" says a boid must fly in the general direction that the flock is moving by averaging the other boids' velocities and directions. Third, the "center rule" says that a boid should minimize exposure to the flock's exterior by moving toward the perceived center of the flock. Flake added a fourth rule, "view", that indicates that a boid should move laterally away from any boid the blocks its view. This boid model seems reasonable if we consider it from another point of view that of it acting according to attraction and repulsion between neighbors in a flock. The repulsion relationship results in the avoidance of collisions and attraction makes the flock keep shape, i.e., copying movements of neighbors can be seen as a kind of attraction. The center rule plays a role in both attraction and repulsion. The swarm behavior of the simulated flock is the result of the dense interaction of the relatively simple behaviors of the individual boids.

The types of algorithms available to solve the collective behavior pattern are Particle Swarm Optimization and Ant Colony Optimization, respectively. Particle Swarm Optimization (PSO) is a swarm intelligence based algorithm to find a solution to an optimization problem in a search space, or model, and predict social behavior in the presence of objectives. The Ant Colony Optimization algorithm (ACO), is a probabilistic technique for solving computational problems which can be reduced to finding good paths through graphs. They are inspired by the behavior of ants in finding paths from the colony to food. Ant systems are also discussed further, where interaction between individuals occurs indirectly by means of modifications to the environment in which they function. By modeling these social interactions, useful algorithms have been devised for solving numerous problems including optimization and clustering.

1.7 Other Paradigms

Some of the major paradigms of computational intelligence are granular computing, chaos theory, and artificial immune systems. Granular computing is an emerging computing paradigm of information processing. It concerns the processing of complex information entities called information granules, which arise in the process of data abstraction and derivation of knowledge from information. Chaos theory is a developing scientific discipline, which is focused on the study of nonlinear systems.

The specific example of a relatively new CI paradigm is the Artificial Immune System (AIS), which is a computational pattern recognition technique, based on how white blood cells in the human immune system detect pathogens that do not belong to the body. Instead of building an explicit model of the available training data, AIS builds an implicit classifier that models everything else but the training data, making it suited to detecting anomalous behavior in systems. Thus, AIS is well suited for applications in anti-virus software, intrusion detection systems, and fraud detection in the financial sector.

Further, fields such as Artificial Life (ALife), robotics (especially multi-agent systems) and bioinformatics are application areas for CI techniques. Alternatively, it can be argued that those fields are a breeding ground for tomorrow's CI ideas. For example, evolutionary computing techniques have been successfully employed in bioinformatics to decipher genetic sequences. Hand in hand with that comes a deeper understanding of the biological evolutionary process and improved evolutionary algorithms.

As another example, consider RoboCup, a project with a very ambitious goal. The challenge is to produce a team of autonomous humanoid robots that will be able to beat the human world championship team in soccer by the year 2050. This is obviously an immense undertaking that will require drawing on many disciplines. The mechanical engineering aspects are only one of the challenges standing in the way of meeting this goal. Controlling the robots is quite another. Swarm robotics, an extension of swarm intelligence into robotics, is a new paradigm in CI that may hold some of the answers. In the mean time, simulated RoboCup challenges, which are held annually, will have to suffice.

1.7.1   Granular Computing

Uncertainty processing paradigms can be considered as conceptual frames. Information granule is a conceptual frame of fundamental entities considered to be of importance in a problem formulation. That conceptual frame is a place where generic concepts, important for some abstraction level, processing, or transfer of results in outer environment, are formulated. Information granule can be considered as knowledge representation and knowledge processing components. Granularity level (size) of information granules is important for a problem description and for a problem-solving strategy. Soft computing can be viewed in the context of computational frame based on information granules, and referred to as granular computing. Essential common features of problems are identified in granular computing, and those features are represented by granularity.

Granular computing has the ability to process information granules, and to interact with other granular or numerical world, eliciting needed granular information and giving results of granular evaluations. Granular computing enables abstract formal theories of sets, probability, fuzzy sets, rough sets, and maybe others, to be considered in the same context, noticing basic common features of those formalisms, providing one more computing level, higher from soft computing, through synergy of considered approaches. Since several computing processes can be present in the same time, with possible mutual communication, a distributed processing model can be conceived. In that model every process, or agent, Figure 1.4, is treated as a single entity.

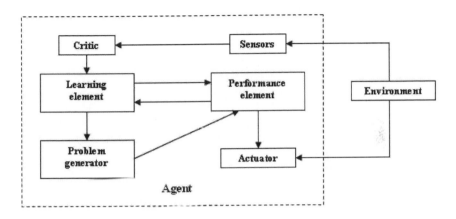

FIGURE 1.4: An Agent

Every agent, as shown in Figure 1.5, acts in its own granular computing environment and communicates with other agents.

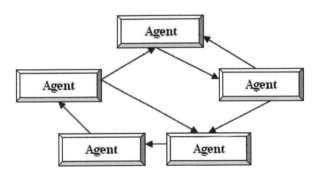

FIGURE 1.5: Granular computing implemented by agents

A formal concept of granular computing can be expressed by four-tuple $<X,F,A,C>$, where: X - is universe of discourse, F - is formal granulation frame, A - is a collection of generic information granules, and, C - is a relevant communication mechanism. Granular computing becomes a layer of computational intelligence, a level of abstraction above soft computing. Granular computing synergically complements different aspects of representing, learning, and optimization. A role of granular computing in development of intelligent systems, and so of computing systems can be significant, as in knowledge integration, and also in development of computing systems more adapted to user, linguistic, and biologically motivated.

1.7.2 Chaos Theory

Chaos Theory is amongst the youngest of the sciences, and has rocketed from its isolated roots in the seventies to become one of the most captivating fields in existence. This theory is applied in the research on several physical systems, and also implemented in areas such as image compression, fluid dynamics, Chaos science assures to continue to yield absorbing scientific information which may shape the face of science in the future. The acceptable definition of chaos theory states, Chaos Theory is the qualitative study of unstable aperiodic behavior in deterministic nonlinear systems. The behavior of chaos is complex, nonlinear, and dynamic complex implies aperiodic (is simply the behavior that never repeats), nonlinear implies recursion and higher mathematical algorithms, and dynamic implies non-constant and non-periodic (time variables). Thus, Chaos Theory is the study of forever changing complex systems based on mathematical concepts of recursion, whether in form of a recursive process or a set of differential equations modeling a physical system.

Newhouse's definition states, A bounded deterministic dynamical system with at least one positive Liaponov exponent is a chaotic system; a chaotic signal is an observation of a chaotic system. The presence of a positive Liaponov exponent causes trajectories that are initially close to each other to separate exponentially. This, in turn, implies sensitive dependence of the dynamics on initial conditions, which is one of the most important characteristics of chaotic systems. What is so incredible about Chaos Theory is that unstable aperiodic behavior can be found in mathematically simple systems. Lorenz proved that the complex, dynamical systems show order, but they never repeat. Our world is a best example of chaos, since it is classified as a dynamical, complex system, our lives, our weather, and our experiences never repeat; however, they should form patterns.

Chaotic behavior has been observed in the laboratory in a variety of systems including electrical and electronic circuits, lasers, oscillating chemical reactions, fluid dynamics, and mechanical and magneto-mechanical devices. Observations of chaotic behavior in nature include the dynamics of satellites in the solar system, the time evolution of the magnetic field of celestial bodies, population growth in ecology, the dynamics of the action potentials in neurons, and molecular vibrations. Everyday examples of chaotic systems include weather and climate. There is some controversy over the existence of chaotic dynamics in the plate tectonics and in economics.

Systems that exhibit mathematical chaos are settled and thus orderly in some sense; this technical use of the word chaos is at odds with common parlance, which suggests complete disorder. A related field of physics called quantum chaos theory studies systems that follow the laws of quantum mechanics. Recently, another field, called relativistic chaos, has emerged to describe systems that follow the laws of general relativity.

As well as being orderly in the sense of being deterministic, chaotic systems usually have well defined statistics. For example, the Lorenz system pictured is chaotic, but has a clearly defined structure. Bounded chaos is a useful term for describing models of disorder.

Chaos theory is applied in many scientific disciplines: mathematics, biology, computer science, economics, engineering, finance, philosophy, physics, politics, population dynamics, psychology, and robotics.

Chaos theory is also currently being applied to medical studies of epilepsy, specifically to the prediction of seemingly random seizures by observing initial conditions.

1.7.3 Artificial Immune Systems

The biological immune system is a highly parallel, distributed, and adaptive system. It uses learning, memory, and associative retrieval to solve recognition and classification tasks. In particular, it learns to recognize relevant patterns, remembers patterns that have been seen previously, and uses combinatorics to construct pattern detectors efficiently. These remarkable information processing abilities of the immune system provide important aspects in the field of computation. This emerging field is sometimes referred to as Immunological Computation, Immunocomputing, or Artificial Immune Systems (AIS). Although it is still relatively new, AIS, having a strong relationship with other biology-inspired computing models, and computational biology, is establishing its uniqueness and effectiveness through the zealous efforts of researchers around the world.

An artificial immune system (AIS) is a type of optimization algorithm inspired by the principles and processes of the vertebrate immune system. The algorithms typically exploit the immune system's characteristics of learning and memory to solve a problem. They are coupled to artificial intelligence and closely related to genetic algorithms.

Processes simulated in AIS include pattern recognition, hypermutation and clonal selection for B cells, negative selection of T cells, affinity maturation, and immune network theory.

1.8   Hybrid Approaches

SC's main characteristic is its intrinsic capability to create hybrid systems that are based on an integration of the techniques to provide complementary learning, reasoning and searching methods to combine domain knowledge and empirical data to develop flexible computing tools and solve complex problems. Several heuristic hybrid soft computing models have been developed for model expertise, decision support, image and video segmentation techniques, process control, mechatronics, robotics, and complicated control and automation tasks. Many of these approaches use a combination of different knowledge representation schemes, decision making models, and learning strategies to solve a computational task. This integration aims at overcoming the limitations of individual techniques through hybridization or the fusion of various soft computing techniques. These ideas have led to the growth of numerous intelligent system architectures. It is very critical to develop and design hybrid intelligent systems to focus primarily on the integration and interaction of different techniques rather than to merge different methods to create ever-new techniques. Techniques already well understood should be applied to solve specific domain problems within the system. Their weaknesses must be addressed by combining them with complementary methods. Nevertheless, developing hybrid intelligent systems is an open-ended concept rather than restricting it to a few technologies. That is, it is evolving those relevant techniques together with the important advances in other new computing methods. Some of the major hybrid combinations are

- Neural Networks controlled by Fuzzy Logic

- Fuzzy logic controllers tuned by Neural Networks

- Neuro Fuzzy Systems

- Neural Networks generated by Genetic Algorithms

- Neural Networks tuned by Genetic Algorithms

- Genetic Algorithms controlled by Fuzzy Logic

- Fuzzy logic controllers tuned by Genetic Algorithms

- Fuzzy logic controllers' learning optimization and by Genetic Algorithms

- Fuzzy Evolutionary Systems

- Fuzzy logic controllers generated by Genetic Algorithms

1.9 Relationship with Other Paradigms

Computational Intelligence is a very young discipline. Other disciplines as diverse as philosophy, neurobiology, evolutionary biology, psychology, economics, political science, sociology, anthropology, control engineering, and many more have been studying intelligence much longer. We first discuss the relationship with philosophy, psychology, and other disciplines which study intelligence; then we discuss the relationship with computer science, which studies how to compute. The science of CI could be described as "synthetic psychology," "experimental philosophy," or "computational epistemology." Epistemology is the study of knowledge. It can be seen as a way to study the old problem of the nature of knowledge and intelligence, but with a more powerful experimental tool than was previously available. Instead of being able to observe only the external behavior of intelligent systems, as philosophy, psychology, economics, and sociology have traditionally been able to do, we are able to experiment with executable models of intelligent behavior. Most importantly, such models are open to inspection, redesign, and experiment in a complete and rigorous way. In other words, you now have a way to construct the models that philosophers could only theorize about.

Just as the goal of aerodynamics isn't to synthesize birds, but to understand the phenomenon of flying by building flying machines, CI's ultimate goal isn't necessarily the full-scale simulation of human intelligence. The notion of psychological validity separates CI work into two categories: that which is concerned with mimicking human intelligence-often called cognitive modeling-and that which isn't. To emphasize the development of CI as a science of intelligence, we are concerned, in this book at least, not with psychological validity but with the more practical desire to create programs that solve real problems. Sometimes it will be important to have the computer to reason through a problem in a human-like fashion. This is especially important when a human requires an explanation of how the computer generated an answer. Some aspects of human cognition you usually do not want to duplicate, such as the human's poor arithmetic skills and propensity for error. Computational intelligence is intimately linked with the discipline of computer science. While there are many non-computer scientists who are researching CI, much, if not most, CI (or AI) research is done within computer science departments. We believe this is appropriate, as the study of computation is central to CI. It is essential to understand algorithms, data structures, and combinatorial complexity in order to build intelligent machines. It is also surprising how much of computer science started as a spin off from AI, from timesharing to computer algebra systems.

The interaction of computational intelligence techniques and hybridization with other methods such as expert systems and local optimization techniques certainly opens a new direction of research toward hybrid systems that exhibit problem solving capabilities approaching those of naturally intelligent systems in the future. Evolutionary algorithms, seen as a technique to evolve machine intelligence, are one of the mandatory prerequisites for achieving this goal by means of algorithmic principles that are already working quite successfully in natural evolution. There exists a strong relationship between evolutionary computation and some other techniques, e.g., fuzzy logic and neural networks, usually regarded as elements of artificial intelligence. Their main common characteristic lies in their numerical knowledge representation, which differentiates them from traditional symbolic artificial intelligence. The term computational intelligence was suggested with the following characteristics:

- numerical knowledge representation

- adaptability

- fault tolerance

- processing speed comparable to human cognition processes

- error rate optimality (e.g., with respect to a Bayesian estimate of the probability of a certain error on future data)

Computational intelligence is considered as one of the most innovative research directions in connection with evolutionary computation, since we may expect that efficient, robust, and easy-to-use solutions to complex real-world problems will be developed on the basis of these complementary techniques.

There are other fields whose goal is to build machines that act intelligently. Two of these fields are control engineering and operations research. These start from different points than CI, namely in the use of continuous mathematics. As building real agents involves both continuous control and CI-type reasoning, these disciplines should be seen as symbiotic with CI. A student of either discipline should understand the other. Moreover, the distinction between them is becoming less clear with many new theories combining different areas. Unfortunately there is too much material for this book to cover control engineering and operations research, even though many of the results, such as in search, have been studied in both the operations research and CI areas. Finally, CI can be seen under the umbrella of cognitive science. Cognitive science links various disciplines that study cognition and reasoning, from psychology to linguistics to anthropology to neuroscience. CI distinguishes itself within cognitive science because it provides tools to build intelligence rather than just studying the external behavior of intelligent agents or dissecting the inner workings of intelligent systems.

1.10 Challenges To Computational Intelligence

A variety of challenges for AI have been formulated and these require a very-large knowledge base and efficient retrieval of structures. The challenge is to present more efficient, knowledge base, knowledge representation and retrieval structures, modeling human brain, using different representations for various purposes. In recent years CI has been extended by adding many other subdisciplines and it became quite obvious that this new field also requires a series of challenging problems that will give it a sense of direction. Without setting up clear goals and yardsticks to measure progress on the way many research efforts are wasted.

Computational Intelligence (CI)-related techniques play an important role in state-of-the-art components and novel devices and services in science and engineering. Some of the major challenges of computational intelligence are concerned with large knowledge bases, bootstraping on the knowledge resources from the Internet etc.

The current state of computational intelligence research can be characterized as in the following.

- The basic concepts of CI have been developed more than 35 years ago, but it took almost two decades for their potential to be recognized by a larger audience.

- Application-oriented research is quite successful and almost dominates the field. Only few potential application domains could be identified, if any, where evolutionary algorithms have not been tested so far. In many cases they have been used to produce good, if not superior, results.

- In contrast, the theoretical foundations are to some extent still weak. To say it pithier: "We know that they work, but we do not know why." As a consequence, inexperienced users fall into the same traps repeatedly, since there are only few rules of thumb for the design and parameterization of evolutionary algorithms.

The challenge is at the meta-level, to find all interesting solutions automatically, especially in difficult cases. Brains are flexible, and may solve the same problem in many different ways. Different applications - recognition of images, handwritten characters, faces, analysis of signals, multimedia streams, texts, or various biomedical data - usually require highly specialized methods to achieve top performance. This is a powerful force that leads to compartmentalization of different CI branches and creation of meta-learning systems competitive with the best methods in various applications will be a great challenge. If we acknowledge that CI should be defined as the science of solving non-algorithmizable problems the whole field will be firmly anchored in computer science and many technical challenges may be formulated.

Summary

This chapter has presented a broad overview of the core computational intelligence techniques, along with some simple examples, describing a range of real world applications that have used these methods. Although these techniques are becoming increasingly common and powerful tools for designing intelligent controllers, there are a number of important practical considerations to bear in mind:

- The ultimate aim is to solve practical problems in the design of controllers, so the techniques considered need to be simple, robust, and reliable.

- Different types of computational intelligence techniques will generally be required for different robot control problems, depending on their different practical requirements.

- The application of a single computational intelligence technique will often be insufficient on its own to provide solutions to all the practical issues.

- Traditional robot control approaches should not be abandoned - they should be considered alongside the computational intelligence techniques.

- Hybrid systems involving combinations of neural computation, fuzzy logic, and evolutionary algorithms, as well as traditional techniques, are a more promising approach for improving the performance of robot controllers.

- A key idea behind computational intelligence is automated optimization, and this can be applied to both the structure and parameters of robot control systems.

Neural network style learning is good for tuning parameters. Evolutionary approaches can be applied to optimize virtually all aspects of these systems. It is hoped that readers will now appreciate the power of computational intelligence techniques for intelligent control, and will be encouraged to explore further the possibility of using them to achieve improved performance in their own applications.

Review Questions

1. Define computational intelligence.

2. Mention the different paradigms of computational intelligence.

3. Explain optimization with an example.

4. Mention a few examples of NP- Complete problems.

5. Differentiate supervised and unsupervised learning.

6. Represent a feed forward neural network diagrammatically.

7. What is the basic concept behind vector quantization?

8. What are fuzzy systems?

9. Mention the fuzzy set operators.

10. Draw the block diagram of a fuzzy controller and explain.

11. What is Evolutionary Computing?

12. Mention the different paradigms of Evolutionary Computing.

13. Explain cultural and co-evolution.

14. Mention some of the best neighborhood topologies in Particle Swarm Optimization.

15. What are Ant systems?

16. What do you mean by Granular computing and Chaos Theory?

17. Mention the challenges involved in computational intelligence.

Chapter 2

Artificial Neural Networks with MATLAB

2.1 Introduction

A neural network is basically a model structure and an algorithm for fitting the model to some given data. The network approach to modeling a plant uses a generic nonlinearity and allows all the parameters to be adjusted. In this way it can deal with a wide range of nonlinearities. Learning is the procedure of training a neural network to represent the dynamics of a plant. The neural network is placed in parallel with the plant and the error h between the output of the system and the network outputs, the prediction error, is used as the training signal. Neural networks have a potential for intelligent control systems because they can learn and adapt, they can approximate nonlinear functions, they are suited for parallel and distributed processing, and they naturally model multivariable systems. If a physical model is unavailable or too expensive to develop, a neural network model might be an alternative. The sections in this chapter provide an understanding of the basic neural networks, their history, and the components of a neural network.

2.2 A Brief History of Neural Networks

The study of the human brain started around thousands of years back. With the advent of advanced electronics, it was only innate to try to harness this thinking process. The first footmark toward artificial neural networks came in 1943 when Warren McCulloch, a neuro-physiologist, and a young mathematician, Walter Pitts, composed a theme on the working of neurons. Initially electrical circuits were used to model simple neural networks.

Reinforcing the concept of neurons and its working was conceptualized by Donald Hebb in his book "The Organization of Behavior" in 1949. It remarked that neural pathways are toned up each time that they are used. As computers upgraded into their infancy in the 1950s, it became affirmable to begin to model the first principles of these theories referring human thought. Nathanial Rochester from the IBM research laboratories led the first effort to model a neural network. Though the first trial failed, the future attempts were successful. It was during this time that conventional computing began to blossom and, as it did, the emphasis in computing left the neural research in the background.

During this period, counsellors of "thinking machines" continued to argue their cases. In 1956 the Dartmouth Summer Research Project on Artificial Intelligence provided a boost to both artificial intelligence and neural networks. One of the major outcomes of this operation was to provoke research in both the intelligent side, AI, as it is known throughout the industry, and in the much lower level neural processing part of the brain.

Later on during the years following the Dartmouth Project, John von Neumann proposed simulating simple neuron functions by applying telegraphy relays or vacuum tubes. Also, Frank Rosenblatt, a neurobiologist of Cornell, began work on the Perceptron in 1958. He was fascinated with the operation of the eye of a fly. The Perceptron, which resulted from this research, was built in hardware and is the oldest neural network still in use today. A single-layer perceptron was found to be useful in classifying a continuous-valued set of inputs into one of two classes. The perceptron computes a weighted sum of the inputs, subtracts a threshold, and passes one of two possible values out as the result. Unfortunately, the perceptron is limited and was proven as such during the "disillusioned years" in Marvin Minsky and Seymour Papert's book Perceptrons.

In 1959, Bernard Widrow and Marcian Hoff of Stanford formulated models they called ADALINE and MADALINE. These models were named for their use of Multiple ADAptive LINear Elements. MADALINE was the first neural network to be applied to a real world problem. It is an adaptive filter, which eliminates echoes on phone lines. This neural network is still in commercial use.

Alas, these earlier successes induced people to magnify the potential of neural networks, particularly in light of the limitation in the electronics then available. This excessive hype, which flowed out of the academic and technological worlds, infected the general literature of the time. Also, a dread harassed as writers began to ponder what effect "thinking machines" would have on man. Asimov's series about robots discovered the effects on man's ethics and values when machines where capable of executing all of mankind's work.

These dreads, fused with unfulfilled, immoderate claims, caused respected voices to critique the neural network research. The result was to halt much of the funding. This period of inferior growth lasted through 1981. In 1982 several events induced a revived interest. John Hopfield's overture was not to merely model brains but to produce useful devices. With clarity and mathematical analysis, he showed how such networks could work and what they could do. However, Hopfield's biggest asset was his personal appeal. He was articulate, likeable, and a champion of a dormant technology.

At the same time, a conference was held in Kyoto, Japan. This conference was the U.S. - Japan Joint Conference on Cooperative/Competitive Neural Networks. Japan subsequently announced their Fifth Generation effort. U. S. periodicals picked up that story, generating a worry that the U. S. could be left behind. Soon funding was flowing once again. By 1985 the American Institute of Physics began an annual meeting — Neural Networks for Computing. By 1987, the Institute of Electrical and Electronic Engineer's (IEEE) first International Conference on Neural Networks drew more than 1,800 attendees.

In 1982, Hopfield showed the usage of "Ising spin glass" type of model to store information in dynamically stable networks. His work paved way for physicians to enter neural modeling. These nets are widely used as associative memory nets and are found to be both continuous valued and discrete valued.

Kohonen's self organizing maps (SOM) evolved in 1972 were capable of reproducing important aspects of the structure of biological neural networks. They make use of data representation using topographic maps, which are more common in nervous systems. SOM also has a wide range of applications in recognition problems.

In 1988, Grossberg developed a learning rule similar to that of Kohonen's SOM, which is widely used in the Counter Propagation Net. This learning is also referred to as outstar learning.

Carpenter and Grossberg invented the Adaptive Resonance Theory (ART) Net. This net was designed for both binary inputs and continuous valued inputs. The most important feature of these nets is that inputs can be presented in any random order.

Nowadays, neural networks discussions are going on everywhere in almost all the engineering disciplines. Their hope appears really brilliant, as nature itself is the proof that this kind of thing works. However, its future, indeed the very key to the whole engineering science, dwells in hardware development. Presently most neural network growth is simply proving that the principal works. This research is producing neural networks that, due to processing limitations, take weeks to learn. To take these prototypes out of the lab and put them into use requires specialized chips. Companies are working on three types of neuro chips - digital, analog, and optical. A few companies are working on creating a "silicon compiler" to render a neural network Application Specific Integrated Circuit (ASIC). These ASICs and neuron-like digital chips appear to be the wave of the near future. Finally, optical chips look very promising, but it may be years before optical chips see the light of day in commercial applications.

2.3   Artificial Neural Networks

Artificial Neural Networks are comparatively crude electronic models based on the neural structure of the brain. The brain in essence acquires knowlwdge from experience. It is natural proof that some problems that are beyond the scope of current computers are indeed solvable by little energy effective packages. This form of brain modeling anticipates a less technical way to produce machine solutions. This new approach to computing also provides a more graceful degradation during system overload than its more orthodox counterparts.

These biologically revolutionized methods of computing are thought to be the next leading progress in the computing industry. Even simple animal brains are able of functions that are presently inconceivable for computers. Computers perform operations, like keeping ledgers or performing complex math. But computers have trouble acknowledging even simple patterns much less generalizing those patterns of the past into actions of the future.

Today, betterments in biological research assure an initial understanding of the natural thinking and learning mechanism. Some of the patterns obtained are very complicated and allow us the ability to recognize individual faces from many different angles. This process of storing information as patterns, utilizing those patterns, and then solving problems encompasses a new field in computing. This field, as mentioned before, does not utilize traditional programming but involves the creation of massively parallel networks and the training of those networks to solve specific problems. This field also utilizes words very different from traditional computing, words like behave, react, self-organize, learn, generalize, and forget.

2.3.1 Comparison of Neural Network to the Brain

The exact operational concepts of the human brain are still an enigma. Only a few aspects of this awesome brain processor are known. In particular, the commonest element of the human brain is a peculiar type of cell, which, unlike the rest of the body, doesn't appear to regenerate. Since this type of cell is the solitary part of the body that is not replaced, it is assumed that these cells provide us with the abilities to remember, think, and apply previous experiences to our every action.

These cells, all 100 billion of them, are known as neurons. Each of these neurons can connect with up to 200,000 other neurons, although 1,000 to 10,000 are typical.

The major power of the human mind comes from the absolute numbers of these basic cellular components and the multiple links between them. It also comes from genetic programming and learning. The individual neurons are perplexed and they have a infinite number of parts, subsystems, and control mechanisms. They carry information through a host of electrochemical pathways. Based on the classificatiion, there are over one hundred different classes of neurons. Collectively these neurons and their links form a process, which is not binary, not stable, and not synchronous. In short, it is nothing like the currently available electronic computers, or even artificial neural networks.

These artificial neural networks try to duplicate only the most basic elements of this complicated, versatile, and powerful organism. They do it in a crude way. But for a technologist who is trying to solve problems, neural computing was never about duplicating human brains. It is about machines and a new way to solve problems.

2.3.2 Artificial Neurons

The fundamental processing element of a neural network is a neuron. This building block of human awareness encompasses a few general capabilities. Basically, a biological neuron receives inputs from other sources, combines them in some way, performs a generally nonlinear operation on the result, and then outputs the final result. Figure 2.1 shows the relationship of these four parts.

Inside human brain there are many variations on this primary type of neuron, further complicating man's attempts at electrically replicating the process of thinking. The x-like extensions of the soma, which act like input channels receive their input through the synapses of other neurons. The soma then processes these incoming signals over time. The soma then turns that processed value into an output, which is sent out to other neurons through the axon and the synapses.

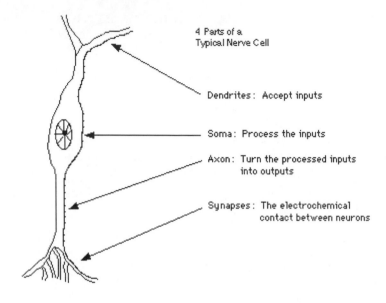

4 Parts of a
Typical Nerve Cell

Dendrites: Accept inputs

Soma: Process the inputs

Axon: Turn the processed inputs
into outputs

Synapses: The electrochemical
contact between neurons

FIGURE 2.1:   A Simple Neuron

The modern observational information has supplied additional evidence that biological neurons are structurally more complex. They are significantly more complex than the existing artificial neurons that are built into today's artificial neural networks. Since biological science furnishes a finer understanding of neurons, and as engineering science advances, network architects can continue to improve their schemes by building upon man's understanding of the biological brain.

Presently, the goal of artificial neural networks is not the pretentious recreation of the brain. Contrarily, neural network researchers are seeking an understanding of nature's capabilities for which people can engineer solutions to problems that have not been solved by traditional computing.Figure 2.2 shows a fundamental representation of an artificial neuron.

In Figure 2.2, several inputs to the network are represented by the mathematical symbol, x(n). Each of these inputs are multiplied by a connection weight represented by w(n). In the simplest case, these products are simply added together, fed through a transfer function to generate an output. This process lends itself to physical implementation on a large scale in a small software package. This electronic implementation is still possible with other network structures, which employ different summing functions as well as different transfer functions.

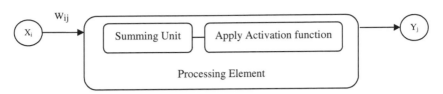

FIGURE 2.2: A Basic Artificial Neuron

A few applications require "monochrome," or binary, answers. These practical applications include the recognition of text, the identification of speech, and the image decrypting of scenes. These practical applications are required to turn real life inputs into discrete values. These potential values are limited to some known set, like the ASCII characters. Due to this limitation of output options, these applications do not, all of the time, use networks framed of neurons that simply sum up, and thereby smooth, inputs. These networks could utilize the binary properties of ORing and ANDing of inputs. These functions, and many others, can be built into the summation and transfer functions of a network.

Other networks work on problems where the resolutions are not just one of several known values. These networks demand to be capable of an infinite number of responses. Example applications of this case constitute the intelligence behind robotic movements. This intelligence works on inputs and then creates outputs, which in reality have a few devices to move. This movement can brace an infinite number of very accurate motions. These networks do indeed require to smooth their inputs which, due to restrictions of sensors, come out as discontinuous bursts. To do that, they might accept these inputs, sum that data, and then produce an output by, for example, applying a hyperbolic tangent as a transfer function. In this way, output values from the network are continuous and satisfy more real life interfaces.

Some other applications might simply sum and compare to a threshold, thereby producing one of two conceivable outputs, a zero or a one. Some functions scale the outputs to match the application, such as the values minus one and one. A few functions still integrate the input data with respect to time, creating time-dependent networks.

2.3.3 Implementation of Artificial Neuron Electronically

Artificial neurons are known as "processing elements" in the presently available software packages. They have many more capabilities than the simple artificial neuron described above. Figure 2.3 is a more detailed schematic of this still simplistic artificial neuron.

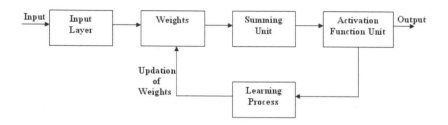

FIGURE 2.3:   A Model of a "Processing Element"

In Figure 2.3, inputs enter into the processing element from the upper left. The initial step is to multiply each of the inputs with their corresponding weighting factor $(w(n))$. Then these altered inputs are fed into the summing function, which usually just sums these products. Eventually, several kinds of operations can be selected. These operations could produce a number of different values, which are then propagated forward; values such as the average, the largest, the smallest, the ORed values, the ANDed values, etc. Moreover, most commercial development products allow software engineers to create their own summing functions through routines coded in a higher-level language. Occasionally the summing function is further elaborated by the addition of an activation function which enables the summing function to operate in a time sensitive manner.

Either way, the output of the summing function is then sent into a transfer function. This function then turns this number into a real output via some algorithm. It is this algorithm that takes the input and turns it into a zero or a one, a minus one or a one, or some other number. The transfer functions that are commonly supported are sigmoid, sine, hyperbolic tangent, etc. This transfer function also can scale the output or control its value via thresholds. The result of the transfer function is usually the direct output of the processing element. An example of how a transfer function works is shown in Figure 2.4.

This sigmoid transfer function takes the value from the summation function, called sum in the Figure 2.4, and turns it into a value between zero and one.

Finally, the processing element is ready to output the result of its transfer function. This output is then input into other processing elements, or to an outside connection, as determined by the structure of

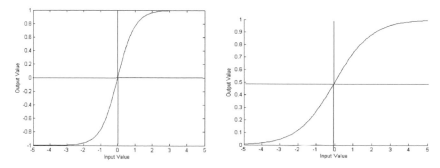

FIGURE 2.4: Sigmoid Transfer Function

the network. All artificial neural networks are constructed from this basic building block - the processing element or the artificial neuron. It is variety and the fundamental differences in these building blocks, which partially cause the implementing of neural networks to be an "art."

2.3.4 Operations of Artificial Neural Network

The "art" of applying neural networks focuses on the infinite number of ways by which the individual neurons can be clustered together. This clustering occurs in the human mind in such a manner that information can be actioned in a dynamic, interactive, and self-organizing way. Biologically, neural networks are constructed in a three-dimensional world from microscopical constituents. These neurons seem capable of almost unclassified interconnections which is not true of any proposed, or existing, man-made network. Integrated circuits, using present technology, are two-dimensional devices with a restricted number of layers for interconnection. This physical reality restrains the types, and scope, of artificial neural networks that can be implemented in silicon.

Currently, neural networks are the simple clustering of the primitive artificial neurons. This clustering occurs by creating layers, which are then connected to one another. How these layers connect is the other part of the "art" of engineering networks to resolve real world problems.

In essence, all artificial neural networks bear a similar structure or topology as shown in Figure 2.5. In this structure some of the neurons interface to the real world to receive its inputs and a few additional neurons provide the real world with the network's output. This output might comprise the particular character that the network thinks that it has scanned or the particular image it thinks is being viewed. The remaining neurons are hidden from view.

Input Layer   **Hidden Layer**   **Output Layer**

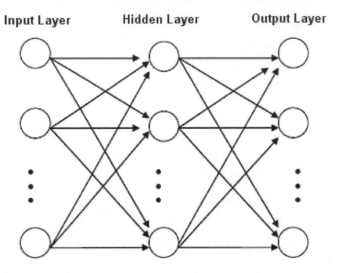

FIGURE 2.5:   A Simple Neural Network Diagram

Many researchers in the early days tried to just connect neurons in a random way, without more success. At present, it is known that even the brains of snails are an organized entity. Among the easiest design techniques to design a structure the most common one is to create layers of elements. It is the grouping of these neurons into layers, the connections between these layers, and the summation and transfer functions that comprise a functioning neural network.

Though there are useful networks, which contain only one layer, or even one element, most applications require networks that contain at least the three normal types of layers - input, hidden, and output. The layer of input neurons receives the data either from input files or directly from electronic sensors in real-time applications. The output layer sends information directly to the outside world, to a secondary computer process, or to other devices such as a mechanical control system. There can be any number of hidden layers between the input and the output layers. These internal hidden layers contain a lot of the neurons in various interconnected structures. The inputs and outputs of each of these hidden neurons merely go away to other neurons.

In almost all networks each neuron in a hidden layer receives the signals from all of the neurons in a layer above it, typically an input layer. Once the neuron performs its function, it passes its output to all of the neurons in the layer below it, providing a feed-forward path to the output. These channels of communication from one neuron to another are important aspects of neural networks. They are referred to as the glue to the system since they interconnect signals. They are the connections, which provide a variable strength to an input. There are two types of these connections. One causes the summing mechanism of the next neuron to add while the other causes it to subtract. In more human terms one excites while the other inhibits.

Some networks want a neuron to inhibit the other neurons in the same layer. This is called lateral inhibition. The most common use of this is in the output layer. For example in text recognition if the probability of a character being a "P" is 0.85 and the probability of the character being an "F" is 0.65, the network wants to choose the highest probability and inhibit all the others. It can do that with lateral inhibition. This concept is also called competition.

Another type of connection is feedback. This is where the output of one layer routes back to a previous layer. An example of this is shown in Figure 2.6.

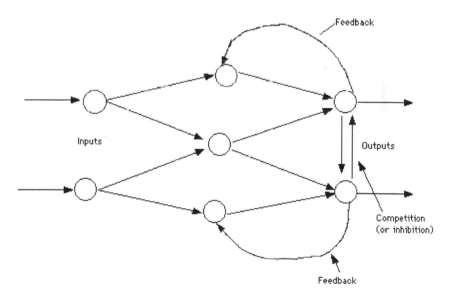

FIGURE 2.6: Simple Network with Feedback and Competition

The way that the neurons are connected to each other has a significant impact on the operation of the network. In the larger, more professional software development packages the user is allowed to add, delete, and control these connections at will. By "tweaking" parameters these connections can be made to either excite or inhibit.

2.3.5 Training an Artificial Neural Network

When a network has been organized and structured for a specific application, it implies that the network is ready to be trained. This operation begins by choosing the initial weights randomly.

Then, the training, or learning, begins. There are two approaches to training - supervised and unsupervised. Supervised training involves a mechanism of providing the network with the desired output either by manually "grading" the network's performance or by providing the desired outputs with the inputs. Unsupervised training is where the network has to make sense of the inputs without outside help.

Neuron connection weights cannot be updated within a single step. The process by which neuron weights are modified occurs over iterations. Training data are presented to the neural network and the results are watched over. The connection weights have to be updated based on these results for the network to learn. The accurate process by which this goes on is influenced by the learning algorithm.

These learning algorithms, which are commonly called learning rules, are almost always expressed as functions. Consider a weight matrix between four neurons, which is expressed as an array of doubles in Java as follows: double weights[ ][ ] = new double[4][4];

This gives us the connection weights between four neurons. Since most of the programming languages index arrays as beginning with zero we shall refer to these neurons as neurons zero through three. Using the above array, the weight between neuron two and neuron three would be contained in the variable weights[2][3]. Therefore, we would like a learning function that would return the new weight between neurons i and j, such as weights[i][j] = weights[i][j] + learningRule(...)

The hypothetical method learningRule calculates the change (delta) that must occur between the two neurons in order for learning to take place. The previous weight value is not discarded altogether, rather a delta value is computed which gives the difference between the weights of the previous iteration and the present one thus modifying the original weight. When the weight of the neural network has been updated the network is presented with the training data again, and the process continues for several iterations. These iterations keep going until the neural network's error rate decreases to an acceptable level.

A common input to the learning rule is the error. The error is the difference between the actual output of the neural network and the anticipated output. If such an error is provided to the training function then the method is called supervised training. In supervised training the neural network is constantly adjusting the weights to attempt to better line up with the anticipated outputs that were provided.

An unsupervised training algorithm can be used when no error is produced. In unsupervised training the neural network is not taught what the "correct output" is. Unsupervised training leaves the neural network to determine the output for itself. The neural network groups the input data when unsupervised training is used. Moreover there is no need for the programmer to have a knowledge of these groups.

Supervised Training

A neural network with supervised training is used when both the inputs and the outputs are provided. The network then processes the inputs and compares its resulting outputs against the desired outputs. The errors produced while comparing, are propagated back through the system, causing the system to adjust the weights, in order to control the network. This process occurs over and over as the weights are continually tweaked. The set of data, which enables the training, is called the "training set." During the training process of a neural network the same set of data is processed several times as the connection weights are refined. The flowchart of supervised training is shown in Figure 2.7.

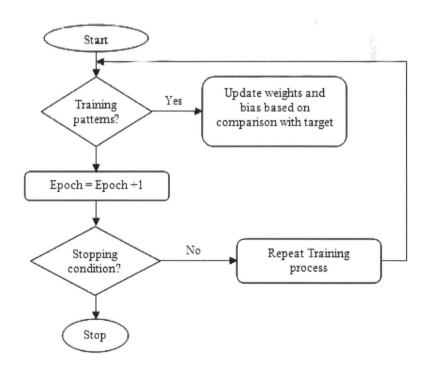

FIGURE 2.7: Flowchart for Supervised Training

The present commercial network development software packages provide functions and methods to monitor the convergence of artificial neural network to predict the correct answer. These functions allow the training process to go on for days; stopping only when the system reaches some statistically desired point, or accuracy. However, some networks never learn. This could be because the input data does not contain the specific information from which the desired output is derived. Networks also don't converge if there is not enough data to enable complete learning. Ideally, there should be enough data so that part of the data can be held back as a test. Several layered networks with multiple nodes are capable of memorizing data. To monitor the network to determine if the system is simply memorizing its data in some non-significant way, supervised training needs to hold back a set of data to be used to test the system after it has undergone its training.

If a network simply can't solve the problem, the designer then has to review the input and outputs, the number of layers, the number of elements per layer, the connections between the layers, the summation, transfer, and training functions, and even the initial weights themselves. Those changes required to create a successful network constitute a process wherein the "art" of neural networking occurs.

Another part of the designer's creativity governs the rules of training. There are many laws (algorithms) used to implement the adaptive feedback required to adjust the weights during training. The most common technique is backward-error propagation, more commonly known as back-propagation. These various learning techniques are explored in greater depth later in this book.

Eventually, training is not exactly a technique. It involves an experience and sensitive analysis, to insure that the network is not overtrained. Initially, an artificial neural network configures itself with the general statistical trends of the data. Later, it continues to "learn" about other aspects of the data which may be spurious from a universal point of view.

Finally when the system has been perfectly trained, and no further learning is required, the weights can, be "frozen." In some systems this finalized network is then turned into hardware so that it can be fast. Other systems don't lock themselves in but continue to learn while in production use.

Unsupervised or Adaptive Training

The other type of training is called unsupervised training. In unsupervised training, the network is provided with inputs but not with desired outputs. The system itself must then decide what features it will use to

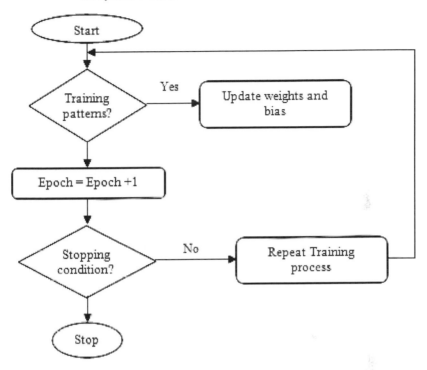

FIGURE 2.8: Flowchart for Unsupervised Training

group the input data. This is often referred to as self-organization or adaptation. The flowchart for unsupervised training is shown in Figure 2.8.

Presently, unsupervised learning is not well understood. This adaptation to the environment is the promise, which would enable science fiction types of robots to continually learn on their own as they encounter new situations and new environments where exact training sets do not exist.

One of the leading researchers into unsupervised learning is Tuevo Kohonen, an electrical engineer at the Helsinki University of Technology. He has developed a self-organizing network, sometimes called an auto-associator, that learns without the benefit of knowing the right answer. It is an unusual looking network in that it contains one single layer with many connections. The weights for those connections have to be initialized and the inputs have to be normalized. The neurons are set up to compete in a winner-take-all fashion.

Kohonen concentrated mainly on the networks that are structured and organized in a different manner compared to the fundamental feed forward networks. Kohonen's research dealt with the clustering of neurons into fields. Neurons within a field are "topologically ordered." Topology is a branch of mathematics that studies how to map from one space to another without changing the geometric configuration. The three-dimensional groupings often found in mammalian brains are an example of topological ordering.

Kohonen has pointed out that the lack of topology in neural network models make today's neural networks just simple abstractions of the real neural networks within the brain. As this research continues, more powerful self learning networks may become possible. But currently, this field remains one that is still in the laboratory.

2.3.6 Comparison between Neural Networks, Traditional Computing, and Expert Systems

Neural networks offer a different way to analyze data, and to recognize patterns within that data, than traditional computing methods. However, they are not a solution for all computing problems. Traditional computing methods work well for problems that can be well characterized. Balancing checkbooks, keeping ledgers, and keeping tabs of inventory are well defined and do not require the special characteristics of neural networks.

Conventional computing machines are perfect for several practical applications. They can process data, track inventories, network results, and protect equipment. These practical applications do not need the special characteristics of neural networks.

Expert systems are an extension of traditional computing and are sometimes called the fifth generation of computing. The first generation computing used switches and wires while the second generation passed off because of the evolution of the transistor. The third generation involved solid-state technology, the use of integrated circuits, and higher level languages like COBOL, Fortran, and "C". End user tools, "code generators," were the fourth generation computing tools. The fifth generation involves artificial intelligence.

Typically, an expert system consists of two parts, an inference engine and a knowledge base. The inference engine is generic. It handles the user interface, external files, program access, and scheduling. The knowledge base contains the information that is specific to a particular problem. This knowledge base allows an expert to define the rules, which govern a process. This expert does not have to understand traditional programming whereas the person has to understand both what he wants a computer to do and how the mechanism of the expert system shell works. The shell which is a part of the inference engine, exactly provides knowledge to the computer regarding the implementation of the expert's needs. To implement this, the expert system itself generates a computer program. The computer program is required in order to build the basic rules for the kind of application. The process of building rules is also complex and does need a person who has a detailed knowledge.

Efforts to make expert systems general have run into a number of problems. As the complexity of the system increases, the system simply demands too much computing resources and becomes too slow. Expert systems have been found to be feasible only when narrowly confined.

Artificial neural networks propose an altogether contrary approach to problem solving and they are sometimes called the sixth generation of computing. These networks are capable of providing algorithms which learn on their own. Neural networks are integrated to provide the capability to solve problems without the intereference of an expert and without the need of complex programming.

Expert systems have enjoyed significant successes. However, artificial intelligence has encountered problems in areas such as vision, continuous speech recognition and synthesis, and machine learning. Artificial intelligence is also a hostage to the speed of the processor that it runs on. Ultimately, it is restricted to the theorctical limit of a single processor. Artificial intelligence is also burdened by the fact that experts don't always speak in rules.

In spite of the benefis of neural networks over both expert systems and more traditional computing in these specific areas, neural nets are not complete solutions. They offer a capability that is not unshakable, such as a debugged accounting system. They learn, and as such, they do continue to make "mistakes." Moreover, whenever a network is developed for a certain application, no method is available to ascertain that the network is the best one to find the optimal solution. Neural systems do precise on their own demands. They do expect their developer to meet a number of conditions such as:

- A data set which includes the information which can characterize the problem

- An adequately sized data set to both train and test the network

- An understanding of the basic nature of the problem to be solved so that basic first-cut decisions on creating the network can be made. These decisions include the activation and transfer functions, and the learning methods

- An understanding of the development tools

- Adequate processing power (some applications demand real-time processing that exceeds what is available in the standard, sequential processing hardware. The development of hardware is the key to the future of neural networks)

Once these conditions are met, neural networks offer the opportunity of solving problems in an arena where traditional processors lack both the processing power and a step-by-step methodology. A number of very complicated problems cannot be solved in the traditional computing environments. Without the massively paralleled processing power of a neural network, spech recognition, image recognition etc., is virtually impractical for a computer to analyze. A traditional computer might try to compare the changing images to a number of very different stored patterns, which is not the optimal solution.

This new way of computing requires skills beyond traditional computing. It is a natural evolution. Initially, computing was only hardware and engineers made it work. Then, there were software specialists - programmers, systems engineers, data base specialists, and designers. This new professional needs to be skilled different than his predecessors of the past. For instance, he will need to know statistics in order to choose and evaluate training and testing situations. It is a stress for the software engineers working to improvise on the skills of neural networks since neural networks offer a distince way to solve problems with their own demands. The greatest demand is that the process is not simply logic. The process demands an empirical skill, an intuitive feel as to how a network might be developed.

2.4   Neural Network Components

At present there has been a general understanding of artificial neural networks among engineers, therefore it is advantageous to do research on them. Merely starting into the several networks, further understanding of the inner workings of a neural network is a major requirement. It is understood that, artificial neural networks are a prominent class of parallel processing architectures that are effective in particular types of composite problems and applications. These architectures ought to be confused with common parallel processing configurations that apply many sequential processing units to standard computing topologies. Alternatively, neural networks are different when compared with conventional Von Neumann computers in the aspect that they inexpertly imitate the underlying properties of a human brain.

As mentioned earlier, artificial neural networks are loosely based on biology. Current research into the brain's physiology has unlocked only a limited understanding of how neurons work or even what constitutes intelligence in general. Investigators and scientists working in biological and engineering fields are focusing and investigating to decipher the important mechanisms as to how a human being learns and reacts to day-to-day experiences. A developed knowledge in neural actioning helps these investigators to produce more effective and more compendious artificial networks. It likewise produces a profusion of modern, and always germinating, architectures. Kunihiko Fukushima, a senior research scientist in Japan, describes the give and take of building a neural network model; "We try to follow physiological evidence as faithfully as possible. For parts not yet clear, however, we construct a hypothesis and build a model that follows that hypothesis. We then analyze or simulate the behavior of the model and compare it with that of the brain. If we find any discrepancy in the behavior between the model and the brain, we change the initial hypothesis and modify the model. We repeat this procedure until the model behaves in the same way as the brain." This common process has created thousands of network topologies.

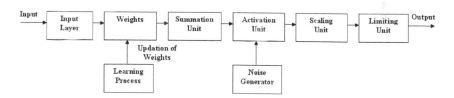

FIGURE 2.9:   Processing Element

Neural computation comprises most machines, not human brains. Neural computation is the process of attempting to build up processing systems that attract the extremely flourishing designs that are happening in biology by nature. This linkage with biological science is the reason that there is a common architectural thread throughout today's artificial neural networks. Figure 2.9 shows a model of an artificial neuron, or processing element, which incarnates a broad variety of network architectures. This figure is adapted from NeuralWare's simulation model used in NeuralWorks Profession II/Plus.

Their processing element model shows that networks designed for prediction can be very similar to networks designed for classification or any other network category. Prediction, classification, and other network categories will be discussed later. The point here is that all artificial neural processing elements have common components.

This section describes the seven major components that make up an artificial neuron. These components are valid whether the neuron is used for input, output, or is in one of the hidden layers.

Component 1. Weighting Factors

An individual neuron usually incurs numerous simultaneous inputs. Each input has its own relative weight that gives the input the energy that it needs on the processing element's summation function. All the weight functions execute the same type of function similar to the varying synaptic intensities of biological neurons. In both cases, a few inputs are given more importance than others so that they have a greater effect on the processing element as they aggregate to produce a neural response.

The weight functions are adaptive coefficients inside the network and they determine the intensity of the input signal as registered by the artificial neuron. Weights are a measure of an input's connection strength, which can be changed in response to various training sets and according to a network's specific topology or through its learning rules.

Component 2. Summation Function

The initial step in the operation of a processing element of a neural network is to compute the weighted sum of all of the inputs. Mathematically, the inputs and the corresponding weights are vectors which can be represented as $(i_1, i_2, \ldots in)$ and $(w_1, w_2, \ldots w_n)$. The total input signal is the dot, or inner, product of these two vectors. This oversimplified summation function is determined by multiplying each component of the $i_t h$ vector by the corresponding component of the w vector and then adding up all the products. Input1 $= i_1 * w_1$, input2 $= i_2 * w_2$, etc., are added as input1 + input2 + ... + input$_n$. The result is a single number, not a multi-element vector.

The dot product of two vectors can be conceived geometrically as a measure of their similarity. If the vectors point in the same direction, the dot product is maximum; if the vectors point in opposite direction (180 degrees out of phase), their dot product is minimum.

The summation function is more complex than just the simple input and weight sum of products. The input and weighting coefficients can be combined in several dissimilar manners before passing on to the transfer function. Likewise a mere product summing, the summation function can select the minimum, maximum, majority, product, or several normalizing algorithms. The specific algorithm for combining neural inputs is determined by the chosen network architecture and paradigm.

Some summation functions have an additional process applied to the result before it is passed on to the transfer function. This process is sometimes called the activation function. The purpose of utilizing an activation function is to allow the summation output to vary with respect to time. Activation functions currently are pretty much confined to research. Most of the current network implementations use an "identity" activation function, which is equivalent to not having one. Additionally, such a function is likely to be a component of the network as a whole rather than of each individual processing element component.

Component 3. Transfer Function

The output of the summation function is the weighted sum, and this is transformed to a working output through an algorithmic process known as the transfer function. The transfer function compares the summation total with a threshold function to determine the output of the nework. If the sum is greater than the threshold value, the processing element generates a signal. If the sum of the input and weight products is less than the threshold, no signal (or some inhibitory signal) is generated. Both cases of obtaining the outputs are substantial.

The threshold, or transfer function, is generally nonlinear. Linear (straight line) functions are limited because the output is simply proportional to the input. Linear functions are not very useful. That was the problem in the earliest network models as noted in Minsky and Papert's book *Perceptrons*.

The transfer function is one in which the result of the summation function may be positive or negative. The output of the network could be zero and one (binary output), plus one and minus one (bipolar), or any other most commonly used numeric combinations. The transfer function that produces such kind of an output is a "hard limiter" or step function. Figure 2.10 shows a few sample transfer functions.

Another type of transfer function, the threshold or ramping function, could mirror the input within a given range and still act as a hard limiter outside that range. It is a linear function that has been clipped to minimum and maximum values, making it nonlinear. However a different choice of a transfer function would be a sigmoid or S-shaped curve. The S-shaped curve goes about a minimum and maximum value at the asymptotes. This curve is termed as a sigmoid when it ranges between 0 and 1, and known as a hyperbolic tangent when it ranges between -1 and 1. Mathematically, the most important and exciting feature of these curves is that both the function and its derivatives are continuous. This option works fairly well and is often the transfer function of choice.

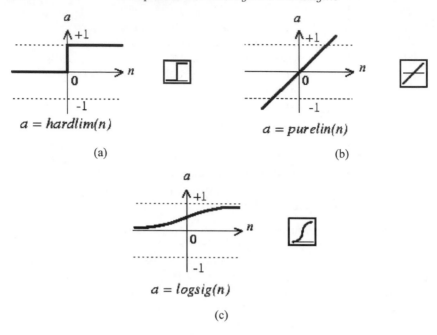

FIGURE 2.10: Sample Transfer Functions (a) Hard Limit Transfer Function (b) Linear Transfer Function (c) Log-Sigmoid Transfer Function

Before applying the transfer function, a uniformly distributed random noise may be added. The origin and volume of this noise is determined by the learning mode of a given network paradigm. This noise is normally referred to as "temperature" of the artificial neurons. The name, temperature, is derived from the physical phenomenon that as people become too hot or cold their ability to think is affected. Electronically, this process is simulated by adding noise. Indeed, by adding different levels of noise to the summation result, more brain-like transfer functions are realized. Most of the researchers use a Gaussian noise source in order to imitate nature's behavior. Gaussian noise is similar to uniformly distributed noise except that the distribution of random numbers within the temperature range is along a bell curve. The use of temperature, that is the noise determined by the network during learning is an ongoing research area and is not being applied to many engineering applications.

Presently NASA is processing a network topology that uses a temperature coefficient in a new feed-forward, back-propagation learning function. But this temperature coefficient is a global term, which is applied to the gain of the transfer function. This should not be confused with the more common term, temperature, which is merely a noise added to individual neurons. In contrast, the global temperature coefficient allows the transfer function to have a learning variable much like the synaptic input weights. This concept is claimed to create a network that has a significantly faster (by several order of magnitudes) learning rate and provides more accurate results than other feedforward, back-propagation networks.

Component 4. Scaling and Limiting

Once the transfer function operation is completed, the output obtained after applying the transfer function is sent through additional processes like scaling and limiting. Scaling just multiplies a scale factor times the transfer function output, and then adds an offset to the result. Limiting is the mechanism which insures that the scaled result does not exceed an upper or lower bound. This limiting is in addition to the hard limits that the original transfer function may have performed. These operations of scaling and limiting are mainly used in topologies to test biological neuron models.

Component 5. Output Function (Competition)

One output signal which is generated from a processing element is taken as a signal to hundreds of other neurons. This process is similar to the biological neuron, where there are many inputs and only one output action. Usually, the output is directly equivalent to the transfer function's output. A few network structures, however, modify the transfer function's output to integrate competition amid neighboring processing elements. Neurons are allowed to compete with each other, inhibiting processing elements unless they have great strength. Competition is usually implemented in one or two steps. First, competition determines which artificial neuron will be active, or provides an output. Second, competitive inputs help to determine which processing element will participate in the learning or adaptation process.

Component 6. Error Function and Back-Propagated Value

Usually the difference between the current output and the desired output is calculated in learning networks and is denoted as the error. An error function is used to transform the error signal in order to match a particular network architecture. Most of the fundamental networks use this error function, but some square the error to find the mean squared error retaining its sign, or some cube the error, or some paradigms modify the raw error to fit their specific purposes.

The artificial neuron's error is then typically propagated into the learning function of another processing element. This error term is sometimes called the current error. Usually the current error is propagated backward to the previous layer. This back-propagated error value can be either the current error, the current error scaled in some manner, or some other desired output depending on the network type. Normally, this back-propagated value, after being scaled by the learning function, is multiplied against each of the incoming connection weights to modify them before the next learning cycle.

Component 7. Learning Function

The aim of the learning function is to update the variable connection weights on the inputs of each processing element based on neural based algorithm. This process of changing the weights of the input connections to achieve some desired result can also be called the adaptation function, as well as the learning mode. There are two types of learning: supervised and unsupervised. Supervised learning requires a teacher. The teacher may be a training set of data or an observer who grades the performance of the network results. Either way, having a teacher is learning by reinforcement. When there is no external teacher, the system must organize itself by some internal criteria designed into the network. This is learning by doing.

2.4.1   Teaching an Artificial Neural Network

Supervised Learning

Majority of artificial neural network solutions have been trained with supervision that is with a teacher. In supervised learning, the actual output of a neural network is compared with the desired output. Weights, which are randomly set to begin with, are then adjusted by the network so that the next iteration, or cycle, will produce a closer match between the desired and the actual output. The learning method tries to minimize the current errors of all processing elements. This global error reduction is created over time by continuously modifying the input weights until an acceptable network accuracy is reached.

With supervised learning, the artificial neural network must be trained before it becomes useful. Training consists of presenting input and output data to the network. This collection of data is often referred to as the training set. A desired output or a target output is provided for every input set of the system. In most applications, actual data must be used. The training phase takes a very long time. In prototype systems, with inadequate processing power, learning can take weeks. This training is considered complete when the neural network reaches a user defined performance level. This level signifies that the network has achieved the desired statistical accuracy as it produces the required outputs for a given sequence of inputs. When no further learning is necessary, the weights are typically frozen for the application. Some network types allow continual training, at a much slower rate, while in operation. This helps a network to adapt to gradually changing conditions.

The training sets should be as ample as possible such that they contain all the necessary and sufficient information that are required for learning. The training sessions should also include a broad variety of data. Instead of learning individually, the system has to learn everything together, to find the best weight settings for the total set of facts available in the problem.

How the input and output data are represented, or encoded, is a major component to successfully instructing a network. Artificial networks only deal with numeric input data. Therefore, the raw data must often be converted from the external environment. Usually it is necessary scale or normalise the input data according to the network's paradigm. This pre-processing of real-world inputs into machine understandable format is already a common work done for standard computers. A lot of training techniques which directly apply to artificial neural network implementations are promptly available. The job of the network designer is only to find the best data format and matching network architecture for a given application by using the instant implementations.

Once a supervised network performs on the training data, it is crucial to ascertain what the network can do with data that it has not seen before. If a system does not produce reasonable outputs for this test set, the training period is not over. Indeed, this testing is critical to insure that the network has not simply memorized a given set of data but has learned the general patterns involved within an application.

Unsupervised Learning

Unsupervised learning also known as clustering is the great promise of the future since it forms natural groups or clusters of patterns. It proves that computers could someday learn on their own in a true robotic sense. Presently, this learning technique is limited to networks known as self-organizing maps. Though these networks are not commonly applied, they have proved that they can provide a solution in a few cases, proving that their promise is not groundless. They have been proven to be more effective than many algorithmic techniques for numerical calculations. They are also being used in the lab where they are split into a front-end network that recognizes short, phoneme-like fragments of speech that are then passed on to a back-end network. The second artificial network recognizes these strings of fragments as words.

This promising field of unsupervised learning is sometimes called self-supervised learning. These networks do not use any external tempts to adjust their weights, instead, they internally monitor their performance. These networks look for regular patterns in the input signals, and makes adaptations according to the function of the network. Even without being told whether it's right or wrong, the network still must have some information about how to organize itself. This information is built into the network topology and learning rules.

An unsupervised learning algorithm may stress cooperation among clusters of processing elements. In such a case, the clusters would work together. If some external input activated any node in the cluster, the cluster's activity as a whole could be increased which is exhibitory. Likewise, if external input to nodes in the cluster was decreased, that could have an inhibitory effect on the entire cluster.

The competition between processing elements could also form a basis for learning. Training of competitive clusters could amplify the responses of specific groups to specific stimuli. As such, it would associate those groups with each other and with a specific appropriate response. Normally, when competition for learning is in effect, only the weights belonging to the winning processing element will be updated.

At the present state of the art, unsupervised learning is not well understood and is still the subject of research. This research is currently of interest to the government because military situations often do not have a data set available to train a network until a conflict arises.

2.4.2   Learning Rates

The rate at which ANNs learn depends upon several controllable factors. Many trade-offs should be taken into account while selecting the learning rates. Evidently, a slower learning rate means more time is spent in accomplishing the off-line learning to produce an adequately trained system. With the faster learning rates, however, the network may not be able to make the fine discriminations possible with a system that learns more slowly.

Commonly, several factors besides time have to be considered when discussing the off-line training task. Some of the most common factors that are to be considered are network complexity, network size, paradigm selection, architecture, type of learning rule or rules employed, and desired accuracy. All these factors play a substantial role in determining the time taken to train a network. If any of these factors is modified then the training time may be extended to an excessive length or the output may be unacceptable and inaccurate.

Almost all learning functions have some provision for a learning rate, or learning constant. Usually this term is positive and between zero and one. If the learning rate is greater than one, it is easy for the learning algorithm to overshoot in correcting the weights, and the network will oscillate. Small values of the learning rate will not correct the current error as quickly, but if small steps are taken in correcting errors, there is a good chance of arriving at the best minimum convergence.

2.4.3 Learning Laws

Numerous learning laws are used commonly to train the neural networks. Most of these laws use the oldest learning law, Hebb's Rule, as the fundamental base. Research into different learning functions keeps going as new ideas routinely come out in commercial publications. The main aim of some researchers and scientists is to model the biological learning. A few other researchers are experimenting with adaptations of biological perceptions to understand how nature handles learning. In both ways, human's understanding of how neural processing actually works is very limited. Learning is surely more complex than the simplifications represented by the learning laws currently developed. A few of the major laws are presented as follows:

Hebb's Rule

The first, and undoubtedly the best-known learning rule was introduced by Donald Hebb to assist with unsupervised training. The rule states, "If a neuron receives an input from another neuron, and if both are highly active (mathematically have the same sign), the weight between the neurons should be strengthened."

Rules for training neural networks are almost always represented as algebraic formulas. Hebb's rule is expressed as:

$$\triangle W_{ij} = \mu a_i a_j$$

The above equation calculates the needed change (delta) in weights from the connection from neuron i to neuron j. The Greek letter mu(μ) represents the learning rate. The activation of each neuron, when provided with the training pattern, is given as a_i and a_j.

To understand the working of Hebb rule, consider a simple neural network with only two neurons. In this neural network these two neurons make up both the input and output layer. There is no hidden layer. Table

TABLE 2.1:   Using Hebb's Rule

Neuron I Output (activation)	Neuron J Output (activation)	Hebb's Rule (R*I*J)	Weight Modification
+1	−1	1*1*−1	−1
−1	+1	1*−1*1	−1
+1	+1	1*1*1	+1

2.1 summarizes some of the possible scenarios using Hebbian training. Assume that the learning rate is one.

It is observed from the above table, that if the activations of neuron 1 was +1 and the activation of neuron J were -1 the neuron connection weight between neuron I and neuron J would be decreased by one.

Hopfield Law

It is similar to Hebb's rule with the exception that it specifies the magnitude of the strengthening or weakening. It states, "if the desired output and the input are both active or both inactive, increment the connection weight by the learning rate, otherwise decrement the weight by the learning rate."

The Delta Rule

This rule is a further variation of Hebb's Rule. It is one of the most commonly used rule. This rule is based on the simple idea of continuously modifying the strengths of the input connections to reduce the difference (the delta) between the desired output value and the actual output of a processing element. This rule changes the synaptic weights in the way that minimizes the mean squared error of the network. This rule is also referred to as the Widrow-Hoff Learning Rule and the Least Mean Square (LMS) Learning Rule. Here since the anticipated output is specified, using the delta rule is considered supervised training. The algebraic expression of delta rule for a single output unit, for several output units and the extended delta rule are illustrated below:

Delta Rule for Single Output Unit

The Delta rule changes the weight of the connections to minimize the difference between the net input to the output unit, yin and the target value t.

The Delta rule is given by,

$$\triangle W_i = \alpha(t - y_{in})x_i$$

where,

x is the vector activation of input units.

y_{in} is the net input to output unit, $-\Sigma$ x.w$_1$

t is the target vector, α is the learning rate

The derivation is as follows,

The mean square error of a particular training pattern is E=Σ_J(t$_j$- y$_{in}$)2

The gradient of E is a vector consisting of a partial derivatives of E with respect to each of the weights. The error can be reduced rapidly by adjusting weight W$_{IJ}$.

Taking partial derivation of E w.r.t. W$_{IJ}$

$$\partial E/\partial W_{IJ} = \partial \Sigma_i (t_j - y_{inj})^2 / \partial W_{IJ}$$
$$= \partial (t_j - y_{inj})^2 / \partial W_{IJ}$$

Since the weight WIJ influences the error only at the input unit yJ. Also,

$$y_{inJ} = \Sigma_i = 1 ton (l_j - y_{inj})^2$$

we get,

$$\partial E/\partial W_{IJ} = \partial F(t_j - y_{inj})^2 / \partial W_{IJ}$$
$$- 2(t_j - y_{inj})(-1)\partial(y_{inj})/\partial W_I J$$
$$\partial E/\partial W_{IJ} = -2(t_j - y_{inj})\partial(y_{inj})/\partial W_{IJ}$$
$$E/\partial W_{IJ} = -2(t_j - y_{inj})x_1$$

Thus the error will be reduced rapidly depending upon the given learning by adjusting the weights according to the Delta rule given by,

$$\triangle W_{IJ} = \alpha(t_J - y_{inJ})x_I$$

Delta Rule for Several Output Units

The derivation of Delta Rule of several output units is similar to that in the previous section. The weights are changed to reduce the difference between net input and target.

The weight correction involving Delta Rule for adjusting the weight from the Ith input unit to the Jth output unit is,

$$\triangle W_{IJ} = \alpha(t_J - y_{inJ})x_I$$

Extended Delta Rule

This can also be called a generalized Delta Rule. The update for the weight from the Ith input unit to the Jth output unit is,

$$\triangle W_{IJ} = \alpha(t_J - y_{inJ})x_I.x_I f^1(y_{in} - J)$$

The derivation is as follows.

The squared error for a particular training pattern is,

$$E = \Sigma_J (t_j - y_j)^2$$

where E is the function of all the weights.

The gradient of E is a vector consisting of the partial derivatives of E with respect to each of the weights. The error can be reduced rapidly by adjusting the weight W_{IJ} in the direction of $-\partial E/\partial\, W_{IJ}$.

Differentiating E partially w.r.t. W_{IJ},

$$\partial E/\partial W_{IJ} = \partial \Sigma_J (t_j - y_j)^2/\partial W_{IJ}$$
$$\partial E/\partial W_{IJ} = \partial \Sigma_J (t_J - y_J)^2/\partial W_{IJ}$$

Since the weight WI J only influences the error at output unit YJ
Since

$$Y_{inJ} = \Sigma i =_1 x_i w_{iJ}$$
$$Y_J = f(y_{in}-_J)$$
$$\partial E/\partial W_{IJ} = 2(t_J - y_J)(-1)\partial y_j \partial/w_{IJ}$$
$$= 2(t_J - y_J)\partial f(y_{inj})/\partial w_{IJ}$$
$$\partial E/\partial W_{IJ} = 2(t_J - y_J)x f^1(y_{inj})$$

Hence the error is reduced rapidly for a given learning rate α by adjusting the weights according to the Delta Rule,

$$\triangle W_{IJ} = \alpha(t_J - y_J)x_I f^1(Y_{inJ})$$

gives the extended Delta Rule.

When using the Delta Rule, it is important to ensure that the input data set is well randomized. Well-ordered or structured presentation of the training set can lead to a network which cannot converge to the desired accuracy. If that happens, then the network is incapable of learning the problem.

Competitive Learning Law or Kohonen's Learning Law

This procedure, developed by Teuvo Kohonen, was inspired by learning in biological systems. In this procedure, the processing elements compete for the opportunity to learn, or update their weights. The processing element with the largest output is declared the winner and has the capability of inhibiting its competitors as well as exciting its neighbors. Only the winner is permitted an output, and only the winner plus its neighbors are allowed to adjust their connection weights.

Further, the size of the neighborhood can vary during the training period. The usual paradigm is to start with a larger definition of the neighborhood, and narrow in as the training process proceeds. Because the winning element is defined as the one that has the closest match to the input pattern, Kohonen networks model the distribution of the inputs. This is good for statistical or topological modeling of the data and is sometimes referred to as self-organizing maps or self-organizing topologies.

For a neuron P to be the winning neuron, its induced local field v_p, for a given particular input pattern must be largest among all the neurons in the network. The output signal of winning neuron is set to one and the signals that lose the competition are set to zero. Hence,

$$N = \begin{cases} 1 & if v_p > v_p for \ all \ q, p \neq q \\ 0 & otherwise \end{cases}$$

This rule is suited for unsupervised network training. The winner-takes-all or the competitive learning is used for learning statistical properties of inputs. This uses the standard Kohonen learning rule.

Let w_{ij} denote the weight of input node j to neuron i. Suppose the neuron has the fixed weight, which are disturbed among its input nodes;

$$\sum_j w_{ij} = 1 for all i$$

A neuron then learns by shifting weights from its inactive to active input modes. If a neuron does not respond to a particular input pattern, no learning takes place in that neuron. If a particular neuron wins the competition, its corresponding weights are adjusted.

Using standard competitive rule, the change w_{ij} is given as,

$$\triangle w_{ij} = \begin{cases} \alpha(x_j - w_{ij}) & if \ neuron \ i \ wins \ the \ competition \\ 0 & if \ neuron \ i \ loses \ the \ competition \end{cases}$$

where α is the learning rate. This rule has the effect of moving the weight vector wi of winning neuron i toward the input pattern x. Through competitive learning, the neural network can perform clustering.

This neighborhood is sometimes entered beyond the simple neuron winner so that it includes the neighboring neurons. Weights are typically initialized at random values and their lengths are normalized during learning in this method. The winner-takes-all neuron is selected either by the dot product or Euclidean norm. Euclidean norm is most widely used because dot product may require normalization.

Out Star Learning Law

Out star learning rule can be well explained when the neurons are arranged in a layer. This rule is designed to produce the desired response t from the layer of n neurons. This type of learning is also called as Grossberg learning. Out star learning occurs for all units in a particular layer and no competition among these units are assumed. However the forms of weight updates for Kohonen learning and Grossberg learning are closely related. In the case of out star learning,

$$\triangle w_{jk} = \begin{cases} \alpha(y_k - w_{jk}) & if \ neuron \ j \ wins \ the \ competition \\ 0 & if \ neuron \ j \ loses \ the \ competition \end{cases}$$

The rule is used to provide learning of repetitive and characteristic properties of input-output relationships. Though it is concerned with supervised learning, it allows the network to extract statistical properties of the input and output signals. It ensures that the output pattern becomes similar to the undistorted desired output after repetitively applying an distorted output versions. The weight change here will be a times the error calculated.

Boltzmann Learning Law

The learning is a stochastic learning. A neural net designed based on this learning is called Boltzmann learning. In this learning the neurons constitute a recurrent structure and they work in binary form. This learning is characterized by an energy function, E, the value of which is determined by the particular states occupied by the individual neurons of the machine given by,

$$E = -\frac{1}{2} \sum_i \sum_j w_{ij} x_j x_i i \neq j$$

where, x_i is the state of neuron i to neuron j. The value i≠j means that none of the neurons in the machine has self feedback. The operation of machine is performed by choosing a neuron at random.

The neurons of this learning process are divided into two groups; visible and hidden. In visible neurons there is an interface between the network and the environment in which it operates but in hidden neurons, they operate independent of the environment. The visible neurons might be clamped onto specific states determined by the environment, called a clamped condition. On the other hand, there is free running condition, in which all the neurons are allowed to operate freely.

Memory Based Learning Law

In memory based learning, all the previous experiences are stored in a large memory of correctly classified input-output examples:$(x_i, t_i)_{i-1}^{N}$ where x_i is the input vector and t_j is the desired response. The desired response is a scalar.

The memory based algorithm involves two parts. They are:

- Criterion used for defining the local neighborhood of the test vector

- Learning rule applied to the training in the local neighborhood

One of the most widely used memory based learning is the nearest neighbor rule, where the local neighborhood is defined as the training example that lies in the immediate neighborhood of the test vector x. The vector,

$$x_n^1 \in f x_1 \dots x_n g$$

is said to be nearest neighbor of x_t if, min $d(x_i, x_t) = d(x_n{}^1, x_t)$.

Where $d(x_i, x_l)$ is the Euclidean distance between the vectors x_i and x_t.

A variant of nearest neighbor classifier is the K-nearest neighbor classifier, which is stated as,

- Identify the K-classified patterns that is nearest to test vector x_t for some integer K.

- Assign xt to the class that is most frequently represented in the K-nearest neighbors to x_t.

Hence K-nearest neighbor classifier acts like an averaging device. The memory-based learning classifier can also be applied to radial basis function network.

2.4.4 MATLAB Implementation of Learning Rules

MATLAB Snippet to Implement Hebb Rule Syntax

```
[dW,LS] = learnh(W,P,Z,N,A,T,E,gW,gA,D,LP,LS)
where
learnh is the Hebb weight learning function
learnh(W,P,Z,N,A,T,E,gW,gA,D,LP,LS) takes several
inputs
W -- S x R weight matrix (or S x 1 bias vector)
P -- R x Q input vectors (or ones(1,Q))
Z -- S x Q weighted input vectors
```

```
N -- S x Q net input vectors
A -- S x Q output vectors
T -- S x Q layer target vectors
E -- S x Q layer error vectors
gW -- S x R gradient with respect to
performance
gA -- S x Q output gradient with respect to performance
D -- S x S neuron distances
LP -- Learning parameters
LS -- Learning state, initially should be = [ ]
dW -- S x R weight (or bias) change matrix
```

Learning occurs according to learnh's learning parameter,
shown here with its default value.
LP.lr - 0.01 -- Learning rate.
For example, consider a random input P and output A for
a layer with a two-element input and three neurons with
a learning rate of 0.5.

```
p = rand(2,1)
a = rand(3,1)
lp.lr = 0.5
dW = learnh([ ],p,[ ],[ ],a,[ ],[ ],[ ],[ ],[ ],lp,[ ])
```

   Output:

```
p =
   0.9501
   0.2311
a =
   0.6068
   0.4860
   0.8913
lp =
   lr: 0.5000
dW =
   0.2883   0.0701
   0.2309   0.0562
   0.4234   0.1030
```

MATLAB Snippet to Implement Delta Rule

```
syntax
```

```
[dW,LS]=learnwh(W,P,Z,N,A,T,E,gW,gA,D,LP,LS)
[db,LS]=learnwh(b,ones(1,Q),Z,N,A,T,E,gW,gA,D,LP,LS)
```
learnwh is the delta learning function, and is also
known as the Widrow-Hoff weight/bias or least mean
squared (LMS)rule.learnwh(W,P,Z,N,A,T,E,gW,gA,D,LP,LS)
takes several inputs
W -- S x R weight matrix (or b, and S x 1 bias vector)
P -- R x Q input vectors (or ones(1,Q))
Z -- S x Q weighted input vectors
N -- S x Q net input vectors
A -- S x Q output vectors
T -- S x Q layer target vectors
E -- S x Q layer error vectors
gW -- S x R weight gradient with respect to performance
gA -- S x Q output gradient with respect to performance
D -- S x S neuron distances
LP -- Learning parameters
LS -- Learning state, initially should be=
[ ] and returns
dW -- S x R weight (or bias) change matrix

Consider a random input P and error E to a layer with
a two-element input and three neurons with a learning
rate of 0.5.
p = rand(2,1)
e = rand(3,1)
lp.lr = 0.5
dW = learnwh([],p,[],[],[],[],e,[],[],[],lp,[])
```

**Output**

```
 p =
0.9501
0.2311
a =
0.6068
0.4860
0.8913
lp =
lr: 0.5000
dW =
0.3620   0.0881
0.2169   0.0528
```
```

```
0.0088    0.0021
```

MATLAB Snippet to Implement Kohonen's Learning Law

Syntax

```
[dW,LS] = learnsom(W,P,Z,N,A,T,E,gW,gA,D,LP,LS)
learnsom is the self-organizing map weight learning
function
learnsom(W,P,Z,N,A,T,E,gW,gA,D,LP,LS) takes several
inputs
W -- S x R weight matrix (or S x 1 bias vector)
P -- R x Q input vectors (or ones(1,Q))
Z -- S x Q weighted input vectors
N -- S x Q net input vectors
A -- S x Q output vectors
T -- S x Q layer target vectors
E -- S x Q layer error vectors
gW -- S x R weight gradient with respect to
performance
gA -- S x Q output gradient with respect to
performance
D -- S x S neuron distances
LP -- Learning parameters
LS -- Learning state
dW -- S x R weight (or bias) change matrix
```

```
For example, consider a random input P, output A,
and weight matrix W, for a layer with a two-element
input and six neurons. The positions and distances
are calculated for the neurons, which are arranged
in a 2-by-3 hexagonal pattern. Four learning
parameters are defined.
p = rand(2,1)
a = rand(6,1)
w = rand(6,2)
pos = hextop(2,3)
d = linkdist(pos)
lp.order_lr = 0.9
lp.order_steps = 1000
lp.tune_lr = 0.02
lp.tune_nd = 1
```

```
ls = [ ]
  [dW,ls] = learnsom(w,p,[],[],a,[],[],[],[],d,lp,ls)

  Output  p =
0.8214
0.4447
a =
0.6154
0.7919
0.9218
0.7382
0.1763
0.4057
w =
0.9355    0.8132
0.9169    0.0099
0.4103    0.1389
0.8936    0.2028
0.0579    0.1987
0.3529    0.6038
pos =
0    1.0000    0.5000    1.5000    0    1.0000
0    0      0.8660    0.8660    1.7321    1.7321

d =
0    1    1    2    2    2
1    0    1    1    2    2
1    1    0    1    1    1
2    1    1    0    2    1
2    2    1    2    0    1
2    2    1    1    1    0
dW =
-0.2189    -0.7071
-0.1909     0.8691
0.8457      0.6291
-0.1426     0.4777
1.3144      0.4235
0.8550     -0.2903
```

MATLAB Snippet to Implement Outstar Learning Law

Syntax

```
[dW,LS] = learnos(W,P,Z,N,A,T,E,gW,gA,D,LP,LS)

learnos(W,P,Z,N,A,T,E,gW,gA,D,LP,LS) takes several
inputs
W -- S x R weight matrix (or S x 1 bias vector)
P -- R x Q input vectors (or ones(1,Q))
Z -- S x Q weighted input vectors
N -- S x Q net input vectors
A -- S x Q output vectors
T -- S x Q layer target vectors
E -- S x Q layer error vectors
gW -- S x R weight gradient with respect to
performance
gA -- S x Q output gradient with respect to
performance
D -- S x S neuron distances
LP -- Learning parameters, none, LP = [ ]
LS -- Learning state, initially should be = [ ] and
returns
dW -- S x R weight (or bias) change matrix
```

For example consider a random input P, output A, and
weight matrix W for a layer with a two-element input
and three neurons with a learning rate of 0.5.

```
p = rand(2,1)
a = rand(3,1)
w = rand(3,2)
lp.lr = 0.5
dW = learnos(w,p,[ ],[ ],a,[ ],[ ],[ ],[ ],[ ],lp,[
])
```

```
 Output

 p =
0.6721
0.8381
a =
0.0196
0.6813
0.3795
w =
0.8318    0.4289
0.5028    0.3046
0.7095    0.1897
```

```
lp =
lr: 0.5000
order_lr: 0.9000
order_steps: 1000
tune_lr: 0.0200
tune_nd: 1
dW =
-0.2729    -0.1715
0.0600     0.1578
-0.1109     0.0795
```

Summary

The design of artificial neural network involves the understanding of the various network topologies, current hardware, current software tools, the application to be solved, and a strategy to acquire the necessary data to train the network. This process further involves the selection of learning rules, transfer functions, summation functions, and how to connect the neurons within the network. Thus this chapter provided an understanding of neural networks, their history, the components such as learning rates, learning laws, error functions, transfer functions etc., This chapter also provided the basic implementation of the learning rules using the Neural Network Toolbox in MATLAB.

Review Questions

1. What are artificial neural networks?

2. Compare ANN with a human brain.

3. How are artificial neurons implemented electronically?

4. Explain supervised and unsupervised training.

5. Mention a few networks that belong to supervised training.

6. Compare Neural networks with traditional computing and expert systems.

7. Describe in detail the major components of a neural network.

8. Mention the different types of transfer function.

9. Define learning rate. What are the constraints on this parameter while training a net?

10. State Hebb rule.

11. Differentiate Hebb and Hopfield learning laws.

12. Derive the algebraic expression of delta rule for single and several outputs.

13. Give an expression to update weights using extended delta rule.

14. Explain competitive learning.

15. State Boltzmann learning law and memory based learning law.

16. Write a MATLAB code to implement the Bolizmann Learning law.

17. Develop a MATLAB code for competitive learning.

Chapter 3

Artificial Neural Networks - Architectures and Algorithms

3.1 Introduction

As stated in the previous chapter artificial neural networks are similar to biological networks based on the concept of neurons, connections, and transfer functions. The architectures of various types of networks are more or less similar. Most of the majority variations prows from the various learning rules like Hebb, Hopfield, Kohonen, etc., and the effect of these rules on the network's topology.

This chapter outlines some of the most common artificial neural networks based on their major class of application. These categories are not meant to be exclusive, they are merely meant to separate out some of the confusion over network architectures and their best matches to specific applications. The single layer and multi layer networks are discussed in this chapter along with the detailed architecture and algorithm of the prediction networks. The chapter also delineates the basic methodology to implement the prediction networks using MATLAB.

Basically, most applications of neural networks fall into the following four categories:

1. Prediction

2. Classification

3. Data Association

4. Data Conceptualization

Table 3.1 shows the differences between these network categories and shows which of the more common network topologies belong to which primary category. A few of these networks grouped according to their specific practical application are being used to solve numerous types of problems. The feedforward back-propagation network is used to solve

TABLE 3.1:   Network Selector Table

Network Category	*Neural Network*	*Applications*
Prediction	1. Perceptron 2. Back Propagation 3. Delta Bar Delta 4. Extended Delta Bar Delta 5. Directed Random search 6. Higher Order Neural Networks 7. Self-organizing map into Backpropagation	Used to pick the best stocks in the market, predict weather, identify people with cancer risks etc.
Classification	1. Learning Vector Quantization 2. Counter-propagation 3. Probabilistic Neural Networks	Used to classify patterns
Data Association	1. Hopfield 2. Boltzmann Machine 3. Hamming Network 4. Bi-directional associative Memory	Used to classify data and also recognizes the data that contains errors.
Data Conceptualization	1. Adaptive Resonance Network 2. Self Organizing Map	Used to analyze the inputs so that grouping relationships can be inferred (e.g. extract from a database the names of those most likely to buy a particular product)

almost all types of problems and indeed is the most popular for the first four categories. Prior to dealing the network types in detail the basic networks based on the number of layers are discussed in the following section. Based on the number of layers artificial neural networks are classified into single layer and multi-layer networks.

3.2   Layered Architectures

Before dealing with the architecture and algorithms of the types of neural networks, let us understand the basic architecture based on the number of layers in the network. The network is classified into single layer and multi layer based on the layers in the architecture.

3.2.1   Single-Layer Networks

A single-layer neural network consists of a set of units organized in a layer. Each unit U_i receives a weighted input x_j with weight w_{ji}. Figure 3.1 shows a single-layer linear model with m inputs and n outputs.

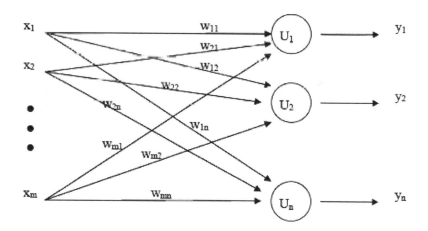

FIGURE 3.1:   A Single-Layer Linear Model

Let $\vec{X} = (x_1, x_2, \ldots x_m)$be the input vector and let the activation function be f, the activation value is just the net sum to a unit. The m × n weight matrix is

$$W = \begin{pmatrix} w_{11} & w_{12}\ldots & w_{1n} \\ w_{21} & w_{22}\ldots & w_{2n} \\ \vdots & & \\ w_{m1} & w_{m2}\ldots & w_{mn} \end{pmatrix}$$

Thus the output y_k at unit U_k is

$$Y_k = (w_{1k}, w_{2k}, ..., w_{mk}) \begin{pmatrix} x_1 \\ x_2 \\ \vdots \\ x_m \end{pmatrix}$$

therefore the output vector $\vec{Y} = (Y_1, Y_2, ... Y_n)T$ is given by

$$\vec{Y} = \vec{W}^T.\vec{X}$$

A simple linear network, with its fixed weights, is limited in the range of output vectors it can associate with input vectors. For example, consider the set of input vectors (x_1, x_2), where each xi is either 0 or 1. No simple linear network can produce outputs as shown in Table 3.2, for which the output is the boolean exclusive-or (XOR) of the inputs. It can be shown that the two weights w1 and w2 would have to satisfy three inconsistent linear equations. Implementing the XOR function is a classic problem in neural networks, as it is a subproblem of other more complicated problems.

TABLE 3.2:   Inputs and Outputs for a Neural Network Implementing the XOR Function

x_1	x_2	Output y
0	0	0
0	1	1
1	0	1
1	1	0

A single-layer model usually uses either the Hebb rule or the delta rule. Refer to Chapter 2 for the algebraic formulas of Hebb and Delta rule. A network using the Hebb rule is guaranteed to be able to learn associations for which the set of input vectors are orthogonal. A disadvantage of the Hebb rule is that if the input vectors are not mutually orthogonal, interference may occur and the network may not be able to learn the associations.

In order to overcome the disadvantages of Hebb rule, the delta rule was proposed. The delta rule updates the weight vector such that the error (difference between the desired and target output) is minimized. Delta rule allows an absolute and accurate method to modify the initial weight vector. The network can learn numerous associations using the delta rule as compared with the Hebb's rule. It is proved that a network using the delta rule can learn associations whenever the inputs are linearly independent.

3.2.2 Multilayer Networks

A network with two or more layers is referred to as a multilayer network. The output from one layer acts as input to the next subsequent layer. The layers that are intermediate and have no external links are referred to as hidden layers (Figure 3.2). Any multilayer system with fixed weights that has a linear activation function is equivalent to a single-layer linear system. Assume a two-layer linear system, in which the input to the first layer is $\vec{X}$, the output $\vec{Y} = W_1\vec{X}$ of the first layer is given as input to the second layer, and the second layer produces output $\vec{Z} = W_2\vec{Y}$.

hidden layer

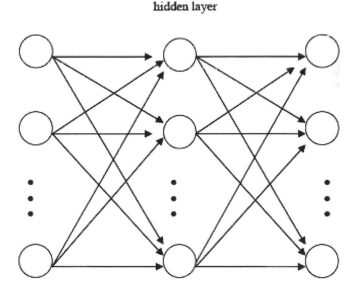

FIGURE 3.2:   A Multilayer Network

Consequently, the system is equivalent to a single-layer network with weight matrix W = W2W1. By induction, a linear system with any number n of layers is equivalent to a single-layer linear system whose weight matrix is the product of the n intermediate weight matrices.

A nonlinear multilayer network is capable of providing more computational potential compared to a single layer network. For example, the difficulties found in the perceptron network can be overcome with the addition of hidden layers; Figure 3.3 demonstrates a multilayer network that symbolizes the XOR function. The threshold is set to zero, and therefore a unit responds if its activation is greater than zero. The weight matrices for the two layers are

$$W_1 = \begin{pmatrix} 1 & -1 \\ -1 & 1 \end{pmatrix}, W_2 = \begin{pmatrix} 1 \\ 1 \end{pmatrix}.$$

We thus get

$$W_1^T \begin{pmatrix} 1 \\ 0 \end{pmatrix} = \begin{pmatrix} 1 \\ 0 \end{pmatrix}, W_2^T \begin{pmatrix} 1 \\ 0 \end{pmatrix} = 1,$$

$$W_1^T \begin{pmatrix} 0 \\ 1 \end{pmatrix} = \begin{pmatrix} 0 \\ 1 \end{pmatrix}, W_2^T \begin{pmatrix} 0 \\ 1 \end{pmatrix} = 1,$$

$$W_1^T \begin{pmatrix} 1 \\ 1 \end{pmatrix} = \begin{pmatrix} 0 \\ 0 \end{pmatrix}, W_2^T \begin{pmatrix} 0 \\ 0 \end{pmatrix} = 0,$$

$$W_1^T \begin{pmatrix} 0 \\ 0 \end{pmatrix} = \begin{pmatrix} 0 \\ 0 \end{pmatrix}, W_2^T \begin{pmatrix} 0 \\ 0 \end{pmatrix} = 0.$$

With input vector (1,0) or (0,1), the output produced at the outer layer is 1; otherwise it is 0.

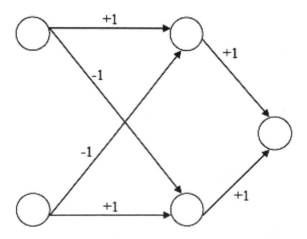

FIGURE 3.3: A Multilayer System Representation of the XOR Function

Multilayer networks have proven to be very powerful. In fact, any boolean function can be implemented by such a network.

No learning algorithm had been available for multilayer networks until Rumelhart, Hinton, and Williams introduced the *generalized delta rule*. At the output layer, the output vector is compared to the expected output.If the error, the difference between the output and the target vector is zero, then no weights are updated. If the difference is non zero, the error is computed from the delta rule and is propagated back through the network. The idea, similar to that of the delta rule, is to adjust the weights to minimize the difference between the real output and the expected output. These networks can learn arbitrary associations by applying differentiable activation functions.

3.3   Prediction Networks

Neural networks are capable of anticipating the most likely things that are to happen. In that respect, there are numerous practical applications in which anticipation or prediction can help in setting priorities. This is the basic idea behind creation of a network group known as prediction networks.

3.3.1   The Perceptron

The perceptron learning rule is a method for finding the weights in a network for pattern classification. Basically the perceptron network consists of a single neuron with a number of adjustable weights. The original perceptron consisted of three units: sensory, associator, and the response units as shown in Figure 3.4. The sensory and associator units had binary activations and an activation of +1, 0 or -1 was used for the response unit.

Architecture

The neuron is the fundamental processor of a neural network. It has three basic elements:

1. A set of connecting links (or *synapses*); each link carries a weight (or *gain*).

2. A *summation* (or *adder*) sums the input signals after they are multiplied by their respective weights.

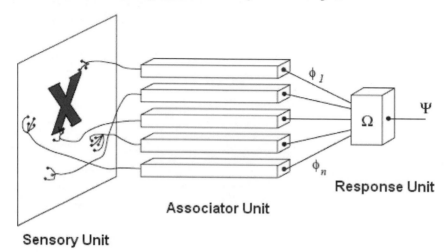

FIGURE 3.4: Original Perceptron

3. An *activation function f(x)*, limits the output of the neuron. Typically the output is limited to the interval [0,1] or alternatively [−1,1].

The architecture of a single layer perceptron is shown in Figure 3.5. The input to the response unit is the output from the associator unit, which is a binary vector. Since the weight adjustment is done only between the associator and the response unit, the network is referred to as a single layer network. The architecture shown consists of only the associator and the response units. The input layer consists of neurons $X_1...X_i...X_n$ which are connected with the output neurons with weighted interconnections. There exists a common bias "1".

Algorithm

The training and testing algorithm for a single layer perceptron is shown below. The algorithm can be implemented in MATLAB.

Parameters

```
x: Input vector (x₁, ... ,xᵢ, ... xₙ)
t: Target vector
w: Weight vector for the output unit (w₁, ... ,wⱼ, ...
wⱼ)
y: Activation function f(y_in)
b: Bias value
```

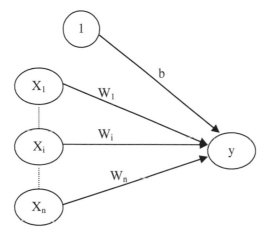

FIGURE 3.5:   Architecture of Single layer Perceptron

θ: Threshold value

Training Algorithm

The pseudocode of the training algorithm for the single layer perceptron is given as follows

Initialize the weights (either to zero or to a small random value)
Set Learning rate α (in the interval [0,1])
While not stopping condition **do**,
      For each training pair s:t
      Assign input unit activations $x_i = s_i$
      Compute the response of output unit:

$$y\_in = b + \sum_i x_i w_i$$

$$y = f(y\_in) = \begin{cases} 1, & if \ y\_in > \theta \\ 0, & if - \theta \le y\_in \le \theta \\ -1, & if \ y\_in < -\theta \end{cases}$$

Update weights and bias according to,
      If y $\ne$ t,
$$w_i(new) = w_i(old) + \alpha t x_i$$
$$b(new) = b(old) + \alpha t$$

    else

$$w_i(new) = w_i(old)$$
$$b(new) = b(old)$$

      End If
    End For
End While

The stopping condition may be no change in weights.

Testing Algorithm

The pseudocode of the testing algorithm for the single layer perceptron is given as follows

```
Set the weights from the training algorithm
    For each input vector x
        Compute the response of output unit:
```
$$y\_in = \sum_i x_i w_i$$

$$y = f(y\_in) = \begin{cases} 1, & if \ y\_in > \theta \\ 0, & if -\theta \le y\_in \le \theta \\ -1, & if \ y\_in < -\theta \end{cases}$$

```
    End For
End
```

Perceptron Training Algorithm for Several Output Classes

The pseudocode of the training algorithm for the single layer perceptron with several output classes is given as follows

```
Initialize the weights (either to zero or to a small
random value)
Set Learning rate α (in the interval [0,1])
While not stopping condition do,
    For each training pair
    Assign input unit activations xᵢ = sᵢ
    Compute the response of output unit:
```
$$y\_in\,j = \sum_i x_i w_i \ \text{for j=1 ... m}$$

$$yj = f(y\_inj) = \begin{cases} 1, & if \ y\_inj > \theta \\ 0, & if -\theta \le y\_inj \le \theta \\ -1, & if \ y\_inj < -\theta \end{cases}$$

```
    Update weights and bias according to,
```

If $y_j \neq t_j$,
  $w_{ij}\ (new)\ =\ w_{ij}\ (old)\ +\ \alpha t_j x_i$
  $b_j\ (new)\ =\ b_j\ (old)\ +\ \alpha t_j$
else
  $wi\ (new)\ =\ wi\ (old)$
  $b\ (new)\ =\ b\ (old)$

  End If
 End For
End While

Perceptron Testing Algorithm for several output classes

The pseudocode of the testing algorithm for the single layer perceptron with several output classes is given as follows

```
Set the weights from the training algorithm
    For each input vector x
        Compute the response of output unit:
```

$$y\_inj - \sum_i x_l w_i$$

$$y = f(y\_inj) = \begin{cases} 1, & if\ \ y\_inj\ >\ \theta \\ 0, & if -\theta \leq y\_inj \leq \theta \\ -1, & if\ \ y\_inj\ <\ -\theta \end{cases}$$

 End For
End

3.3.2 MATLAB Implementation of a Perceptron Network

Consider a 4-input and 1-output problem, where the output should be "one" if there are odd number of 1s in the input pattern and "zero" otherwise.

Enter the input and the target vectors

```
clear
inp=[0 0 0 0 0 0 0 0 1 1 1 1 1 1 1 1;0 0 0 0 1 1 1 1
1 0 0 0 0 1 1 1 1;...
0 0 1 1 0 0 1 1 0 0 1 1 0 0 1 1;0 1 0 1 0 1 0 1 0
1 0 1 0 1 0 1];
out=[0 1 1 0 1 0 0 1 1 0 0 1 0 1 1 0];
```

Create a network using the newff function in MATLAB Neural Network toolbox. The function newff creates a feedforward network. It requires four inputs and returns the network object. The first input is an R by 2 matrix of minimum and maximum values for each of the R elements of the input vector. The second input is an array containing the sizes of each layer. The third input is a cell array containing the names of the transfer functions to be used in each layer. The final input contains the name of the training function to be used.

```
network=newff([0 1;0 1; 0 1; 0 1],[6 1],
'logsig','logsig');
```

Before training a feedforward network, the weights and biases must be initialized. The newff command will automatically initialize the weights, but it may be required to reinitialize them. This can be done with the command init.

```
network=init(network);
```

The function sim simulates a network. The function sim takes the network input, and the network object net, and returns the network outputs.

```
y=sim(network,inp);
figure,plot(inp,out,inp,y,'o'),
title('Before Training');
axis([-5 5 -2.0 2.0]);
```

There are seven training parameters associated with traingd: epochs, show, goal, time, min_grad, max_fail, and lr. Here the number of epochs and the learning rate are defined as follows

```
network.trainParam.epochs = 500;
network.trainParam.lr = 0.05;
```

Training the network. Simulate the network and plot the results (Figure 3.6)

```
network=train(network,inp,out);
y=sim(network,inp);
figure,plot(inp,out,inp,y,'o'),
title('After Training');
axis([-5 5 -2.0 2.0]);
```

Updated weights and bias values

```
Layer1_Weights=network.iw1
Layer1_Bias=network.b1
```

```
Layer2_Weights=network.lw2
Layer2_Bias=network.b2
Actual_Desired=[y' out']
```

Output:
```
TRAINLM, Epoch 0/500, MSE 0.376806/0,
Gradient 0.527346/1e-010
TRAINLM, Epoch 25/500, MSE 5.86095e-005/0,
Gradient 0.0024019/1e-010
TRAINLM, Epoch 34/500, MSE 1.67761e-012/0,

Gradient 2.89689e-011/1e-010
TRAINLM, Minimum gradient reached, performance goal
was not met.
```

Layer1_Weights =

```
   -9.5620     9.4191     6.5063    11.2360
    8.7636    10.0589   -10.4526    -8.9109
    9.8375   -11.4056    14.3788    -5.5974
    7.4334    10.0157   -13.7152     5.9788
  -10.6286    14.6436    10.6738   -15.2093
    8.7772    -6.0128     6.6699     9.0275
```

Layer1_Bias =

```
 -11.8265
   -5.8238
    4.2346
   -7.7644
   -5.0030
   -6.0732
```

Layer2_Weights =

```
    -28.4686 17.7464 -10.1867 -26.2315 28.9088
27.1797
```

Layer2_Bias =

```
   -4.4138
```

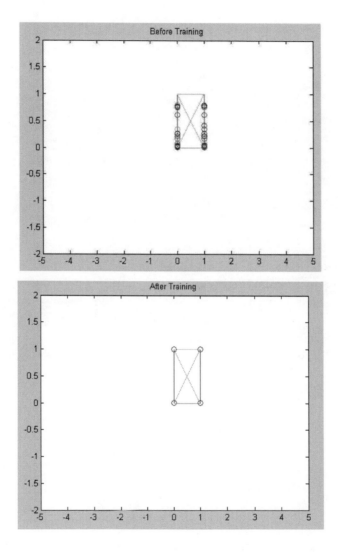

FIGURE 3.6: Pattern before Training and after Training, Performance and Goal at the end of 34 Epochs

3.3.3   Feedforward Back-Propagation Network

The feed-forward, back-propagation architecture was developed in the early 1970's by Rumelhart, Hinton and Williams. Presently, the back-propagation architecture is the commonest, most efficient, and easy model for complex, multi-layered networks. The network is used in a wide range of applications ranging from image processing, signal processing, face recognition, speech recognition, etc. This network architecture has bred a large class of network types with several different topologies and various training methods. The major capability is to obtain nonlinear solutions to undefined problems.

Architecture

The typical back-propagation network has an input layer, an output layer, and at least one hidden layer. There is no theoretical limit on the number of hidden layers but typically there is just one or two. Some work has been done which indicates that a minimum of four layers (three hidden layers plus an output layer) are required to solve problems of any complexity. Each layer is fully connected to the succeeding layer, as shown in Figure 3.7.

The in and out layers indicate the flow of information during recall. Recall is the process of putting input data into a trained network and receiving the answer. Back-propagation is not used during recall, but only when the network is learning a training set.

The number of layers and the number of processing element per layer are important decisions. These parameters to a feedforward, back-propagation topology are also the most ethereal. There are only general rules picked up over time and followed by most researchers and engineers applying this architecture of their problems.

Rule A: The number of the processing elements in the hidden layer should increase as the complexity in the input and output relationship increases.

Rule B: If the process being modeled is separable into multiple stages, then additional hidden layer(s) may be required. If the process is not separable into stages, then additional layers may simply enable memorization and not a true general solution.

Rule C: The amount of training data available sets an upper bound for the number of processing elements in the hidden layers. To calculate this upper bound, use the number of input output pair examples in the training set and divide that number by the total number of input

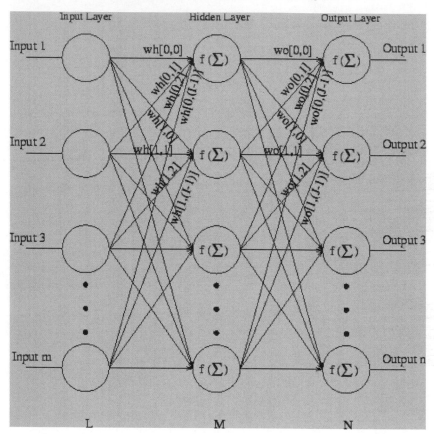

FIGURE 3.7: Back-Propagation Network

and output processing elements in the network. Then divide that result again by a scaling factor between five and ten. Larger scaling factors are used for relative noisy data. A highly noisy input data requires a factor of twenty or fifty, while very non-noisy input data with an exact relationship to the output unit might drop the factor to approximately two. Therefore it is necessary that the hidden layers possess a few processing elements.

As soon as the network is created, the process of teaching commences. This teaching makes use of a learning rule which is the variant of the Delta Rule. The rule starts with determining the error which is the difference between the actual outputs and the desired outputs. Based on this error, the connection weights are increased in proportion to the error times for global accuracy. While doing this for an individual node the inputs, the output, and the desired output should be present at the same processing element. The composite section of this learning rule is for the system to determine which input contributed the most to an incorrect output and how does that element get changed to correct the error. An inactive node would not contribute to the error and would have no need to update its weights.

To figure out this problem, training inputs are put on to the input layer of the network, and desired outputs are compared at the output layer. During the learning process, a forward sweep is made through the network, and the output of each element is computed layer by layer. The difference between the actual output and the target output is back-propagated to the previous layer(s), which is modified by the derivative of the transfer function, and the connection weights are normally adjusted using the Delta Rule. This process proceeds for the previous layer(s) until the input layer is reached.

There are many variations to the learning rules for back-propagation network. Different error functions, transfer functions, and even the modifying method of the derivative of the transfer function can be used. The concept of momentum error was introduced to allow for more prompt learning while minimizing unstable behavior. Here, the error function, or delta weight equation, is modified so that a portion of the previous delta weight is fed through to the current delta weight. This acts, in engineering terms, as a low-pass filter on the delta weight terms since general trends are reinforced whereas oscillatory behavior is canceled out. This allows a low, normally slower, learning coefficient to be used, but creates faster learning.

One more common technique that is used for the network to converge is to only update the weights after many pairs of inputs and their desired outputs are presented to the network, rather than after every presentation. This kind of network is referred to as cumulative back-propagation because the delta weights are not accumulated until the complete set of pairs is presented. The number of input-output pairs that are presented during the accumulation is referred to as an epoch. This epoch may correspond either to the complete set of training pairs or to a subset.

Algorithm of a Back Propagation Network

This section discusses the training and application algorithm of a feed forward back propagation network.

Training Algorithm

The training algorithm of back propagation involves four stages

1. Initialization of weights

2. Feed forward

3. Back propagation of errors

4. Updation of the weights and biases

During first stage, which is the initialization of weights, some small random values are assigned.

During feed forward stage each input unit (X_i) receives an input signal and transmits this signal to each of the hidden units $z_1 \ldots z_p$. Each hidden units then calculates the activation function and sends its signal z_j to each output unit. The output unit calculates the activation function to form the response of the net for the given input pattern.

During back propagation of errors, each output unit compares its computed activation y_k with its target value t_k to determine the associated error for that pattern with that unit. Based on the error, the factor $\delta_k(k=1, \ldots m)$ is computed and is used to distribute the error at output unit y_k back to all units in the previous layer. Similarly the factor δ_j $(j=1 \ldots p)$ is computed for each hidden unit z_j.

During final stage, the weight and biases are updated using the δ factor and the activation.

Parameters

The various parameters used in the training algorithm is as follows:

x : Input training vector x= $(x_1, \ldots x_i, \ldots x_n)$

t : Output target vector t= $(t_1, \ldots t_k, \ldots t_m)$

δ_k= error of output unit y_k

δ_j= error of output unit z_j

α = learning rate

v_{oj}= bias on hidden unit j

z_j = hidden unit j

w_{oK}=bias on output unit k

y_k = output unit k

The pseudocode of the training algorithm of the back propagation network is as follows. The algorithm is given with the various phases:

```
Initialization stage:
Initialize weights v and w to small random values
Initialize Learning rate
While not stopping condition for phase I Do
      For each training pair
```

Feed forward stage:
Each input unit receives the input signal
xi and transmits to all units in the hidden
layer
Each hidden unit (z_j, j=1,...,p) sums its
weighted input signals

$$z_{inj} = v_{oj} + \sum_{i=1}^{n} x_i v_{ij}$$

applying activation function
zj = f(z_{inj})
and sends this signal to all units.
Each output unit (y_k, k=1,...,m) sums its
weighted input signals

$$y_{ink} = w_{ok} + \sum_{j=1}^{p} z_j W_{jk}$$

applies its activation function to
calculate output signals
y_k= f(y_{ink})

Back propagation of errors:
Each output unit (y_k, k=1,..., m) receives a target pattern
corresponding to an input pattern, error information term
is calculated as
$$\delta k = (t_k - y_k)f(y_{ink})$$

Each hidden unit (z_j, j=1,....,p) sums its
delta inputs from units in the layer above

$$\delta_{inj} = \sum_{k-1}^{m} \delta j W_{jk}$$

The error information is calculated as
$$\delta_j = \delta_{inj} f(z_{inj})$$

Updation of weights and biases:
Each output unit (y_k, k=1,...,m) updates
its bias and weights (j=0,,p)
The weight correction term is given by
$$\Delta W_{jk} = \alpha \delta_k z_j$$
The bias correction term is given by
$$\Delta W_{ok} = \alpha \delta_k$$
$$W_{jk(new)} = W_{jk(old)} + \Delta W_{jk}, W_{ok(new)} = W_{ok(old)} + \Delta W_{ok}$$

Each hidden unit (z_j, j=1,...,p) updates
its bias and weights (i=0,...,n)

The weight correction term

$$\triangle V_{ij} = \alpha \delta_j x_i$$

The bias correction term

$$\triangle V_{oj} = \alpha \delta_j$$

$$V_{ij(new)} = V_{ij(old)} + \triangle V_{ij}, V_{oj(new)} = W_{oj(old)} + \triangle V_{oj}$$

End For

Test the stopping condition

End while

End

The stopping condition may be the minimization of the errors, number of epochs etc.

Choice of Parameters

Parameters that are used for training should be selected such that the network operates efficiently. The initial value assignments are discussed in the following section.

Initial Weights: Initial weights determine whether the error gradient reaches a global (or only a local) minima in order for the network to converge. If the initial weight is too large, the initial input signals to each hidden or output unit will fall in the saturation region where the derivative of the sigmoid has a very small value (f(net)=0). If initial weights are too small, the net input to a hidden or output unit will approach zero, which then causes extremely slow learning. In,order to obtain optimum results the initial weights and biases are set to random numbers between -0.5 and 0.5 or between -1 and 1.

Selection of Learning Rate: A high learning rate leads to rapid learning but the weights may oscillate, while a lower learning rate leads to slower learning. Methods suggested for adopting learning rate are as follows:

- Start with a high learning rate and steadily decrease it. Changes in the weight vector must be small in order to reduce oscillations or any divergence.

- A simple suggestion is to increase the learning rate in order to improve performance and to decrease the learning rate in order to worsen the performance.

- Another method is to double the learning rate until the error value worsens.

Learning in Back Propagation

There are two types of learning.

i. Sequential learning or pre-pattern method

ii. Batch learning or pre-epoch method

In sequential learning a given input pattern is propagated forward, the error is determined and back propagated, and the weights are updated.

In Batch learning the weights are updated only after the entire set of training network has been presented to the network. Thus the weight update is only performed after every epoch.

If p= patterns in one epoch, then

$$\triangle w = \frac{1}{p} \sum_{p-1}^{\alpha} \triangle w_p$$

In some cases, it is advantageous to accumulate the weight correction terms for several patterns and make a single weight adjustment (equal to the average of the weight correction terms) for each weight rather than updating the weights after each pattern is presented. This procedure has a "smoothing effect". In some cases the smoothing may increase the chances of convergence to the local minimum.

Time taken to train a net

The need for applying back propagation network is to accomplish a balance between memorization and generalization; it is not needfully appropriate to proceed training until the error reaches minimum value. The two disjoint sets of data used during training are:

• Set of training patterns

• Set of training - testing patterns.

Training patterns have a great influence on weight adjustments. As the error in a neural network increases the network starts to store and memorise the training patterns.

Number of Training Pairs

The number of training matches likewise plays a crucial part on training of the nets. A simple thumb rule is utilized to ascertain the number of training pairs.

Consider a net trained to classify the fraction (1-**e**/2) of the trained patterns correctly. This means it will also classify (1-**e**) of the testing pattern correctly by using the following condition.

If there are enough training patterns the net will be able top generalize as desired. Enough training pattern is given by w/p=e, where

p=number of training patterns
and w=number of weights to be trained.
The value of e lies between 0 to 1/8. The necessary condition is given by
p>| w |/(1-a)
where a= expected accuracy for the test set.
Sufficient condition is given as
p>|w|/(1-a)log(n/(1-a))
where n= number of nodes.

Number of Hidden Units

In general, the choice of hidden units depends on many factors such as:
- numbers of input and output units
- number of training patterns
- noise in the target vectors
- complexity of the function or classification to be learned
- architecture
- type of hidden unit activation function
- training algorithm
- regularization

In a number of situations, there is no specific way to determine the best number of hidden units without training several networks and estimating the generalization error of each. If there are a few hidden units, then the training error and generalization error are very high due to underfitting and there occurs high statistical bias. If there are too many hidden units, then the training error is too low but still has high generalization error due to overfitting and high variance.

Momentum Factor

The weight updation in a BPN is proceeded in a direction which is a combination of present gradient and the former gradient. This kind of an approach is advantageous while a few training data differ from a majority of the data. A low value of learning rate is applied to avoid great disturbance in the direction of learning whenever identical and unusual pair of training patterns is presented. The speed of convergence increases as if the momentum factor is added to the weight updation rule. The weights from one or more former training patterns must be saved in order to use momentum. For the BPN with momentum, the new weights for training step t+2, is based on t and t+1. It is found that momentum allows the net to perform large weight adjustments as long as the correction proceeds in the same general direction for several patterns. Hence applying momentum, the net does not go forward in the direction of the gradient, just travels in the direction of the combination of the present gradient and the former gradient for which the weight updation is made. The main purpose of the momentum is to speed up the convergence of error propagation algorithm. This method makes the current weight adjustment with the fraction of the recent weight adjustment.

The weight updating formula for BPN with momentum is,

$$w_{jk}(t+1) = w_{jk}(t) + \alpha\delta_k z_j + \mu[w_{jk}(t) - W_{jk}(t-1)]$$
$$v_{jk}(t+1) = v_{jk}(t) + \alpha\delta j x_i + \mu[v_{ij}(t) - v_{ij}(t-1)]$$

μ is called the momentum factor. It ranges from $0<\mu<1$.

Application Algorithm

The application procedure for BPN is shown below:

```
Initialize weights from (training algorithm).
      For each input vector
          For i =1,...., n ;
              set activation of input unit, xᵢ;
          End For
          For j =1, ., p ;
```
$$Z_{inj} = V_{aj} + \sum_{i-1}^{n} X_i V_{ij}$$
```
          End For
          For k=1, ., m ;
```
$$Y_{inj} = W_{ok} + \sum_{j-1}^{p} Z_j W_{jk}$$
```
          End For
      End For
  End
```

3.3.4 Implementation of BPN Using MATLAB

MATLAB has a suite of inbuilt functions designed to build neural networks, the Neural Networks Toolbox. Basically a programmer has to go through three phases to implement neural networks: Design, Training, and Testing.

- Design phase:

  - Define the number of nodes in each of the three layers (input, hidden, output)
  - Define the transfer functions
  - Initialize the training routine that is to be used.
  - Optional parts of the design: Error function (Mean Square Error is the default), plot the progress of training, etc.

- Training phase:

  - Once the network has been designed, the network is designed by optimizing the error function
  - This process determines the "best" set of weights and biases for the users data set.

- Testing phase

  - Test the network to see if it has found a good balance between memorization (accuracy) and generalization

The BPN network can be implemented in MATLAB using the Neural Network Toolbox as follows.

```
clc;
clear all;
close all;
```

The first step in training a feedforward network is to create the network object. The function newff creates a feedforward network. It requires four inputs and returns the network object. The first input is an R by 2 matrix of minimum and maximum values for each of the R elements of the input vector. The second input is an array containing the sizes of each layer. The third input is a cell array containing the names of the transfer functions to be used in each layer. The final input contains the name of the training function to be used.

```
% Creating a network
net=newff([-1 2; 0
5],[3,1],'tansig','purelin','traingd');
```

Before training a feedforward network, the weights and biases must be initialized. The newff command will automatically initialize the weights, but it may be required to reinitialize them. This can be done with the command init.

```
Initialising weights
net = init(net);
```

The function `sim` simulates a network. The function `sim` takes the network input p, and the network object net, and returns the network outputs a.

```
% Simulating the network
p = [1 2 3 4 5;6 7 8 9 0];
a = sim(net,p)
```

There are seven training parameters associated with `traingd`: epochs, show, goal, time, min_grad, max_fail, and lr. The learning rate lr is multiplied times the negative of the gradient to determine the changes to the weights and biases. The larger the learning rate, the bigger the step. If the learning rate is made too large, the algorithm becomes unstable. If the learning rate is set too small, the algorithm takes a long time to converge. The training status is displayed for every show iteration of the algorithm. (If show is set to NaN, then the training status never displays.) The other parameters determine when the training stops. The training stops if the number of iterations exceeds epochs, if the performance function drops below goal, if the magnitude of the gradient is less than mingrad, or if the training time is longer than time seconds.

```
% Parameter setting
net.trainParam.show = 50;
net.trainParam.lr = 0.05;
net.trainParam.epochs = 300;
net.trainParam.goal = 1e-5;
```

Now the network is ready for the training phase

```
% Training the network
t = [1 -1 1 1 -1 1];
  [net,tr]=train(net,p,t);
```

The output obtained is shown in Figure 3.8
    The observations are shown below

```
TRAINGD, Epoch 0/300, MSE 1.27137/1e-005, Gradient
2.30384/1e-010
TRAINGD, Epoch 50/300, MSE 0.730344/1e-005, Gradient
0.17989/1e-010
TRAINGD, Epoch 100/300, MSE 0.684018/1e-005,
Gradient 0.0927535/1e-010
```

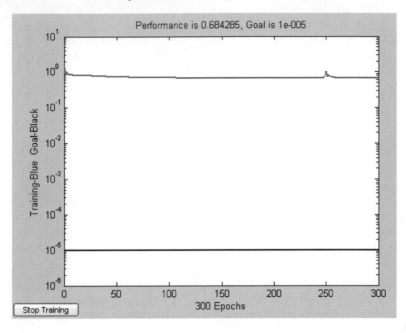

FIGURE 3.8:   Performance and Goal of BPN Network

```
TRAINGD, Epoch 150/300, MSE 0.671868/1e-005,
Gradient 0.0528921/1e-010
TRAINGD, Epoch 200/300, MSE 0.666293/1e-005,
Gradient 0.0428047/1e-010
TRAINGD, Epoch 250/300, MSE 1.0139/1e-005, Gradient
3.96228/1e-010
TRAINGD, Epoch 300/300, MSE 0.684285/1e-005,
Gradient 1.13772/1e-010
TRAINGD, Maximum epoch reached
```

Limitations of Back Propagation Network

A few limitations are also available while using the feedforward back-propagation network. Back-propagation demands more of supervised training, with numerous input-output examples. In addition, the internal mapping operations are not clear to understand, and also there is no assurance that the system will meet to an acceptable solution. From time to time, the learning process grinds to a halt in a local minima, limiting the optimal solution.

This kind of stucking operation occurs when the network system finds an error that is lower than the circumventing possibilities but does not finally get to the smallest possible error. Numerous learning applications add a term to the computations to bump or jog the weights preceding superficial barriers and find the actual minimum rather than a temporary fault.

Applications

- speech synthesis from text

- robot arms

- evaluation of bank loans

- image processing

- knowledge representation

- forecasting and prediction

- multi-target tracking

3.3.5 Delta Bar Delta Network

The architecture of a delta bar delta network is similar to that of a back-propagation network. The major difference between the two networks lie in the algorithmic method of learning. Delta bar delta was developed by Robert Jacobs in order to overcome the limitations of BPN by improving the learning rate of standard feedforward, back-propagation networks.

The back-propagation learning process is based on a steepest descent method that minimizes the networks anticipation error during the process where the connection weights to each artificial neuron are updated. The momentum term is assigned initially and the learning rates are assigned on a layer-by-layer basis. The back-propagation approach allows the learning rate to decrease step by step as more numbers of training sets are presented to the network. Though this technique is eminent in working out many practical applications, the convergence rate of the procedure is too slow.

The learning method used in the delta bar delta network uses a weight that has its own self-adapting coefficient. Here no momentum factor is used. The persisting operations of the network, such as back propagating the error to the previous layer etc., are identical to the normal back-propagation architecture. Delta bar delta is a heuristic approach to training artificial networks. Here the preceding error values can be used to infer future calculated error values. A knowledge of the probable errors enables the system to update the weights in an intelligent way. However, this approach is complicated in that empirical evidence suggests that each weight may have quite different effects on the overall error. Jacobs then suggested the common sense notion that the back-propagation learning rules should account for these variations in the effect on the overall error. To put differently, every connective weight of a network should have its own learning rate. The claim is that the step size appropriate for one connection weight may not be appropriate for all weights in that layer. Additionally the learning rates ought to be varied over time. By assigning a learning rate to each connection weight and permitting this learning rate to change continuously over time, more degrees of freedom are introduced to reduce the time to convergence.

Rules which directly apply to this algorithm are straight forward and easy to implement. The learning rates are altered based on the present error information found with standard back-propagation. During weight updation, if the local error has the same sign for several consecutive time steps, the learning rate for that connection is linearly increased. Incrementing the learning rate linearly prevents the learning rates from becoming too large. When the local error changes signs frequently, the learning rate is decreased geometrically. Decrementing geometrically ensures that the connection learning rates are always positive. Further, they can be decreased more rapidly in regions where the change in error is large.

Learning Rule of the Delta Bar Delta Network

Depending upon the learning rate the following heuristics are followed in order to train the network:

i. When the sign of the derivative of a weight is the same on several consecutive steps, the corresponding learning rate should be increased.

ii. When the sign of the derivative of a weight alternates on consecutive steps, the corresponding learning rate should be decreased.

The delta bar delta rule is a variation of the generalized delta rule and implements the heuristics mentioned above. In fact, it consists of two rules: one for weight adjustment and the other for control of the learning rates.

The weights are modified as:

$$\vec{W}(t+1) = \vec{W}(t) - \sum_{i=1}^{n} \eta_i(t) E_{i,i} \bigtriangledown_{\vec{w}} F(\vec{w})$$

where $\vec{W}(t)$ is the weight vector's value at time step t
$\eta_i(t)$ is the learning rate at time step t

$E_{i,i}$ is an n x n matrix, with every component = 0,
except with the component with
row=column=i, which is =1, $(1 \leq i \leq n)$
$F(\vec{w})$ is the minimum of $\vec{w}$
Every learning rate is adjusted according to

$$\eta(t+1) = \eta(t) + \triangle\eta(t)$$

where

$$\eta(t) = \begin{cases} k, & if\bar{\delta}(t-1)\delta(t)>0 \\ -\phi\eta(t), if\bar{\delta}(t-1)\delta(t)<0 \\ 0, & else \end{cases}$$

where

$$\delta(t) - \frac{\partial F(t)}{\partial W(t)} \text{ and } \bar{\delta}(t) = (1-\theta)\delta(t) + \theta\bar{\delta}(t-1)$$

Here k > 0 , $\phi \in [0,1]$ and $\theta \in [0,1]$ are constants.k, ϕ and are θ the same for every learning rate. $\bar{\delta}$ is the exponential average of the current and past derivatives of w.

By allowing different learning rates for each connection weight in a network, there is no need for the steepest descent search that is to search direction of the negative gradient. Therefore the connection weights are updated by partially differentiating the error with respect to the weight itself. The weight updations satisfy the locality constraint, that is, they need data only from the processing elements to which they are connected.

3.3.6 Extended Delta Bar Delta

To overcome the limitations of delta bar delta network, Ali Minai and Ron Williams developed the extended delta bar delta algorithm to enhance the delta bar delta by applying an exponential decay to the learning rate increase. As discussed in the section on back-propagation, momentum is a factor used to smooth the learning rate. It is a term added to the standard weight updation, which is proportional to the previous weight change. In this way, by applying the momentum factor good general trends are reinforced, and oscillations are dampened.

The learning rate and the momentum rate for each weight have separate constants controlling their increase and decrease. Once again, the sign of the current error is used to indicate whether an increase or decrease is appropriate. As the learning rate and momentum rate increases, they are modified as exponentially decreasing functions. Thus, greater increases will be applied in areas of small slope or curvature than in areas of high curvature. This is a partial answer to the jump and oscillation problem of delta bar delta.

Further to prevent the jumps and oscillations in the weights, ceilings are placed on the individual connection learning rates and momentum rates along with a memory with a recovery feature. As the training set is presented in each epoch, the accumulated error is evaluated. If the error is less than the previous minimum error, the weights are saved in memory as the current best. Similarly, if the current error exceeds the minimum previous error, modified by the tolerance parameter, than all connection weight values return stochastically to the stored best set of weights in memory. Furthermore, the learning and momentum rates are decreased to begin the recovery process.

In order to overcome the shortcomings of the delta bar delta learning algorithm, the Extended Delta-Bar-Delta (EDBD) algorithm was suggested which was based on the following considerations:

1. Even when the direction of the slope is not changing, an increase in steepness often signals the approach of a minimum. Since, for steepest descent, q is a multiplicative factor and the actual step size is proportional to the slope, so is the effect of any learning rate increment. Using a constant learning rate increment thus means that step size increases more rapidly on steep slopes than on shallow ones. Clearly, this is the opposite of what is required. Intuitively, the descent on steep slopes should be more careful lest the minimum be overshot beyond recovery. At the very least, it should not be any more reckless than it is in flat areas! Also, as pointed out by Jacobs, the direction of steepest descent points toward the minimum only when the slope in all directions is equal. If slopes are not equal, the orientation of the descent vector can be improved (under certain topological assumptions) by taking disproportionately large steps in flat directions and disproportionately smaller steps in steep directions. However, this must not be overdone, or most of the time will be spent spiraling instead of descending.

2. In DBD, momentum leads to divergence because it magnifies learning rate increments. Since the step size is largest on steep slopes preceding a minimum, this is also precisely when momentum is at its most dangerous, and should be kept in check. On flat areas, however, it provides added speed-up and should be used to the utmost.

3. Increasing learning rate (or momentum) without bound can lead to divergence even with small increments, depending, of course, upon the error surface. Thus, growth should be capped or tapered off to preclude this event. Since such preclusion is impossible to

guarantee for all situations, some sort of memory and recovery should be included in the algorithm.

Based on these points, the following changes were made to the delta bar delta learning algorithm:

1. The learning rate increase was made an exponentially decreasing function of $|\bar{\delta}(t)|$ instead of being a constant. This meant that learning rate would increase faster on very flat areas than on areas of greater slope. This allowed the use of fairly large increments in flat areas without risking wild jumps from steep slopes. Also, using an exponential ensured that deviations from the steepest descent direction would be significant only when different slopes were markedly disproportionate and compensating distortion was needed.

2. Momentum was used as a standard part of the algorithm, but was varied just like the learning rate. Thus, momentum was increased on plateaus and decreased exponentially near minima. The DBD criiterion was used for this purpose too, but the increment factor was again a decreasing exponential function of $|\bar{\delta}(t)|$.

3. To prevent either the learning rate or momentum from becoming too high, a ceiling was defined for both at which they were hard-limited. This further facilitated the use of large increments, since their effect was not unbounded.

4. Memory and recovery were incorporated into the algorithm. Thus, the best result seen until the current time step was saved. A tolerance parameter ? was used to control recovery. If the error became greater than ? times the lowest seen so far, the search was restarted at the best point with attenuated learning rate and momentum. To prevent thrashing, this was done stochastically, so there was a small probability P that the search would restart at a totally new point.

The equations for the EDBD algorithm can be written as follows:

$$\triangle W_{ij}(t) = -\eta_{ij}(t)\frac{\partial E(t)}{\partial W_{ij}} + \mu_{ij}(t)\triangle W_{ij}(t-1)$$

$$\eta_{ij}(t+1) = IN[\eta_{max}, \eta_{ij}(t) + \triangle\eta_{ij}(t)]$$

$$\mu_{ij}(t+1) = MIN[\mu_{max}, \mu_{ij}(t) + \triangle\mu_{ij}(t)]$$

where

$$\triangle\eta_{ij}(t) = \begin{cases} k_l exp(-\gamma_l \mid \bar{\delta}_{ij}(t) \mid) \\ -\phi_l\eta_{ij}(t) \\ 0 \end{cases}$$

$$\triangle\mu_{ij}(t) = \begin{cases} k_m exp(-\gamma_m \mid \bar{\delta}_{ij}(t) \mid) & if\bar{\delta}_{ij}(t-1)\delta_{ij}(t)>0 \\ -\phi_m\eta_{ij}(t) & if\bar{\delta}_{ij}(t-1)\delta_{ij}(t)<0 \\ 0 & otherwise \end{cases}$$

Overall, it appears that the EDBD algorithm in its current form is quite successful, at least for networks learning in real-valued, continuous problem domains. Success with small problems notwithstanding, we think that EDBD should be evaluated separately for binary domain learning, and experiments are underway in that direction. Also, since learning rate variability and momentum confer an ability to escape local minima, it might be worthwhile to investigate some sort of annealing procedure, whereby learning rate variability and/or momentum decreases with time or decreasing error. This should make the search even more robust and effective in terms of solution quality.

3.3.7 Directed Random Search Network

All the network architectures that were discussed in the previous sections were based on learning rules, or algorithmic paradigms. These algorithms applied a gradient descent technique to update the weights. Unlike thse networks, the directed random search makes use of a standard feedforward recall structure. The directed random search updates the weights in a random manner. To regularize this random weight updation, a direction component is added to insure that the weights tend toward an antecedently successful search direction. Totally all the processing elements available in the network are regulated individually.

The directed random search algorithm has numerous advantages. The problem has to be well understood while using directed random search so that the optimal results occur when the initial weights are within close proximity to the best weights. The algorithm learns at a very fast rate since the algorithm cycles through its training much more quickly than calculus-bases techniques like the delta rule and its variations. No error terms are computed for the intermediate processing elements and only one output error is computed. The given problem should accumulate a small network since the training becomes too long and complex if the number of connections are large.

In order to maintain the weights within a compact region that is the area in which the algorithm works best, an upper bound is required on the weight's magnitude. Therefore, the weight's bounds are set reasonably high, the network is still allowed to seek what is not exactly known - the true global optimum. The second key parameter to this learning rule involves the initial variance of the random distribution of the weights. In most of the commercial packages there is a vendor recommended number for this initial variance parameter. Yet, the setting of this number is not all that important as the self-adjusting feature of the directed random search has proven to be robust over a wide range of initial variances.

There are four key components to a random search network. They are the random step, the reversal step, a directed component, and a self-adjusting variance.

Random Step: A random value is added to each weight. Then, the entire training set is run through the network, producing a "prediction error." If this new total training set error is less than the previous best prediction error, the current weight values (which include the random step) becomes the new set of "best" weights. The current anticipation error is then saved as the new, best prediction error.

Reversal Step: If the random step's results are worse than the previous best, then the same random value is subtracted from the original weight value. A set of wieghts are produced such that they are in the opposite direction when compared to the previous random step. If the total "prediction error" is less than the previous best error, the current weight values of the reversal step are stored as the best weights. The current prediction error is also saved as the new, best prediction error. If both the forward and reverse steps fail, a completely new set of random values are added to the best weights and the process is then begun again.

Directed Component: In order to increase the speed of convergence, a set of directed components are created, based on the outcomes of the forward and reversal steps. These components speculate the history of success or failure for the previous random steps. The directed components, which are initialized to zero, are added to the random components at each step in the procedure. Directed components provide a common way for the network to train in the direction. Addition of directed components provide a dramatic performance improvement to convergence.

Self-adjusting Variance: An initial variance parameter is specified to control the initial size (or length) of the random steps which are added to the weights. An adaptive mechanism changes the variance parameter based on the current relative success rate or failure rate. The learning rule assumes that the current size of the steps for the weights is in the right direction if it records several consecutive successes, and it then expands to try even larger steps. Conversely, if it detects several consecutive failures it contracts the variance to reduce the step size.

Directed random search brings out good solutions for small and medium sized networks in a reasonable amount of time. The training is automatic, requiring little, user interaction. The number of connection weights imposes a practical limit on the size of a problem that this learning algorithm can effectively solve. If a network has more than 200 connection weights, a directed random search can require a relatively long training time and still end up yielding an acceptable solution.

3.3.8   Functional Link Artificial Neural Network (FLANN) or Higher-Order Neural Network

The network can be called as functional link artificial neural network or higher order artificial neural network. These neural networks, are an extended version of the standard feedforward, back-propagation network. Here the input layer processing elements include nodes to provide the network with a more complete understanding of the input. Basically, the inputs are transformed in a well-understood mathematical way so that the network does not have to learn some basic math functions. These functions do enhance the network's understanding of a given problem. These mathematical functions translate the inputs via higher-order functions such as squares, cubes, or sines. It is from the very name of these functions, higher-order or functionally linked mappings, that the two names for this same concept were derived.

This method has been shown to dramatically improve the learning rates of practical applications. An additional advantage to this extension of back-propagation is that these higher order functions can be applied to other derivations - delta bar delta, extended delta bar delta, or any other enhanced feedforward, back-propagation networks.

There are two basic methods to add additional input nodes to the input layer in FLANN.

- The cross-products of the input terms can be added into the model and this method is called as the output product or tensor model, where each component of the input pattern multiplies the entire input pattern vector. A reasonable way to do this is to add all interaction terms between input values. For example, for a back-propagation network with three inputs (A, B, and C), the cross-products would include: AA, BB, CC, AB, AC, and BC. This example adds second-order terms to the input structure of the network. Third-order terms, such as ABC, could also be added.

- The second method for adding additional input nodes is the functional expansion of the base inputs. Thus, a back-propagation model with A, B, and C might be transformed into a higher-order neural network model with inputs: A, B, C, SIN(A), COS(B), LOG(C), MAX(A,B,C), etc. In this model, input variables are individually acted upon by appropriate functions. Many different functions can be used. The overall effect is to provide the network with an enhanced representation of the input. It is even possible to combine the tensor and functional expansion models together.

No new information is added, but the representation of the inputs is enhanced. Higher-order representation of the input data can make the network easier to train. The joint or functional activations become instantly available to the model. While adding nodes in some applications, a hidden layer is no longer required. There are a few limitations to the network model. Many more input nodes must be processed to use the transformations of the original inputs. With higher-order systems, the problem is worsened. Eventually, because of the finite processing time of computers, it is important that the inputs are not expanded more than required to get an accurate solution.

Pao draws a distinction between truly adding higher order terms in the sense that some of these terms represent joint activations versus functional expansion which increases the dimension of the representation space without adding joint activations. While most developers recognize the difference, researchers typically treat these two aspects in the same way. Pao has been awarded a patent for the functional-link network, so its commercial use may require royalty licensing.

Architecture

The function approximation capability of FLANNs can be understood with an example as follows: Consider a MLP with d = 2 input units and h = 3 sigmoidal hidden units in the lone hidden layer as an example. The output function calculated for this neural network is

$$Y_k = G_k \left(\sum_{j=1}^{h} W_{jk} * \Phi_j(x) \right)$$

where W_{jk} is the weight connecting hidden unit j with output unit k and G_k the activation function employed by the output layer neurons.

The hidden layer units calculate a projection of original input space into an intermediate one by means of hidden layer weights, W_{ij} .

$$(X_1, ..., X_i, ...X_d) \longrightarrow (\Phi_1(X), ..., \Phi_j(x), ...\Phi_h(x))$$

$$\Phi_j(x) = G_j \left(\sum_{j=1}^{d} W_{ij} * x_i \right)$$

G_j is the activation function used at hidden layer neurons.

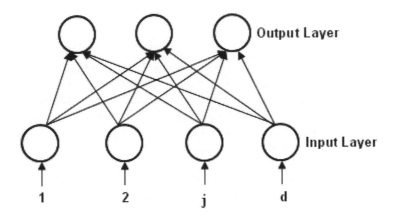

FIGURE 3.9: Architecture of FLANN

In this hidden space, linear discrimination, to be carried out by the output weights becomes easier than in the original input space. By contrast, linear networks used in FLANN take hidden units to the input layer and work with a single layer of weights. The output function is modified as

$$Y_k = G_k \left(\sum_{j=1}^{d'} W_{ik} * \Psi_i(x) \right)$$

where d' is the dimension of the new input space $(\Psi(x), ... \Psi_i(x), ... \Psi d'(x))$. The new units $\Psi_i(x)$ instead of being learnable arbitrary functions of the original attributes, are now fixed polynomial and trigonometric terms constructed out of the original attributes.

The FLANN network architecture is shown in Figure 3.9. The corresponding linear mapping in this polynomial space is, in fact, nonlinear in the original input space. FLANNs do not have any hidden layer and the computational cost, in fact, moves from the hidden layer to selection of suitable expanded inputs for the input layer. The expanded inputs are chosen using an evolutionary technique, which makes use of Genetic Algorithms and gradually evolves inputs of the FLANN to achieve the desired model.

The order of the polynomial used to obtain expanded inputs can be gradually increased so that minimum number of inputs is used and the complexity of the neural network is minimized. The evolutionary algorithm is reproduced below in brief:

1. The evolutionary algorithm begins from original input attributes. This makes sense since some problems can be solved linearly, perhaps after rejecting some noisy or irrelevant attributes.

2. Each input vector is encoded by means of a binary chromosome of length equal to the number of available polynomial terms. A bit 1 specifies that corresponding polynomial term is fed into the network.

3. Instead of choosing initial random population, it starts from a pool of single feature networks. For example if the system to be modeled has five inputs, the following initial pool of chromosomes is taken:

$$[(10000), (01000), (00100), (00010) and (00001)].$$

4. Roulette-wheel selection and single-point crossover have been employed in the GA used. The crossover and mutation probability is fixed at 0.9 and 0.05 respectively.

5. The maximum number of generations has been fixed at 50.

6. If the error reached by the best individual in the population on the validation set is not satisfactory, then the order of the polynomial terms is raised by one. For example, system having two attributes x1 and x2 will yield five polynomials of degree two.

7. The number of terms grows very quickly with the degree of product or trigonometric polynomial.

8. The process is repeated with increased number of polynomial terms. The best individual obtained in previous evolution run is also included in the new population. The algorithm is run till the error goal is achieved or the degree of polynomial becomes prohibitively high. As FLANNs do not require any hidden layer; the architecture becomes simple and training does not involve full backpropagation. Thus, nonlinear modeling can be accomplished, by means of a linear learning rule, such as delta rule. The computational complexity is also reduced and the neural net becomes suitable for on-line applications. Further, it reaches its global minima very easily. As FLANNs involve linear mapping in polynomial

space, they can easily map linear and nonlinear terms. Notwith-standing advantages accrued, unlike MLP, these networks lack universal approximation capability.

Summary

The design and development of an application based on neural network requires a lot of hard work as data is fed into the system, performances are monitored, processes tweaked, connections added, rules modified, and so on until the network achieves the desired results. These desired results are statistical in nature. The network is not always right. It is for that reason that neural networks are finding themselves in applications where humans are also unable to always be right. This chapter provides an understanding about the single layer and multi layer networks along with the detailed architecture and algorithm of the prediction networks. The chapter also delineates the basic methodology to implement a few the prediction networks using MATLAB.

Review Questions

1. Based on applications, how are neural networks categorized?

2. Differentiate single layer and multi layer neural networks.

3. Mention the networks related with data association.

4. Mention a few networks used for data conceptualization.

5. Explain the architecture and algorithm of back propagation network.

6. How is the learning rate selected in BPN?

7. Mention the use of momentum factor in BPN.

8. Write a note on directed random search.

9. Write a MATLAB program to implement the back propagation algorithm using the Neural Network Toolbox.

10. What is the learning rate that should be used in the implementation of back propagation algorithm.

11. Mention the rule used in training a back propagation network.

12. State the perceptron algorithm for several output classes.

13. Differentiate bias and threshold value.

14. Mention the Neural Network Toolbox functions that are used to implement perceptron algorithm.

Chapter 4

Classification and Association Neural Networks

4.1 Introduction

The previous chapter described networks that attempt to make predictions of the future. But understanding trends and impacts of those trends might have is only one of several types of applications. This chapter discusses the major class of neural network based on applications such as data classification and data association. Another network type described in this chapter is data conceptualization. Implementations of these networks using MATLAB Neural Network Tool box are also discussed. Customers might exist within all classifications, yet they might be concentrated within a certain age group and certain income levels. In reality, additional data might stretch and twist the region, which comprises the majority of expected buyers. This process is known as data conceptualization, which includes networks such as Adaptive Resonance Theory network and Self Organizing Feature Maps.

4.2 Neural Networks Based on Classification

The second class of applications is classification. A network that can classify could be used in the medical industry to process both lab results and doctor-recorded patience symptoms to determine the most likely disease. Some of the network architectures used for data classification are Learning Vector Quantization (LVQ), Counter-propagation Network (CPN), and Probabilistic Neural Network (PRNN). This section discusses the architecture, algorithm, and implementation of these networks.

4.2.1   Learning Vector Quantization

The vector quantization technique was originally evoked by Tuevo Kohonen in the mid 80's. Both Vector quanization network and self-organizing maps are based on the Kohonen layer, which is capable of sorting items into appropriate categories of similar objects. Such kind of networks find their application in classification and segmentation problems.

Topologically, the network contains an input layer, a single Kohonen layer, and an output layer. An example network is shown in Figure 4.1. The output layer has as many processing elements as there are distinct categories, or classes. The Kohonen layer consists of a number of processing elements classified for each of the defined classes. The number of processing elements in each class depends upon the complexity of the input-output relationship. Every class has the same number of elements throughout the layer. It is the Kohonen layer that learns and performs relational classifications with the aid of a training set. However, the rules used to classify vary significantly from the back-propagation rules. To optimize the learning and recall functions, the input layer should contain only one processing element for each separable input parameter. Higher-order input structures could also be used.

Learning Vector Quantization sorts its input data into groups that it determines. Fundamentally, it maps an n-dimensional space into an m-dimensional space. The meaning is that the network has n-inputs and produces m-outputs. During training the inputs are classified without disturbing the inherent topology of the training set. Generally, topology preserving maps preserve nearest neighbor relationships in the training set such that the input patterns which have not been previously learned will be categorized by their nearest neighbors in the training data.

While training, the distance of the training vector to each processing element is computed and while doing so, the processing element with the shorter distance is declared the winner. Always, there is only one winner for the entire layer. This winner fires only one output processing element, which determines the class or category the input vector belongs to. If the winning element is in the expected class of the training vector, it is reinforced toward the training vector. If the winning element is not in the class of the training vector, the connection weights entering the processing element are moved away from the training vector. This later operation is referred to as repulsion. On this training method, individual processing elements allotted to a particular class migrate to the region associated with their specific class.

During the recall mode, the distance of an input vector to each processing element is computed and again the nearest element is declared

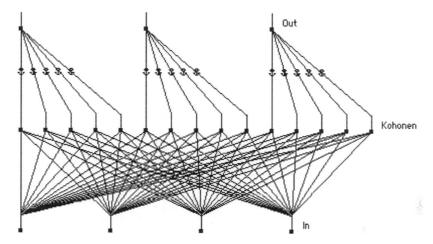

FIGURE 4.1: An Example Learning Vector Quantization Network

the winner. That in turn generates one output, signifying a particular class found by the network.

There are some limitations with the Learning Vector Quantization architecture. Apparently, for complex classification problems with similar objects or input vectors, the network requires a large Kohonen layer with many processing elements per class. This can be overcome with better choices, or higher-order representation for, the input parameters.

The learning mechanisms have some disadvantages which are addressed by variants to the paradigm. Usually these variants are applied at different stages of the learning process. They pervade a conscience mechanism, a boundary adaptation algorithm, and an attraction function at different points while training the network.

In the basic form of the Learning Vector Quantization network a few processing elements tend to win too often while others, do nothing. Those processing elements that are close tend to win and those that are far away do not involve. To overcome this defect, a conscience mechanism is added so that a processing element which wins too often develops a blameable conscience and is penalized. The actual conscience mechanism is a distance bias which is added to each processing element. This distance bias is proportional to the difference between the win frequency of an element and the average processing element win frequency. As the network progresses along its learning curve, this bias proportionality factor needs to be decreased.

A boundary modification algorithm is used to refine a solution once a relatively good solution has been found. This algorithm effects the cases when the winning processing element is in the wrong class and the second best processing element is in the right class. A further limitation is that the training vector must be near the midpoint of space joining these two processing elements. The winning processing element is moved away from the training vector and the second place element is moved toward the training vector. This procedure refines the boundary between regions where poor classifications commonly occur.

In the early training of the Learning Vector Quantization network, it is sometimes desirable to turn off the repulsion. The winning processing element is only moved toward the training vector if the training vector and the winning processing element are in the same class. This option is particularly helpful when a processing element must move across a region having a different class in order to reach the region where it is needed.

Architecture

The architecture of an LVQ neural net is shown in Figure 4.2.

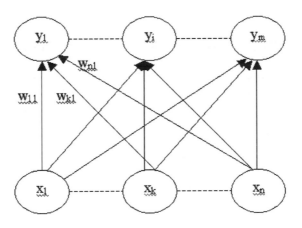

FIGURE 4.2:   Architecture of LVQ

The architecture is similar to the architecture of a Kohonen self organizing neural but without a topological structure assumed for the output units, In LVQ net, each output unit has a known class, since it uses supervised learning, thus differing from Kohonen SOM, which uses unsupervised learning. The architecture may resemble competitive network architecture, but this is a competitive net where the output is known; hence it is a supervised learning network.

Methods of initialization of reference networks

   i. Take first 'm' training vectors and use them as weight vectors; the remaining vectors are used for training.

   ii. Initialize the reference vectors randomly and assign the initial weights and class randomly.

  iiii. K-means clustering method can be adapted.

Training Algorithm

The algorithm for the LVQ net is to find the output unit that has a matching pattern with the input vector. At the end of the process, if x and w belong to the same class, weights are moved toward the new input vector and if they belong to a different class, the weights are moved away from the input vector. In this case also similar to Kohonen self-organizing feature map, the winner unit is identified. The winner unit index is compared with the target, and based upon the comparison result, the weight updation is performed as shown in the algorithm given below. The iterations are further continued by reducing the learning rate.

Parameters Used in the Pseudcode

The various parameters used in the training of the LVQ network is given below.

x: Training vector $(x_1,...,x_i,...,x_n)$

T: Category or class for the training vector

w_j: Weight vector for the jth output unit $(w_{1j},...,w_{ij},...w_{nj})$

C_j: Category or class represented by j^{th} output unit

| | x-wj | |: Euclidean distance between input vector and weight vector for the jth output unit

The pseudocode of the LVQ algorithm is as follows:


```
Initialize weights (reference) vectors.
Initialize learning rate
While not stopping condition  do
    For each training input vector x
        Compute J using squared Euclidean distance
        D(j)= ∑ (w_{ij} − x_i)²
        Find j when D(j) is minimum
        Update w_J as follows:
        If T=C_J , then
        w_{J(new)} = w_{J(old)}+ α (x−w_{J(old)})
        If T≠C_J , then
        w_{J(new)} = w_{J(old)}+ α (x−w_{J(old)})
```

End For
```
Reduce the learning rate.
Test for the stopping condition.
```
End While

The stopping condition may be fixed number of iterations or the learning rate reaching a sufficiently small value.

Variants of LVQ

Kohonen developed two variant techniques, LVQ2 and LVQ3, which are more complex than initial LVQ but allow for important performance in classification. In the LVQ algorithm, only the reference vector that is closest to the input vector is updated. In the LVQ2, LVQ3 algorithms, two vectors learn if several conditions are satisfied. The vectors are winner up and runner up. The technique followed is, if the input is approximately the same distance from the winner up and runner up, then each should learn.

LVQ2

In this case, the winner and runner up represent different classes. The runner up class is the same as the input vector. The distances between the input vector to the winner and the input vector to runner are approximately equal. The fundamental condition of LVQ2 is formed by a window function. Here, x is current input vector, y_c is reference vector that is closest to x, y_r is the reference vector that is next closest to x, d_c is distance from x to yc and d_r is distance from x to y_r. The window is defined as : the input vector x falls in the window if,

$$(d_c/d_r) > (1 - \epsilon) \text{ and } (d_c/d_r) < (1 + \epsilon)$$

where ϵ is the number of training samples (ϵ =0.35)

The updation formula is given by

$$y_c(t + 1) = y_c(t) + \alpha(t)[x(t) - y_c(t)]$$
$$y_r(t + 1) = y_r(t) + \alpha(t)[x(t) - y_r(t)]$$

LVQ3

The window is defined as

$$\min(d_{c1}/d_{c2}, d_{c2}/d_{c1}) > (1 - \epsilon)(1 + \epsilon)(\epsilon = 2)$$

Considering the two closest vectors y_{c1} and y_{c2} .

LVQ3 extends the training algorithm to provide for training if x, yc1 and yc2 belong to the same class. The updates are given as

$$y_{c1}(t+1) = y_c1(t) + \beta(t)[x(t) - y_{c1}(t)]$$
$$y_{c2}(t+1) = y_{c2}(t) + \beta(t)[x(t) - y_{c2}(t)]$$

The value of β is a multiple of the learning rate $\alpha(t)$ that is used if y_{c1} and y_{c2} belong to different classes, i.e., $\beta = $ m $\alpha(t)$; for $0.1 < $ m < 0.5.

This change in β indicates that the weights continue to approximate class distributions and prevents codebook vectors from moving away from their placement if the learning continues.

4.2.2   Implementation of LVQ in MATLAB

An LVQ network can be created with the function newlvq available in MATLAB Neural Network tool box as follows:

```
net = newlvq(PR,S1,PC,LR,LF)
```

where: PR is an R-by-2 matrix of minimum and maximum values for
          R input elements.
        S1 is the number of first layer hidden neurons.
        PC is an S2 element vector of typical class percentages.
        LR is the learning rate (default 0.01).
        LF is the learning function (default is learnlv1).

Example

Enter the input and the target vectors

```
clear all;
close all;
```

The input vectors P and target classes Tc below define a classification problem to be solved by an LVQ network.

```
inp = [-3 -2 -2 0 0 0 0 +2 +2 +3;0 +1 -1 +2 +1 -1 -2
+1 -1 0];
target_class = [1 1 1 2 2 2 2 1 1 1];
```

The target classes are converted to target vectors T. Then, an LVQ network is created (with inputs ranges obtained from P, four hidden neurons, and class percentages of 0.6 and 0.4) and is trained.

```
T = ind2vec(target_class);
```

The first-layer weights are initialized to the center of the input ranges with the function midpoint. The second-layer weights have 60% (6 of the 10 in Tc above) of its columns with a 1 in the first row, (corresponding to class 1), and 40% of its columns will have a 1 in the second row (corresponding to class 2).

```
network = newlvq(minmax(inp),4,[.6 .4]);
network = train(network,inp,T);
```

To view the weight matrices

```
network.IW(1,1) ; % first layer weight matrix
```

The resulting network can be tested.

```
Y = sim(network,inp)
Yc = vec2ind(Y)
```

Output: The network has classified the inputs into two basic classes, 1 and 2.

```
Y =
1       1       1       0       0       0       0       1       1       1       1
0       0       0       1       1       1       1       0       0       0       0
Yc =
1       1       1       2       2       2       2       1       1       1       1

Weight matrix of layer 1
ans =
      2.6384         -0.2459
     -2.5838          0.5796
     -0.0198         -0.3065
      0.1439          0.4845
```

Thus the above code implements the learning vector quantization algorithm to classify a given set of inputs into two classes.

4.2.3 Counter-Propagation Network

The Counter Propagation Network was proposed and developed by Robert Hecht-Nielsen to synthesize complex classification problems instead of reducing the number of processing elements and training time. The learning process of Counter propagation is more or less similar to LVQ with a small difference such that the middle Kohonen layer plays the role of a look-up table. This look-up table finds the closest fit to a given input pattern and outputting its equivalent mapping.

The first counter-propagation network comprised of a bi-directional mapping between the input and output layers. Essentially, while data is presented to the input layer to generate a classification pattern on the output layer, the output layer in turn would accept an additional input vector and generate an output classification on the network's input layer. The network got its name from this counter-posing flow of information through its structure. Most developers use a uni-flow variant of this formal representation of counter-propagation. Counter propagation networks have only one feedforward path from input layer to output layer.

An example network is shown in Figure 4.3. The uni-directional counter-propagation network has three layers. If the inputs are not already normalized before they enter the network, a fourth layer is sometimes required. The main layers include an input buffer layer, a self-organizing Kohonen layer, and an output layer which uses the Delta Rule also known as the Grossberg out star layer to modify its incoming connection weights.

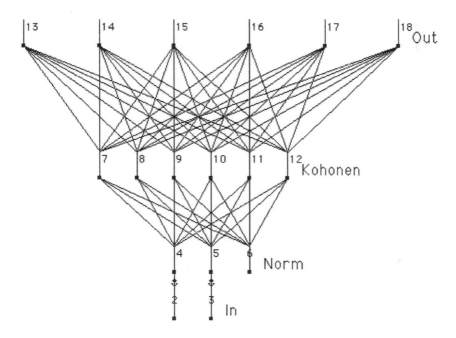

FIGURE 4.3:   An Example Counter-Propagation Network

Depending upon the parameters that define a problem the input layer's size varies. If the input layer has very few processing elements, then the network may not generalize and if the input layer has a large number of processing elements then the processing time is very high.

Generally for fine operation of a network, the input vector must be normalized. Normalization refers to the process of adding that is for every combination of input values, the total "length" of the input vector must add up to one. The normalization process can be done with a preprocessor before presenting the data to the network. In specific applications, a normalization layer is added between the input and Kohonen layers. The normalization layer requires one processing element for each input, plus one more for a balancing element. This normalization layer assures that all input sets sum up to the same total.

Normalization of the inputs is necessary to insure that the Kohonen layer finds the correct class for the problem. Without normalization, larger input vectors bias many of the Kohonen processing elements such that weaker value input sets cannot be properly classified. Due to the competitive nature of the Kohonen layer, the bigger value input vectors overcome the smaller vectors. Counter-propagation uses a standard Kohonen layer which self-organizes the input sets into classification zones. It follows the classical Kohonen learning law. This layer acts as a closest neighbor classifier such that the processing elements in the competitive layer autonomously update their connection weights to divide up the input vector space in approximate correspondence to the frequency with which the inputs occur. There should be as many processing elements as possible in the Kohonen layer equivalent to the output classes. The Kohonen layer generally has a lot more elements than classes simply because additional processing elements provide a finer resolution between similar objects.

The output layer for counter-propagation fundamentally consists of processing elements which learn to produce an output when a specific input is applied. Because the Kohonen layer is a competitive layer, only a single winning output is produced for a given input vector. This layer renders a method of decoding the input to a purposeful output class. The delta rule is used to back-propagate the error between the desired output class and the actual output generated with the training set. The weights in the output layer are alone updated while the Kohonen layer is unaffected.

As only one output from the competitive Kohonen layer is active at a time and all other elements are zero, the only weight adjusted for the output processing elements are the ones connected to the winning element in the competitive layer. In this way the output layer learns to reproduce a definite pattern for each active processing element in the competitive layer. If numerous competitive elements belong to the same class, then the output processing element will acquire weights in response to those competitive processing elements and zero for all others.

The major limitation of this architecture is the competitive Kohonen layer learns without any supervision. Therefore it cannot predict the type of class it is reporting to. This infers that it is possible for a processing element in the Kohonen layer to learn two or more training inputs, which belong to different classes. During this process, the output of the network will be multi-valued for any inputs. To overcome this difficulty, the processing elements in the Kohonen layer can be pre-conditioned to learn only about a specific class.

Counter propagation network is classified into two types. They are

1. Full counter propagation network

2. Forward only counter propagation network

In this section, the training algorithm and application procedure of full CPN is described.

Training Phases of Full CPN

The full CPN is achieved in two phases.

The first phase of training is called as In star modeled training. The active units here are the units in the x-input ($x = x_1, \ldots, x_i, \ldots x_n$), z-cluster ($z = z_1, \ldots, z_j, \ldots, z_p$) and y-input ($y = y_1, \ldots, y_k, \ldots, y_m$) layers.

Generally in CPN, the cluster unit does not assume any topology, but the winning unit is allowed to learn. This winning unit uses our standard Kohonen learning rule for its weight updation. The rule is given by

$$v_{ij(new)} = v_{ij(old)} + \alpha(x_i - v_{ij(old)})$$
$$= (1 - \alpha)v_{ij(old)} + \alpha x_i; i = 1 \text{ to } n$$
$$w_{jk(new)} = w_{kj(old)} + \beta(y_k - w_{jk(old)})$$
$$= (1 - \beta)w_{kj(old)} + \beta y_k; k = 1 \text{ to } m$$

In the second phase, we can find only the J unit remaining active in the cluster layer. The weights from the winning cluster unit J to the output units are adjusted, so that vector of activation of units in the y output layer, y*, is approximation of input vector x. This phase may be called the out star modeled training. The weight updation is done by the Grossberg learning rule, which is used only for out star learning. In out star learning, no competition is assumed among the units, and the learning occurs for all units in a particular layer. The weight updation rule is given as,

$$U_{ij(new)} = u_{jk(old)} + \alpha(y_k - u_{jk(old)})$$
$$= (1 - a)\alpha_{jk(old)} + ay_k; k = 1 \text{ to m}$$
$$t_{ji(new)} = t_{ji(old)} + \alpha(x_i - t_{ji(old)})$$
$$= (1 - b)t_{ji(old)} + bx_i; i = 1 \text{ to n}$$

The weight change indicated is the learning rate times the error.

Training Algorithm

The parameters used are

x- Input training vector x=$(x_1, ...x_i, ...x_n)$
y- target output vector Y=$(y_1,...y_k,...y_m)$
z_j- activation of cluster unit Z_j
x*- approximation to vector x
y*- approximation to vector y
v_{ij}- weight from x input layer to Z-cluster layer
w_{jk}- weight from x input layer to Z-cluster layer
t_{ji}- weight from x input layer to X-cluster layer
u_{jk}- weight from x input layer to Y-cluster layer
α , β- learning rates during Kohonen learning
a, b- learning rates during Grossberg learning

The algorithm uses the Euclidean distance method or dot product method for calculation of the winner unit. The winner unit is calculated during both first and second phase of training. In the first phase of training for weight updation Kohonen learning rule is used and for second phase of training Grossberg learning rule is used.

The pseudocode of the the full CPN is given by,

```
Initialize weights v and w
Initialize learning rates α and β
While not stopping condition for phase I Do
    For each training input pair x:y
        Assign X input layer activations to
        vector x
        Assign Y input layer activations to
        vector y
        Calculate winning cluster D(j)=
        ∑ (xᵢ-vᵢⱼ)²; i = 1 to n
        i
```

```
Update weights
```
v_{ij}(new)=v_{ij}(old)+ α(x_i-v_ij(old)); i = 1 to
n
w_{jk}(new)=w_{jk}(old)+ β(y_k-w_{jk}(old)); k = 1
to m
End For
```
Reduce learning rates α and β
Test stopping condition for Phase I training
```
End while
While not stopping condition for phase II **Do**
   For each training input pair x:y
```
        Assign X input layer activations to
        vector x
        Assign Y input layer activations to
        vector y
        Calculate winning cluster D(j)=
```
$$\sum_i (x_i-v_{ij})^2; \quad i = 1 \text{ to } n$$
```
        Update weights from unit zⱼ to the output
        layers
        (α and β are constant in this phase)
```
$U_{ij(new)}$= $u_{jk(old)}$+ α (y_k-$u_{jk(old)}$); k=1 to m
$t_{ji(new)}$ = $t_{ji(old)}$+ α (x_i-$t_{ji(old)}$); i=1 to n
End For
```
Reduce learning rates α and β
Test for stopping condition for phase II
training
```
  End while

The winning unit selection is done either by dot product or Euclidean distance.

The dot product is obtained by calculating the net input.

$$Z_{inj} \sum_i x_i u_{ij} + \sum_k y_k w_{kj}$$

The cluster unit with the largest net input is winner. Here the vectors should be normalized. In Euclidean distance,

$$D_j = \sum i(x_i - v_{ij})^2$$

The square of whose distance from the input vector is smallest is the winner.

In case of tie between the selections of the winning unit, the unit with the smallest index is selected.

The stopping condition may be the number of iteration or the reduction in the learning rate up to a certain level.

Application Procedure

In the training algorithm, if only one Kohonen neuron is activated for each input vector, this is called the Accretive mode. If a group of Kohonen neurons having the highest outputs is allowed to present its outputs to the Grossberg layer, this is called the interpolative mode. This mode is capable of representing more complex mappings and can produce more accurate results.

The application procedure of the full CPN is

```
Initialize weights
For each input pair x:y
    Assign X input layer activation to vector x;
    Assign Y input layer activation to vector y;
    Find the cluster unit ZJ close to the input
    pair
    Compute approximations to x and y:
      x_i* = t_Ji
      y_k* = u_Jk
End For
```

Initializing the Weight Vectors

All the network weights must be set to initial value before training starts. It is a common practice with neural networks to randomize the weights to small numbers. The weight vectors in CPN should be distributed according to the density of the input vectors that must be separated, thereby placing more weight vectors in the vicinity of the large number of input vectors. This is obtained by several techniques.

One technique called, Convex combination method, sets all the weights to the same value $1/\sqrt{n}$, where n is the number of inputs and hence, the number of components in each weight vector. This makes all the weight vectors of unit length and coincident. This method operates well but slows the training process, as the weight vectors have to adjust to a moving target.

Another method adds noise to the input vectors. This causes them to move randomly, eventually capturing a weight vector. This method also works, but it is even slower than convex combination.

A third method starts with randomized weights but in the initial stages of the training process adjusts all the weights, not just those associated with the winning Kohonen neuron. This moves the weight vectors around, to the region of the input vectors. As training starts, weight adjustments are restricted to those Kohonen neurons that are nearest to the winner. This radius of adjustment is gradually decreased until only those weights are adjusted that are associated with the winning Kohonen neuron.

There is another method, which gives each Kohonen neuron a "conscience". If it has been winning more than its fair share of the time, say $1/k$, where k is the number of Kohonen neurons, it raises the threshold that reduces its chances of winning, thereby allowing the other neurons an opportunity to be trained. Thus the weights may be initialized.

4.2.4 Probabilistic Neural Network

The developement of probabilistic neural network is based on Parzen's windows and these networks were designed by Donald Specht based on his papers, Probabilistic Neural Networks for Classification, Mapping or Associative Memory and Probabilistic Neural Networks, released in 1988 and 1990, respectively. Pattern classification is a major application area of this network and the network is based on Bayes theory. Bayes theory, developed in the 1950's, allows the relative likelihood of events and applies a priori information to ameliorate prediction. The probability density functions required by the Bayes theory is constructed using Parzen Windows.

The learning process of the probabilistic neural network is a supervised learning and during this learning process distribution functions are developed within a pattern layer. The distribution functions are used to estimate the likeliness of an input feature vector. The learned patterns are grouped in a manner with the a priori probability, also called the relative frequency, to determine the most likely class for a given input vector. All the classes or categories are assumed to be equally likely if the knowledge about the a priori probability is available. The feature vectors determine the class based on the shortest Euclidean distance between the input feature vector and the distribution function of a class.

The basic straucture of a probabilistic neural network is shown in Figure 4.4. The fundamental architecture has three layers, an input layer, a pattern layer, and an output layer. The input layer has many elements since separable parameters are required to describe the objects to be classified. The pattern layer organizes the training set in such a way that every input vector is delineated by an independent processing element. The output layer also called the summation layer has numerous processing elements according to the classes that are to be recognized. Each and every processing element in the summation layer combines with processing elements in the pattern layer, and if they are associated to the same class then that category for output is developed. Based on the application, occasionally a fourth layer is added to normalize the input vector if the inputs are not already normalized before they enter the network.

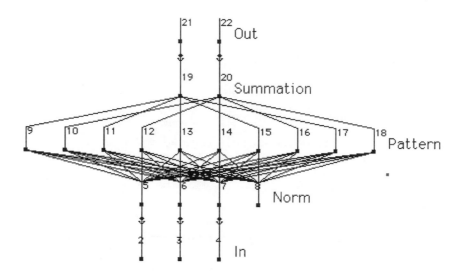

FIGURE 4.4:   A Probabilistic Neural Network Example

The pattern layer constitutes a neural implementation of a Bayes classifier, where the class dependent Probability Density Functions (PDF) are approximated using a Parzen estimator. By using the Parzen estimator to determine the PDF the expected risk in classifying the training set incorrectly is minimized. Using the Parzen estimator, the classification gets closer to the true underlying class density functions as the number of training samples increases, so long as the training set is an adequate representation of the class distinctions.

The pattern layer consists of a processing element corresponding to each input vector in the training set. Each output class should consist of equal number of processing elements otherwise a few classes may be inclined falsely leading to poor classification results. Each processing element in the pattern layer is trained once. An element is trained to return a high output value when an input vector matches the training vector. In order to obtain more generalization a smoothing factor is included while training the network. In such a case, it is not necessary for the training vectors to be in any predefined order within the training set, since the category of a particular vector is specified by the desired output of the input. The learning function merely selects the first untrained processing element in the correct output class and updates its weights to match the training vector.

The pattern layer classifies the input vectors based on competition, where only the highest match to an input vector wins and generates an output. Hence only one classification category is generated for any given input vector. If there is no relation between input patterns and the patterns programmed into the pattern layer, then no output is generated.

In order to fine tune the classification of objects the Parzen estimator is added. This process is carried out by adding the frequency of occurrence for each training pattern built into a processing element. Fundamentally, the probability distribution of happening for each example in a class is multiplied into its respective training node. Similarly, a more accurate expectation of an object is added to the features, which makes it recognizable as a class member.

Compared to the feedforward back propagation network, training of the probabilistic neural network is much more simpler. Anyway, the pattern layer can be quite huge if the distinction between categories is varied. There are many proponents for this type of network, since the groundwork for optimization is founded in well known, classical mathematics.

Since the probabilistic networks classify on the basis of Bayesian theory, it is essential to classify the input vectors into one of the two classes in a Bayesian optimal manner. This theory provides a cost function to comprise the fact that it may be worse to misclassify a vector that is actually a member of class A than it is to misclassify a vector that belongs to class B. The Bayes rule classifies an input vector belonging to class A as,

$$P_A C_A f_A(x) > P_B C_B f_B(x)$$

where, PA - Priori probability of occurrence of patterns in class A
CA - Cost associated with classifying vectors
fA(x) - Probability density function of class A

The PDF estimated using the Bayesian theory should be positive and integratable over all x and the result must be 1.

The probabilistic neural net uses the following equation to estimate the probability density function given by,

$$f_A(x) = \frac{1}{(2\Pi)^{n/2}\sigma^n} \frac{1}{m_n} \sum_{i=1}^{m_A} exp[-2\frac{(x - x_A)^r(x - x_{Ai})}{\sigma^2}]$$

where x_{Ai} - i_{th} training pattern from class A

      n - Dimension of the input vectors

      σ - Smoothing parameter (corresponds to standard deviations of Guassian distribution)

The function f_A (x) acts as an estimator as long as the parent density is smooth and continuous. f_A (x) approaches the parent density function as the number of data points used for the estimation increases. The function $f_A(x)$ is a sum of Guassian distributions. The disadvantage of using this parent density function along with Bayes decision rule is that the entire training set must be stored and the computation needed to classify an unknown vector is proportional to the size of the training set.

Architecture

The architecture of probabilistic neural net is shown in the Figure 4.5. The architecture is made up of four types of units.

- Input units

- Pattern units - Class A and Class B.

- Summation units

- Output units

The weights between the summation unit and the output unit are,

$$VA = 1$$
$$VB = -PBCBmA/PACAmB$$

Training Algorithm

The training algorithm for the probabilistic neural net is given as,

```
For each training input pattern, x (p), p=1,0,P
    Create pattern unit Zp:
    Weight vector for unit Zp:     wp=x(p)
    (unit Zp is either a ZA unit or ZB unit)
    Connect the pattern unit to summation unit
    If x(p) belongs to class A,
        Then connect pattern unit Zp to summation
        unit SA
    Else
        connect pattern unit Zp to summation unit SB
    End If
End For
```

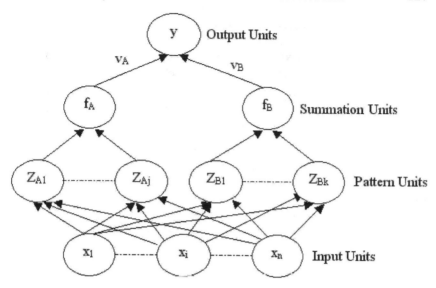

FIGURE 4.5: Architecture of Probabilistic Neural Net

Application Algorithm

The pseodocode of the application algorithm for classifying is given as,

```
Initialize weights from training algorithm.
For input pattern to be classified
      Patterns units:
      Calculate net input,
```

$$z_{inj} = x.w_j = x^T W_j$$

```
      Calculate the output,
```

$$Z = \exp[(z_inj-1)/\sigma^2]$$

```
      Summation units:
      The weights used by the summation unit for class
      B is,
```

$$v_B = -P_B\, C_B\, m_A\, /P_A\, C_A\, m_B$$

```
      Output unit:
      Sums the signals from fA and fB
      Input vector is classified as class A if the
      total input to decision unit is positive
End For
```

4.2.5   Implementation of the Probabilistic Neural Net Using MATLAB

The probabilistic neural net can be implemented in MATLAB using the following function

$$net = newpnn(P, T, spread)$$

where,
P - matrix of Q input vectors
T - matrix of Q target class vectors
spread - Spread of radial basis functions, default = 0.1

The function returns a new probabilistic neural network. If spread is near zero, the network will act as a nearest neighbor classifier. As spread becomes larger, the designed network will take into account several nearby design vectors.

Example

Enter the input and the target vectors

clear all;
close all;

The input vectors P and target classes Tc below define a classification problem to be solved by an LVQ network.

inp = [1 2 3 4 5 6 7 8 9;9 8 7 6 5 4 3 2 1];
target_class = [1 1 1 2 2 2 3 3 3];

The target classes are converted to target vectors T. Then, an LVQ network is created (with input ranges obtained from P, four hidden neurons, and class percentages of 0.6 and 0.4) and is trained.

T = ind2vec(target_class);

The network is created and simulated, using the input to make sure that it does produce the correct classifications. The function vec2ind is used to convert the output Y into a row Yc to make the classifications clear.

network = newpnn(inp,T);

Y = sim(network,inp)

Yc = vec2ind(Y)

Output


```
Y =
        (1,1)      1
        (1,2)      1
        (1,3)      1
        (2,4)      1
        (2,5)      1
        (2,6)      1
        (3,7)      1
        (3,8)      1
        (3,9)      1
        (3,9)      1
Yc =
        1     1     1     2     2     2     3     3     3
```

Thus the probabilistic neural net can be used for classification of pattern, from each of the two classes that has been presented. This type of probabilistic neural net can be used to classify electrocardiogram's output as normal or abnormal.

4.3   Data Association Networks

The classification networks discussed in the previous sections are related to data association networks. In data association, classification is also done. For instance, a recognition unit can classify each of its scanned inputs into several groups. Yet, some data with errors are also available. The data association networks recognize these error occurrences as merely defective data and that this defective data can span all classifications.

4.3.1   Hopfield Network

The first data association network was proposed by John Hopfield in 1982 at the National Academy of Sciences and it was named as the Hopfield network. The network is conceived based on energy functions. Here in Hopfield network, the processing elements update or change their state only if the overall "energy" of the state space is minimized. Basic applications for these kind of networks have included associative, optimization problems, like the shortest path algorithm.

The Figure 4.6 shows the basic model of a Hopfield network. Earlier the original network used processing elements in binary formats which later on changed to bipolar format. Therefore the restriction in binary processing elements was during the quantization, that is the output is quantized to a zero or one.

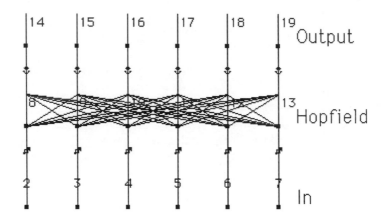

FIGURE 4.6:   A Hopfield Network Example

The basic model of Hopfield network includes three layers; an input buffer layer, a middle Hopfield layer, and an output layer. All the layers have equal number of processing elements. The inputs of the Hopfield layer are connected to the outputs of the corresponding processing elements in the input buffer layer through variable connection weights. The outputs of the Hopfield layer are connected back to the inputs of every other processing element except itself. While training, the network applies the data from the input layer through the learned connection weights to the Hopfield layer. The Hopfield layer takes some time to saturate, and the current state of that layer is passed on to the output layer. This state matches a pattern that is predefined and programmed into the network.

During learning the network requires the training pattern to be presented to the input and the output layer at the same time. The recursive and oscillating nature of the Hopfield layer allows the network to adjust all connection weights. The learning rule is the Hopfield Law, which states that "if the desired output and the input are both active or both inactive, increment the connection weight by the learning rate, otherwise decrement the weight by the learning rate". Evidently, any non-binary implementation of the network must have a threshold mechanism in the transfer function, or matching input-output pairs could be too rare to train the network properly.

The network can be used as a content addressable memory. While applied as a content addressable memory, there are a few limitations like

- The amount of patterns that can be stored and precisely recalled is severely bounded. If too many patterns are stored, the network may converge to a wrong pattern or may not converge at all. Therefore the storage capacity should be limited to approximately fifteen percent of the number of processing elements in the Hopfield layer.

- The Hopfield layer may become unstable if there are common patterns. A pattern is considered unstable if it is applied at time zero and the network converges to some other pattern from the training set. This disadvantage is overcome by altering the pattern sets to be orthogonal with each other.

Architecture

The architecture of the discrete Hopfield net is shown in the Figure 4.7. It consists of "n" number of x input neurons and "y" output neurons. It should noted that apart from receiving a signal from input, the y1 neuron receives signal from its other output neurons also. This is the same for the all other output neurons. Thus, there exists a feedback output being returned to each output neuron. That is why the Hopfield network is called a feedback network.

Training Algorithm

Discrete Hopfield net is described for both binary as well as bipolar vector patterns. The weight matrix to store the set of binary input patterns s(p),p=1,...P, where

$$s(p){=}(s1(p),...si~(p),...sn(p))$$

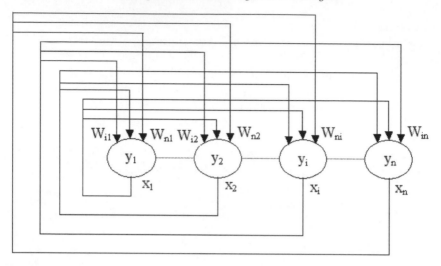

FIGURE 4.7: Architecture of Hopfield Net

can be determined with the help of Hebb rule as discussed earlier. The weight matrix can be determined by the formula

$$W_{ij} = \sum_p (2s_i(p) - 1) \text{ for } i \neq j \text{ and } W_{ii}=0$$

For bipolar input patterns, the weight matrix is given by,

$$W_{ij} = \sum_p s_i(p) \text{ for } i \neq j \text{ and } W_{ii}=0$$

Application Algorithm

The weights to be used for the application algorithm are obtained from the training algorithm. Then the activations are set for the input vectors. The net input is calculated and applying the activations, the output is calculated. This output is broadcasted to all other units. The process is repeated until the convergence of the net is obtained. The pseudocode of the application algorithm of a discrete Hopfield net is given as follows:

```
Initialize weights to store pattern (use Hebb rule)
While activations of the net are not converged do
     For each input vector x
   Assign initial activations to external input vector
```

```
   x, yi=xi (i=1,n)
      For each unit yi
         Compute the net input
```
$$Y_{inj} = X_i + \sum_i Y_i \ W_{ji}$$
```
         Determine activation (output signal)
```
$$Y_j = \begin{cases} 1, & if \ Y_{inj} > \theta \\ Y_j, & if \ Y_{inj} = \theta \\ 1, & if \ Y_{inj} < \theta \end{cases}$$
```
   Broadcast the value of yi to all other units
         End For
      End For
Test for convergence
End While
```

The value of threshold θ_i is usually taken to be zero. The order of update of the unit is random but each unit must be updated at the same average rate.

4.3.2 Implementation of Hopfield Network in MATLAB

The following code stores the vector (1 1 1 0 1 1). Assume that there are mistakes in the first and the second component of the stored vector such that (0 0 1 0 1 1).

```
clc;
clear all;
```

The stored vector is
```
x=[1 1 1 0 1 1 ];
```

```
t=[0 1 1 0 0 1 ];
```

The weight matrix is determined by the formula
$$W_{ij} = \sum_p (2s_i(p) - 1)(2s_j(p) - 1) \ \text{for} \ i \neq j \ \text{and} \ W_{ii} = 0$$
```
w=(2*x'-1)*(2*x-1);
for i=1:6
    w(i,i)=0;
end
stop=1;
y=[ 0 1 1 0 0 1 ];

while stop
```

```
update=[1 2 3 4 5 6 ];
for i=1:6
```

   Compute the net input according to the formula
   $$Y_{ini}=X_i+\sum_j Y_i W_{ji}$$

```
   yin(update(i))=t(update(i))+y*w(1:6,update(i));
   if(yin(update(i))>0)
       y(update(i))=1;
   end
end
if y = = x
   disp('Convergence has reached);
   stop=0;
   disp(y)
end
end
```

Output
```
Convergence has reached
The converged output is
   y = 1 1 1 0 1 1
```

4.3.3   Boltzmann Machine

Boltzmann machine is like the Hopfield network functionally and operation wise. The only difference is that simulated annealing technique is used in Boltzmann machine while finding the original pattern. Simulated annealing is used to search the pattern layer's state space to find a global minimum. Therefore the machine is able to span to an improved set of values over time as data iterates through the system.

The Boltzmann learning rule was first proposed by Ackley, Hinton, and Sejnowski in 1985. Similar to Hopfield network, the Boltzmann machine has an associated state space energy based upon the connection weights in the pattern layer. The processes of learning a training set involve the minimization of this state space energy.

The simulated annealing schedule is added to the learning process. This technique is done similar to physical annealing, temperatures start at higher values and decrease over time. This high value of temperature adds a noise factor into each processing element in the pattern layer and a low value of temperature tries to settle the oscillations of the network. An optimal solution can be determined while adding more iterations at lower temperatures.

Therefore the learning process at high temperatures is a random pro-

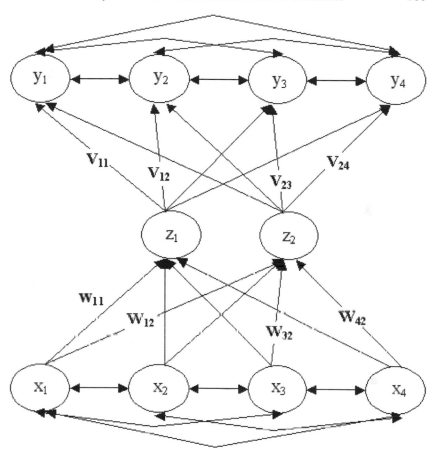

FIGURE 4.8: Architecture of Boltzmann Machine

cess and at low temperatures is a deterministic process. During random learning process, the processing elements tend to change its state thereby increasing the overall energy of the system.

Architecture

The architecture of Boltzmann machine is shown in Figure 4.8.

The architecture of the Boltzmann machine looks like a two-dimensional array. It consists of a set of units and a set of bidirectional connections between a pair of units. The units in each row and column are fully interconnected. The weights on each connection are -p and there is also a self-connection with weight b. The units are labeled as $V+_{i,j}$.

The objective of the neural network is to maximize the consensus function,

$$C = \sum_{i} \left[\sum_{i} W_{ij} X_i Y_j \right]$$

The net finds the maximum by making each unit change its state.

The change in consensus if unit Xi were to change its state is given by,

$$\Delta C(i) = [1 - 2x_i][W_{ij} + \sum W_{ij} X_j]$$

where x_i is the current state.

$$(1 - 2x_i) = \begin{cases} +1, if fx_i -' on' \\ -1, if fx_i -' off' \end{cases}$$

The probability of the net accepting a change in state for unit X_i is,

$$A(i, t) = \frac{1}{1 + exp\left(\frac{-\Delta e(i)}{T}\right)}$$

To obtain the optimal solution, the parameter T is to be reduced gradually. The weights -p indicate the penalties for violating the condition that almost one unit be "on" in a row or column. The self-connection weights b indicate the incentives or bonus to a unit to turn 'on', without making more than one unit to be "on" in a row or column. The net will function as desired if p > b.

Application Algorithm

The pseudocode of the application algorithm of the Boltzmann machine is as follows, here the weights between unit V_{ij} and $V_{I,J}$ are denoted by W(i,j:I,J). Also,

$$W(i, j : I, J) = \begin{cases} -p, if i = I or j = J (not both) \\ b, \qquad otherwise \end{cases}$$

```
Initialize weights, control parameters,
activation of units.
While not stopping condition do
    For n² times
    Chose integers I and J at random between 1 n.
    Compute the change in consensus:
```
$$\Delta \ C= \ [1-2V_{I,J}]\,[W(i,j:I,J) + \sum_{i,j} \sum_{I,J} W(i,j:I,J)U_{IJ}]$$

```
Compute the probability of acceptance of the
changes,
```
$$A(T) = \frac{1}{1+exp(-\Delta e/T)}$$

```
Determination of whether or not to accept
the change.
                If random number N between 0 to 1.
                If N<A
                    change is accepted
                    V_{I,J}=1-V_{I,J}
                Else If N ≥ A
                    change is rejected
                End If
End For
Reduce the control parameter,
        T(new)=0.95T(old)
Test for stopping condition.
End While
```

The stopping condition may be change of state for a specified number of epochs or if T reached a specified value. Thus, Boltzmann machine can be applied to the traveling salesman problem.

4.3.4 Hamming Network

The Hamming network was developed by Richard Lippman in the mid 1980's. The Hamming network is more or less similar to the Hopfield network, the only difference is that the Hamming net implements a maximum likehood classifier. The Hamming distance is defined as the number of bits which differ between two corresponding, fixed-length input vectors among which one input vector is the noiseless pattern and the other input is the vector corrupted by real world noisy effects. Thus the output categories are defined by a noiseless training set. During training the input vectors are alloted to a class in which the Hamming distance is minimum.

The basic model of the Hamming network is shown in Figure 4.9. The network has three layers: input layer, a middle Hopfield layer, and an output layer. The input layer has numerous nodes which are binary in nature. The Hopfield layer also known as the category layer, has nodes that are equivalent to the classes. The output layer matches the number of nodes in the category layer.

The network is a simple feedforward architecture with the input layer fully connected to the category layer. Each processing element in the category layer is connected back to every other element in that same layer, as well as to a direct connection to the output processing element. The output from the category layer to the output layer is done through competition.

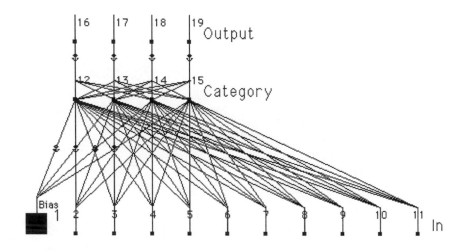

FIGURE 4.9:   A Hamming Network

The learning of a Hamming network is similar to the Hopfield network. The desired training pattern is presented to the input layer and the desired class to which the input vector belongs to is presented to the output layer. The connection weights are updated based on the recursive behavior of the neural network.

The connection weights are first set in the input to category layer. Matching scores are generated to determine the best match between the desired input and the desired output. The category layer's recursive connection weights are trained in the same manner as in the Hopfield network. In normal feedforward operation an input vector is applied to the input layer and must be presented long enough to allow the matching score outputs of the lower input to category subnet to settle. This will initialize the input to the Hopfield function in the category layer and allow that portion of the subnet to find the closest class to which the input vector belongs. This layer is competitive, so only one output is enabled at a time.

Hamming Distance

The Hamming distance between two vectors is the number of components in which the vectors differ. It can also be defined as the number of differences between two binary or bipolar vectors (x,y). It can be denoted as H(x,y). The average Hamming distance is given as,

$$= (1/n) \ H(x,y)$$

where n is the number of components in each vector.

In Hamming net, the measure of similarity between the input vector and the stored exemplar is minus the Hamming distance between the vectors.

Consider two bipolar vectors x and y,

If "a" - Number of components in which the vectors agree and "d"- Number of components in which the vectors differ (hamming distance)

Then, x,y=a-d

If "n" is the number of components, then,

$$n=a+d \ \text{or} \ d=n-a$$

As a result

$$x,y=a-d=a-(n-a)$$
$$x,y=2a-n \ (\text{or}) \ 2a=x.y+n$$

From the results it is understood that if the weights are set to one-half of the exemplar vector and the bias to one half of the number of components, the net will find the unit with the closest exemplar simply by finding the unit with the largest net input.

Architecture

The architecture is show in Figure 4.10. Assuming input vectors with 4-tuples and the output to be classified to one of the two classes given.

The architecture consists of n-input modes in the lower net, with each input node connected to the m-output nodes. These output nodes are connected to the upper net (i.e., Max-Net-acting as subnet for Hamming net) which calculates the best exemplar match to the input vector. The 'm' in the output nodes represent the number of exemplar vectors stored in the net. It is important to note that the input vector and the exemplar vector are bipolar. For a given set of exemplar vectors, the Hamming net finds the exemplar vector that is closest to the bipolar input vector x. The number of components in which the input vector and the exemplar vector agree is given by the net input.

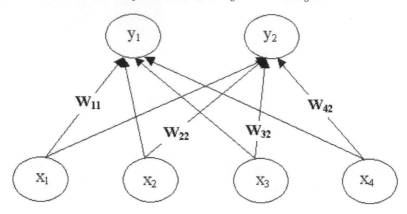

FIGURE 4.10: Architecture of Hamming Net

Application Procedure

The parameters used are,
    n - number of input nodes (input vectors)
    m - number of output nodes(exemplar vectors)
    e(j)-j_{th} exemplar vector
The pseudocode of the application procedure is as follows:

Initialize weights for storing m exemplar vectors
    w_{ij}=e$_i$(j)/2, (i=1..n, j=1..m)
Initializing the bias
    b_j=n/2(j=1..m)
For each vector x
Compute the net input to each unit Y$_j$
    Y_{inJ}=b$_j$+ $\sum\limits_{i} x_i w_{ij}$ j=1..m

    Initialize activations for Max Net
    y$_j$(0)=y$-_{inj}$ (j=1,..m)
    Max net iterates to find the best match exemplar
End For

The Hamming network is more advantageous when compared with the Hopfield network. Hamming applies the optimum minimum error classifier to classify the input vectors when input bit errors are stochastic. Hamming networks use a few processing elements when compared to Hopfield network. The category layer needs only one element for the

entire category instead of individual elements for each input node. The
Hamming network is faster and more accurate than the Hopfield network
since it takes less time to settle.

4.3.5   Bi-Directional Associative Memory

Bart Kosko developed a model to generalize the Hopfield model and
this termed as the bi-directional associative memory (BAM). In BAM,
the patterns are presented as a pair of bipolar vectors and when a noisy
input is presented the nearest pattern associated with the input pattern
is determined.

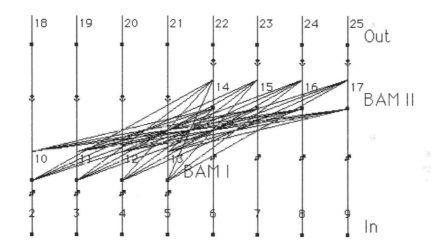

FIGURE 4.11:   Bi-Directional Associative Memory Example

The basic model of bi-directional associative memory network is shown
in Figure 4.11. The BAM network has several inputs equal to the output
processing nodes. The hidden layers are made up of separate associated
memories and represent the size of the input vectors. The lengths of the
units need not be equal. The middle layers are fully connected to each
other. The input and output layers are for implementation purposes the
means to enter and retrieve information from the network. Kosko original
work targeted the bi-directional associative memory layers for optical
processing, which would not need formal input and output structures.

Vectors are stored as associated pairs in the middle layer. The middle
layers swing to and fro until a stable state is attained whenever noisy
patterns are presented. The stable state represents the nearest learned

association and gives the best output similar to the input pattern presented initially. Similar to the Hopfield network, the bi-directional associative memory network is liable to incorrectly find a trained pattern when complements of the training set are used as the unknown input vector.

Architecture

The layers in BAM are referred as X-layer and Y-layer instead of input and output layer, because the weights are bidirectional and the algorithm alternates between updating the activations of each layer.

Three forms of BAM are

1. Binary

2. Bipolar

3. Continuous

But the architecture for the three types remains the same. The architecture of BAM is shown in Figure 4.12.

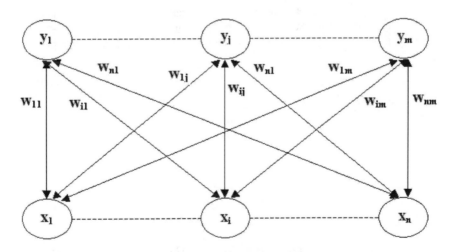

FIGURE 4.12:   Architecture of Bi-Directional Memory Network

The hetero associative BAM network has "n" units in X-layer and "m" units in Y-layer. The connections between the layers are bi-directional, i.e., if the weight matrix for signals sent from the X-layer to Y-layer is W, then weight matrix for signals sent from Y-layer to X-layer is W_T.

The BAM architecture resembles a single layer feedforward network which consists of only one layer of weighted interconnections. There exists "n" number of input neurons in the input layer and "m" number of output neurons in the output layer. The training process is based on the Hebb learning rule. This is a fully interconnected network, wherein the inputs and outputs are different. The most important feature of a bidirectional associative memory is that there exists weighted interconnections from both X-layer to Y-layer and vice versa. The link between one layer to another and its reciprocate is shown in Figure 4.12 by the arrow pointing on both sides. The weights are adjusted between X-layer to Y-layer and also from Y-layer to X-layer.

Types of Bi-Directional Associative Memory Net

There exists two types of BAM. They are

- Discrete BAM

- Continuous BAM

Discrete BAM

The training patterns can be either binary or bipolar. In both the formats the weights are found from the sum of the Hebb outer product of the bipolar form of the training vector pairs. Step activation function with a threshold value is used as the activation function. Generally the bipolar vectors improve the performance of the net.

The weight matrix to store the set of input and target vectors s (p): t (p), p=1,...,P, where

$$s (p) = (s1 (p), \ldots s_i (p), \ldots s_n (p))$$

and can be determined with the help of Hebb rule.

For binary input vectors, weight matrix can be determined by the formula.

$$W_{ij} = \sum_p (2s_i (p) - 1) (2t_j (p) - 1)$$

For bipolar input vectors, the weight matrix is given by,

$$W_{ij} = \sum_p s_i (p) t_j (p)$$

Activation Function

The activation function for discrete BAM depends on whether binary or bipolar vectors are used. The activation function is the appropriate step function. For binary input vectors, the activation function for the Y-layer is

$$y_j = \begin{cases} 1, & if \ y_{inj} > 0 \\ y_j, & if \ y_{inj} = 0 \\ 1, & if \ y_{inj} < 0 \end{cases}$$

and the activation layer for the X-layer is

$$x_i = \begin{cases} 1, & if \ x_{inj} > 0 \\ x_j, & if \ x_{inj} = 0 \\ 1, & if \ x_{inj} < 0 \end{cases}$$

For bipolar vector, the activation function for Y-layer is

$$y_j = \begin{cases} 1, & if \ y_{inj} > \theta_j \\ x_j, & if \ y_{inj} = \theta_j \\ 1, & if \ y_{inj} < \theta_j \end{cases}$$

and the activation layer for the X-layer is

$$x_j = \begin{cases} 1, & if \ x_{inj} > \theta_j \\ x_j, & if \ x_{inj} = \theta_j \\ 1, & if \ x_{inj} < \theta_j \end{cases}$$

If the net input is equal to the threshold value, the activation function decides to leave the activation of that unit at its previous value. In the above θ_i, indicates the threshold value.

Continuous BAM

The continuous BAM was introduced by Kosko in 1988. A continuous bidirectional associative memory has the capability to transfer the input smoothly and continuously into the respective output range between [0,1]. The continuous BAM uses logistic sigmoid function as the activation function for all units.

For binary input vectors $(s(p),t(p)),p=1,2,...P$, the weights are determined by the formula.

$$W_{ij} = \sum_p (2s_i(p) - 1)(2t_j(p) - 1)$$

Y-layer

The logistic sigmoid activation function is given by

$$f(y_{inj}) = \frac{1}{1 + exp(-y_{inj})}$$

If bias is included in calculating

$$y_{inj} = b_j + \sum x_i w_{ij}$$

X-layer

The logistic sigmoid activation function is given by

$$f(x_{inj}) = \frac{1}{1 + exp(-y_{inj})}$$

If bias is included in calculating

$$x_{inj} = b_j + \sum x_i w_{ij}$$

The memory storage capacity of the BAM is min (n, m) where,

n is the number of units in x-layers
m is the number of units in y-layers
according to Haines and Hecht-Neilsen. It should be noted that this could
be extended to min (2n, 2m) if appropriate non-zero threshold value is
chosen for each unit.

Application Algorithm

From the training process, obtain the final weights. The input patterns are presented to both X-layer and Y-layer. The net input and the activations to the Y-layer are calculated. Then the signals are sent to X-layer and here its net input and activations are found. In this manner, the bi-directional associative memory is tested for its performance. The pseudocode of the bi-directional memory net is given as follows:

Initialize the weight to store a set of P vectors
Initialize all activations to 0

```
For each testing input
      Assign activation of X-layer to current pattern
      Input pattern y is presented to the Y-layer
              While activations are not converged do
              Compute the net input  y_inj = Σ w_ij xi
                                            i
              Compute activations.  y_i = f(y_-inj)
              Send signals to the X-layer
              Compute the net input  x_ini = Σ w_ij Yi
                                            i
              Compute the activations  x_i = f(x_-ini)
              Send signals to the Y-layer
```

```
        Test for convergence
        End While
End For
```

The stopping condition may be that the activation vectors x and y have reached equilibrium. The activation function applied in Steps 6 and 7 is based on the discussions made earlier.

4.4 Data Conceptualization Networks

In most of the neural network applications the data that is presented as the training set seems to vary. A few applications require grouping of data that may, or may not be, clearly definable. In such cases it is required to identify a group as optimal as possible. Such kind of networks are grouped as data conceptualization networks.

4.4.1 Adaptive Resonance Network

The last unsupervised learning network we discuss differs from the previous networks in that it is recurrent; as with networks in the next chapter, the data is not only fed forward but also back from output to input units.

Background

In 1976, Grossberg introduced a model for explaining biological phenomena. The model has three crucial properties:

1. a normalization of the total network activity. Biological systems are usually very adaptive to large changes in their environment. For example, the human eye can adapt itself to large variations in light intensities.

2. contrast enhancement of input patterns. The awareness of subtle differences in input patterns can mean a lot in terms of survival. Distinguishing a hiding panther from a resting one makes all the difference in the world. The mechanism used here is contrast enhancement.

3. short-term memory (STM) storage of the contrast-enhanced pattern. Before the input pattern can be decoded, it must be stored

in the short-term memory. The long-term memory (LTM) implements an arousal mechanism (i.e., the classification), whereas the STM is used to cause gradual changes in the LTM.

The system consists of two layers, F1 and F2, which are connected to each other via the LTM (Figure 4.13). The input pattern is received at F1, whereas classification takes place in F2. As mentioned before, the input is not directly classified. First a characterization takes place by means of extracting features, giving rise to activation in the feature representation field.

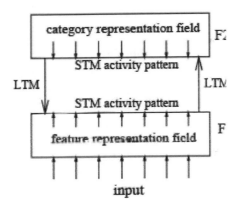

FIGURE 4.13: The ART Architecture

The expectations, residing in the LTM connections, translate the input pattern to a categorization in the category representation field. The classification is compared to the expectation of the network, which resides in the LTM weights from F2 to F1. If there is a match, the expectations are strengthened, otherwise the classification is rejected

ART1: The Simplified Neural Network Model

The architecture of ART1 is a very simplified model and consists of two layers of binary neurons (with values 1 and 0), called the comparison layer denoted as F1 and the recognition layer denoted as F2 (Figure 4.14). Every individual neuron in the comparison layer is connected to all neurons in the recognition layer through the continuous-valued forward long term memory (LTM) W^f, and vice versa via the binary-valued backward LTM W^b. A gain unit and a reset unit are also available. There are two gain units and are denoted as G1 and G2. Every neuron in the F1 layer receives three inputs: a component of the input pattern, a component of the feedback pattern, and a gain G1. The neuron fires if and only if two-third of the input is high. The neurons in the recognition layer each compute the inner product of their incoming (continuous-valued) weights and the pattern sent over these connections. The winning neuron then inhibits all the other neurons via lateral inhibition. Gain 2 is the logical "or" of all the elements in the input pattern x. Gain 1 equals gain 2, except when the feedback pattern from F2 contains any 1; then it is forced to zero. Finally, the reset signal is sent to the active neuron in F2 if the input vector x and the output of F1 differ by more than some vigilance level.

Architecture

The ART 1 has computational units and supplemental units. Its architecture is shown in Figure 4.13.

Computational Units

The computational unit comprises of F_1 and F_2 units and the reset unit. The $F_1(a)$ input unit is connected to the $F_2(b)$ interface unit. The input and the interface units are connected to reset mechanism unit. By means of top-down and bottom-up weights, the interface layer units are connected to the cluster units and the reciprocity is also achieved.

Supplemental Units

There are a few limitations of the computational unit. All the units of the computational unit are expected to react very often during the learning process. Moreover the F_2 unit is inhibited during some specific conditions and then again should be returned back when required. Therefore, in order to overcome these limitations two gain control units G_1 and G_2 act as supplemental units. These special units receive signals from and send their signal to, all the units present in occupational structure. In Figure 4.14, the excitatory signals are indicated by "+" and inhibitory signals by "−". The signal may be sent, wherever any unit in interface or cluster layer has three sources from which it can receive a signal. Each of these units also receives two excitatory signals in order to be "on". Hence, due to this, the requirement is called the two-thirds rule. This rule plays a role in the choice of parameters and initial weights. The reset unit R also controls the vigilance matching.

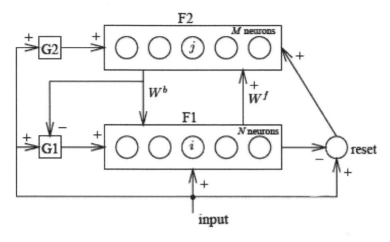

FIGURE 4.14: The ART1 Neural Network

Operation

The network starts by clamping the input at F1. Because the output of F2 is zero, G1 and G2 are both on and the output of F1 matches its input. The pattern is sent to F2, and in F2 one neuron becomes active. This signal is then sent back over the backward LTM, which reproduces a binary pattern at F1. Gain 1 is inhibited, and only the neurons in F1 which receive a 'one' from both x and F2 remain active. If there is a substantial mismatch between the two patterns, the reset signal will inhibit the neuron in F2 and the process is repeated.

Training Algorithm

The parameters used in the training algorithm are

n: Number of components in the input vector

m: Maximum number of clusters that can be formed

b_{ij}: bottom-up weights (from F_1 (b) to F_2 unit)

t_{ij}: top-bottom weights (from F_2 to F_1 (b) units)

ρ: vigilance parameter

s: binary input vector

x: activation vector for interface layer (F_1 (b) layer (binary))

$||x||$: norm of vector x (sum of the components x_i)

The binary input vector is presented to F_1 (a) input layer and is then received by F_1(b), the interface layer. The F_1(b) layer sends the activation signal to F_2 layer over weighted interconnection path. Each F_2 unit calculates the net input. The unit with the largest net input will be the winner that will have the activation d=1. All the other units will have the activation as zero. That winning unit alone will learn the current input pattern. The signal sent from F_2 to F_1 (b) through weighted interconnections is called as top-bottom weights. The "X" units remain "on" only if they receive non-zero weights from both the F_1 (a) to F2 units.

The norm of the vector $\|x\|$ will give the number of components in which top-bottom weight vector for the winning unit t_{ji} and the input vector S are both '1'. Depending upon the ratio of norm of x to norm of S ($\|x\|/\|S\|$), the weights of the winning cluster unit are adjusted. The whole process may be repeated until either a match is found or all neurons are inhibited. The ratio ($\|x\|/\|s\|$) is called Match ratio.

At the end of each presentation of a pattern, all cluster units are returned to inactive states but are available for further participation.

The pseudocode of the training algorithm of ART 1 network is as follows.

Initialize parameters L >1 and $0 <_i \rho \le 1$
```
Initialize weights 0 <b_{ij} (0) < L/(L-1+n) <t_{ji}(0) = 1
```

While not stopping condition **do**
    For each training input
        Assign activations of all F_2 units to zero
        Assign activations of F1 (a) units to input vector s
Compute the norm of s: $\| s \| = \sum_i S_i$
        Send input signal from F_1 (a) to $F_1 1$(b) layer $x_i = s_i$
        For each F_2 node that is not inhibited
            If $y_J \ne -1$
                $y_j = \sum_i b_{ij} x_i$
        End If
        While reset **do**
            Find J such that $y_J \ge y_j$ for all nodes j
                If $y_j = -1$
                    All nodes are inhibited
            End If
        Recompute activation x of F_1 (b) $x_i = s_i t_{Ji}$
        Compute the norm of vector x: $\| x \| = \sum_i X_i$

            Test for reset

If $\| x \|/\| s \| < \rho$

$y_J = -1$, (inhibit node J)

End If

End While

End For

Update the weights for node J

$$b_{ij \ (new)} = \frac{Lx_i}{L - 1 + jj \ xjj}$$

$$t_{ji \ (new)} = x_i$$

End For

Test for stopping condition

End While

The stopping condition may be no weight changes, no units reset or maximum number of epochs searched.

In winner selection, if there is a tie, take J to be the smallest such index. Also t_{ji} is either 0 or 1, and once it is set to 0 during learning, it can be never set back to 1, and once it is set to 0 during learning, it can be never set back to 1 because of stable learning method.

The parameters used have the typical values as shown below.

Parameter	Range	Typical value
L	$L > 1$	2
ρ	$0 > \rho \leq 1$	0.9
b_{ij}	$0 < b_{ij} \ (0) <$	
	$(L/(L-1+n)) \ 1/(1+n)$	
		(Bottom-up weights)
t_{ji}	$t_{ji} \ (0) = 1$ (top down weights)	

4.4.2   Implementation of ART Algorithm in MATLAB

The top-down weights for an ART network after a few iterations are given as tji=[1 1 0 0;1 0 0 1;1 1 1 1] and the bottom up weights are bij=[.57 0 .3;0 0 .3;0 .57 .3;0 .47 .3]. The following MATLAB code illustrates the steps of ART algorithm to find the new weight after the vector [1 0 1 1] is presented.

```
clc;
clear all;

% Step 1: Initialization
% The bottom up weights
b=[.57 0 .3;0 0 .3;0 .57 .3;0 .47 .3 ];
```

```
% The top down weights
t=[1 1 0 0;1 0 0 1;1 1 1 1];

% Vigilance parameter (0<ρ ≤ 1) set to 0.9
p=0.9;

% Initialize L (L>1) to 2
L=2;
```

% **Step 2:** Start training

% **Step 3:** Present the new vector
```
x=[1 0 1 1];
```

% **Step 4:** Set the activations to the input vector
```
s=x;
```

% **Step 5:** Compute the norm of s according to the
formula $\| s \| = \sum_i s_i$

```
norm_s=sum(s);
```

% Step 6: Send input signal from
   F_1 (a) to F_1(b) layer $x_i = s_i$
% Step 7: Calculate the net input $y_j = \sum_i b_{ij} x_i$

```
y=x*b;
stop=1;
while stop
```
% Step 8: While reset do
```
      for i=1:3
```
         % **Step 9:** Find J such that $y_J \geq y_j$ for
               all nodes j
```
         if y(i)==max(y)
             J=i;
         end
      end
```
         % **Step 10:** Recomputing activation x of F_1(b)
                  $x_i = s_i t_{Ji}$
```
      x=s.*t(J,:);
```
         % **Step 11:** Compute the norm of vector

$$\text{x:} \quad \| \text{ x } \| = \sum_i \text{X}_i$$

```
      norm_x=sum(x);
      % Step 12: Test for reset
      if norm_x/norm_s >= p
         % Step 13: Updating the weights
         b(:,J)=L*x(:)/(L-1+nx);
         t(J,:)=x(1,:);
         stop=0;
      else
         y(J)=-1;
         stop=1;
      end
      if y+1 == 0
         stop=0;
      end
end
      disp('Top down weights')
      disp(t);
      disp('Bottom up weights')
      disp(b);
```

Output

```
The updated weights are:
Top down weights
     1        1        0        0
     1        0        0        1
     1        0        1        1

Bottom up weights
     0.5700   0          0.5000
     0        0          0
     0        0.5700     0.5000
     0        0.4700     0.5000
```

4.4.3 Self-Organizing Map

The Kohonen network was developed by Teuvo Kohonen in the early 1980's, based on clustering data. In this network if two input vectors are close, they will be mapped to processing elements that are close together in the two-dimensional Kohonen layer that represents the features or clusters of the input data. Here, the processing elements constitute a two-dimensional map of the input data.

The basic use of the self-organizing map is to picture topologies and hierarchical structures of multidimensional input spaces. The self-organizing network has been used to create area-filled curves in two-dimensional space created by the Kohonen layer. The Kohonen layer can also be used for optimization problems by providing the connection weights to settle out into a minimum energy pattern.

The major advantage of this network is that this network learns based on unsupervision. When the topology is combined with other neural layers for prediction or categorization, the network first learns in an unsupervised manner and then switches to a supervised mode for the trained network to which it is attached.

The basic architectural model of a self-organizing map network is shown in Figure 4.15. The self-organizing map has typically two layers: input layer and the Kohonen layer. The input layer is fully connected to a two-dimensional Kohonen layer. The output layer shown here is used in a categorization problem and represents three classes to which the in-

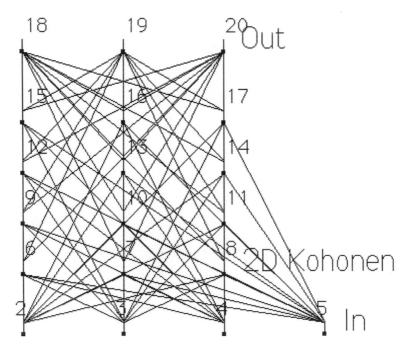

FIGURE 4.15:   An Example Self-Organizing Map Network

put vector can belong. This output layer typically learns using the delta rule and is similar in operation to the counter-propagation paradigm.

The processing elements in the Kohonen layer measures the Euclidean distance of the weights from the presented input patterns. During recall, the Kohonen element with the minimum distance is the winner and outputs a one to the output layer. Since this a competitive network, once the winning unit is chosen all the other processing elements are forced to zero. Hence the winning element is the nearest element to the input value and this represents the input value in the two-dimensional map.

During the training process, the Kohonen processing element with the smallest distance adjusts its weight to be closer to the values of the input data. The neighbors of the winning element also adjust their weights to be closer to the same input data vector.

The processing elements naturally represent approximately equal information about the input data set. Where the input space has sparse data, the representation is compacted in the Kohonen space, or map. Where the input space has high density, the representative Kohonen elements spread out to allow finer discrimination. In this way the Kohonen layer is thought to mimic the knowledge representation of biological systems.

Self-Organized Maps allow a network to develop a feature map. Self-Organized learning can be characterized as displaying "global order emerging from local interactions". One example of self-organized learning in a neural network is the SOM algorithm. There are three important principles from which the SOM algorithm is derived:

1. Self-amplification

2. Competition

3. Co-operation

These principles are defined as follows:

1. **Self-amplification:** units, which are on together, tend to become more strongly connected. Thus, positive connections tend to be self-amplifying. This is the Hebbian learning principle.

2. **Competition:** Units enter into a competition according to which one responds "best" to the input. The definition of "best" is typically according to either (i) the Euclidean distance between the unit's weight vector and the input, or (ii) the size of the dot product between the unit's weight vector and the input. Provided the vectors are normalized, a minimum Euclidean distance is equivalent to a maximum dot product so it doesn't matter which you

choose. The best-matching unit is deemed to be the winner of the competition.

3. **Co-operation:** In the SOFM, each unit in the "competing layer" is fully connected to the input layer. Further, each competing unit is given a location on the map. Most often, a two dimensional map is used so the units are assigned locations on a 2-D lattice. (maps of one dimension or more than two dimensions are also possible). Whenever a given unit wins the competition, its neighbors are also given a chance to learn. The rule for deciding who are the neighbors may be the "nearest neighbor" rule, i.e., only the four nearest units in the lattice are considered to be in the neighborhood, or it could be "two nearest neighbors", or the neighborhood could be defined as a shrinking function of the distance from each other unit and the winner. Whatever the basis for determining neighborhood membership, the winner and all its neighbors do some Hebbian learning, while units not in the neighborhood do not learn for a given pattern.

Architecture

The architecture of the Kohonen SOM is shown in Figure 4.16. All the units in the neighborhood that receive positive feedback from the winning unit participate in the learning process. Even if a neighboring unit's weight is orthogonal to the input vector, its weight vector will still change in response to the input vector. This simple addition to the competitive process is sufficient to account for the order mapping.

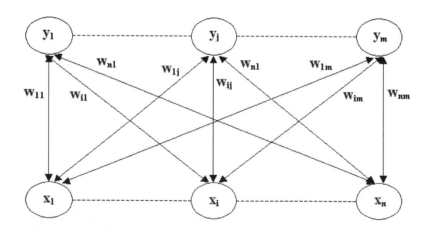

FIGURE 4.16:   Architecture of Kohonen SOM

Training Algorithm

The weights and the learning rate are set initially. The input vectors to be clustered are presented to the network. When the input vectors are presented, the winner unit is calculated either by Euclidean distance method or sum of products method based on the initial weights. Based on the winner unit selection, the weights are updated for that particular winner unit using competitive learning rule as discussed earlier. An epoch is said to be completed once all the input vectors are presented to the network. By updating the learning rate, several epochs of training may be performed.

The pseudocode of the training algorithm of the SOM network is shown below:

```
Initialize topological neighborhood parameters
Initialize learning rate
Initialize weights
While not stopping condition do
    For each input vector x
            For each I, compute squared Euclidean
            distance
                    D(j)= ∑ (wᵢⱼ xᵢ)², i=1 to n and j=1 to m
            Find index J, when D (j) is minimum
            For all units J, with specified neighborhood
            of J,
                    and for all i,
            update the weights as
                    wᵢⱼ(new) =wᵢⱼ(old) + α [xᵢ-wᵢⱼ(old)]
    End For
    Update the learning rate
    Reduce the radius of topological neighborhood
    at specified times
    Test the stopping condition
End While
End
```

The map function occurs in two phases:

- Initial formation of perfect (correct) order

- Final convergence

The second phase takes a longer duration than the first phase and requires a small value of learning rate. The learning rate is a slowly decreasing function of time and the radius of the neighborhood around a cluster unit also decreases as the clustering process goes on. The initial weights are assumed with random values. The learning rate is updated by, α (t+1) =0.5 α (t).

4.5   Applications Areas of ANN

Artificial neural networks are going through the change that happens once a concept departs the academic environment and is thrown into the harsher world of users who merely wish to get a job arranged. Several networks that are being designed recently are statistically quite exact but they still leave a defective impression with users who anticipate computers to solve their problems absolutely. These networks could be 85% to 90% accurate. Regrettably, a couple of applications tolerate that level of error.

While researchers continue to work on improving the accuracy of their "creations," some explorers are finding uses for the current technology.

In reviewing this state of the art, it is hard not to be overcome by the bright promises or tainted by the unachieved realities. Presently, neural networks are not the user interface which translates spoken works into instructions for a machine, but someday they will be. Someday, VCRs, home security systems, CD players, and word processors will simply be activated by voice. Touch screen and voice editing will replace the word processors of today while bringing spreadsheets and data bases to a level of usability pleasing to most everyone. But for now, neural networks are simply entering the marketplace in niches where their statistical accuracy is valuable as they await what will surely come.

Many of these niches indeed involve applications where answers are nebulous. Loan approval is one. Financial institutions make more money by having the lowest bad loan rate they can achieve. Systems that are "90% accurate" might be an improvement over the current selection process. Indeed, some banks have proven that the failure rate on loans approved by neural networks is lower than those approved by some of their best traditional methods. Also, some credit card companies are using neural networks in their application screening process.

This latest method of looking into the future by examining past experiences has rendered its own independant problems. One of the major problems is to provide a reason behind the computer-generated answer, such as to why a particular loan application was denied. As mentioned throughout this chapter, the inner workings of neural networks are "black boxes." The explanation of a neural net and its learning has been difficult. To alleviate this difficult process, a lot of neural net tool developers have provided programs which explain which input through which node dominates the decision making process. From that information, experts in the application should be able to infer the reason that an exceptional piece of data is crucial.

Besides this filling of niches, neural network work is progressing in other more promising application areas. The following section of this chapter goes through some of these areas and briefly details the current work. This is done to help stimulate within the reader the various possibilities where neural networks might offer solutions, possibilities such as language processing, character recognition, image compression, pattern recognition, among others.

4.5.1 Language Processing

Human language users perform differently from their linguistic competence, that is from their knowledge of how to communicate correctly using their language. Natural language processing is an application reinforced by the use of association of words and concepts, implemented as a neural network. A single neural network architecture is capable of processing a given sentence, and outputs a host of language processing predictions: part-of-speech tags, chunks, named entity tags, semantic roles, semantically similar words, and the likelihood that the sentence makes sense grammatically and semantically using a language model.

Several researchers belonging to various universities are researching how a computer could be programmed to respond to spoken commands using the artificial neural networks. Natural language processing (NLP) has become the main-stream of research with neural networks (NNs), which are powerful parallel distributed learning/processing machines and they play a major role in several areas of NLP.

Presently, as reported by the academic journals, most of the hearing-capable neural networks are trained to only one talker. These one-talker, isolated-word recognizers can recognize only a few hundred words. But when there is a pause between each word, then the neural neworks can recognize more number of words.

A few investigators are touting even bigger potentialities, but due to the expected reward the true progress, and methods involved, are being closely held. The most highly touted, and demonstrated, speech-parsing system comes from the Apple Corporation. This network, according to an April 1992 Wall Street Journal article, is capable of recognizing almost any person's speech through a limited vocabulary.

4.5.2   Character Recognition

The recognition of optical characters is known to be one of the earliest applications of Artificial Neural Networks, which partially emulate human thinking in the domain of artificial intelligence. Recognition of either handwritten or printed characters is a major area in which neural networks are providing optimal solutions. Multilayer neural networks are used to recognize characters which is a vital application in areas like banking, etc. The main issue in character recognition is the trade-off between cost and benefits such as accuracy and speed. Neural networks provide a method for combining independently trained characters to achieve higher performance at relatively low cost.

The greatest amount of recent research in the domain of character recognition is targeted at scanning oriental characters into a computer. Presently, these characters require four or five keystrokes each. This complicated process stretches the task of identifying a page of text into hours of drudgery.

4.5.3   Data Compression

In neural networks, when the number of hidden units is less when compared to he input and output units, then the neurons of the middle layer are capable of data compression. Researches have been proved that neural networks can do real-time compression and decompression of data. These networks are auto associative in that they can reduce eight bits of data in the input layer to three in the hidden layer and then reverse that process upon restructuring to eight bits again in the output layer. While compressing there is no loss in the information. Some of the major applications of data compression are multispectral lossless image compression pattern recognition lossy or lossless compression video compression handwritten numeral classification edge detection and magnetic resonance image compression.

4.5.4   Pattern Recognition

Pattern recognition is a very old application of neural networks and it has been studied in relation to many different (and mainly unrelated) applications, such as classifying patterns by shape, identifying fingerprints, identifying tumors, handwriting recognition, face recognition, coin recognition, etc. The majority of these applications are concerned with problems in pattern recognition, and make use of feed-forward network architectures such as the multi-layer perceptron and the radial basis function network. According to the perspective of pattern recognition, neural networks can be looked upon as an extension of the several traditional techniques which have been developed over several decades.

4.5.5 Signal Processing

The role of neural networks in signal processing is getting distributed, with practical applications in areas such as filtering, parameter estimation, signal detection, pattern identification, signal reconstruction, system identification, signal compression, and signal transmission. The signals concerned include audio, video, speech, image, communication, geophysical, sonar, radar, medical, musical, and many others. The main characteristics of neural networks applied to signal processing are their asynchronous parallel and distributed processing, nonlinear dynamics, global interconnection of network elements, self-organization, and high-speed computational capability. Neural networks are capable of providing effective means for resolving several problems encountered in signal processing, especially, in nonlinear signal processing, real-time signal processing, adaptive signal processing, and blind signal processing.

4.5.6 Financial

Earlier financial experts used charts as the main source to navigate the large amount of financial data that was available. A few experts study the long term investments of companies while a few others try to anticipate the approaching economy or stock market in general. All these processes involved a large amount of risk in the work. In order to aid people in predicting particular markets,numerous computer programs are available. Traditionally, these programs are expensive, need complex programming, use surveys of financial experts to define the "game rules", and are still limited in their ability to think like people. Still, the task is difficult even if the solution is obtained. In spite of the implications of the effective market hypothesis, many traders continue to make, buy, and sell decisions based on historical data. These decisions are made under the premise that patterns exist in that data, and that these patterns provide an indication of future movements. If such patterns exist, then it is possible in principle to apply automated pattern recognition techniques such as neural networks to the discovery of these patterns.

A neural network is a new kind of computing tool that is not limited by equations or rules. Neural networks function by finding correlations and patterns in the financial data provided by the user. These patterns become a part of the network during training. A separate network is needed for each problem you want to solve, but many networks follow the same basic format.

Summary

This chapter discusses the major class of neural network based on applications such as data classification and data association. Another network type described in this chapter is data conceptualization. Implementations of these networks using MATLAB Neural Network Tool box are also illustrated. The future of ANN holds even more promises. Neural networks need faster hardware. They need to become part of hybrid systems, which also utilize fuzzy logic and expert systems. It is then that these systems will be able to hear speech, read handwriting, and formulate actions. They will be able to become the intelligence behind robots that never tire nor become distracted. It is then that they will become the leading edge in an age of "intelligent" machines.

Review Questions

1. Explain the architecture and algorithm of LVQ.

2. What are the variants of LVQ? How do they differ from LVQ?

3. Mention the different types of counter propagation network.

4. What are in star and out star models?

5. Explain the application procedure of CPN.

6. Describe the architecture and algorithm of probabilistic neural network.

7. Explain the application procedure of Discrete Hopfield network.

8. Write a note on Boltzmann machine.

9. Derive an expression to determine the Hamming distance.

10. What are the different forms of BAM? How are BAM nets classified?

11. Describe the major application areas of ANN.

12. Write a MATLAB program to implement competitive learning rule.

13. Mention the functionality of supplemental and computational units of the ART network.

14. Assume the stored pattern to be [1 1 1 1 0 1 1 1 1]. Let there be mistakes in the 2nd and 5th positions. Develop a Hopfield algorithm in MATLAB and determine the converged output.

15. What is the MATLAB Neural Network toolbox function used to find the weights in the layers of a LVQ network?

Chapter 5

MATLAB Programs to Implement Neural Networks

Neural network computations are naturally expressed in matrix notation, and there are several toolboxes in the matrix language MATLAB, for example the commercial neural network toolbox, and a toolbox for identification and control. In the wider perspective of supervisory control, there are other application areas, such as robotic vision, planning, diagnosis, quality control, and data analysis (data mining). The strategy in this chapter is to aim at all these application areas, but only present the necessary and sufficient neural network material for understanding the basics. A few MATLAB simulated examples such as Coin detection, Pattern Recall, Pattern Classification, and Simulink models using different Neural Network architectures are illustrated in this chapter.

5.1 Illustration 1: Coin Detection Using Euclidean Distance (Hamming Net)

A set of coins is taken as input image and they are detected using Hamming Network. The edges are detected and the minimal distance to non-white pixels are calculated using Euclidean distance formula.

```
% - - - - - - - - - - - - - - - - - - - - - -
% Main Program
% - - - - - - - - - - - - - - - - - - - - - -


close all;
clear all;
clc;
```

```
I = imread('Image1','jpeg');
flg=isrgb(I);


if flg==1
   I=rgb2gray(I);
end


[h,w]=size(I);
figure;imshow(I);


c = edge(I, 'canny',0.3);       % Mcanny edge
detection
figure; imshow(c);               % binary edges


se = strel('disk',2);               %
I2 = imdilate(c,se);           %
imshow(I2);                        %


d2 = imfill(I2, 'holes');        %
figure, imshow(d2);            %


Label=bwlabel(d2,4);


a1=(Label==1);
a2=(Label==2);
a3=(Label==3);
a4=(Label==4);
a5=(Label==5);
a6=(Label==6);


 D1 = bwdist( a1); % computing minimal euclidean
                    % distance to non-white pixel
 figure, imshow(D1,[]),
 [xc1 yc1 r1]=merkz(D1);
 f1=coindetect(r1)
```

```
D2 = bwdist( a2); % computing minimal euclidean
                          % distance to non-white pixel
figure, imshow(D2,[]),
%[xc2 yc2 r2]=merkz(D2);
f2=coindetect(r2)


D3 = bwdist( a3); % computing minimal euclidean
                          % distance to non-white pixel
figure, imshow(D3,[]),
%[xc3 yc3 r3]=merkz(D3);
f3=coindetect(r3)


D4 = bwdist( a4); % computing minimal euclidean
                          % distance to non-white pixel
figure, imshow(D4,[]),
%[xc4 yc4 r4]=merkz(D4);
f4=coindetect(r4)


D5 = bwdist( a5); % computing minimal euclidean
                          % distance to non-white pixel
figure, imshow(D5,[]),
%[xc5 yc5 r5]=merkz(D5);
f5=coindetect(r5)


D6 = bwdist( a6); % computing minimal euclidean
                          % distance to non-white pixel
figure, imshow(D6,[]),
%[xc6 yc6 r6]=merkz(D6);
f6=coindetect(r6)


% - - - - - - - - - - - - - - - - - - - - - - -
% End of Main Program
% - - - - - - - - - - - - - - - - - - - - - - -
```

```
% - - - - - - - - - - - - - - - - - - - - - - -
% Sub-Program Used
% - - - - - - - - - - - - - - - - - - - - - -


function f=coindetect(rad);


if rad >100


f=1;
elseif (68<rad) & (rad<69)
f=5;
elseif (76<rad) & (rad<77)
f=10;
elseif (85<rad) & (rad<86)
f=25;
elseif (95<rad) & (rad<96)
f=50;
else
f=0;
end


function [centx,centy,r]=merkz(D);


[w h]=size(D');


mx=max(max(D));
r=mx;
for i=1:h
    for j=1:w
        if D(i,j)==mx;
            centx=j;
            centy=i;
        end
    end
end
```

FIGURE 5.1:   Input Image

Observations

The input image shown in Figure 5.1 is converted to gray scale (Figure 5.2) and the edges are detected (Figure 5.3). The holes of the image are shown in Figure 5.4. The minimal distance to non-white pixels are computed using Euclidean distance formula and shown in Figure 5.5.

FIGURE 5.2:   Input Image in Gray Scale

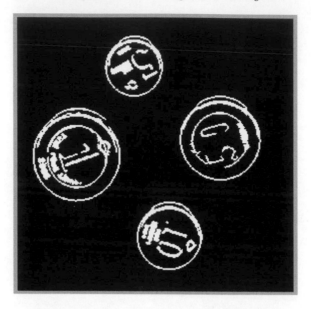

FIGURE 5.3: Edge Detection

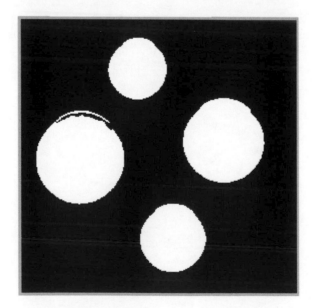

FIGURE 5.4: Fill in the Holes of an Intensity Image

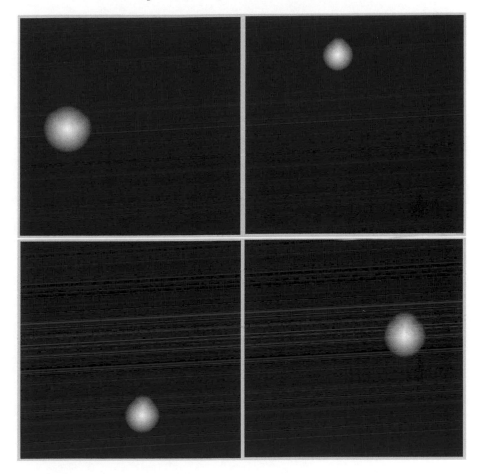

FIGURE 5.5: Computation of Minimal Euclidean Distance to Non-White Pixel

---

5.2   Illustration 2: Learning Vector Quantization - Clustering Data Drawn from Different Regions

An LVQ net with two input units and four target classes is considered, with 16 classification units and weights assigned in random. The learning rate is chosen between [0,1]. An input vector is drawn from different regions and accordingly the class is assigned. The winning unit is found based on competition. In this example the training vectors are drawn from the following regions:

Class 1   0.0 <=x1<0.5   0.0<=x2<0.5
Class 2   0.5 <=x1<1.0   0.0<=x2<0.5
Class 3   0.0 <=x1<0.5   0.5<=x2<1.0
Class 4   0.5 <=x1<1.0   0.5<=x2<1.0

Based on the input value given by the user the class is chosen. The weight vectors are assigned as follows according to the class chosen from the input vector,

Weight vector for Class 1 = $[0.2\ 0.2\ 0.6\ 0.6; 0.2\ 0.6\ 0.8\ 0.4]$
Weight vector for Class 2 = $[0.4\ 0.4\ 0.8\ 0.8; 0.2\ 0.6\ 0.4\ 0.8]$
Weight vector for Class 3 = $[0.2\ 0.2\ 0.6\ 0.6; 0.4\ 0.8\ 0.2\ 0.6]$
Weight vector for Class 4 = $[0.4\ 0.4\ 0.8\ 0.8; 0.4\ 0.8\ 0.2\ 0.6]$

```
% MATLAB CODE

% Training with inputs lying in the specified regions
% Different learning rates and different geometries for
    input
clc;
n=2;
m=4;
p=1;
ep=1;
J=0;

prompt1={'ENTER LEARNING RATE','ENTER TRAINING INPUT x1',
         'ENTER TRAINING INPUT x2'};
dlgTitle1='PATTERN CLUSTERING';
answer1=inputdlg(prompt1);
disp(answer1);
learn_rate=str2double(answer1(1));
disp(learn_rate);
x1=str2double(answer1(2));
disp(x1);
x2=str2double(answer1(3));
disp(x2);

if((x1 <=0.0)&&(x1<0.5)&&(x2>=0.0)&&(x2<0.5))
    T=1;
    X=[x1 x2];
    W=[0.2 0.2 0.6 0.6;0.2 0.6 0.8 0.4];
    disp('Class 1');
elseif((x1>=0.5)&&(x1<1.0)&&(x2>=0.0)&&(x2<0.5))
```

```
   T=2;
   X=[x1 x2];
   W= [0.4 0.4 0.8 0.8;0.2 0.6 0.4 0.8];
   disp('Class 2');
elseif((x1>=0.0)&&(x1<0.5)&&(x2>=0.5)&&(x2<1.0))
   T=3;
   X=[x1 x2];
   W= [0.2 0.2 0.6 0.6;0.4 0.8 0.2 0.6];
   disp('Class 3');
elseif((x1>=0.5)&&(x1<1.0)&&(x2>=0.5)&&(x2<1.0))
   T=4;
   X=[x1 x2];
   W= [0.4 0.4 0.8 0.8;0.4 0.8 0.2 0.6];
   disp('Class 4');
end
for c=1:p
   fprintf('Input\n')
   disp(X)
   for j-1:m
     D(j)=0;
     for i=1:n
         D(j)=D(j)+(X(c,i)-W(i,j)^2);
         disp(D(j));
     end
     fprintf('Distance D(%g)',j);
     disp(D(j));
   end
end
Temp=D(1)
disp(Temp);
J=1;
for j=2:m
   if Temp>D(j);
     Temp=D(j);
     J=j;
   end
end
   disp(J);
   if(J==T)
     for i=1:n
         W(i,J)=W(i,J)+learn_rate*
             (X(c,i)-W(i,j));
     end
```

```
else
  for i=1:n
       W(i,J)=W(i,J)+learn_rate*
          (X(c,i)-W(i,j));
  end
end
disp(W);
```

```
prompt={'INDEX J','CLASS','UPDATED WEIGHTS
w11','w12','w13','w14','w21','w22','w23','w24'};
J1=num2str(J);
class1=num2str(J);
w11=num2str(W(1,1));
w12=num2str(W(1,2));
w13=num2str(W(1,3));
w14=num2str(W(1,4));
w21=num2str(W(2,1));
w22=num2str(W(2,2));
w23=num2str(W(2,3));
w24=num2str(W(2,4));
def=J1,class1,w11,w12,w13,w14,w21,w22,w23,w24;
dlgTitle='Result of Training'
lineNo=1;
answer=inputdlg(prompt,dlgTitle,lineNo,def);
```

Observations:

The learning process performed by the LVQ is shown in Figure 5.6 and 5.7. From the output we find that the vectors are classified as belonging to class 4.

5.3   Illustration 3: Character Recognition Using Kohonen Som Network

The objective of this illustration is to identify each of a large number of black-and-white rectangular pixel displays as one of the 26 capital letters in the English alphabet. The character images were based on 20 different fonts and each letter within these 20 fonts was randomly distorted to produce a file of 20,000 unique stimuli. Each stimulus was converted into 16 primitive numerical attributes (statistical moments

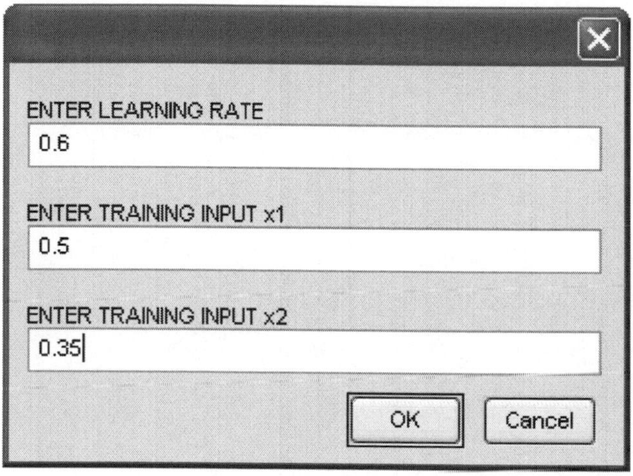

FIGURE 5.6: Input Format

and edge counts), which were then scaled to fit into a range of integer values from 0 through 15. The first 16000 items were trained and the resulting model is used to predict the letter category for the remaining 4000. Some of the parameters initialized in this program are follows:

The Number of Instances: 20000
Number of Attributes: 17 (Letter category and 16 numeric features)
Attribute Information:

1.	lettr capital letter	(26 values from A to Z)
2.	position of box	(integer)
3.	y-box vertical position of box	(integer)
4.	width width of box	(integer)
5.	high height of box	(integer)
6.	onpix total # on pixels	(integer)
7.	x-bar mean x of on pixels in box	(integer)
8.	y-bar mean y of on pixels in box	(integer)
9.	x2bar mean x variance	(integer)
10.	y2bar mean y variance	(integer)
11.	xybar mean x y correlation	(integer)
12.	x2ybr mean of x * x * y	(integer)
13.	xy2br mean of x * y * y	(integer)
14.	x-ege mean edge count left to right	(integer)
15.	xegvy correlation of x-ege with y	(integer)
16.	y-ege mean edge count bottom to top	(integer)
17.	yegvx correlation of y-ege with x	(integer)

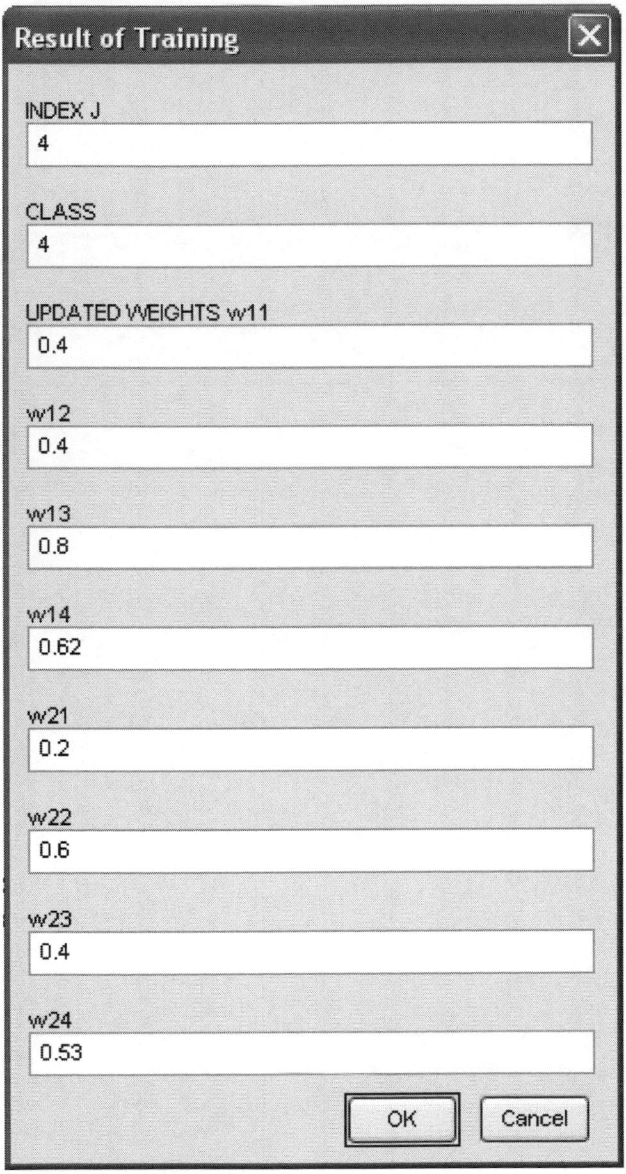

FIGURE 5.7: Output Format

TABLE 5.1: Class Distribution

A	B	C	D	E	F	G	H
789	766	736	805	768	775	773	734
I	J	K	L	M	N	O	P
755	747	739	761	792	783	753	803
Q	R	S	T	U	V	W	X
783	758	748	796	813	764	752	787
Y	Z						
786	734						

Missing Attribute Values: None
Class Distribution:

```
% MATLAB CODE
% ANALOG DATA

clear all;
clc;
m1=26;
alpha = input('Enter the value of alpha = ');
per1 = input('Enter the percentage of traing vectors ');
per2 = input('Enter the percentage of testing vectors ');
x1 = load('d:\finalpgm\data160rand.txt');
 [patt n] = size(x1);
x2=x1;
maxi=max(x1,[],1);
value= x2(:,1);

for j = 2:n
       input(:,(j-1)) = x2(:,j)/maxi(j);
end

 [pattern n] = size(input);
ci = 1;
for i = 1:m1
       while (i ~= value(ci));
              ci = ci + 1;
              if(ci>patt)
                     ci = 1;
              end
       end
       w(i,:) = input(i,:);
       ci = 1;
end
```

```
countw = ones(1,m1);
alphacond = 0.000001*alpha;
ep = 0;
patterntrain = round(pattern*per1/100);
for i = 1:patterntrain
        for j = 1:m1
            if(value(i)==j)
                    countw(j) = countw(j)+1;
                    w(j,:) = ((countw(j)-1)*w(j,:)+
                        input(i,:))/countw(j);
            end
        end
end
tic;
while(alpha>alphacond)
        clc;
        ep = ep+1
        for p = 1:patterntrain;
            data = input(p,:);
            for i = 1:m1
                    d(i) = sum(power((w(i,:)-
                        data(1,:)),2));
            end
              [mind mini] = min(d);

                    w(mini,:) = w(mini,:)+alpha*
                        (data(1,:)-w(mini,:));
                end
        alpha = alpha*0.9;

end
t = toc;
count = 0;
patterntest = round(pattern*per2/100);
for p = 1:patterntest
        data = input(p,:);
            for i = 1:m1
                    d(i) = sum(power((w(i,:)-
                        data(1,:)),2));
            end
          [mind mini] = min(d);
        output(p) = mini;
        if(mini==value(p))
```

```
                    count = count+1;
              end
      end
      fprintf('\nPercentage of TRAING Vectors : % f',per1);
      fprintf('\nPercentage of TESTING Vectors : % f',per2);
      fprintf('\nTime Taken for TRANING : % f in secs',t);

      eff = count*100/patterntest;
      fprintf('\nEfficiency = % f',eff);

      % DIGITAL DATA

      clear all;
      clc;
      m1=26;
      alpha = input('Enter the value of alpha = ');
      per1 = input('Enter the percentage of traing vectors ');
      per2 = input('Enter the percentage of testing vectors ');
      x1 = load('d:\finalpgm\data160rand.txt');
       [patt n] =size(x1);
      x2-x1;
      maxi=max(x1,[],1);
      value = x2(:,1);

      for j = 2:n
              input(:,(j-1)) = x2(:,j)/maxi(j);
      end
       [pattern n] = size(input);
      for i = 1:pattern
              for j = 1:16
                    if(input(i,j)>0.5)
                            input(i,j) = 1;
                    else
                            input(i,j) = 0;
                    end
              end
      end
      ci = 1;
      for i = 1:m1
              while (i ~= value(ci));
                    ci = ci + 1;
                    if(ci>patt)
                            ci = 1;
```

```
                end
            end
        w(i,:) = input(i,:);
        ci = 1;
end
countw = ones(1,m1);
alphacond = 0.000001*alpha;

ep = 0;
patterntrain = round(pattern*per1/100);
for i = 1:patterntrain
        for j = 1:m1
            if(value(i)==j)
                    countw(j) = countw(j)+1;
                    w(j,:) = ((countw(j)-1)*w(j,:)+
                        input(i,:))/countw(j);
            end

        end
end
tic;

while(alpha>alphacond)
        clc;
        ep = ep+1
        for p = 1:patterntrain;
            data = input(p,:);
            for i = 1:m1
                    d(i) = sum(power((w(i,:)-
                        data(1,:)),2));
            end
             [mind mini] = min(d);
                    w(mini,:) = w(mini,:)+alpha*
                        (data(1,:)-w(mini,:));
        end
        alpha = alpha*0.9;

end
t = toc;
count = 0;
patterntest = round(pattern*per2/100);
for p = 1:patterntest
        data = input(p,:);
```

TABLE 5.2: Results for Analog Data

Training Vectors	Test Vectors	Time (Secs)	Efficiency (%)
20	80	2.01	51.56
40	60	3.98	73.95
60	40	5.78	89.06
80	20	9.23	93.75
95	5	9.36	100

```
for i = 1:m1
    d(i) = sum(power((w(i,:)-
        data(1,:)),2));
end
[mind mini] = min(d);
output(p) = mini;
if(mini==value(p))
    count = count+1;
end
end

% RESULTS:
fprintf('\nPercentage of TRAING Vectors : % f',per1);
fprintf('\nPercentage of TESTING Vectors : % f',per2);
fprintf('\nTime Taken for TRANING : % f in secs',t);

eff = count*100/patterntest;
fprintf('\nEfficiency = % f',eff);
```

Observations:

Tables 5.2 and 5.3 and show the results of KOHONEN for analog and digital inputs with optimum efficiency. It can be inferred that the efficiency increases with number of training vectors.

TABLE 5.3: Results for Digital Data

Training Vectors	Test Vectors	Time (Secs)	Efficiency (%)
20	80	1.98	70.31
40	60	3.87	79.16
60	40	6.03	82.81
80	20	7.76	84.37
95	5	9.29	100

5.4   Illustration 4: The Hopfield Network as an Associative Memory

This illustration demonstrates the implementation of the Hopfield network, which acts as an associative memory for pattern classification.

```
% - - - - - - - - - - - - - - - - - - - - - - - -
% Main Program
% - - - - - - - - - - - - - - - - - - - - - - - -

clear all
% Load input data


X = [-1 -1 -1 -1 -1 -1 -1 -1 -1 -1 -1 -1 1 1 1 -1 -1
-1 -1 -1 1 1 -1
1 1 -1 -1 -1 -1 1 1 -1 1 1 -1 -1 -1 -1 1 1 -1 1 1 -1
-1 -1 -1 1 1 -1 1 1 -1 -1
-1 -1 1 1 -1 1 1 -1 -1   -1 -1 -1 1 1 1 -1 -1 -1
-1 -1 -1

                   -1 -1 -1 -1 -1 -1;

-1 -1 -1 1 1 1 -1 -1 -1 -1 -1 -1 1 1 1 -1 -1 -1 -1
-1 -1 1 1 1 -1 -1 -1 -1 -1 -1 1 1 1 -1 -1 -1 -1 -1
-1 1 1 1 1 -1 -1 -1 -1 -1 -1 1 1 1 -1 -1 -1 -1 -1 -1 1
1 1 -1 -1 -1 -1 -1 -1 1 1 1 -1 -1 -1 -1 -1 -1 1 1 1
-1 -1 -1;
-1 -1 1 1 1 1 1 1 -1 -1 -1 1 -1 -1 -1 -1 -1 1 -1 -1 -1
-1 -1 -1 -1 -1 1 -1 -1 -1 -1 -1 -1 -1 1 -1 -1 -1 -1
-1 -1 -1 1 -1 -1 -1 -1 -1 -1 -1 1 -1 -1 -1 -1 -1 -1
-1 1 -1 -1 -1 -1 -1 -1 -1 1 -1 -1 -1 -1 -1 -1 -1 1 1
1 1 1 1 -1;
-1 1 1 -1 -1 -1 1 1 -1 -1 1 1 -1 -1 -1 1 1 -1 -1 1 1
-1 -1 -1 1 1 -1 -1 1 1 -1 -1 -1 1 1 -1 -1 1 1 1 1 1
1 1 -1 -1 -1 -1 -1 -1 -1 1 1 -1 -1 -1 -1 -1 -1 -1 1
1 -1 -1 -1 -1 -1 -1 1 1 -1 -1 -1 -1 -1 -1 -1 1 1
-1;
1 1 1 1 1 1 -1 -1 -1 1 1 -1 -1 -1 -1 -1 -1 -1 1 1 1 -1
-1 -1 -1 -1 -1 1 1 1 1 1 1 1 -1 -1 -1 1 1 -1 -1 1 1
-1 -1 -1 1 1 -1 -1 1 1 -1 -1 -1 1 1 -1 -1 1 1 -1 -1
-1 1 1 -1 -1 1 1 -1 -1 -1 1 1 1 1 1 1 -1 -1 -1];
```

```
X = X([1 2 3 4 5],:);
numPatterns = size(X,1);
numInputs = size(X,2);
% Plot the fundamental memories.
figure
plotHopfieldData(X)

% STEP 1. Calculate the weight matrix using Hebb's
postulate
W = (X'* X- numPattern)s*eye(numInputs)/numInputs;

% STEP 2. Set a probe vector using a predefined
noiselevel. The probe
% vector is a distortion of one of the fundamental
memories
noiseLevel = 1/3;
patInd = ceil(numPatterns*rand(1,1));
xold -
(2*(rand(numInputs,1)>noiseLevel)-1).*X(patInd,:)';

% STEP 3. Asynchronous updates of the elements of the
% probe vector until it converges. To guarantee convergence,
% the algorithm performs at least numPatterns=81 iterations
% even though convergence generally occurs before this
figure
converged = 0;
x=xold;
while converged==0,
        p=randperm(numInputs);
        for n=1:numInputs
            r = x(p(n));
            x(p(n)) = hsign(W(p(n),:)*x, r);
            plotHopfieldVector(x);
            pause(0.01);
        end

% STEP 4. Check for convergence
                if (all(x==xold))
                    break;
                end
                xold = x;
end
```

```
% - - - - - - - - - - - - - - - - - - -
% End of Main
% - - - - - - - - - - - - - - - - - - -


% - - - - - - - - - - - - - - - - - - - -
% Subprograms used
% - - - - - - - - - - - - - - - - - - - -


% Update the elements asynchronously.
function y = hsign(a, r)
y(a>0) = 1;
y(a==0) = r;
y(a<0) = -1;


% - - - - - - - - - - - - - - - - - - - -
% End of hsign
% - - - - - - - - - - - - - - - - - - - -


% Plot the fundamental memories.
function plotHopfieldData(X)
numPatterns = size(X,1);
numRows = ceil(sqrt(numPatterns));
numCols = ceil(numPatterns/numRows);
for i=1:numPatterns
                  subplot(numRows, numCols, i);
                  axis equal;
                  plotHopfieldVector(X(i,:))
end


% - - - - - - - - - - - - - - - - - - - -
% End of plotHopfieldData
% - - - - - - - - - - - - - - - - - - - -


% Plot the sequence of iterations for the probe vector.
function plotHopfieldVector(x)
cla;
numInputs = length(x);
numRows = ceil(sqrt(numInputs));
numCols = ceil(numInputs/numRows);
for m=1:numRows
```

```
      for n=1:numCols
          xind = numRows*(m-1)+n;
          if xind >numInputs
              break;
          elseif x(xind)==1
              rectangle('Position', [n-1 numRows-m 1 1],
                  'FaceColor', 'k');
          elseif x(xind)==-1
              rectangle('Position', [n-1 numRows-m 1 1],
                  'FaceColor', 'w');
          end
      end
end
set(gca, 'XLim', [0 numCols], 'XTick', []);
set(gca, 'YLim', [0 numRows], 'YTick', []);

% - - - - - - - - - - - - - - - - - - - - - - -
% End of plotHopfieldVector
% - - - - - - - -      - - - - - - - - - - - - -
```

Observations

Figure 5.8 shows the input patterns, of which "2" is selected for pattern classification. A noisy pattern is introduced as shown in Figure 5.9. The net was able to classify the correct pattern after 81 iterations as shown in Figure 5.10.

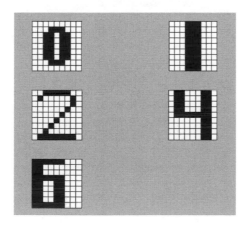

FIGURE 5.8:   Input Patterns

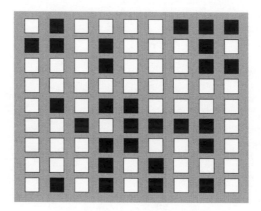

FIGURE 5.9:   Noisy Pattern

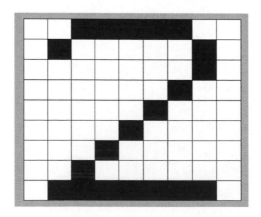

FIGURE 5.10:   Network Converged after 81 Iterations

5.5 Illustration 5: Generalized Delta Learning Rule and Back-Propagation of Errors for a Multilayer Network

This illustration shows the implementation of the generalized learning rule and the back propagation of errors in a back propagation network.

```
% - - - - - - - - - - - - - - - - - - - - - - - -
% Main Program
% - - - - - - - - - - - - - - - - - - - - - - - -

% Load data
load housing.txt
X = housing(:,1:13);
t = housing(:,14);

% Scale to zero mean, unit variance and introduce
bias on input.
xmean = mean(X);

xstd = std(X);
X =
(X-ones(size(X,1),1)*xmean)./(ones(size(X,1),1)*xstd);
X = [ones(size(X,1),1) X];
tmean = mean(t);
tstd = std(t);
t = (t-tmean)/tstd;

% Iterate over a number of hidden nodes
maxHidden = 2;
for numHidden=1:maxHidden

% Initialise random weight vector.
% Wh are hidden weights, wo are output weights.
randn('seed', 123456);
Wh = 0.1*randn(size(X,2),numHidden);
wo = 0.1*randn(numHidden+1,1);

% Do numEpochs iterations of batch error back propagation.
numEpochs = 2000;
numPatterns = size(X,1);
```

```
% Set eta.
eta = 0.05/numPatterns;
for i=1:numEpochs
              % Calculate outputs, errors, and gradients.
              phi = [ones(size(X,1),1) tanh(X*Wh)];
              y = phi*wo;
              err = y-t;
              go = phi'*err;
              Gh = X'*((1-phi(:,2:numHidden+1).^2).*
                 (err*wo(2:numHidden+1)'));
              % Perform gradient descent.
              wo = wo - eta*go;
              Wh = Wh - eta*Gh;
              % Update performance statistics.
              mse(i) = var(err);
end


plot(1:numEpochs, mse, '-')
hold on
end


fsize=15;
set(gca,'xtick',[0:500:2000],'FontSize',fsize)
set(gca,'ytick',[0:0.5:1],'FontSize',fsize)
xlabel('Number of Epochs','FontSize',fsize)
ylabel('Mean Squared Error','FontSize',fsize)
hold off


% - - - - - - - - - - - - - - - - - - - - - - - -
% End of Main
% - - - - - - - - - - - - - - - - - - - - - - - -
```

Observations

The performance statistics of the generalized delta learning rule and backpropagation of errors for a multilayer network is shown in Figure 5.11. It is found that the Mean Squared Error decreases as the number of epochs increases.

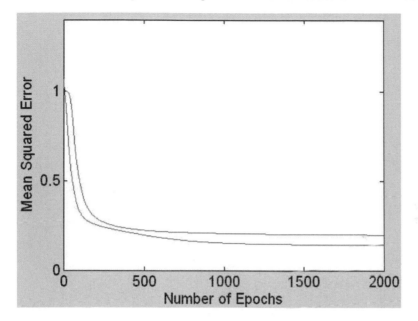

FIGURE 5.11: Illustration of Generalized Delta Rule and Back Propagation of Errors

5.6 Illustration 6: Classification of Heart Disease Using Learning Vector Quantization

The following program illustrates the use of MATLAB to classify a normal person from a sick person using LVQ. This is supervised version of Kohonen learning rule. The original data is available in 13 attributes and a class attribute. Depending upon the value of attribute the network will classify a normal person from a sick person.

The datasets chosen for the project are heart disease Cleveland database. All attribute names and values of the attribute are given so that the reader can understand the aim of the program.

```
% PROGRAM FOR DIGITAL(BIPOLAR) INPUT DATA
clc;
tic;
m=4;
x5=load('heartdata00.txt'); % loading input datas from
the file c3=size(x5)*[1;0];% calculating size of matrix.
u=size(x5)*[0;1];
```

```
prompt1={'Percentage of datas to be trained',
   'Percentage of datas to be tested',
'Enter the value of learning rate'};
dlgTitle1='HEART DISEASE DIAGNOSIS USING LEARNING VECTOR
QUANTISATION NETWORK';
lineNo=1;
answer1=inputdlg(prompt1,dlgTitle1,lineNo);
per=str2double(answer1(1));
per1=str2double(answer1(2));
al=str2double(answer1(3));
pe=per/100;
v=round(pe*c3);
% Separating theinputattributes and the target from the
input file for s1=1:m
     for i=1:c3
          if(x5(i,u)==(s1-1))
               for(j=1:u)
                    temp=x5(s1,j);
                    x5(s1,j)=x5(i,j);
                    x5(i,j)=temp;
               end
          end
     end
end
for i=1:c3
     for j=1:u
          if((j==u))
               t(i)=x5(i,j);
          end
     end
end
for i=1:c3
     for j=1:u-1
          x(i,j)=x5(i,j);
     end
end
for i=1:c3
     for j=1:u-1
          if x(i,j)==0
               x(i,j)=.05;
          end
     end
end
```

```
% Normalizing the datas
q2=size(x)*[0;1];
p2=size(x)*[1;0];
y=max(x,[],1);
z=min(x,[],1);
for i=1:q2
    if y(i)~=z(i)
        e(i)=y(i)-z(i);
    else
        e(i)=1;
        z(i)=0;
    end
end
for i=1:p2
    for j=1:q2
        x(i,j)=(x5(i,j)- z(j))/(e(j));
    end
end
% Initialising the weight matrix
for i=1:u-1
    for j=1:4
        w(i,j)=x(j,i);
    end
end
% Converting the normalized data into bipolar form
for i=1:p2
    for j=1:q2
        if x(i,j)>.3
            x(i,j)=1;
        else
            x(i,j)=-1;
        end
    end
end
q=size(x)*[0;1];
p=size(x)*[1;0];
N=0;
% Stopping condition.
while (al>0.0000000001)
    N=N+1;
% Calculating the distance by using Euclidean distance
    method
    for i=5:v
```

```
       for k=1:4
            d(k)=0;
            for j=1:u-1
                  d(k)=d(k)+[x(i,j)-w(j,k)]^2;
            end
       end
       b=min(d);
% Finding the winner
       for l=1:4
                if (d(l)==b)
                    J=l;
                end
       end
% Weight updation
       for f=1:q
                if(t(J)==t(i))
                      w(f,J)=w(f,J)+al*[x(i,f)-w(f,J)];
                else
                      w(f,J)=w(f,J)-al*[x(i,f)-w(f,J)];
                end
       end
     end
% Reducing the learning rate
     al=al/2;
end
% LVQ Testing
pe1=per1/100;
v1=round(pe1*c3);
for i=1:v1
     for j=1:u-1
          x1(i,j)=x(i,j);
     end
end
p1=size(x1)*[0;1];
q1=size(x1)*[1;0];
count=0;
if (x1(i,j)>.3)
     x1(i,j)=1;
else
     x1(i,j)=-1;
end
for i=1:v1
     t1(i)=t(i);
```

```
end
for i=1:q1
      for k=1:m
            d1(k)=0;
            for j=1:p1
                  d1(k)=d1(k)+[(x1(i,j)-w(j,k))]^2;
            end
      end
      c1=min(d1);
      for a=1:m
            if(d1(a)==c1)
                  O1=a-1;
            end
      end
      if (O1==t1(i))
            count=count+1;
      end
end
% calculting the efficiency.
eff=round(count*100/(v1));
sec=toc;
% Result display
clc;
prompt={'Total number of datas available ','
                  Number of training inputs presented',
                  'Number of testing inputs presented',
                  'Number of recognized datas',
                  'Number of iterations performed',
                  'Efficiency','Execution time'};
c31=num2str(c3) ;
v2=num2str(v) ;
vs=num2str(v1) ;
count1=num2str(count);
N1=num2str(N) ;
eff1=num2str(eff) ;
sec1=num2str(sec);
def=c31,v2,vs,count1,N1,eff1,sec1;
dlgTitle='Result';
lineNo=1;
answer=inputdlg(prompt,dlgTitle,lineNo,def);

% PROGRAM FOR ANALOG INPUT DATA
clc;
```

```
tic;
m=4;
x5=load('heartdata00.txt');% loading input datas from
    the file
c3=size(x5)*[1;0];% calculating size of matrix.
u=size(x5)*[0;1];
prompt1={'Percentage of datas to be trained',
'Percentage of datas to be tested','Enter the value
of learning rate'};
dlgTitle1='HEART DISEASE DIAGNOSIS USING
                LEARNING VECTOR QUANTIZATION NETWORK';
lineNo=1;
answer1=inputdlg(prompt1,dlgTitle1,lineNo);
per=str2double(answer1(1));
per1=str2double(answer1(2));
al=str2double(answer1(3));
pe=per/100;
v=round(pe*c3);
% Seperating the input attributes and the target from the
input file for s1=1:m
     for i=1:c3
          if(x5(i,u)==(s1-1))
               for(j=1:u)
                    temp=x5(s1,j);
                    x5(s1,j)=x5(i,j);
                    x5(i,j)=temp;
               end
          end
     end
end
for i=1:c3
     for j=1:u
          if((j==u))
               t(i)=x5(i,j);
          end
     end
end
for i=1:c3
     for j=1:u-1
          x(i,j)=x5(i,j);
     end
end
for i=1:c3
```

```
        for j=1:u-1
            if x(i,j)==0
                x(i,j)=.05;
            end
        end
end
% Normalizing the datas
q2=size(x)*[0;1];
p2=size(x)*[1;0];
y=max(x,[],1);
z=min(x,[],1);
for i=1:q2
        if y(i)~=z(i)
            e(i)=y(i)-z(i);
        else
            e(i)=1;
            z(i)=0;
        end
end
for i=1:p2
        for j-1:q2
            x(i,j)=(x5(i,j)- z(j))/(e(j));
        end
end
% Initialising then weight matrix
for i=1:u-1
        for j=1:4
            w(i,j)=x(j,i);
        end
end
q=size(x)*[0;1];
p=size(x)*[1;0];
N=0;
% Stopping condition
while (al>0.0000000001)
        N=N+1;
% Calculating the distance by using Euclidean distance
method
        for i=5:v
            for k-1:4
                d(k)=0;
                for j=1:u-1
                    d(k)=d(k)+[x(i,j)-w(j,k)]^2;
```

```
                  end
           end
           b=min(d);
% Finding the winner
           for l=1:4
                  if (d(l)==b)
                        J=l;
                  end
           endv % Weight updation
           for f=1:q
                  if(t(J)==t(i))
                        w(f,J)=w(f,J)+al*[x(i,f)-w(f,J)];
                  else
                        w(f,J)=w(f,J)-al*[x(i,f)-w(f,J)];
                  end
           end
     end
% Reducing the learning rate
     al=al/2;
end
% LVQ Testing
pe1=per1/100;
v1=round(pe1*c3);
for i=1:v1
     for j=1:u-1
           x1(i,j)=x(i,j);
     end
end
p1=size(x1)*[0;1];
q1=size(x1)*[1;0];
count=0;
for i=1:v1
     t1(i)=t(i);
end
for i=1:q1
     for k=1:m
           d1(k)=0;
           for j=1:p1
                  d1(k)=d1(k)+[(x1(i,j)-w(j,k))]^2;
           end
     end
     c1=min(d1);
     for a=1:m
```

```
            if(d1(a)==c1)
                O1=a-1;
            end
        end
        if (O1==t1(i))
            count=count+1;
        end
end
% calculting the efficiency.
eff=round(count*100/(v1));
sec=toc;
% Result display
clc;
prompt={'Total number of datas available ',
        ' Number of training inputs presented',
        'Number of testing inputs presented',
        'Number of recognized datas',
        'Number of iterations performed',
        'Efficiency','Execution time'};
c31=num2str(c3) ;
v2=num2str(v) ;
vs=num2str(v1) ;
count1=num2str(count);
N1=num2str(N) ;
eff1=num2str(eff) ;
sec1=num2str(sec);
def=c31,v2,vs,count1,N1,eff1,sec1;
dlgTitle='Result';
lineNo=1;
answer=inputdlg(prompt,dlgTitle,lineNo,def);
```

TABLE 5.4: Efficiency and Time for Analog Input

Training Vectors	Test Vectors	Time (sec)	Recognised Vectors	Efficiency
83(20%)	331(80%)	0.141	271	81.87%
166(40%)	248(60%)	0.109	196	79.03%
248(60%)	166(40%)	0.078	144	86.747%
331(80%)	83(20%)	0.031	77	92.77%

TABLE 5.5:   Efficiency and Time for Digital Input

Training Vectors	Test Vectors	Time (sec)	Recognised Vectors	Efficiency
83(20%)	331(80%)	0.141	282	85.19%
166(40%)	248(60%)	0.109	202	81.46%
248(60%)	166(40%)	0.062	154	92.77%
331(80%)	83(20%)	0.047	82	98.92%

Observations:

Tables 5.4 and 5.5 show the results of LVQ for analog and digital inputs with optimum efficiency and time. The learning rate alpha was varied from 0.1 to 0.9 and it was found that maximum efficiency was obtained for alpha value 0.1. The following parameters were used in the program:

Total number of instances : 414
Total number of attributes : 13
Number of classes : 4

The efficiency is higher with more number of training vectors and the time taken for complete execution of the program is less if the number of training vectors is more. The net identifies the pattern faster if the numbers of training patterns are more.

5.7   Illustration 7: Neural Network Using MATLAB Simulink

A system is described by a first order difference equation

$$y(k + 2) = 0.3y(k + 1) + 0.6y(k) + f(u(k))$$

where $f(u(k)) = u^3 + 0.3u^2 - 0.4u$ and $u(k)$ is random noise.

Generate data in appropriate range for $f(u)$ and fit a neural network on the data. Simulate the system response for exact $f(u)$ and the neural network approximation. Initial condition could be zero. Input can be assumed to be noise. The Simulink model is shown in Figure 5.12.

The neural network approximation can be obtained in various ways, using the following commands

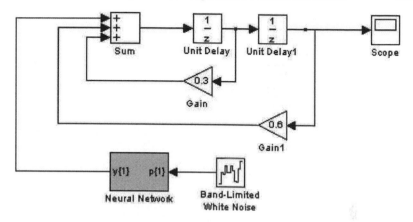

FIGURE 5.12: Simulink Model of the Function

```
% Generate parabolic data
u=-10:0.1:10; y=u.^3+0.3*u.^2-0.4*u;
P=u; T=y;
% Define network
net=newff([-10 10], [10,1],
'tansig','purelin','trainlm');

% Define parameters
net.trainParam.show = 50;
net.trainParam.lr = 0.05;
net.trainParam.epochs = 300;
net.trainParam.goal = 1e-3;

% Train network
net1 = train(net, P, T);

% Simulate result
a= sim(net1,P);

% Plot result
plot(P,T,P,a-T)
Title('Cubic function: y=u^3+0.3u^2-0.4u')
xlabel('Input u');ylabel('Output y')
```

Observations:

The given cubic function is simulated and plotted as shown in Figure 5.13.

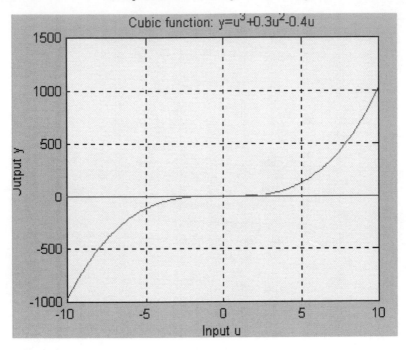

FIGURE 5.13: Result of Simulation

Summary

This chapter provided different MATLAB programs for the user to understand and write programs according to the required applications.

Review Questions

1. Consider humps function in MATLAB. It is given by $y = \frac{1}{(x-3)^2} + 0.01 + \frac{1}{(x-0.9)^2} + 0.04 - 6$; Find a neural network to fit the data generated by humps-function between $[0,2]$ using MATLAB. Fit a multilayer perceptron network on the data. Try different network sizes and different teaching algorithms.

2. Consider a surface described by $z = \cos(x) \sin(y)$ defined on a square $-2 \leq x \leq 2, -2 \leq y \leq 2$. Plot the surface z as a function

of x and y. Design a neural network, which will fit the data using MATLAB.

3. Consider Bessel functions J α (t), which are solutions of the differential equation $t^2 y - ty + (t^2 - \alpha^2)y = 0$. Implement backpropagation network in MATLAB to approximate first order Bessel function J1 , $= \alpha 1$, when t ϵ [0,20].

4. Analyze a biomedical data set (e.g., Iris data set, breast cancer data set, etc.) using Kohonen SOM network. Read the data from the ASCII file and normalize the data. Once the data is normalized then train the feature map and analyze the data using MATLAB.

5. Write a MATLAB program to solve the Traveling Salesman Problem using Hopfield network.

6. Using the feedforward neural network simulate and analyze the behavior of PID tuning in control systems in MATLAB.

Chapter 6

MATLAB-Based Fuzzy Systems

6.1 Introduction

Fuzzy Logic was initiated in 1965 by Lotfi A. Zadeh, professor for computer science at the University of California in Berkeley. Basically, Fuzzy Logic (FL) is a multivalued logic that defines intermediate values between traditional evaluations like true/false, yes/no, high/low, etc. These intermediate values can be formulated mathematically and processed by computers, in order to apply a more human like way of thinking. Based on Aristotle and other later mathematicians, the so called "Laws of Thought" was posited. One of these, the "Law of the Excluded Middle," states that every proposition must either be True or False. Even when Parminedes proposed the first version of this law (around 400 B.C.) there were strong and immediate objections: for example, Heraclitus proposed that things could be simultaneously True and not True. It was Plato who laid the foundation for what would become fuzzy logic, indicating that there was a third region (beyond True and False) where these opposites "tumbled about." Other, more modern philosophers echoed his sentiments, notably Hegel, Marx, and Engels. An alternative approach to the bivalued logic of Aristotle was proposed by Lukasiewicz. Fuzzy Logic has been developed as a profitable tool for the controlling and steering of systems and complex industrial processes, as well as for household and entertainment electronics.

In viewing the evolution of fuzzy logic, three principal phases may be discerned. The first phase, from 1965 to 1973, was concerned in the main with fuzzification, that is, with generalization of the concept of a set, with two-valued characteristic function generalized to a membership function taking values in the unit interval or, more generally, in a lattice. The basic issues and applications that were addressed were, for the most part, set-theoretic in nature, and logic and reasoning were not at the center of the stage. The second phase, 1973 to 1999, began with two key concepts: (a) the concept of a linguistic variable; and (b) the concept of a fuzzy if-then rule. Today, almost all applications of fuzzy set theory and fuzzy logic involve the use of these concepts. Nowadays, fuzzy logic is applied in various applications in two different senses:

A narrow sense — In a narrow sense fuzzy logic, abbreviated as FL, is a logical system which is a generalization of multivalued logic

A wide sense — In a wide sense fuzzy logic is abbreviated as FL, is a union of FLn, fuzzy set theory, possibility theory, calculus of fuzzy if-then rules, fuzzy arithmetic, calculus of fuzzy quantifiers, and related concepts and calculi

The distinctive feature of FL is that in FL everything is, or is allowed to be, a matter of degree. Possibly the most salient growth during the second phase of the evolution was the rapid growth of fuzzy control, alongside the boom in fuzzy logic applications, especially in Japan.

There were many other major developments in fuzzy-logic-related basic and applied theories, among them the genesis of possibility theory and possibilistic logic, knowledge representation, decision analysis, cluster analysis, pattern recognition, fuzzy arithmetic, fuzzy mathematical programming, fuzzy topology and, more generally, fuzzy mathematics. Fuzzy control applications proliferated but their dominance in the literature became less pronounced.

An important development making the beginning of the third phase was "From Computing with Numbers to Computing with Words" in 1999. Basically, development of computing with words and perceptions brings together earlier strands of fuzzy logic and suggests that scientific theories should be based on fuzzy logic rather than on Aristotelian, bivalent logic, as they are at present. The concept of Precisiated Natural Language (PNL) is the key constituent in the area of computing words. PNL gives room to a major enlargement of the purpose of natural languages in technological hypotheses. It may well turn out to be the case that, in coming years, one of the most important application-areas of fuzzy logic, and especially PNL, will be the Internet, centering on the conception and design of search engines and question-answering systems.

From its inception, fuzzy logic has been and to some degree still an object of skepticism and controversy. The disbelief about fuzzy logic is a manifestation of the reality that, in English, the term 'fuzzy' is usually used in a uncomplimentary sense. Merely, fuzzy logic is hard to accept as abandoning bivalence breaks with centuries-old tradition of basing scientific theories on bivalent logic.

It may take some time for this to happen, but eventually abandonment of bivalence will be viewed as a logical development in the evolution of science and human thought. This chapter will discuss the basic fuzzy sets, operations on fuzzy sets, relations between fuzzy sets, composition, and fuzzy arithmetic. A few MATLAB programs are also illustrated on topics such as membership functions, fuzzy operations, fuzzy arithmetic, relations, and composition.

6.2 Imprecision and Uncertainty

Fuzziness should not be confused with other forms of imprecision and uncertainty. There are several types of imprecision and uncertainty and fuzziness is just one aspect of it. Imprecision and uncertainty may be in the aspects of measurement, probability, or descriptions. Imprecision in measurement is associated with a lack of precise knowledge. Sometimes there are measurements that are inaccurate, inexact, or of low confidence.

Imprecision as a form of probability is associated with an uncertainty about the future occurrence of events or phenomena. It concerns the likelihood of non-deterministic events (stochastic uncertainty). An example is the statement "It might rain tomorrow" which exhibits a degree of randomness.

Imprecision in description is the type of imprecision addressed by fuzzy logic. It is the ambiguity, vagueness, qualitativeness, or subjectivity in natural language (linguistic, lexical, or semantic uncertainty). It is the ambiguity found in the definition of a concept or the meaning of terms such as "tall building" or "low scores". It is also the ambiguity in human thinking, that is, perceptions and interpretations. Examples of statements that are fuzzy in nature are "Hemoglobin count is very low." and "Teddy is rather heavy compared to Ike."

The nature of fuzziness and randomness are therefore quite different. They are different aspects of imprecision and uncertainty. The former conveys subjective human thinking, feelings, or language, and the latter indicates an objective statistic in the natural sciences.

Fuzzy models and statistical models also possess philosophically different kinds of information: fuzzy memberships comprise similarities of objects to imprecisely defined properties, while probabilities convey information about relative frequencies. Thus, fuzziness deals with deterministic plausability and not non-deterministic probability.

6.3 Crisp and Fuzzy Logic

Fuzzy logic forms a bridge between the two areas of qualitative and quantitative modeling. Although the input-output mapping of such a model is integrated into a system as a quantitative map, internally it can be considered as a set of qualitative linguistic rules. Since the pioneering work of Zadeh in 1965 and Mamdani in 1975, the models formed by fuzzy logic have been applied to many varied types of information processing including control systems.

The term Fuzzy Logic implies that in some manner the methodological analysis is vague or ill-defined. The basic idea behind the development of fuzzy logic rose from the requirement to design the type of vague or ill-defined systems. These ill-defined systems cannot operate on traditional binary valued logic therefore the fuzzy methodological analysis is used based on mathematical theory.

The commonest binary valued logic and set theory is defined as "an element belongs to a set of all possible elements and given any specific subset, whether that element is or is not a member of it." For instance, a tiger belongs to the set of all animals. Human reasoning can be enhanced in this way. Likewise many occurrences can be described with different types of sets. Consider, the basic statement

IF a fruit is red AND mature THEN the fruit is ripe

applies to many people across the world with complete precision. The above rule is formed using operators such as intersection operator AND, which manipulates the sets.

Eventually, not all parameters can be described using binary valued sets. For instance, classifications of a person into males and females are easy, but it is problematic to classify them as being tall or not tall. The set of tall people is far more difficult to define, since there is no exact precise value to define tall. Such a kind of problem is frequently twisted so that it can be delineated using the well-known existing technique. The heights, e.g., 1.80 m, can be defined as tall, as shown in Figure 6.1 (a). This kind of crisp reasoning would not produce smooth results, since a person of height 1.79 m or a person of height 1.8 m would produce different results.

Fuzzy logic was suggested by Zadeh as a method for mimicking the ability of human reasoning using a small number of rules and still producing a smooth output via a process of interpolation. It forms rules that are based upon multi-valued logic and so introduced the concept of set membership. Using fuzzy logic a component can be as decided as belonging to a set, this kind of allotment is carried out by membership functions. For example, a person of height 1.79 m would belong to both tall and not tall sets with a particular degree of membership. Equally the membership grade increases or decreases proportionate with the height of a person, as shown in Figure 6.1 (b). The output of a fuzzy logical thinking system would produce like results for similar inputs. The fuzzy logic theory is just a prolongation of traditional logic where partial set membership could exist, rule conditions could be satisfied partially,

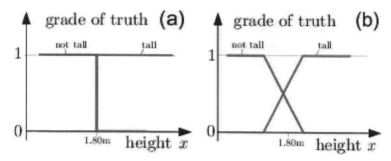

FIGURE 6.1: The Difference between the Grade of Truth in (a) Binary Valued Logic 0,1 and (b) Fuzzy Logic [0,1]

and system outputs are calculated by interpolation. Hence, the output is smooth over the equivalent binary-valued rule base. This property is especially crucial to control system applications.

A fuzzy logic control system is one that has at least one system component that uses fuzzy logic for its internal knowledge representation. Although it is possible for fuzzy systems to communicate information using fuzzy sets, most applications have a single fuzzy system component communicating with conventional system components via deterministic values. In this case, and also in this chapter, fuzzy logic is used purely for internal knowledge representation and, externally, can be considered as any other system component.

6.4 Fuzzy Sets

Mathematical theory of sets have been extended to create fuzzy sets. Set theory was first proposed by a German mathematician, Georg Cantor (1845 to 1918). His hypothesis of sets encountered more opposition during his lifetime, but later most mathematicians considered that it is possible to express almost all concepts of mathematics in the language of set theory.

Conventional Sets - A set comprises some collection of objects, which can be handled at large as a whole component. According to Cantor a set is an item from a given universe. The terms set, collection, and class are synonyms, just as the terms item, element, and member. Nearly anything called a set in ordinary conversation is an acceptable set in the mathematical sense, as explained in the following example.

Illustration 1 (sets):

The following are well-defined lists or collections of objects, and therefore entitled to be called sets:
(a) The set of non-negative integers less than 4. This is a finite set with four members: 0, 1, 2, and 3.
(b) The set of live dinosaurs in the basement of a British museum. This set has no members and is called an empty set.
(c) The set of measurements greater than 10V even though this set is infinite, it is possible to determine whether a given measurement is a member or not.

A set can be specified by its members, they characterize a set completely. The list of members A=0,1,2,3 specifies a finite set. It is not possible to list all elements of an infinite set; we must instead state some property which characterizes the elements in the set, for instance the predicate x >10. That set is defined by the elements of the universe of discourse, which make the predicate true. So there are two ways to describe a set: explicitly in a list or implicitly with a predicate. In the classical set theory a set can be represented by enumerating all its elements using

$$A = fa_1, a_2 \ldots, a_n g$$

The grade of membership for all its members thus describes a fuzzy set. An item's grade of membership is normally a real number between 0 and 1, often denoted by the Greek letter μ. The higher the number, the higher the membership. Zadeh regards Cantor's set as a special case where elements have full membership, i.e., $\mu = 1$. If the elements of the above equation ai (i=1, ... ,n) of A are together a subset of the universal base set X, the set A can be represented for all elements xϵ X by its characteristic function

$$\mu_A(x) \begin{cases} 1 & if \ x\epsilon \ A \\ 0 & otherwise \end{cases} \tag{6.1}$$

In classical set theory μ_A(x) has only values 0 ("false") and 1 ("true"), so two values of truth. Such sets are also called crisp sets.

Non-crisp sets are called fuzzy sets, for which a characteristic function can also be defined. This function is a generalization of Equation (6.1) and called a *membership function*. The membership of a fuzzy set is described by this membership function $\mu_A(x)$ of A, which associates to each element $x_o\epsilon X$ a grade of membership $\mu_A(x_o)$. In contrast to classical set theory a membership function $\mu_A(x)$ of a fuzzy set can have in the normalized closed interval [0,1] an arbitrary grade of truth. Therefore,

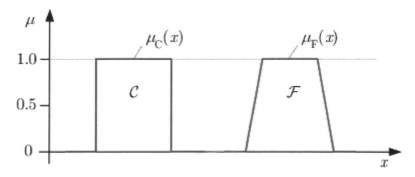

FIGURE 6.2: Membership Functions of a Crisp Set C and a Fuzzy Set F

each membership function maps elements of a given universal base set X, which is itself a crisp set, into real numbers in [0,1]. The notation for the membership function $\mu_A(x)$ of a fuzzy set A

$$A : X \longrightarrow [0,1] \tag{6.2}$$

is used. Each fuzzy set is completely and uniquely defined by one particular membership function. Consequently symbols of membership functions are also used as labels of the associated fuzzy sets. That is, every fuzzy set and its membership function are referred by the same capital letter. Since crisp sets and the associated characteristic functions may be viewed, respectively, as special cases of fuzzy sets and membership functions, the same notation is used for crisp sets as well, as shown in Figure 6.2

The base set X is introduced as a universal set. In practical applications, physical or similar quantities are considered that are defined in some interval. When such quantities are described by sets, a base set can be generalized seamless to a crisp base set X that exists in a defined interval. This is a generalization of fuzzy sets. Base sets are not always crisp sets. Another generalization is that the base set is itself a fuzzy set.

6.5 Universe

The constituents of a fuzzy set are acquired from a universe of discourse also referred to as universe. The universe comprises the complete elements that can inherit consideration. The following example delineates the application of universe.

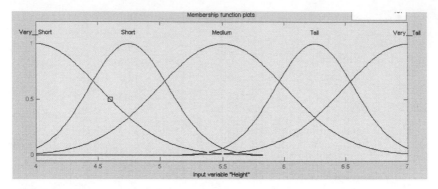

FIGURE 6.3: The sets very short, medium, and very tall are derived from short and tall

Illustration 2: (Universe)

(a) The set of people could have all human beings in the world as its universe. Alternatively it could be the numbers between 4 and 6 feet: these would represent their height as shown in Figure 6.3.

In case we are dealing with a non-numerical quantity, for instance taste, which cannot be measured against a numerical scale, we cannot use a numerical universe. The elements are then said to be taken from a *psychological continum*; an example of such a universe could be {bitter, sweet, sour, salt, hot, ... }

6.6   Membership Functions

The membership function $\mu_A(x)$ describes the membership of the elements x of the base set X in the fuzzy set A, whereby for $\mu_A(x)$ a large class of functions can be taken. Reasonable functions are often piecewise linear functions, such as triangular or trapezoidal functions.

The grade of membership $\mu_A(x_o)$ of a membership function $\mu_A(x)$ describes for the special element x=x_o, to which grade it belongs to the fuzzy set A. This value is in the unit interval [0,1]. Of course, xo can simultaneously belong to another fuzzy set B, such that $\mu_B(x_o)$ characterizes the grade of membership of x_o to B. This case is shown in Figure 6.4.

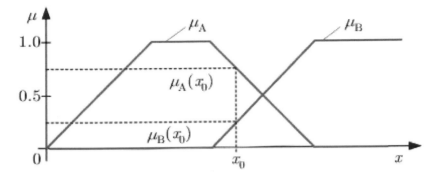

FIGURE 6.4: Membership Grades of x_o in the Sets A and B: $\mu_A(x_o)$ =0.75 and $\mu_B(x_o)$ =0.25

The membership for a 50-year old in the set "young" depends on one's own view. The grade of membership is a precise, but subjective measure that depends on the context. A fuzzy membership function is different from a statistical probability distribution. This is illustrated following egg-eating example.

Illustration 3: (Probability vs. Possibility)

Consider the statement "Hans ate X eggs for breakfast", where $X \in U = \{1,2,3, \dots ,8\}$. We may associate a probability distribution p by observing Hans eating breakfast for 100 days.

U=[1 2 3 4 5 6 7 8]
p=[.1 .8 .1 0 0 0 0 0]

A fuzzy set expressing the grade of ease with which Hans can eat eggs may be the following possibility distribution Π

U=[1 2 3 4 5 6 7 8]
p=[1 1 1 1 .8 .6 .4 .2]

Where the possibility for X = 3 is 1, the probability is only 0.1. The example shows, that a possible event does not imply that it is probable. However, if it is probable it must also be possible. You might view a fuzzy membership function as the personal distribution, in contrast with a statistical distribution based on observations.

6.6.1   Types of Membership Functions

In principle any function of the form $A{:}X{\longrightarrow}[0,1]$ describes a membership function associated with a fuzzy set A that depends not only on the concept to be represented, but also on the context in which it is used. The graphs of the functions may have different shapes and may have specific properties. Whether a particular shape is suitable can be determined only in the application context. In certain cases, however, the meaning semantics captured by fuzzy sets is not too sensitive to variations in the shape, and simple functions are convenient. In many practical instances fuzzy sets can be represented explicitly by families of parameterized functions, the most common being the following:

1. Triangular Function

2. Γ -Function

3. S-Function

4. Trapezoidal Function

5. Gaussian Function

6. Exponential Function

The membership function definitions for the above mentioned common membership functions are given in the following sections:

Triangular Function

The membership definition for a triangular function is given as:

$$A(x) = \begin{cases} 0, & if\ x \leq a \\ \frac{x-a}{b-a}, & if\ x\ \epsilon[a,b] \\ \frac{c-x}{c-b}, & if\ x\ \epsilon[b,c] \\ 0, & if\ x\ \geq a \end{cases} \tag{6.3}$$

where m is a modal value, and a and b denote the lower and upper bounds, respectively, for nonzero values of A(x). Sometimes it is more convenient to use the notation explicitly highlighting the membership functions' parameters; in this case,

A(x;a,m,b)= **max**{**min**[(x-a)/(m-a),(b-c)/(b-m)],**0**} (6.4)

Γ **Function**

The membership definition for a -function is given as:

$$A(x) = \begin{cases} 0, & if\ x \leq a \\ 1 - e_{-k(x-a)^2} & if\ x{<}a \end{cases} \tag{6.5}$$

or

$$A(x) = 0 \begin{cases} 0, & if\ x \leq a \\ \frac{k(x-a)^2}{1+k(x-a)^2}, & if\ x{>}a \end{cases}$$

where k>0.

S-Function

The membership definition for a S-function is given as:

$$A(x) = \begin{cases} \mathbf{0}, & if\ x \leq a \\ \mathbf{2}(\frac{x-a}{b-a})^2, & if\ x\ \epsilon[a,m] \\ \mathbf{1{-}2}(\frac{x-a}{b-a})^2, & if\ x\ \epsilon[m,b] \\ \mathbf{1}, & if\ x{>}b \end{cases} \tag{6.6}$$

The point m=a+b/2 is known as the crossover of the S-function.

Trapezoidal function

The membership definition for a trapezoidal function is given as:

$$A(x) = \begin{cases} \mathbf{0}, & if\ x \leq a \\ \frac{x-a}{b-a}, & if\ x\ \epsilon[a,b] \\ \mathbf{1}, & if\ x\ \epsilon[b,c] \\ \frac{d-x}{d-c}, & if\ x\ \epsilon[c,d] \\ \mathbf{0}, & if\ x \geq b \end{cases} \tag{6.7}$$

Using equivalent notation, we obtain

$$A(x;a,m,n,b)\mathbf{max\{min}[(x\text{-}a)/(m\text{-}a),\mathbf{1},(b\text{-}x)/(b\text{-}m)],\mathbf{0\}} \tag{6.8}$$

Gaussian Function

The membership definition for a Gaussian function is given as:

$$A(x) = -e^{\frac{(x-c)^2}{2\sigma^2}}, where\ 2\sigma^2{>}0. \tag{6.9}$$

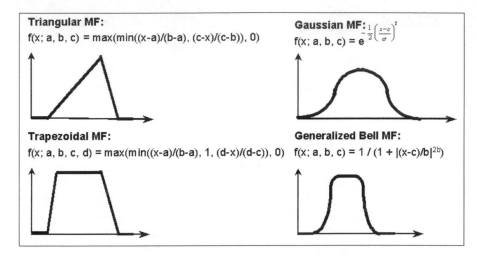

FIGURE 6.5: Membership Functions

Exponential Function

The membership definition for a exponential function is given as:

$$A(x) = \frac{1}{1 + (k - m)^2}, k > 1 \tag{6.10}$$

or

$$A(x) = \frac{(k - m)^2}{1 + (k - m)^2}, k > 0$$

Figure 6.5 shows a few membership functions.

6.6.2   Membership Functions in the MATLAB Fuzzy Logic Toolbox

The only condition a membership function must really satisfy is that it must vary between 0 and 1. The function itself can be an arbitrary curve whose shape can be defined as a function that suits us from the point of view of simplicity, convenience, speed, and efficiency. A classical set might be expressed as

$$A = \{xjx > 6\} \tag{6.11}$$

A fuzzy set is an extension of a classical set. If X is the universe of discourse and its elements are denoted by x, then a fuzzy set A in X is defined as a set of ordered pairs.

$$X = \{X, \mu X(X) j X \epsilon X\}$$

μ A(x) is called the membership function (or MF) of x in A. The membership function maps each element of X to a membership value between 0 and 1.

The Fuzzy Logic Toolbox includes 11 built-in membership function types as shown in Figure 6.6. These 11 functions are, in turn, built from several basic functions: piecewise linear functions, the Gaussian distribution function, the sigmoid curve, and quadratic and cubic polynomial curves. By convention, all membership functions have the letters mf at the end of their names. The simplest membership functions are formed using straight lines. Of these, the simplest is the triangular membership function, and it has the function name trimf, which is a collection of three points forming a triangle. The trapezoidal membership function, trapmf, has a flat top and is just a truncated triangle curve. These straight-line membership functions have the advantage of simplicity.

Two membership functions are built on the Gaussian distribution curve: a simple Gaussian curve and a two-sided composite of two different Gaussian curves. The two functions are gaussmf and gauss2mf. The generalized bell membership function is specified by three parameters and has the function name gbellmf. The bell membership function has one more parameter than the Gaussian membership function, so it can approach a non-fuzzy set if the free parameter is tuned. Because of their smoothness and concise notation, Gaussian and bell membership functions are popular methods for specifying fuzzy sets. Both Gaussian and bell membership curves hold the advantage of representing a smooth and nonzero output at all points. Though these curves attain smoothness, they are incapable of specifying asymmetric membership functions, which play a crucial role in certain essential applications.

The sigmoidal membership function is either open left or right. Asymmetric and closed (i.e., not open to the left or right) membership functions can be synthesized using two sigmoidal functions. In addition to the basic *sigmf*, the difference between two sigmoidal functions, *dsigmf*, and the product of two sigmoidal functions *psigmf* are also available.

Various membership functions are polynomial based curves in the fuzzy logic toolbox. Three related membership functions are the Z, S, and Pi curves, all named because of their shape. The function zmf is the asymmetrical polynomial curve open to the left, smf is the mirror-image function that opens to the right, and pimf is zero on both extremes with a rise in the middle. In that respect, there is an absolute wide selection to pick out the preferred membership function. A user can also create his own membership functions using the fuzzy logic toolbox.

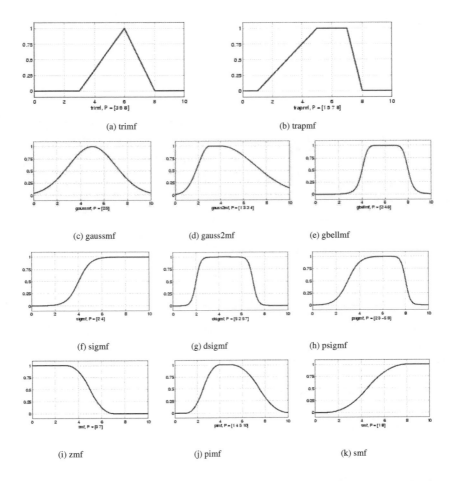

(a) trimf

(b) trapmf

(c) gaussmf

(d) gauss2mf

(e) gbellmf

(f) sigmf

(g) dsigmf

(h) psigmf

(i) zmf

(j) pimf

(k) smf

FIGURE 6.6:   Membership Functions in MATLAB Fuzzy Logic Toolbox. (a) Triangular Membership Function (b) Trapezoidal Membership Function (c) Simple Gaussian Membership Function (d) Two Sided Composite Gaussian Membership Function (e) Generalised Bell Membership Function (f) Sigmoidal Membership Function (g) Difference Sigmoidal Membership Function (h) Product Sigmoidal Membership Function (i) Z curve Membership Function (j) Pi Curve Membership Function (k) S Curve Membership Function

6.6.3 MATLAB Code to Simulate Membership Functions

Triangular Membership Function

```
y = trimf(x,params)
y = trimf(x,[a b c])
```

The triangular curve is a function of a vector, x, and depends on three scalar parameters a, b, and c, as given in Eq. (6.3) and more compactly, by

$$f(x; a, b, c) = \max\left(\min\left(\frac{x-a}{b-a}, \frac{c-x}{c-b},\right), 0\right)$$

The parameters a and b locate the "feet" of the triangle and the parameter c locates the peak.

Code

```
% Triangular membership function
x=0:0.1:10;
y=trimf(x,[3 6 8]);
plot(x,y);
xlabel('trimf, P=[3 6 8]');
title('Triangular Membership Function');
```

Output

The Triangular Membership Function plotted from the above code is shown in Figure 6.7.

Trapezoidal Membership Function

Syntax

```
y = trapmf(x,[a b c d])
```

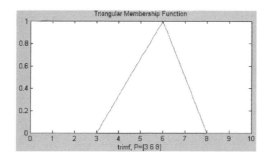

FIGURE 6.7: Triangular Membership Function

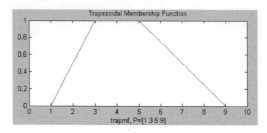

FIGURE 6.8:   Trapezoidal Membership Function

The trapezoidal curve is a function of a vector, x, and depends on four scalar parameters a, b, c, and d, as given by Eq. (6.7) or, more compactly, by

$$f(x; a, b, c, d) = \max\left(\min\left(\tfrac{x-a}{b-a}, 1, \tfrac{d-x}{d-c},\right), 0\right)$$

The parameters a and d locate the "feet" of the trapezoid and the parameters b and c locate the "shoulders."

Code

```
% Trapezoidal membership function
x=0:0.1:10;
y=trapmf(x,[1 3 5 9]);
plot(x,y)
xlabel('trapmf, P=[1 3 5 9]')
title('Trapezoidal Membership Function');
```

Output

The Trapezoidal Membership Function plotted from the above code is shown in Figure 6.8

Gaussian Membership Function Syntax

```
y = gaussmf(x,[sig c])
```

The symmetric Gaussian function depends on two parameters ? and c as given by Eq. (6.9). The parameters for gaussmf represent the parameters ? and c listed in order in the vector [sig c].

Code

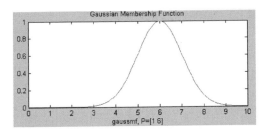

FIGURE 6.9: Gaussian Membership Function

```
% Gaussian membership function
x=0:0.1:10;
y=gaussmf(x,[1 6]);
plot(x,y)
xlabel('gaussmf, P=[1 6]')
title('Gaussian Membership Function');
```

Output

The Gaussian Membership Function plotted from the above code is shown in Figure 6.9.

Generalized Bell Shaped Membership Function

Syntax

```
y = gbellmf(x,params)
```

The Generalized bell function depends on three parameters a, b, and c as given by

$$f(x; a, b, c) = \left(\frac{1}{1+|\frac{x-c}{a}|^{2b}} \right)$$

where the parameter b is usually positive. The parameter c locates the center of the curve. The parameter vector params, the second argument for gbellmf, is the vector whose entries are a, b, and c, respectively.

Code

```
% Generalized Bell Shaped membership function
x=0:0.1:10;
y=gbellmf(x,[1 3 7]);
plot(x,y)
xlabel('gbellmf, P=[1 3 7]')
```

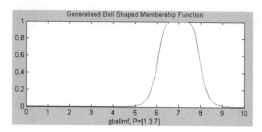

FIGURE 6.10:   Generalized Bell Shaped Membership Function

```
title('Generalized Bell Shaped Membership Function');
```

Output

The Generalized Bell Shaped Membership Function plotted from the above code is shown in Figure 6.10.

Sigmoidal Shaped Membership Function

Syntax

```
colorboxlightgray y = sigmf(x,[a c])
```

The sigmoidal function, sigmf(x,[a c]), as given below by, f(x,a,c) is a mapping on a vector x, and depends on two parameters a and c.

$$f(x; a, c) = \frac{1}{1 + e^{-a(x-c)}}$$

Depending on the sign of the parameter a, the sigmoidal membership function is inherently open to the right or to the left, and thus is appropriate for representing concepts such as "very large" or "very low." Various traditional membership functions can be built by taking either the product (psigmf) or difference (dsigmf) of two different sigmoidal membership functions.

Code

```
% Sigmoidal Shaped membership function
x=0:0.1:10;
y=sigmf(x,[3 6]);
plot(x,y)
xlabel('sigmf, P=[3 6]')
title('Sigmoidal Shaped Membership Function');
```

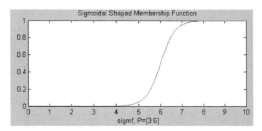

FIGURE 6.11:   Sigmoidal Shaped Membership Function

Output

The Sigmoidal Shaped Membership Function plotted from the above code is shown in Figure 6.11.

Z Curve Membership Function

Syntax

```
y = zmf(x, [a b])
```

The spline-based function of x is so named because of its Z-shape. The parameters a and b locate the extremes of the sloped portion of the curve as given by Eq. (6.6)

Code

```
% Z Curve membership function
x=0:0.1:10;
y=zmf(x,[3 6]);
plot(x,y)
xlabel('zmf, P=[3 6]')
title('Z Curve Membership Function');
```

Output

The Z Curve Membership Function plotted from the above code is shown in Figure 6.12.

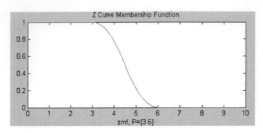

FIGURE 6.12:   Z Curve Membership Function

Pi Curve Membership Function

Syntax

```
y = pimf(x,[a b c d])
```

The spline-based curve given by Eq. (6.6) is so named because of its Π shape. This membership function is evaluated at the points determined by the vector x. The parameters a and d locate the "feet" of the curve, while b and c locate its "shoulders."

Code

```
% Pi Curve membership function
x=0:0.1:10;
y=pimf(x,[1 5 6 10]);
plot(x,y)
xlabel('pimf, P=[1 5 6 10]')
title('Pi Curve Membership Function');
```

Output

The Pi Curve Membership Function plotted from the above code is shown in Figure 6.13.

S Curve Membership Function

Syntax

```
y = smf(x,[a b])
```

The spline-based curve given by Eq. (6.6) is a mapping on the vector x, and is named because of its S-shape. The parameters a and b locate

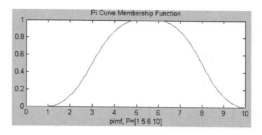

FIGURE 6.13: Pi Curve Membership Function

the extremes of the sloped portion of the curve.

Code
```
% S Curve membership function
x=0:0.1:10;
y=smf(x,[1 8]);
plot(x,y)
xlabel('smf, P=[1 8]')
title('S Curve Membership Function');
```

Output
The S Curve Membership Function plotted from the above code is shown in Figure 6.14.

6.6.4 Translation of Parameters between Membership Functions Using MATLAB

To translate parameters between membership functions a function "mf2mf" is available in Matlab Fuzzy Logic Toolbox. The general syntax of this function is

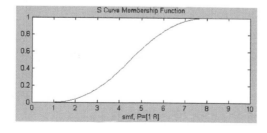

FIGURE 6.14: S Curve Membership Function

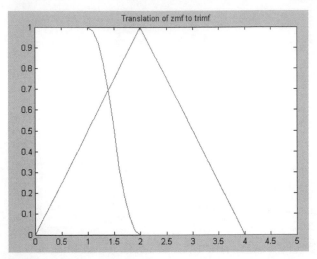

FIGURE 6.15: Illustration of 'mf2mf' Function

outParams = mf2mf (inParams, inType, outType)

where inParams - the parameters of the membership function to be transformed inType - a string name for the type of membership function to be transformed outType - a string name for the new membership function The following code translates the polynomial Z curve membership function into triangular membership function.

```
x=0:0.1:5;
mfp1 = [1 2 3];
mfp2 = mf2mf(mfp1,'zmf','trimf');
plot(x,zmf(x,mfp1),x,trimf(x,mfp2))
```

The resultant plot is shown in Figure 6.15.

6.7   Singletons

Strictly speaking, a fuzzy set D is a collection of ordered pairs

$$A=\{(x,\mu(x))\} \qquad (6.11)$$

Item x belongs to the universe and $\mu(x)$ is its grade of membership in A. A single pair $(x,\mu(x))$ is called a fuzzy singleton; therefore the entire set can be considered as the union of its constituent singletons. It is often convenient to think of a set A just as a vector

$$a = (\mu(x_1), \mu(x_2), \mu(x_3), .., \mu(x_n)) \tag{6.12}$$

It is understood then, that each position i (1,2,3,..,n) corresponds to a point in the universe of n points.

6.8 Linguistic Variables

Just like an algebraic variable takes numbers as values, a linguistic variable takes words or sentences as values. The set of values that it can take is called its term set. Each value in the term set is a fuzzy variable defined over a base variable. The base variable defines the universe of discourse for all the fuzzy variables in the term set. In short, the hierarchy is as follows: linguistic variable $\longrightarrow$ fuzzy variable $\longrightarrow$ base variable.

Illustration 4 (term set)

Let x be a linguistic variable with the label "*Age*". Terms of this linguistic variable, which are fuzzy sets, could be "*old*", "*young*", "*very young*" from the term set.

T = Old, Very Old, Not So Old, More or Less Young, Quite Young, Very Young

Each term is a fuzzy variable defined on the base variable, which might be the scale from 0 to 100 years.

6.9 Operations on Fuzzy Sets

Fuzzy set operations are a generalization of crisp set operations. There is more than one possible generalization. The most widely used operations are called standard fuzzy set operations. Let U be a domain and A,B be the fuzzy sets on U. There are three basic operations on fuzzy sets:

Standard Fuzzy Complement: defined as a fuzzy set on U for which A'(x) = 1 - A(x) for every x in U.

Standard Fuzzy Intersection: defined as a fuzzy set on U for which (A B)(x) = min [A(x), B(x)] for every x in U.

Standard Fuzzy Union: defined as a fuzzy set on U for which (A Y B)(x) = max [A(x), B(x)] for every x in U.

Functions that meet the criteria of fuzzy intersections are known as t-norms and those that meet the criteria of fuzzy unions are known as t-conorms. A largest fuzzy set is obtained while performing standard fuzzy intersection, while a smallest fuzzy set is formed while performing standard fuzzy union operation.

x is within X where $A(x) = A \char`\^ (x)$ are called "equilibrium points"of A. Functions that qualify as fuzzy intersections and fuzzy unions are usually referred to in the literature as "t-norms"and "t-conorms", respectively. The standard fuzzy intersection is the weakest fuzzy intersection (the largest fuzzy set is produced), while the standard fuzzy union is the strongest fuzzy union (the smallest fuzzy set is produced).

Fuzzy intersections and fuzzy unions do not cover all operations that aggregate fuzzy sets, but they are the only ones that are associative. The remaining aggregating operations must be defined as functions of n arguments for each n >2. Aggregation operations that, for any given membership grades a1, a2, ... , an, produce a membership grade that lies between min(a1,a2, ... ,an) and max(a1,a2, ... ,an) are called "averaging operations". For an given fuzzy sets, each of the averaging operations produces a fuzzy set that is larger than any fuzzy intersection and smaller than any fuzzy union.

If each x in X has to be a member of A, B and/or C, we have

$$B = A \ \mathbf{I} \ C$$

which means that if a person is not young and they are not old, they must be middle aged.

Similarly,

$$A \ \mathbf{Y} \ (A \ \mathbf{I} \ B) = A$$

proof:

max[A,min(A,B)] = max[A,A] = A if A(x) ≤ B(x)
max[A,min(A,B)] = max[A,B] = A if A(x) >B(x)
so it is true for all x.

For any fuzzy set A defined on a finite universal set X, we define its "scalar cardinality", |A|, by the formula

$$|A| = sum \ (x) \ [A(x)] \ for \ all \ x \ in \ X$$

Some authors refer to |A|as the "sigma count"of A.

For any pair of fuzzy sets defined on a finite universal set X, the "degree of subset hood", S(A,B), of A in B is defined by

$$S(A,B) = (|A|\text{- sum(x) max}[0,A(x)\text{-}B(x)])/|A|$$
$$= |A \textbf{ I } B|/|A|$$

Properties: A and B are fuzzy sets. The following properties hold for a,b in the range [0,1].

(i) a+A is a subset of aA

(ii) a ≤ b implies that bA is a subset of aA and b+A is a subset of a+A

(iii) a(A **I** B) = aA **I** aB and a(A**Y**B) = aA **Y**aB

(iv) a+(A **I** B) = a+A **I** a+B and a+(A **Y** B) = a+A **Y** a+B

(v) a(Aˆ) = (1-a)+Aˆ

This means that the alpha-cut of the compliment of A is the

(1-a) strong alpha-cut of A complimented.

a(A˝) is not equal to aAˆ

a+(Aˆ) is not equal to a+Aˆ

6.9.1 Fuzzy Complements

A(x) defines the degree to which x belongs to A. Let cA denote a fuzzy complement of A of type c. cA(x) is the degree that x belongs to cA. (A(x) is therefore the degree to which x does not belong to cA.)

$$c(A(x)) = cA(x)$$

Axioms for fuzzy complements

Axiom c1. c(0)=1 and c(1)=0 (boundary conditions)
Axiom c2. for all a,b in [0,1], if a<=b then c(a)>=c(b) (monotonicity)
Axiom c3. c is a continuous function (continuity)
Axiom c4. c is "involutive", which means that c(c(a))=a for all a in [0,1] (Involution)

These axioms are not independent since it can be shown that if a complement function c obeys Axioms c2 and c4, then it also must obey c1 and c3. In addition, it must be a one-to-one function, which means that the function is also "bijective". It therefore follows that all involutive complements for a special subclass of all continuous complements, which in turn forms a special subclass of all fuzzy complements. The Standard Fuzzy Complement is one type of involutive fuzzy complement.

Classes of Fuzzy complements:

1. "Sugeno class" defined by

c-lambda(a) = (1-a)/(1+lambda*a)
where lambda is in the range (-1,inf). When lambda=0, this becomes the Standard Fuzzy Complement.

2. "Yager class" defined by

c-w(a) = (1-a**w)**(1/w)
where w is in the range (0,inf). When w=1, we again have the Standard Fuzzy Complement.

Theorems of Fuzzy Complements:

Theorem 6.2: Every fuzzy complement has at most one equilibrium. This means that c(a)-a=0 for at most one a in [0,1]. (The fuzzy complements are assumed to obey all 4 axioms throughout.)

Theorem 6.3: Assume that a given fuzzy complement c has an equilibrium e, which by Theorem 6.2 is unique. Then

$$a \leq c(a) \text{ if } a \leq e$$
$$a \geq c(a) \text{ if } a \geq e$$

Theorem 6.4: If c is a continuous fuzzy complement, then c has a unique equilibrium. A dual point, da, with respect to a is defined by

$$c(da) - da = a - c(a)$$

Theorem 6.5: If a complement c has an equilibrium e, then the equilibrium is its own dual point

$$de = e$$

Theorem 6.6: For each a in [0,1], da = c(a) if c(c(a)) = a, that is, when the complement is involutive.

This means that if the complement is not involutive, whether the dual point does not exist, or it does not coincide with the complement point.

Theorem 6.7 (First Characterization Theorem of Fuzzy Complements): Let c be a function from [0,1] to [0,1]. Then, c is an involutive fuzzy complement if there exists a continuous function g from [0,1] on R such that g(0)=0, g is strictly increasing, and c(a) = g˜(g(1)-g(a)), g˜is the inverse of g for all a in [0,1].

Functions g are usually called increasing generators. Each function that qualifies as an increasing generator determines an involutive fuzzy complement by the equation above.

For a Standard Fuzzy Complement, g(a)=a.
For the Sugeno class of complements,

$$g\text{-lambda}(a) = \ln(1+\text{lambda}*a)\ /\text{lambda}\ (\text{lambda} > -1)$$

For the Yager class of complements, g-w(a) = a**w (w >0)

Combining to give

$$g\text{-lambda,w}(a) = \ln(1+\text{lambda}*(a**w))/\text{lambda}\ (\text{lambda} > -1, w > 0)$$

which yields

$$c\text{-lambda,w}(a) = ((1-a**w)/(1+\text{lambda}*(a**w)))**(1/w)$$

which contains the Sugeno class and the Yager class as special subclasses. As one more example

$$g\text{-gamma}(a) = a/(\text{gamma}+(1-\text{gamma})*a)\ (\text{gamma} > 0)$$

produces the class of involutive fuzzy complements

$$c\text{-gamma}(a) = ((\text{gamma}**2)*(1-a))/(a+(\text{gamma}**2)*(1-a))(\text{gamma} > 0)$$

Involutive fuzzy complements can also be produced by "decreasing generators".

Theorem 6.8 (Second Characterization Theorem of Fuzzy Complements): Let c be a function form [0,1] to [0,1]. Then c is an involutive fuzzy complement if there exists a continuous function f from [0,1] to R such that f(1)=0, f is strictly decreasing, and c(a) = f˜(f(0)-f(a)), f˜is the inverse of f for all a in [0,1].
For a Standard Fuzzy Complement, f(a) = -k*a + k for all k >0.
For the Yager class of complements, f-w(a) = 1-a**w (w >0).

6.9.2   Fuzzy Intersections: t-norms

The intersection of two fuzzy sets, A and B, is specified by a binary operation on the unit interval; that is, a function of the form

$$i: [0,1] \times [0,1] \longrightarrow [0,1]$$

every element x in the universal set is assigned the membership grade of x in the fuzzy sets A and B, and yields the membership grade of the element in the set constituting the intersection of A and B.

$$(A \mathbf{I} B)(x) = i[A(x),B(x)]$$

The operator, i, in the above equation obeys a few specific properties in order to ensure that fuzzy sets produced by i are intuitively acceptable as meaningful fuzzy intersections of a given pair of fuzzy sets. The class of t-norms is now generally accepted as equivalent to the class of fuzzy intersections. Therefore, "t-norms" and "fuzzy intersections" may be used interchangably.

Note that the function i is independent of x; it depends only upon the values if A(x) and B(x) (or a and b that are in [0,1], respectively).

A "fuzzy intersection/t-norm" i is a binary operation on the unit interval that satisfies at least the following axioms for all a, b, and d in the range [0,1].

Axioms for fuzzy intersections:

Axiom i1: i(a,1) = a (boundry condition)
Axiom i2: b <= d implies i(a,b) <= i(a,d) (monotonicity)
Axiom i3: i(a,b) = i(b,a) (commutativity)
Axiom i4: i(a,i(b,d)) = i(i(a,b),d) (associativity)

The above four axioms are called the "axiomatic skeleton for fuzzy intersections/t-norms".

It is often desirable to restrict the class of fuzzy intersections (t-norms) by considering additional requirements. Three of the most important are:

Axiom i5: i is a continuous function (continuity)
Axiom i6: i(a,a) <a (subidempotency)
Axiom i7: a1 <a2 and b1 <b2 implies i(a1,b1) ¡ i(a2,b2) (strict monotomicity)

Since the requirement in Axiom i6 is weaker than "idempotency", the requirement that i(a,a)=a, it is called "subidempotency".

A continuous t-norm that satisfies subidempotency is called an "Archimedean t-norm"; if it also satisfies strict monotomicity, it is called a "strict Archimedean t-norm".

Theorem 6.9: The Standard Fuzzy Intersection is the only idempotent t-norm.

The following are examples of some t-norms that are frequently used as fuzzy intersections (each defined for all a,b in [0,1]).

Standard intersection: i(a,b) = min(a,b)

Algebraic product: i(a,b) = ab

Bounded difference: i(a,b) = max(0,a+b-1)

Drastic intersection: i(a,b) = a when b=1

b when a=1

0 otherwise

The drastic intersection is also denotes Imin(a,b). The full range of all fuzzy intersections, i(a,b) is specified in the nest theorem.

Theorem 6.10: for all a,b in [0,1],

$$I_min(a,b) \leq i(a,b) \leq \min(a,b)$$

As described above, a "decreasing generator" is a continuous and strictly decreasing function f from [0,1] to R such that f(1)=0. The pseudo-inverse of a decreasing generator f, denoted f(-1), is a function from R to [0,1] given by

$$f(-1)(a) = \begin{cases} 1, & for \ a \ in \ (-inf, 0) \\ f \sim (a), & for \ a \ in[0, f \ (0)] (f \sim is \ the \ inverse) \\ 0, & for \ a \ in[f(0), inf] \end{cases}$$

Some examples are

f1(a)=1-a**p(p>0) for any a in [0,1]

$$f1(-1)(a) = \begin{cases} 1, & for \ a \ in(-inf, 0) \\ (1 - a) * *(1/p), & for \ a \ in[0, 1] \\ 0, & for \ a \ in(1, inf) \end{cases}$$

and

f2(a) = - ln(a) for any a in [0,1] with f2(0) = inf.

$$f2(-1)(a) = \begin{cases} 1, & for\ a\ in(-inf, 0) \\ exp(-a) & for\ a\ in(0, inf) \end{cases}$$

A decreasing generator and its pseudo-inverse satisfy f(-1)(f(a))=a for any a in [0,1] and

$$f(f(-1)(a)) = \begin{cases} 0, & for\ a\ in(-inf, 0) \\ a, & for\ a\ in(0, f(0)) \\ f(0) & for\ a\ in(f(0), inf) \end{cases}$$

An "increasing generator" is a continuous and strictly increasing function g from [0,1] to R such that g(0)=0. The pseudo-inverse of an increasing generator g, denoted g(-1), is a function from R to [0,1] defined by

$$g(-1)(a) = \begin{cases} 0, & for\ a\ in\ (-inf, 0) \\ g \sim (a), & for\ a\ in[0, g\ (1)](g \sim is\ the\ inverse\ of\ g) \\ 1, & for\ a\ in[g(1), inf] \end{cases}$$

Some examples are

g1(a)=**p(p>0)for any a in [0,1]

$$g1(-1)(a) = \begin{cases} 0, & for\ a\ in(-inf, 0) \\ a ** (1/p), & for\ a\ in[0, 1] \\ 1, & for\ a\ in(1, inf) \end{cases}$$

and

g2(a) = -ln(1-a) for any a in [0,1] with g2(1)=inf

$$g2(-1)(a) = \begin{cases} 0, & for\ a\ in(-inf, 0) \\ 1 - exp(-a) & for\ a\ in(0, inf) \end{cases}$$

An increasing generator and its pseudo-inverse satisfy g(-1)(g(a))=a for any a in [0,1] and

$$g(g(-1)(a)) = \begin{cases} 0, & for\ a\ in(-inf, 0) \\ a, & for\ a\ in(0, g(1)) \\ g(1) & for\ a\ in(g(1), inf) \end{cases}$$

Lemma 6.1: Let f be a decreasing generator. Then a function g defined by

$$g(a) = f(0) - f(a)$$

for any a in $[0,1]$ is an increasing generator with $g(1)=f(0)$, and its pseudo-inverse $g(-1)$ is given by

$$g(-1)(a) = f(-1)(f(0)-a)$$

for any a in R.

Lemma 6.2: Let g be an increasing generator. Then a function f defined by

$$f(a) = g(1) - g(a)$$

for any a in $[0,1]$ is a decreasing generator with $f(0)=g(1)$, and its pseudo-inverse $f(-1)$ is given by

$$f(-1)(a) = g(-1)(g(1)-a)$$

for any a in R.

Theorem 6.11 (Characterization Theorem of t-Norms): Let i be a binary operation on the unit interval. Then, i is an Archimedean t-norm if there exists a decreasing generator f such that

$$i(a,b) = f(-1)(f(a)+f(b)) \text{ for all } a,b \text{ in } [0,1].$$

The following theorem shows you how to generate new t-norms from existing t-norms.

Theorem 6.13: Let i be a t$\longrightarrow$ norm and let g : $[0,1]$ $[0,1]$ be a function such that g is strictly increasing and continuous in $(0,1)$ and $g(0)=0$, $g(1)=1$. Then the function i-g defined by

$$i\text{-}g(a,b) = g(-1)(i(g(a),g(b))) \text{ for all } a,b \text{ in } [0,1], \text{ where } g(-1) \text{ is the}$$
pseudo-inverse of g, is also a t-norm.

There are other methods for obtaining t-norms from given t-norms, which are based on various ways of combining several t-norms into one t-norm.

6.9.3   Fuzzy Unions: t-conorms

The general fuzzy union of two fuzzy sets A and B is specified by a function

$$\mathbf{Y} : [0,1] \times [0,1] \longrightarrow [0,1]$$

or

$$(A\mathbf{Y}B)(x) = \mathbf{Y}[A(x), B(x)]$$

The union function satisfies exactly the same properties of functions that are known in the literature as t-conorms.

A fuzzy union/t-conorm u is a binary operation on the unit interval that satisfies at least the following axioms for all a,b,d in [0,1]:

Axioms for fuzzy union:

Axiom u1: $\mathbf{Y}$ (a,0) = a (boundary condition)
Axiom u2: b $\leq$ d implies $\mathbf{Y}$ (a,b) $\leq$ $\mathbf{Y}$ (a,d) (monotonicity)
Axiom u3: $\mathbf{Y}$ (a,b) = $\mathbf{Y}$ (b,a) (commutativity)
Axiom u4: $\mathbf{Y}$ (a, $\mathbf{Y}$ (b,d)) = $\mathbf{Y}$ ($\mathbf{Y}$ (a,b),d) (associativity)

These are called the "axiomatic skeleton for fuzzy unions/t-conorms".
The most important additional requirements for fuzzy unions are expressed by the following axioms:

Axiom u5: $\mathbf{Y}$ is a continuous function (continuity)
Axiom u6: $\mathbf{Y}$ (a,a) >a (superidempotency)
Axiom u7: a1 <a2 amd b1 <b2 implies $\mathbf{Y}$ (a1,b1) <$\mathbf{Y}$ (a2,b2) (strict monotonicity)

Any continuous and superidempotent t-conorm is called "Archimedean"; if it is also strictly monotomic, it is called "strictly Archimedean".

Theorem 6.14: The standard fuzzy union is the only idempotent t-conorm. (i.e., it is the only one where u(a,a) = a).

Here are some examples to t-conorms that are frequently used as fuzzy unions (each defined for all a and b in [0,1]):

Standard union: $\mathbf{Y}$ (a,b) = max(a,b)
Algebraic sum: $\mathbf{Y}$ (a,b) = a + b $-$ ab
Bounded sum: $\mathbf{Y}$ (a,b) = min(1,a+b)
Drastic union: $\mathbf{Y}$ (a,b) = a when b=0, b when a=0, 1 otherwise

The full range of fuzzy unions is defined by the following theorem.

Theorem 6.15: For all a,b in [0,1]

$$\max(a,b) \leq (a,b) \leq U_{m}ax(a,b)$$

where $U_{m}ax$ denotes the drastic union.

Theorem 6.16 (Characterization Theorem of t-Conorms):

Let u be a binary operation on the unit interval. Then, u is an Archemedean t-cornom if there exists an increasing generator such that

$$\mathbf{Y}\ (a,b) = g(-1)(g(a) + g(b))\ for\ all\ a,b\ in\ [0,1].$$

Theorem 6.17: Let u-w denote the class of Yager t-conorms. Then

$$\max(a,b) \leq \text{u-w}(a,b) \leq U_{m}ax(a,b)\ \text{for all a,b in [0,1] and all w} > 0.$$

New t-conorms can generated from existing t-conorms.

Theorem 6.18: Let u be a t-conorm and let g: $[0,1] ->[0,1]$ be a function such that g is strictly increasing and continuous on $(0,1)$ and $g(0)=0$, $g(1)=1$. Then, the function u-g defined by

As with t-norms, t-conorms can also be constructed by combining existing t-conorms.

6.9.4 Combinations of Operations

In classical set theory, the operations of intersection and union are dual with respect to the complement in that they satisfy the De Morgan laws:

- The complement of the intersection of A and B equals the union of the complement of A and the complement of B.

- The complement of the union of A and B equals the intersection of the complement of A and the complement of B.

Evidently, duality is gratified only by certain specific combinations of t-norms, t-conorms, and fuzzy complements. The t-norm i and the t-conorm u are "dual with respect to a fuzzy complement c" if

$$c(i(a, b)) = u(c(a), c(b))$$

and

$$c(u(a, b)) = i(c(a), c(b))$$

These equations define the De Morgan laws for fuzzy sets. Let the triple (i,u,c) denote that i and u are dual with respect to c, and let any such triple be called a "dual triple".

The following t-norms and t-conorms are dual with respect to the Standard Fuzzy Complement cs (i.e., dual triples):

$$(\min(a,b),\max(a,b),cs)$$
$$(ab,a+b-ab,cs)$$
$$(\max(0,a+b-1),\min(1,a+b),cs$$
$$(I_{m}in(a,b),U_{m}ax(a,b),cs)$$

Theorem 6.19: The triples (min,max,c) and (Imin,Umax,c) are dual with respect to any fuzzy complement c.

Theorem 6.20: Given a t-norm i and an involutive fuzzy complement c, the binary operation u on [0,1] defined by

$$u(a,b) = c(i(c(a),c(b)))$$

for all a,b in [0,1] is a t-conorm such that (i,u,c) is a dual triple.

Theorem 6.21: Given a t-conorm u and an involutive fuzzy complement c, the binary operation i on [0,1] defined by

$$i(a,b) = c(u(c(a),c(b)))$$

for all a,b in [0,1] is a t-norm such that (i,u,c) is a dual triple.

Theorem 6.22: If an involutive fuzzy complement c and an increasing generator g of c are known, then the t-norm and t-conorm generated by g are dual with respect to c.

Theorem 6.23: Let (i,u,c) be a dual triple generated by Theorem 6.22. Then, the fuzzy operations i, u, c satisfy the law of excluded middle [u(a,c(a))=1] and the law of contradiction [i(a,c(a))=0]

Theorem 6.24: Let (i,u,c) be a dual triple that satisfies the law of excluded middle and the law of contradiction. Then (i,u,c) does not satisfy the distributive laws. This means that

i(a,u(b,d)) is not equal to u(i(a,b),i(a,d)) for all a,b,d in [0,1]

6.9.5 MATLAB Codes for Implementation of Fuzzy Operations

Complement

Two options are listed here to compute the standard complement using MATLAB. The first option is to compute the complement of the membership matrix A by subtracting A from the ones array (with all elements equal 1) with the following syntax:

```
Acomp=ones(length(A(:,1)),length(A(1,:)))-A
```

```
A =
        0    0.3000   0.4000
   0.2000    0.5000   0.3000
   0.8000    0    0
   0.7000    1.0000   0.9000
```

```
>>Acomp=ones(length(A(:,1)),length(A(1,:)))-A
```

```
Acomp =
1.0000 0.7000 0.6000
0.8000 0.5000 0.7000
0.2000 1.0000 1.0000
0.3000 0 0.1000
```

```
Using the abilities of MATLAB this may be made
shorter using the syntax:
```

```
Acomp=1-A
```

```
A =
   0    0.3000    0.4000
0.2000    0.5000    0.3000
0.8000    0    0
0.7000    1.0000    0.9000
```

```
>>Acomp=1-A
```

```
Acomp =
1.0000   0.7000   0.6000
0.8000   0.5000   0.7000
0.2000   1.0000   1.0000
0.3000   0   0.1000
```

Union

The standard fuzzy union of two membership matrices A and B is calculated as described in Section 6.9.3. The function max(A,B) of MAT-LAB returns an array of the same size as A and B by taking the maxima between two elements with the same indices. The fuzzy union of the matrices A and B, is calculated using the following syntax:

un = max(A,B) , where un returns a matrix with the maximum of A and B matrices

A =
0   0.3000   0.4000
0.2000   0.5000   0.3000
0.8000   0   0
0.7000   1.0000   0.9000

B =
1.0000   0.2000   0.2000
0.1000   0.4000   0.9000
0.1000   1.0000   0
0.7000   0.4000   0.3000

>>un=max(A,B)

un =
1.0000   0.3000   0.4000
0.2000   0.5000   0.9000
0.8000   1.0000   0
0.7000   1.0000   0.9000

Intersection

The standard fuzzy intersection of the matrices A and B is calculated as described in Section 6.9.2. In MATLAB it is computed by using the function min(A,B) taking the minima between the elements with the same indices.

The fuzzy intersection of the matrices A and B, is calculated using the following syntax:

in = min(A,B) , where in returns a matrix with the minimum of A and B matrices

A =
0   0.3000   0.4000
0.2000   0.5000   0.3000
0.8000   0   0
0.7000   1.0000   0.9000

```
B =
1.0000   0.2000   0.2000
0.1000   0.4000   0.9000
0.1000   1.0000   0
0.7000   0.4000   0.3000

>>in=min(A,B)
in =
0   0.2000   0.2000
0.1000   0.4000   0.3000
0.1000   0   0
0.7000   0.4000   0.3000
```

6.9.6   Aggregation Operations

Aggregation operations on fuzzy sets are operations by which several fuzzy sets are aggregated in some standard manner to produce a single fuzzy set. Formally, an "aggregation operation" on n fuzzy sets (n>=2) is defined by a function

$$h: [0,1]^{**}n \longrightarrow [0,1]$$

or

$$A(x) = h(A1(x), A2(x), \ldots, An(x))$$

The following axioms should be satisfied in order to aggregate fuzzy sets. These axioms express the essence of the notion of aggregation:

Axioms for aggregation operations:

Axiom h1: h(0,0, ... ,0)=0 and h(1,1, ... ,1)=1 (boundary conditions)

Axiom h2: For any pair (a1,a2, ... ,an) and (b1,b2, ... ,bn) of n-tuples such that ai, bi are in [0,1] for all i, if ai ≤ bi for all i, then h(a1,a2, ... ,an) ≤ h(b1,b2, ... ,bn) (i.e. h is monotonic increasing in all its arguments).

Axiom h3: h is a continuous function.

Aggregation operations on fuzzy sets are usually expected to satisfy two additional axiomatic requirements.

Axiom h4: h is a "symmetric" function in all its arguments; that is
h(a1,a2, ... an) = h(a-p(1),a-p(2), ... ,a-p(n))
for any permutation p of the indices.
Axiom h5: h is an "idempotent" function; that is,
h(a,a, ... ,a) = a
for all a in [0,1].

Any aggregation operation h that satisfies Axioms h2 and h5 also satisfies
the inequalities

$$\min(a1,a2, ... ,an) \leq h(a1,a2, ... ,an) \leq \max(a1,a2, ... ,an)$$

for all n-tuples (a1,a2, ... ,an) in [0,1]**n. It is also true that any aggregation operation that satisfies the above inequalities also satisfies Axiom
h5. This means that all aggregation operations between the standard
fuzzy intersection and the standard fuzzy union are idempotent, and
they are the only ones that are idempotent. These aggregation operations are usually called "averaging operations".

One class of averaging operations that covers the entire interval between the min and max operations consist of "generalized means". They
are defined by the formula

$$\text{h-x}(a1,a2, ... ,an) = ((a1^{**}x + a2^{**}x + ... + an^{**}x)/n)^{**}(1/x)$$

where x is any real number except 0 and all ai's can't be zero. If x = -1,
h is the harmonic mean, as x $\longrightarrow$ 0, h approaches the geometric mean,
and for x = 1, h is the arithmetic mean.

Another class of aggregation operations that covers the entire interval between the min and max operators is called the class of "ordered
weighted averaging operations"; the acronym OWA is often used in the
literature to refer to these operations. Let

$$w = (w1,w2,...,wn)$$

be a "weighting vector" such that wi is in [0,1] for all i and

$$\text{sum}(i=1,n)wi = 1.$$

Then, an OWA operation associated with w is the function

$$\text{hw}(a1,a2, ...,an) = w1b1 + w2b2 + ... + wnbn,$$

where bi for any i is the i-th largest element in a1,a2, ... ,an (the b vector
is a permutation of the a vector such that the elements are ordered from
largest to smallest).

Theorem 6.25: Let h : $[0,1]^{**}n \longrightarrow$ R+ (positive reals) be a function that satisfies Axioms h1 and h2 and the property

h(a1+b1,a2+b2, ... ,an+bn) = h(a1,a2, ... ,an)+h(b1,b2, ... ,bn)

where ai,bi,ai+bi are in [0,1] for all i, then

h(a1,a2, ... ,an) = sum(i=1,n)wi*ai

where wi >0 for all i.

Theorem 6.26: Let h: $[0,1]^{**}n \longrightarrow [0,1]$ be a function that satisfies Axioms h1 and h3 and the properties

h(max(a1,b1), ... ,max(an,bn))=max(h(a1, ... ,an),h(b1, ... ,bn))

hi(hi(ai))=hi(ai)

where hi(ai)=h(0, ... ,0,ai,0, ... ,0) for all i. Then,

h(a1, ... ,an)=max(min(w1,a1), ... ,min(wn,an))

where wi is in [0,1] for all i. (This is a "weighted quasi average", in which the min and max operations replace, respectively, the arithmetic produce and sum.)

Theorem 6.27: Let h: $[0,1]^{**}n \longrightarrow [0,1]$ be a function that satisfies Axioms h1 and h3 and the properties

h(min(a1,b1),...,min(an,bn))=min(h(a1, ... ,an),h(b1, ... ,bn))

hi(ab) = hi(a)*hi(b) and hi(0)=0

for all i, where hi(ai) = h(1,...,1,ai,1,...,1). Then, there exist

numbers x1,x2, ... ,xn in [0,1] such that

h(a1,a2, ... ,an) = min(a1**x1,a2**x2, ... ,an**xn)

A special kind of aggregation operations are binary operations h on [0,1] that satisfy the properties of "monotonicity", "commutativity" and "associativity" of t-norms and t-conorms, but replace the boundary conditions with weaker boundary conditions

h(0,0)=0 and h(1,1)=1.

These aggregation operations are called "norm operations".

When a norm operation also has the property h(a,1)=a, it becomes a t-norm; when it also has the property h(a,0)=a, it becomes a t-conorm. Otherwise, it is an "associative averaging operation". Hence, norm operations cover the whole range of aggregating operations, from $I_{m}in$ to $U_{m}ax$.

6.10   Fuzzy Arithmetic

Fuzzy Numbers

To qualify as a "fuzzy number", a fuzzy set A on R must possess at least the following three properties:

(i) A must be a normal fuzzy set;

(ii) a (alpha-cut of A; {x |A(x) ≥ a}) must be a closed interval for every a in (0,1];

(iii) the support of A, 0+A (strong 0-cut of A; {x |A(x) >0 }), must be bounded.

Therefore, every fuzzy number is a convex fuzzy set. Although triangular and trapezoidal shapes of membership functions are most often used for representing fuzzy numbers, other shapes may be preferable in some applications, and they need not be symmetrical. This can include symmetric or asymmetric "bell-shaped" membership functions, or strictly increasing or decreasing functions (e.g., sigmoids) that capture the concept of a "large number" or a "small number".

The following theorem shows that membership functions of fuzzy numbers may be, in general, piecewise-defined functions.

Theorem 4.1: Let A be a member of the fuzzy sets. Then, A is a fuzzy number if there exists a closed interval [a,b], which is not empty, such that

$$A(x) = \begin{cases} 1, & for\ x\ in[a,b] \\ l(x), & for\ x\ in[-inf,a] \\ r(x) & for\ x\ in[b,inf] \end{cases}$$

where l is a function from (-inf,a) to [0,1] that is monotomic increasing, continuous from the right, and such that l(x)=0 for x in (-inf,w1); r is a function from (b,inf) to [0,1] that is monotonic decreasing, continuous from the left, and such that r(x)=0 for x in (w2,inf)

This implies that every fuzzy number can be represented in the form of Theorem 4.1. This form means that fuzzy numbers can be defined in a piecewise manner. The sigmoid increasing or decreasing functions can be defined in this form by extending the function to -infinity and infinity by just setting l(x) and/or r(x) to 0.

Using fuzzy numbers, we can define the concept of a fuzzy cardinality for fuzzy sets that are defined on finite universal sets. Given a fuzzy set

A defined on a finite universal set X, its fuzzy cardinality, $|A\sim|$, is a fuzzy number defined by the formula

$$|A\sim|(|aA|) = a$$

for all a in Lambda(A).

6.10.1 Arithmetic Operations on Intervals

Fuzzy arithmetic is based on two properties of fuzzy numbers:

1. Each fuzzy set, and thus each fuzzy number, can fully and uniquely be represented by its alpha-cuts.

2. Alpha-cuts of each fuzzy number are closed intervals of real numbers for all alpha in [0,1]. Therefore, arithmetic operations on fuzzy numbers can be defined in terms of arithmetic operations on their alpha-cuts (i.e., arithmetic operations on closed intervals), which is treated in the field of "interval analysis".

Let # denote any of the four arithmetic operations on closed intervals: "addition" |, "subtraction" −, "multiplication" *, and "division" /. Then,

$$[a, b]\#[d, e] = \{f\#g \vert a \le f \le b, d \le g \le e\}$$

is a general property of all arithmetic operations on closed intervals, except [a,b]/[d,e] is not defined when 0 is in the interval [d,e]. Therefore, the result of an arithmetic operation on closed intervals is again a closed interval.

The four arithmetic operations on closed intervals are defined as follows:

$$[a,b] + [d,e] = [a+d,b+e]$$
$$[a,b] - [d,e] = [a-e,b-d]$$
$$[a,b] * [d,e] = [\min(ad,ae,bd,be),\max(ad,ae,bd,be)]$$

and, provided that 0 is not in [d,e]

$$[a,b] / [d,e] = [a,b] * [1/e,1/d]$$
$$= [\min(a/d,a/e,b/d,b/e),\max(a/d,a/e,b/d,b/e)]$$

Note that a real number r may also be regarded as a special (degerated) interval [r,r]. When one of the intervals in the above equations is degenerated, we obtain special operations; when both of them are degenerated, we obtain the standard arithmetic of real numbers.

Letting A=[a1,a2], B=[b1,b2], C=[c1,c2], 0=[0,0], 1=[1,1], useful properties of arithmetic operations are as follows:

1. A+B = B+A; A*B = B*A (commutativity)

2. (A+B)+C = A+(B+C); (A*B)*C = A*(B*C) (associativity)

3. A = 0+A = A+0; A = 1*A = A*1 (identity)

4. A*(B+C) is a subset of A*B + A*C (subdistributivity)

5. If b*c >= 0 for every b in B and c in C, then A*(B+C) = A*B + A*C (distributivity)
   Furthermore, if A = [a,a], then a*(b+c) = a*B + a*C

6. 0 is a subset of A-A and 1 is a subset of A/A

7. If A is a subset of E and B is a subset of F, then

   A+B is a subset of E+F

   A-B is a subset of E-F

   A*B is a subset of E*F

   A/B is a subset of E/F (inclusion monotomicity)

6.10.2 Arithmetic Operations on Fuzzy Numbers

We assume in this section that fuzzy numbers are represented by continuous membership functions. If # is any of the four basic arithmetic operations on the Fuzzy sets A and B, we define a fuzzy set A#B by defining its alpha-cut, a(A#B), as

$$a(A\#B) = aA \# aB$$

for any a in (0,1]. Therefore,

$$A\#B = \text{Union(all } a \text{ in } [0,1]) \ a(A\#B)$$

Since a(A#B) is a closed interval for each a in [0,1], and A and B are fuzzy numbers, A#B is also a fuzzy number. As an example, consider two triangular shaped fuzzy numbers A and B defined as follows:

$$A(x) = \begin{cases} \mathbf{0}, & for\ x\ \leq \mathbf{-1}\ and\ x \mathbf{>3} \\ (x+1)/\mathbf{2}, & for\ \mathbf{-1} < x \leq \mathbf{1} \\ (\mathbf{3}-x)/\mathbf{2} & for\ \mathbf{1} < x \leq \mathbf{3} \end{cases}$$

$$B(x) = \begin{cases} \mathbf{0}, & for\ x\ \leq \mathbf{1}\ and\ x \mathbf{>5} \\ (x-\mathbf{1})/\mathbf{2}, & for\ \mathbf{1} < x \leq \mathbf{3} \\ (\mathbf{5}-x)/\mathbf{2} & for\ \mathbf{3} < x \leq \mathbf{5} \end{cases}$$

their alpha-cuts are

$$aA = [2a\text{-}1, 3\text{-}2a]$$
$$aB = [2a\text{+}1, 5\text{-}2a]$$

Using the operations definitions, we obtain

a(A+B) = [4a,8-4a] for a in [0,1]
a(A-B) = [4a-6,2-4a] for a in [0,1]
a(A*B) = [-4*a* *2+12*a-5,4*a* *2-16*a+15] for a in [0,.5]
[4*a* *2-1,4*a* *2-16*a+15] for a in [.5,1]
a(A/B) = [(2a-1)/(2a+1),(3-2a)/(2a+1)] for a in [0,.5]
[(2a-1)/(5-2a),(3-2a)/(2a+1)] for a in [.5,1]

The resulting fuzzy numbers are then:

$$(A+B)(x) = \begin{cases} 0, & for\ x \leq 0\ and\ x>8 \\ x/4, & for\ 0< x \leq 4 \\ (8-x)/4, & for 4 <x \leq 8 \end{cases}$$

$$(A-B)(x) = \begin{cases} 0, & for\ x \leq -6\ and\ x>2 \\ (x+6)/4, & for -6< x < -2 \\ (2-x)/4, & for -2 <x \leq 2 \end{cases}$$

$$(A*B)(x) = \begin{cases} 0, & for\ x< -5\ and\ x \geq 2 \\ [3 - sqrt(4-x)]/2, & for\ -5 \leq x< -0 \\ [sqrt(1-x)/2], & for\ 0 \leq x<3 \\ [4 - sqrt(1+x)]/2, & for\ 3 \leq x<15 \end{cases}$$

$$(A/B)(x) = \begin{cases} 0, & for\ x< -1\ and\ x \geq 2 \\ (x+1)/(2-2x), & for\ -1 \leq x<0 \\ (5x+1)/(2x+2)], & for\ 0 \leq x<1/3 \\ (3-x)]/(2x+2), & for\ 1/3 \leq x<3 \end{cases}$$

Another technique for developing fuzzy arithmetic based on the extension principle is one in which standard arithmetic operations on real numbers are extended to fuzzy number

$$(A\#B)(z) = supremum(z=x\#y)\ min[A(x),B(y)]$$

for all z in the set of real numbers. More specifically, we define for all z in R:

(A+B)(z) = supremum(z=x+y) min[A(x),B(y)]
(A-B)(z) = supremum(z=x-y) min[A(x),B(y)]
(A*B)(z) = supremum(z=x*y) min[A(x),B(y)]
(A/B)(z) = supremum(z=x/y) min[A(x),B(y)]

Theorem 4.2: Let # be an element of $+,-,*,/$, and let A,B denote continuous fuzzy numbers. Then, the fuzzy set A#B defined above is a continuous fuzzy number.

6.10.3 Fuzzy Arithmetic Using MATLAB Fuzzy Logic Toolbox

MATLAB Fuzzy Logic toolbox provides a function `'fuzarith'` to perform fuzzy arithmetic. The general syntax of the function is

C = fuzarith (X, A, B, operator)

where A, B and X - vectors of same dimension operator - is one of the following strings: "sum', "sub', "prod", and "div" C - column vector with the same length as X.

Illustration

The following program illustrates the fuzzy arithmetic operation implemented in MATLAB. Here two membership functions triangular and generalized bell shaped functions are considered and the four basic fuzzy arithmetic operations are performed on them.

```
x=0:0.1:10;
A=trimf(x,[3 6 8]);
figure;
subplot(3,2,1);
plot(x,A);
title('Triangular Membership Function');
B=gbellmf(x,[1 3 7]);
subplot(3,2,2);
plot(x,B)
title('Generalized Bell Shaped Membership Function');

 SUM = fuzarith(x, A, B, 'sum');
 subplot(3,2,3);
 plot(x, SUM);
 title('Fuzzy addition A+B');
 DIFF = fuzarith(x, A, B, 'sub');
 subplot(3,2,4);
 plot(x, DIFF);
 title('Fuzzy Difference A-B');
 PROD = fuzarith(x, A, B, 'prod');
 subplot(3,2,5);
```

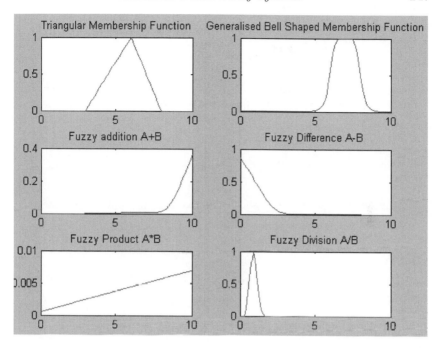

FIGURE 6.16: Illustration of Fuzzy Arithmetic

```
plot(x, PROD);
title('Fuzzy Product A*B');
DIV = fuzarith(x, A, B, 'div');
subplot(3,2,6);
plot(x,DIV);
title('Fuzzy Division A/B');
```

The result of the above program is plotted in Figure 6.16.

6.11 Fuzzy Relations

Fuzzy relations are fuzzy subsets of X x Y , i.e., mapping from X $\longrightarrow$ Y . Relations map elements of one universe, X to those of another universe, say Y , through the Cartesian product of the two universes. This relation is expressed as

$$R= \{ ((x,y),\mu_R(x,y))—(x,y)\epsilon \ X \ x \ Y \}$$

The operations of fuzzy relations are similar to those of fuzzy sets and are defined below.

Let R and S denote two fuzzy relations in the Cartesian space X x Y.

1. Union operation: The union of two fuzzy relations R and S in the Cartesian space X x Y is defined as $\mu_{RYS}(x,y) = max(\mu_R(x,y), \mu_s(x,y))$

2. Intersection operation: The intersection of two fuzzy relations R and S in the Cartesian space X x Y is defined as $\mu_{RIS}(x,y) = max(\mu_R(x,y), \mu_s(x,y))$

3. Complement operation: The complement operation of the fuzzy relation R is defined as $\mu_R(x,y) = 1 - \mu_R(x,y)$

Let us describe the relationship between the type of dirt in the clothes x and the washing time y.

The type of dirt is a linguistic variable characterized by a crisp set X with three linguistic terms as

$$X = greasy, \ moderate \ greasy, \ not \ greasy$$

and similarly the washing time as

$$Y = high, \ medium, \ low$$

One knows that a crisp formulation of a relation $X \longrightarrow Y$ between the two crisp sets would look like Table 6.1.

The zeros and ones describe the grade of membership to this relation. This relation is now a new kind of crisp set that is built from the two crisp base sets X and Y. This new set is now called R and can be expressed also by the rules:

(1) IF the type of dirt is greasy THEN the washing time is high.

(2) IF the type of dirt is moderate greasy THEN the washing time is medium.

(3) IF the type of dirt is not greasy THEN the washing time is low.

TABLE 6.1:　Relation between the two crisp sets

	High	*Medium*	*Low*
Greasy	1	0	0
Moderate Greasy	0	1	0
Not Greasy	0	0	1

TABLE 6.2: Modified form of the Table 6.1

	High	*Medium*	*Low*
Greasy	1	0.5	0
Moderate Greasy	0.3	1	0.4
Not Greasy	0	0.2	1

As can be seen from this example, a relation, which is called a rule or rule base, can be used to provide a model.

This crisp relation R represents the presence or absence of association, interaction, or interconnection between the elements of these two sets. This can be generalized to allow for various degrees of strength of association or interaction between elements. Degrees of association can be represented by membership grades in a fuzzy relation in the same way as degrees of the set membership are represented in a fuzzy set. Applying this to the above washing example, Table 6.1 can be modified as shown in Table 6.2.

where there are now real numbers in [0,1]. This table represents a fuzzy relation and models the connectives in a fuzzy rule base. It is a two-dimensional fuzzy set and the question is, how this set can be determined from its elements.

In the above example, the linguistic terms were treated as crisp terms. For example, when one represents the type of dirt on a spectrum scale, the type of dirt would be described by their spectral distribution curves that can be interpreted as membership functions. The washing times can also be treated as fuzzy terms, the above relation is a two-dimensional fuzzy set over two fuzzy sets. For example, taking from the washing machine example the relation between the linguistic terms greasy and high, and represent them by the membership functions as shown in Figure 6.17a.

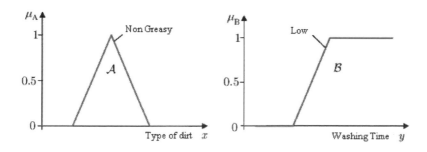

FIGURE 6.17a: Membership Functions

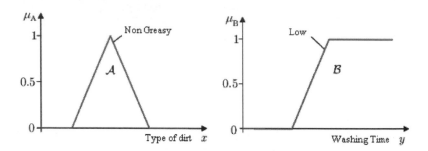

FIGURE 6.17b: Membership Function of the Relation after Applying the min Operation

This expression can be re-written in mathematical form using elementary connective operators for the membership functions by

$$\mu_R(x, y) = min\{\mu_A(x), \mu_B(y)\}$$

or

$$\mu_R(x, y) = \mu_A(x), \mu_B(y) \qquad (6.13)$$

Figure 6.17b shows a 3-dimensional view of these two fuzzy terms and Figure 6.17c the result of the connective operation according to Equation (6.13). This result combines the two fuzzy sets by an operation that is a Cartesian product

$$R: X \times Y \longrightarrow [0,1] \qquad (6.14)$$

From this example it is obvious that the connective operation in a rule for the operation is simply performed by a fuzzy intersection in two dimensions.

Combining rules into a rule base the example from above may help when it is rewritten as

1. IF the type of dirt is greasy THEN the washing time is high,

   (OR)

2. IF the type of dirt is moderate greasy THEN the washing time is medium,

   (OR)

3. IF the type of dirt is not greasy THEN the washing time is low

which describes the union of three rules in a linguistic way. For the complete rule base R one can combine the relations formed for each individual rule with a fuzzy union operator, which is the fuzzy OR.

According to the standard mathematical notation for IF-THEN and AND

$$A_r1(x_1)^A_{r2}(x_2)^{\cdot}..A_m(x_n) \longrightarrow B_r(u) \tag{6.15}$$

Where x1,x2,.xn are the several input variables, r = 1, ... R is the rule number, and Ari and Br are words from natural language.

Applying the union operator by writing the rule base with max/min operators as follows:

$$\mu_R(x_1, x2, ...x_n, u) = maxmin\{\mu_{Pr}(x_1, x_2, ...x_n), \mu_{Br}(u)\} \tag{6.16}$$

where $\mu_{Pr}(x_1, x_2, ...x_n)$ is the premise of the r_{th} rule. This representation is the standard max/min representation of a rule base that is used for fuzzy controllers. Instead of the max/min representation a so called max-prod representation is also usual, where the algebraic product

$$\mu_R(x_1, x2, ...x_n, u) = maxmin\mu_{Pr}(x_1, x_2, ...x_n), \mu_{Br}(u) \tag{6.17}$$

is used to build the relation between the premise and the conclusion.

6.12 Fuzzy Composition

Composition of fuzzy relations plays a crucial role in the study of approximate reasoning. Composition is defined as the operation of combining fuzzy relations in different product spaces with each other. This allows the fuzzy relations to operate with a given fact to produce an output that represents the decision in a fuzzy way, known as fuzzy reasoning, which is a special case of the more general operation called fuzzy composition.

Two relations of the form given in Equation (6.15)

$$R: X \times Y \longrightarrow [0,1]$$

$$S: Y \times Z \longrightarrow [0,1]$$

can be composed to one relation

$$T: X \times Z \longrightarrow [0,1]$$

using the max and min operators for union and intersection, one can express the composition operation T = RoS by the corresponding membership functions. Based upon the mathematical properties, composition can be performed either by min-max composition or the max product composition.

Min-Max Composition: Let $\bar{R}_1 \epsilon X \times Y$ and $\bar{R}_2 \epsilon Y \times Z$ denote two fuzzy relations, the min-max composition on these relations is defined as $\bar{R}_1 \circ \bar{R}_2 = max\{min\{\mu_{R1}(x,y), \mu_{R2}(y,z)\}\}x\epsilon X, y\epsilon Y, z\epsilon Z$

Max Product Composition: Let $R_1 \epsilon X \times Y$ and $R_2 \epsilon Y \times Z$ denote two fuzzy relations, the max product composition on these relations is defined as $R_1 R_2 = max\mu_{R1}(x,y) * \mu_{R2}(y,z)x\epsilon X, y\epsilon Y, z\epsilon Z$

Consider $\bar{R}_1 \epsilon X \times Y and \bar{R}_2 \epsilon Y \times Z$ be defined as :

$$\bar{R}_1 = \begin{matrix} 1 & 0.4 & 0.5 \\ 0.3 & 0 & 0.7 \\ 0.2 & .6 & 0.9 \end{matrix}$$

$$\bar{R}_2 = \begin{matrix} 0.1 & 0.4 \\ 0.9 & 0.4 \\ 0 & 0.8 \end{matrix}$$

The min-max composition cam be computed as

$$\bar{R}_1 \circ \bar{R}_2(x_1, z_1) = max\{min(1, 0.1), min(0.4, 0.9), min(0.5, 0)\}$$
$$= max\{0.1, 0.4, 0\} = 0.4$$
$$\bar{R}_1 \circ \bar{R}_2(x_1, z_2) = max\{min(1, 0.5), min(0.4, 0.4), min(0.5, 0.8)\}$$
$$= max\{0.5, 0.4, 0.5\} = 0.5$$
$$\bar{R}_1 \circ \bar{R}_2(x_2, z_1) = max\{min(0.3, 0.1), min(0, 0.9), min(0.7, 0)\}$$
$$= max\{0.1, 0.0\} = 0.1$$
$$\bar{R}_1 \circ \bar{R}_2(x_2, z_2) = max\{min(0.3, 0.1), min(0, 0.9), min(0.7, 0)\}$$
$$= max\{0.1, 0.0\} = 0.1$$
$$\bar{R}_1 \circ \bar{R}_2(x_3, z_1) = max\{min(0.2, 0.1), min(0.6, 0.9), min(0.9, 0)\}$$
$$= max\{0.1, 0.6, 0\} = 0.6$$
$$\bar{R}_1 \circ \bar{R}_2(x_3, z_2) = max\{min(0.2, 0.5), min(0.6, 0.4), min(0.9, 0.8)\}$$
$$= max\{0.2, 0.4, 0.8\} = 0.8$$

Therefore the result obtained by min-max composition is

$$\bar{R}_1 o \bar{R}_2 = \begin{matrix} 0.4 & 0.5 \\ 0.1 & 0.1 \\ 0.6 & 0.8 \end{matrix}$$

The max product composition can be computed as follows:

$$\bar{R}_1 \circ \bar{R}_2(x_1, z_1) = max\{(1 * 0.1), (0.4 * 0.9), (0.5 * 0)\}$$
$$= max\{0.1, 0.36, 0\} = 0.36$$
$$\bar{R}_1 \circ \bar{R}_2(x_1, z_2) = max\{(1 * 0.5), (0.4 * 0.4), (0.5 * 0.16, 0.4)\} = 0.5$$
$$\bar{R}_1 \circ \bar{R}_2(x_1, z_2) = max\{(1 * 0.5), (0.4 * 0.4), (0.5 * 0.8)\}$$
$$= max\{0.5, 0.16, 0.4\} = 0.5$$
$$\bar{R}_1 \circ \bar{R}_2(x_2, z_1) = max\{(0.3 * 0.1), (0 * 0.9), (0.7 * 0)\}$$
$$= max\{0.03, 0, 0\} = 0.03$$
$$\bar{R}_1 \circ \bar{R}_2(x_2, z_2) = max\{(0.3 * 0.1), (0 * 0.9), (0.7 * 0)\}$$
$$= max\{0.03, 0, 0\} = 0.03$$
$$R_1 R_2(x_3, z_1) = max\{(0.2 * 0.1), (0.6 * 0.9), (0.9 * 0)\}$$
$$= max\{0.02, 0.54, 0\} = 0.54$$
$$\bar{R}_1 \circ \bar{R}_2(x_3, z_2) = max\{(0.2 * 0.5), (0.6 * 0.9), (0.9 * 0.8)\}$$
$$= max\{0.1, 0.24, 0.72\} = 0.72$$

Therefore the result obtained by max-product composition is

$$\bar{R}_1 \circ \bar{R}_2 = \begin{matrix} 0.36 & 0.5 \\ 0.03 & 0.03 \\ 0.54 & 0.72 \end{matrix}$$

In the same manner as relations can be composed, the one-dimensional facts can be composed with the rule base to realize the reasoning operation. This can now be precisely re-formulated for the general case of a rule base.

If for the rule base

$$R{:}X \times Y \longrightarrow [0,1]$$

its membership function is described by Equations (6.16) or (6.17) and if there is a fact described by the fuzzy set

$$A'{:}X \longrightarrow [0,1]$$

and its membership function $\mu_{A'}$ the result

$$B'{:}A'oR{:}Y \longrightarrow [0,1]$$

of the fuzzy reasoning is represented by the membership function

$$\mu_B(y) = \max_{x \epsilon X}\{min\{\mu_{A'}(x), \mu_R(x, y)\}\} \qquad (6.18)$$

6.12.1 MATLAB Code to Implement Fuzzy Composition

Fuzzy composition is illustrated through a Matlab code as follows. Here the relations

$$R{:}X \times Y \longrightarrow [0,1]$$

$$S{:}Y \times Z \longrightarrow [0,1]$$

are denoted as R and S respectively. The composed relation T:X × Z $\longrightarrow$ [0,1]

Illustration 1: Given R: x × y and S: × z as follows.

$$R = \begin{matrix} 0.1 & 0.3 & 0.5 & 0.6 \\ 0.3 & 0.5 & 0.8 & 1 \\ 0.2 & 0.9 & 0.7 & 0.1 \end{matrix}$$

and

$$S = \begin{matrix} 0 & 0.3 \\ 0.9 & 0.8 \\ 0.4 & 0.7 \\ 0.2 & 0.6 \end{matrix}$$

Compute the max min composition T: x × y using a MATLAB program.

% Max Min Composition

```
R=[.1 .3 .5 .6 ; .3 .5 .8 1; .2 .9 .7 .1]
S=[0 .3;0.9 0.8;0.4 0.7; 0.2 0.6]
[x,y]=size(R);
[y1,z]=size(S);
if y==y1
T=zeros(x,z);
for i=1:x,
for j=1:z,
T(i,j)=max(min([R(i,:); S(:,j)']));
end
end
else
display('Matrix dimension does not match')
end
```

Observations:

Max-Min composition

$$R = \begin{matrix} 0.1000 & 0.3000 & 0.5000 & 0.6000 \\ 0.3000 & 0.5000 & 0.8000 & 1.0000 \\ 0.2000 & 0.9000 & 0.7000 & 0.1000 \end{matrix}$$

$$S = \begin{matrix} 0 & 0.3000 \\ 0.9000 & 0.8000 \\ 0.4000 & 0.7000 \\ 0.2000 & 0.6000 \end{matrix}$$

$$T = \begin{matrix} 0.4000 & 0.6000 \\ 0.5000 & 0.7000 \\ 0.9000 & 0.8000 \end{matrix}$$

Illustration 2: Given R: x × y and S: y1 × z as follows.

$$R = \begin{matrix} 0.1 & 0.3 & 0.5 & 0.6 \\ 0.3 & 0.5 & 0.8 & 1 \\ 0.2 & 0.9 & 0.7 & 0.1 \end{matrix}$$

and

$$S = \begin{matrix} 0 & 0.3 \\ 0.9 & 0.8 \\ 0.4 & 0.7 \\ 0.2 & 0.6 \end{matrix}$$

Compute the max product composition T: x × y using a MATLAB program.

```
% Max Product composition
R=[.1 .3 .5 .6 ; .3 .5 .8 1; .2 .9 .7 .1]
S=[0 .3;0.9 0.8;0.4 0.7; 0.2 0.6]
[x,y]=size(R);
[y1,z]=size(S);
if y==y1
T=zeros(x,z);
for i=1:x,
for j=1:z,
T(i,j)=max(R(i,:).*S(:,j)');
end
end
else
display('Matrix dimension does not match')
end
```

Observations

Max Product composition

$$R = \begin{matrix} 0.1000 & 0.3000 & 0.5000 & 0.6000 \\ 0.3000 & 0.5000 & 0.8000 & 1.0000 \\ 0.2000 & 0.9000 & 0.7000 & 0.1000 \end{matrix}$$

$$S = \begin{matrix} 0 & 0.3000 \\ 0.9000 & 0.8000 \\ 0.4000 & 0.7000 \\ 0.2000 & 0.6000 \end{matrix}$$

$$T = \begin{matrix} 0.2700 & 0.3600 \\ 0.4500 & 0.6000 \\ 0.8100 & 0.7200 \end{matrix}$$

Illustration 3: Given a fuzzy relation

$$\begin{matrix} 0.1 & 0.8 & 0.5 & 0.6 & 0.7 \\ 0.8 & 1 & 0.4 & 0 & 0.2 \\ 0 & 0.1 & 1 & 0.5 & 0.2 \\ 0.7 & 0.5 & 0.2 & 1 & 0.3 \\ 0.2 & 0.9 & 0.5 & 0.3 & 1 \end{matrix}$$

Write a MATLAB program to compute R^2 and R^3.

Here we make use of the rule to compute $R^{n-1} = RoRoR.......... R_2$ and R_3.

% MATLAB code to compute R^2 using max min composition

```
R=[0.1 0.8 0.5 0.6 0.7; 0.8 1 0.4 0 0.2; 0 0.1 1 0.5
0.2; 0.7 0.5 0.2 1 0.3; 0.2 0.9 0.5 0.3 1]

  [x,y]=size(R);
    R2=zeros(x,y);
      for i=1:x, % for each row of R matrix
       for j=1:y, % for each column of R matrix
        R2(i,j)=max(min([R(i,:); R(:,j)']));
       end
      end
```

Observations:

Max-Min composition
Input relation:

```
R =
   0.1000     0.8000     0.5000     0.6000     0.7000
   0.8000     1.0000     0.4000     0          0.2000
   0          0.1000     1.0000     0.5000     0.2000
   0.7000     0.5000     0.2000     1.0000  0.3000
0.2000     0.9000     0.5000     0.3000     1.0000


Output:
R2 =
0.8000     0.8000     0.5000     0.6000     0.7000
0.8000     1.0000     0.5000     0.6000     0.7000
0.5000     0.5000     1.0000     0.5000     0.3000
0.7000     0.7000     0.5000     1.0000     0.7000
0.8000     0.9000     0.5000     0.5000     1.0000
```

% MATLAB code to compute R^3 using max min composition

```
R=[0.1 0.8 0.5 0.6 0.7; 0.8 1 0.4 0 0.2; 0 0.1 1 0.5
0.2; 0.7 0.5 0.2 1 0.3; 0.2 0.9 0.5 0.3 1]
[x,y]=size(R);
R2=zeros(x,y);
R3=zeros(x,y);
   for i=1:x, % for each row of R matrix
      for j=1:y, % for each column of R matrix
 R2(i,j)=max(min([R(i,:); R(:,j)']));
      end
   end
   for i=1:x, % for each row of R matrix
for j=1:y, % for each column of R matrix
  R3(i,j)=max(min([R2(i,:); R(:,j)']));
end
end
```

Observations: Max-Min composition
Input relation:

```
R =
   0.1000     0.8000     0.5000     0.6000     0.7000
   0.8000     1.0000     0.4000     0          0.2000
   0          0.1000     1.0000     0.5000     0.2000
   0.7000     0.5000     0.2000     1.0000     0.3000
   0.2000     0.9000     0.5000     0.3000     1.0000
Output:
```

```
R3 =
    0.8000      0.8000      0.5000      0.6000      0.7000
    0.8000      1.0000      0.5000      0.6000      0.7000
    0.5000      0.5000      1.0000      0.5000      0.5000
    0.7000      0.7000      0.5000      1.0000      0.7000
    0.8000      0.9000      0.5000      0.6000      1.0000
```

```
% MATLAB code to compute R^2 using max product composition
R=[0.1 0.8 0.5 0.6 0.7; 0.8 1 0.4 0 0.2; 0 0.1 1 0.5
 0.2; 0.7 0.5 0.2 1 0.3; 0.2 0.9 0.5 0.3 1]
[x,y]=size(R);
    R2=zeros(x,y);
  for i=1:x, % for each row of R matrix
    for j=1:y, % for each column of R matrix
% R2(i,j)=max(min([R(i,:); R(:,j)']));
 end
    end
```

Observations:

```
Max-Product composition
Input relation:

  R =

      0.1000      0.8000      0.5000      0.6000      0.7000
    0.8000      1.0000      0.4000      0       0.2000
    0       0.1000      1.0000      0.5000      0.2000
    0.7000      0.5000      0.2000      1.0000      0.3000
    0.2000      0.9000      0.5000      0.3000      1.0000
Output:
R2 =
    0.6400      0.8000      0.5000      0.6000      0.7000
    0.8000      1.0000      0.4000      0.4800      0.5600
    0.3500      0.2500      1.0000      0.5000      0.2000
    0.7000      0.5600      0.3500      1.0000      0.4900
    0.7200      0.9000      0.5000      0.3000      1.0000
```

% MATLAB code to compute R^3 using max product composition

```
R=[0.1 0.8 0.5 0.6 0.7; 0.8 1 0.4 0 0.2;0 0.1 1 0.5
0.2;0.7 0.5 0.2 1
0.3; 0.2 0.9 0.5 0.3 1]
[x,y]=size(R);
R2=zeros(x,y); R3=zeros(x,y);
for i=1:x, % for each row of R matrix
for j=1:y, % for each column of R matrix
R2(i,j)=max(R(i,:).*R(:,j)');
End
end
for i=1:x, % for each row of R matrix
for i=1:x, % for each row of R matrix
for j=1:y, % for each column of R matrix
R3(i,j)=max(R2(i,:).*R(:,j)');
end
end
```

Observations:

```
Max-Product composition
Input relation:
R =


0.1000     0.8000     0.5000     0.6000     0.7000
0.8000     1.0000     0.4000     0     0.2000
0     0.1000     1.0000     0.5000     0.2000
0.7000     0.5000     0.2000     1.0000     0.3000
0.2000     0.9000     0.5000     0.3000     1.0000
```
Output:
```
R3 =
      0.6400     0.8000     0.5000     0.6000     0.7000
      0.8000     1.0000     0.4000     0.4800     0.5600
      0.3500     0.2800     1.0000     0.5000     0.2450
      0.7000     0.5600     0.3500     1.0000     0.4900
      0.7200     0.9000     0.5000     0.4320     1.0000
```

Summary

Fuzzy Logic provides a completely different, unorthodox way to approach a problem. This method focuses on what the system should do rather than trying to understand how it works. This chapter discussed the basic fuzzy sets, operations on fuzzy sets, relations between fuzzy sets, composition and fuzzy arithmetic. A few MATLAB programs were also illustrated on topics such as membership functions, fuzzy operations, fuzzy arithmetic, relations, and composition.

Review Questions

1. State "Law of excluded middle".

2. Differentiate crisp and fuzzy logic.

3. Give a few examples of sets.

4. Mention a few examples of crisp and fuzzy sets.

5. Define universe of discourse.

6. What is a membership function? State the different types of membership functions.

7. Define Singleton and Linguistic variables.

8. Define term set with an example.

9. What are the major operations performed by a fuzzy set?

10. State the axioms for fuzzy complement.

11. State the axioms for fuzzy intersection.

12. State the axioms for fuzzy union.

13. What are aggregation operators? State the axioms for aggregation operations.

14. Mention the three properties to qualify a fuzzy set as a "fuzzy number".

15. Explain fuzzy composition with a suitable example.

Chapter 7

Fuzzy Inference and Expert Systems

7.1 Introduction

Fuzzy inference is the process of formulating the mapping from a given input to an output using fuzzy logic. The mapping then provides a basis from which decisions can be made, or patterns discerned. The process of fuzzy inference involves all the topics such as fuzzification, defuzzification, implication, and aggregation. Expert control/modeling knowledge, experience, and linking the input variables of fuzzy controllers/models to output variable (or variables) are mostly based on Fuzzy Rules. A fuzzy expert system consists of four components namely, the fuzzifier, the inference engine, and the defuzzifier, and a fuzzy rule base. This chapter focuses on these rules, expert system modeling, fuzzy controllers, and implementation of fuzzy controllers in MATLAB.

7.2 Fuzzy Rules

For any fuzzy logic operation, the output is obtained from the crisp input by the process of fuzzification and defuzzification. These processes involve the usage of rules, which form the basis to obtain the fuzzy output. A fuzzy if-then rule is also known as fuzzy rule, or fuzzy conditional statement or fuzzy implication. It is generally of the form

IF (x *is* A) AND (y *is* B) AND THEN (z *is* Z)

where x,y,z etc. represent the variables and A,B,Z are the linguistic values in the universe of discourse. Here the IF part is referred to as the antecedent or premise and the THEN part is referred to as consequent or conclusion. AND is the Boolean operator which connects two or more antecedents. These fuzzy rules are multi-valued.

An individual fuzzy rule-based can possess more than one rule. Based on several set of rules an overall decision can be made from the individual consequents. This process of obtaining the overall decision is known as aggregation of rules.

Fuzzy rules are most commonly applied to control systems. The common types of fuzzy rules applied to control systems are the Mamdani fuzzy rules and Takagi–Sugeno (TS) fuzzy rules.

7.2.1   Generation of Fuzzy Rules

The fuzzy rules are formed or generated in a canonical method. The canonical rules can be formed by integrating the linguistic variables using assignment, conditional, or unconditional statements. Assignment statements assign a value to a particular variable in the universe. Conditional statement, involve, set of IF-THEN rules. The unconditional statements are of the form in which no condition is to be satisfied. The fuzzy rules are generated based on conditional statements and are also referred to as the canonical rule based system. The rules are usually formed as follows:

IF antecedent THEN consequent

A few examples of canonical rules are illustrated in the following:

IF $\underbrace{temperature\ is\ hot}_{antecedant}$ THEN $\underbrace{pressure\ is\ medium}_{consequent}$

IF $\underbrace{height\ is\ tall}_{antecedant}$ THEN $\underbrace{weight\ is\ heavy}_{consequent}$

7.2.2   Disintegration of Rules

Most of the practical applications do not involve rules like the above mentioned with one antecedent part. These applications involve a compound rule structure. Such rules can be disintegrated into smaller rules and from which simple canonical rules can be formed. These rules have more than one antecedent part connected by conjunction and disjunction connectives. Conjunction connective uses intersection operation involving the "AND" connective as follows

IF antecedent$_1$ AND antecedent$_2$ AND ...
AND antecedent$_n$ THEN consequent

Similarly the disjunction connective uses union operation involving the "OR" connective as follows

$$IF\ antecedent_1\ OR\ antecedent_2\ OR\ ...$$
$$OR\ antecedent_n\ THEN\ consequent$$

Likewise complex rules can be broken into simpler forms and connected using the "AND" or "OR" connectives as shown in the following rule.

IF $\underbrace{\textit{type of dirt is High}}_{antecedent\ 1}$ $\underbrace{\textit{AND}}_{connective}$ $\underbrace{\textit{volume of clothes is Small}}_{antecedent\ 2}$

THEN $\underbrace{\textit{washing Time is medium}}_{consequent}$

IF $\underbrace{\textit{type of dirt is High}}_{antecedent\ 1}$ $\underbrace{\textit{AND}}_{connective}$ $\underbrace{\textit{volume of clothes is Large}}_{antecedent\ 2}$

THEN $\underbrace{\textit{washing Time is High}}_{consequent}$

These rules can be decomposed to a set of relations.

7.2.3 Aggregation of Rules

The rule based system involves several rules and each rule provides an output or consequent. The consequent part also known as conclusion is unique for every rule that has been executed based on the input parameters. An overall conclusion has to be obtained from the individual consequents. This method of obtaining the overall conclusion from the set of rules is referred to as aggregation of rules.

Fuzzy rules can be aggregated by using the "AND" or "OR" connectives. The process of aggregating the rules using "AND" connective is known as conjunctive aggregation and the process of aggregating the rules using "OR" connective is known as disjunctive aggregation.

Conjunctive aggregation:

Consequent $=$ $Consequent_1$ AND $Consequent_2$ AND ... AND $Consequent_r$

Disjunctive aggregation:

Consequent $= Consequent_1$ OR $Consequent_2$ OR ... OR $Consequent_r$

Using these operators a final decision is made on the output of the fuzzy set.

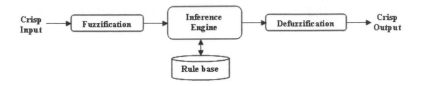

FIGURE 7.1: Fuzzy Expert System Model

7.3 Fuzzy Expert System Model

A fuzzy expert system shown in Figure 7.1 consists of four components namely, the fuzzifier, the inference engine, the defuzzifier, and a fuzzy rule base.

During fuzzification, crisp inputs are converted into linguistic values and are related to the input linguistic variables. Subsequently as the fuzzification process is completed, the inference engine refers to the fuzzy rule base containing fuzzy IF-THEN rules to deduct the linguistic values for the intermediate and output linguistic variables. When the output linguistic measures are obtainable, the defuzzifier produces the final crisp values from the output linguistic values.

7.3.1 Fuzzification

The process by which the input values from sensors are scaled and mapped to the domain of fuzzy variables is known as fuzzification. The fuzzy variables also known as linguistic variables are determined based on intuition (from knowledge) or inference (known facts). These linguistic variables can be either continuous or discrete theoretically, but in practice it should be discrete. Fuzzification is a two step process: Assign fuzzy labels and Assign numerical meaning to each label.

Assign fuzzy label:

Each crisp input is assigned a fuzzy label in the universe of discourse. For example for the input parameter height fuzzy labels can be "tall", "short", "Normal", "Very Tall", and "Very short". Every crisp input can be assigned multiple labels. As the number of labels increases the resolution of the process is better. In some cases, assigning large number of labels leads to a large computational time and thus making the fuzzy system unstable. Therefore in general the number of labels for a system

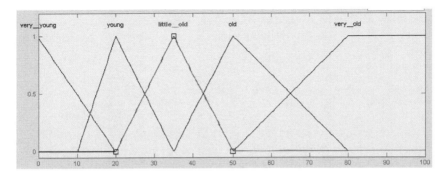

FIGURE 7.2: Sample fuzzification of crisp inputs

is limited to an odd number in the range [3, 9], such that the surface is balanced and symmetric.

Assign numerical meaning:

Here membership functions are formed to assign a numerical meaning to each label. The range of the input value that corresponds to a specific label can be identified by the membership function. Though there are different membership function shapes, triangular and trapezoidal membership functions are commonly used to avoid time and space complexity.

For each fuzzy set and for each linguistic variable, the grade of membership of a crisp measure in each fuzzy set is ascertained. As an example, the numerical variable age which has a given value of 25.0 was fuzzified using the triangular membership functions defined for each fuzzy set for linguistic variable age. As a result of fuzzification (Figure 7.2), linguistic variable age has linguistic values of "young", "Little old" etc.

In a fuzzy expert system application, each input variable's crisp value is first fuzzified into linguistic values before the inference engine proceeds in processing with the rule base.

7.3.2 Fuzzy Rule Base and Fuzzy IF-THEN Rules

Fuzzy expert systems use fuzzy IF-THEN rules. A fuzzy IF-THEN rule is of the form

$$IF\ X_1 = A_1\ and\ X_2 = A_2...and\ X_n = A_n\ THEN\ Y = B$$

where X_i and Y are linguistic variables and A_i and B are linguistic terms. An example of a fuzzy IF-THEN rule is

$$IF pressure = \text{``low''} THEN volume = \text{``big''}$$

In a fuzzy expert system, the fuzzy inference engine stores the set of fuzzy IF-THEN rules and these rules are referred while processing inputs.

7.3.3 Fuzzy Inference Machine

The fuzzy inference machine combines the facts obtained from the fuzzification with the rule base and conducts the fuzzy reasoning process. The fuzzy inference engine is the kernel of the fuzzy logic controller. This reasoning mechanism infers the fuzzy control actions with the aid of fuzzy implication and fuzzy logic rules. These fuzzy logic rules are in the fuzzy rule base and the membership functions of the fuzzy sets that are formed are contained in the data base. The rule base and the data base together are referred to as the knowledge base.

The rule base represents the entire process in the form of a set of production rules. It includes information such as

- process input and output variables

- contents of rule antecedent and rule consequent

- range of linguistic values

- derivation of the set of rules

The data base provides the necessary information for proper functioning of the fuzzy inference system. It includes information such as

- membership functions representing the meaning of the linguistic values

The operation of the fuzzy inference engine is explained with the aid of the following flowchart shown in Figure 7.3. The first step is the fuzzification process in which the crisp values are converted to their equivalent linguistic variables. The antecedent parts are combined to obtain the firing strength of each rule, based on which the consequents are generated. Finally the consequents are aggregated to obtain an overall crisp output from the individual consequents.

The most commonly used fuzzy inference methods are Mamdani's inference method, Takagi–Sugeno (TS) inference method and the Tsukamoto inference method. All these methods are similar to each other but differ only in their consequents. Mamdani fuzzy inference method uses fuzzy sets as the rule consequent while TS method uses functions of input variables as the rule consequent and the Tsukamoto inference method uses fuzzy set with a monotonical membership function as the rule consequent. These methods are discussed in detail in section 7.3.

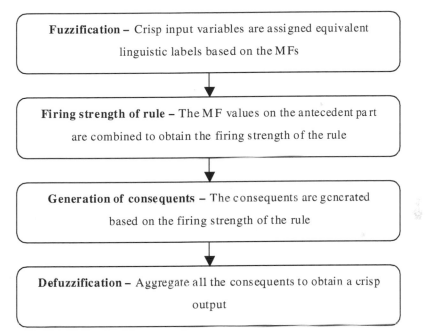

FIGURE 7.3: Flowchart of the Fuzzy inference engine operation

7.3.4 Defuzzification

As a result of applying the previous steps, one obtains a fuzzy set from the reasoning process that describes, for each possible value u, how reasonable it is to use this particular value. In other words, for every possible value u, one gets a grade of membership that describes to what extent this value u is reasonable to use. Using a fuzzy system as a controller, one wants to transform this fuzzy information into a single value u' that will actually be applied. This transformation from a fuzzy set to a crisp number is called a defuzzification. The fuzzy results generated cannot be used as such to the applications, hence it is necessary to convert the fuzzy quantities into crisp quantities for further processing. This can be achieved by using defuzzification process by using the methods as follows:

Max-Membership principle

- The max membership principle method finds the defuzzified value at which the membership function is a maximum

- This method of defuzzification is also referred to as the height

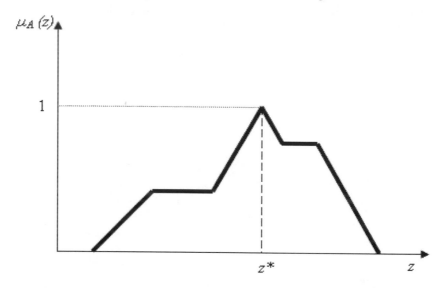

FIGURE 7.4:   Max membership principle

method.

- The defuzzified value can be determined from the following expression $\mu_A Z^* \geq \mu_A(Z)$

- Computes the defuzzified value at a very fast rate

- Very accurate only for peaked output membership functions

- Graphical representation of max-membership defuzzification shown in Figure 7.4

Center of gravity method (COG)

- The COG method of defuzzification was developed by Sugeno in 1985

- This method is also known as center of area or centroid method

- Most commonly used method

- Defined as z* = where z* $\frac{\int \mu_A(z)z dz}{\int \mu_A(z)z dz}$ is the defuzzified output, $\mu A(z)$ is the aggregated membership function and z is the output variable

- Capable of producing very accurate results

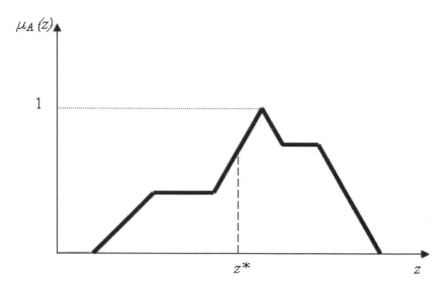

FIGURE 7.5: Centroid defuzzification method

- Major disadvantage - computationally difficult for complex membership functions

- Graphical representation of COG method is shown in Figure 7.5

Weighted Average method

- In the weighted average method, the output is obtained by the weighted average of the each membership function output of the system

- This method can be applied only for symmetrical output membership functions

- Each membership function is weighted by its largest membership function- Defined as where z* $\frac{\sum \mu_A(z)z}{\sum \mu_A(z)}$ is the defuzzified output, A(z) is the aggregated membership function z and is the weight associated with the membership function

- The defuzzified value obtained in this method is very close to that obtained by COG method

- Overcomes the disadvantage of COG method - Less computationally intensive

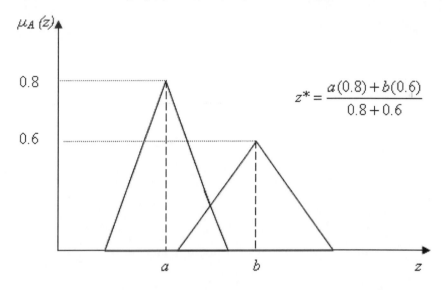

FIGURE 7.6:   Weighted Average method

- Graphical representation of Weighted Average method is shown in Figure 7.6

Mean Max membership

- The defuzzified result of the Mean max method represents the mean value whose membership function is the maximum. (Mean of the maximum membership function)

- This method is commonly referred to as middle of maxima method

- The maximum membership can be either a single point or a range of values

- Defined as $z^* = \frac{a+b}{2}$ where a and b denote the end points of the maximum membership

- Graphical representation of Mean Max method is shown in Figure 7.7

Center of sums

- Center of sums (COS) method computes the sum of the fuzzy sets

- This method is used to determine the defuzzified value by computing the algebraic sum of individual fuzzy sets

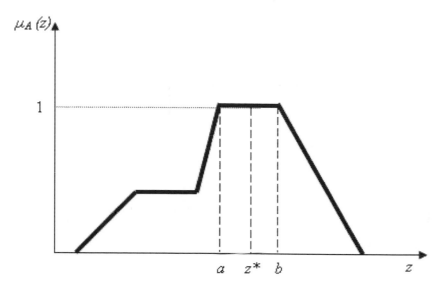

FIGURE 7.7: Mean Max method

- More similar to the center of gravity method but COS involves algebraic sum of individual output membership functions instead of their union

- Faster compared to all other defuzzification methods

- The major drawback is that the intersecting areas are added twice

- Defined as where $z^* = \dfrac{\int z \sum\limits_{i=1}^{n} \mu_{A_i}(z)dz}{\int z \sum\limits_{i=1}^{n} \mu_{A_i}(z)dz}$ is the defuzzified output, μ A(z) is the aggregated membership function and z is the output variable

- Graphical representation of COS method is shown in Figure 7.8

Center of largest area

- Center of Largest area method of computing defuzzifciation is used when the universe is non-convex and if it contains atleast two convex fuzzy subsets

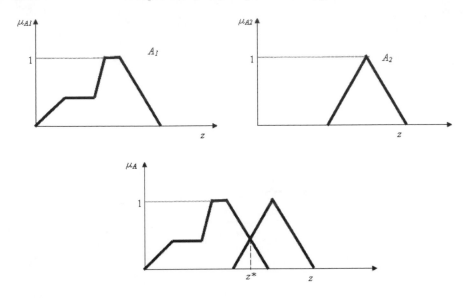

FIGURE 7.8:   Center of sums method

- The convex fuzzy subset with the largest area is chosen to determine the defuzzified value.

- Complex method since it involves finding convex regions and then the areas

- Defined as $z* = \dfrac{\int \mu_{Am}(z)z\ dz}{\int \mu_{Am}(z)z\ dz}$ where Am is the convex subregion that has the largest area

- The major drawback is that if the largest membership functions have equal areas, then there occurs an ambiguity in choosing the defuzzified value

- Graphical representation of Center of largest area method is shown in Figure 7.9

First or Last of maxima

- In first of maxima, the first value of the overall output membership function with maximum degree is considered.

- First of maxima is defined as $z* = \inf_{z \epsilon Z}\{z \epsilon Z | \mu_A(Z) = height(A)\}$

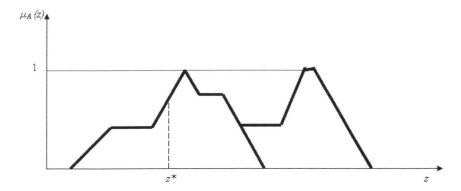

FIGURE 7.9:   Center of Largest area method

- In last of maxima, the last value of the overall output membership function with maximum degree is considered.

- Last of maxima is defined as $z* = \sup_{z \epsilon Z}\{z\epsilon Z|\mu_A(Z) = height(A)\}$

where height(A) $= \sup_{z\epsilon Z} \mu_A(z)$

7.3.5   Implementation of Defuzzification using MAT-LAB Fuzzy Logic Toolbox

Syntax
```
out = defuzz(x,mf,type)
```
`defuzz(x,mf,type)` returns a defuzzified value out, of a membership function mf positioned at associated variable value x, using one of several defuzzification strategies, according to the argument, type. The variable type can be one of the following:
`centroid`: centroid of area method
`bisector`: bisector of area method
`mom`: mean of maximum method
`som`: smallest of maximum method
`lom`: largest of maximum method

If type is not one of the above, it is assumed to be a user-defined function. x and mf are passed to this function to generate the defuzzified output.

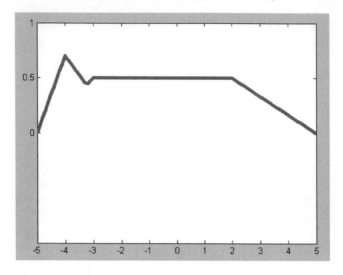

FIGURE 7.10:   Region to be defuzzified

Illustration

Assume that the following region is to be defuzzified. This example illustrates the methods of defuzzification and compares the different methods to find the best one.

```
x = -5:0.1:5;
m1 = trimf(x,[-5 -4 -2 3]);
m2 = trapmf(x,[-5 -3 2 5]);
m4 = max(0.5*m2,0.7*m1);
plot(x,m4,'Linewidth',4);
set(gca,'YLim',[-1 1],'YTick',[0 .5 1])
```

Output

Figure 7.10 shows the plot of trapezoidal membership function (region to be defuzzified).

Centroid method

Centroid defuzzification returns the center of area under the curve. If the area is a plate of equal density, the centroid is the point along the x axis about which this shape would balance. The following code returns the centroid value as -0.5896 shown in Figure 7.11.

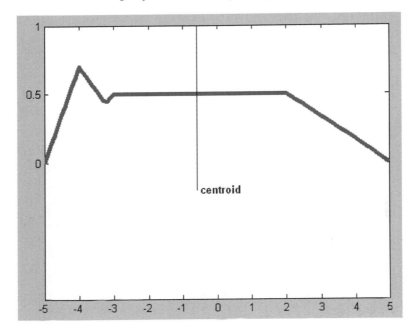

FIGURE 7.11: Centroid defuzzification

```
x1 = defuzz(x,m4,'centroid')
h1 = line([x1 x1],[-0.2 1.2],'Color','k');
t1 = text(x1,-0.2,' centroid','FontWeight','bold');
```

Bisector method

The bisector is the vertical line that will divide the region into two sub-regions of equal area. It is sometimes, but not always coincident with the centroid line. Here the defuzzified value is -0.6000 as shown in Figure 7.12.

```
x2 = defuzz(x,m4,'bisector')
h2 = line([x2 x2],[-0.4 1.2],'Color','k');
t2 = text(x2,-0.4,' bisector','FontWeight','bold');
```

Mean of Maximum methods

MOM, SOM, and LOM stand for Middle, Smallest, and Largest of Maximum, respectively. These three methods key off the maximum value assumed by the aggregate membership function. The mean of maximum method is used to compute the defuzzified value at -4 as shown in Figure 7.13.

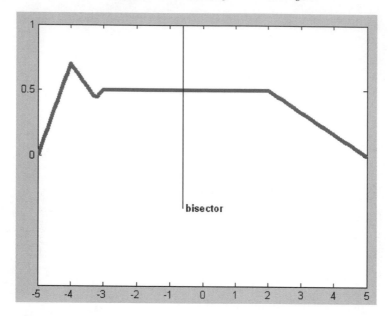

FIGURE 7.12:   Bisector method

texttttx3 = defuzz(x,mf1,'mom')
h3 = line([x3 x3],[-0.7 1.2],'Color','k');
t3 = text(x3,-0.7,'MOM','FontWeight','bold');

Choosing the best method

From the figures above, the centroid method is the best method since it is capable of producing best and accurate results. Thus the MATLAB code was given to illustrate the methods of defuzzification.

7.4   Fuzzy Inference Methods

The design or modeling of a fuzzy inference method is based on two different approaches.

- Composition based inference

- Individual rule based inference

In ***composition based inference*** all the rules are combined and fired with the fuzzy value through composition operation. The fuzzy relations

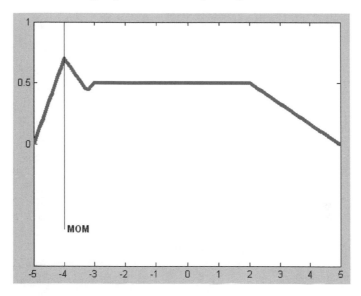

FIGURE 7.13:   MOM methods

representing the meaning of each individual rule are aggregated into one fuzzy relation describing the meaning of the overall set of rules. Then the inference or firing operation is performed through fuzzy composition between the fuzzified crisp input and the fuzzy relation representing the meaning of the overall set of rules. As a result of composition the fuzzy set describing the fuzzy value of the overall control output is obtained.

In the ***individual rule based inference*** each rule is fired individually with the crisp value to get "n" fuzzy sets which are combined over an overall fuzzy set. Initially each single rule is fired. Firing refers to computing the degree of match between the crisp input and the fuzzy sets describing the meaning of the rule antecedent. Once the firing is completed, clipping is performed. Clipping refers to cutting of the membership function at the point at which the degree to which the rule antecedent has been matched by the crisp output. Finally the clipped values are aggregated thus forming the value of the overall control output.

Among these two approaches, individual rule based inference method is commonly used since it is computationally efficient and also saves memory. The most commonly used fuzzy inference methods are Mamdani's inference method, Takagi–Sugeno (TS) inference method and the Tsukamoto inference method. All these methods are similar to each other but differ only in their consequents. Mamdani fuzzy inference method uses fuzzy sets as the rule consequent while TS method uses functions of input variables as the rule consequent and the Tsukamoto inference method uses fuzzy set with a monotonical membership function as the rule consequent.

7.4.1   Mamdani's Fuzzy Inference Method

Among the above mentioned inference methods, the Mamdani model is the commonly used method due to its simple min-max structure. The model was proposed by Mamdani (1975) as an attempt to control a steam engine and boiler combination by synthesizing a set of linguistic control rules obtained from experienced human operators. An example of a Mamdani inference system is shown in Figure 7.14. The following flowchart in Figure 7.15 explains the operations involved in computing the FIS using the Mamdani model.

The steps involved in the flowchart are explained below:

Generate Fuzzy Rules: The crisp values obtained from the physical world are converted to its equivalent linguistic variables. A set of linguistic statements form a fuzzy rule. The decision is made by the fuzzy inference system using these fuzzy rules. The fuzzy rules are of the form

IF ($input_1$ is Linguistic $variable_1$) AND OR ($input_2$ is Linguistic $variable_2$) AND OR ... THEN (output is Linguistic $variable_n$)

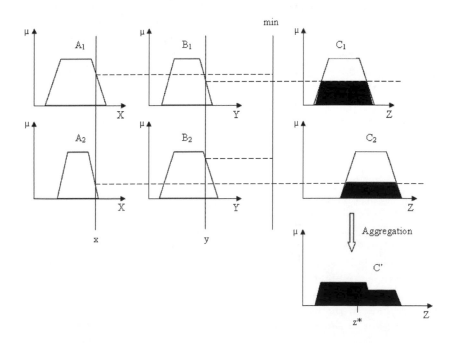

FIGURE 7.14:   The Scheme of Mamdani Fuzzy Inference

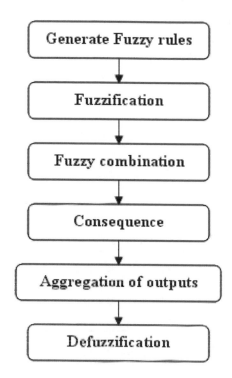

FIGURE 7.15: Operations Involved in the Mamdani Inference Method

The input and output variables are defined by means of membership functions. The process of converting the crisp value obtained from the external world with the aid of sensors to its equivalent fuzzy linguistic value is referred to as fuzzification.

Fuzzification: The process of converting crisp values to fuzzy linguistic variables is known as fuzzification. Fuzzification is a two step process as discussed earlier involving steps such as assigning fuzzy labels and then assigning a numerical meaning to each label.

Each crisp input is assigned a fuzzy label in the universe of discourse. For example for the input parameter height fuzzy labels can be "tall", "short", "Normal", "very tall" and "very short".

Membership functions are formed to assign a numerical meaning to each label. The range of the input value that corresponds to a specific label can be identified by the membership function. Though there are different membership function shapes, triangular and trapezoidal membership functions are commonly used to avoid time and space complexity.

*Fuzzy combination:*The Boolean operators such as "AND" and "OR" are used as connectives in the fuzzy rules. These operators are known as fuzzy combination operators since they are used to combine more than one antecedent part.

The fuzzy AND is expressed as

$$\mu_{AIB} = min(\mu_A(x), \mu_B(x))$$

Though the fuzzy AND is similar to the Boolean AND the difference is that, the Boolean AND can perform with only 0 and 1, while the fuzzy AND performs for the numbers between 0 and 1.

The fuzzy OR is expressed as

$$\mu_{AYB} = max(\mu_A(x), \mu_B(x))$$

Though the fuzzy OR is similar to the Boolean OR the difference is that, the Boolean OR can perform with only 0 and 1, while the fuzzy OR performs for the numbers between 0 and 1.

Consequence: Using the combination operators AND/OR the fuzzified inputs are combined and the rule strengths are determined. Then the output membership functions are clipped at the rule strength to obtain the consequence. The rule consequent is correlated with the truth value of the antecedent by cutting the consequent membership function at the level of the antecedent truth. This process is referred to as clipping or alpha cut. The top most membership function is cut, therefore some information loss occurs. In order to preserve the unique shape of the membership function scaling is preferred over clipping. In scaling the degree of the membership function of the rule consequent is multiplied by the truth value of the antecedent, thus reducing the loss of information.

Aggregation of Outputs: The process of unification of the outputs of all the rules is known as aggregation. The clipped or scaled membership functions are combined into a single fuzzy set. Each individual fuzzy rule yields a consequence, from which the overall output is to be computed. All the consequences are aggregated by using the "AND" or "OR" connectives. The process of aggregating the rules using "AND" connective is known as conjunctive aggregation and the process of aggregating the rules using "OR" connective is known as disjunctive aggregation.

Defuzzification: The final step in the Mamdani Inference method is defuzzification. In order to obtain a crisp output number several defuzzification methods can be used. The aggregated output from the previous step acts as the input to the defuzzification module and outputs a single crisp number. Though there are several defuzzification methods such as max membership principle, weighted average, centroid, center of sums, max mean, etc., and the most commonly used method is the centroid method. This method computes a defuzzification value which slices the aggregate set into two equal parts. Mathematically the defuzzified value is computed as

$$\frac{\int \mu_A(z)zdz}{\int \mu_A(z)zdz}$$

where z* is the defuzzified output, $\mu_A(z)$ is the aggregated membership function and z is the output variable.

7.4.2 Takagi–Sugeno Fuzzy Inference Method

The Takagi–Sugeno model also known as TS method was proposed by Takagi and Michio Sugeno in 1985 in order to develop a systematic approach to generate fuzzy rules. The Sugeno type fuzzy inference is similar to the Mamdani inference, they differ from each other in their rule consequent. The TS method was developed to function as an efficient model for systems whose input output relations are well defined. In the TS method also known as the parametric method the consequents are linear parametric equations represented in terms of the inputs of the system. The general form of a TS rule is

$$IF\ antecedent_1\ AND\ antecedent_2\ THEN\ \underbrace{output\ =\ f(x,y)}_{consequent}$$

Here *output* = $f\,(x,y)$ is a crisp function in the consequent. This mathematical function can either be linear or nonlinear. Most commonly linear functions are used and adaptive techniques are used for nonlinear equations. The membership function of the rule consequent is a single spike or a singleton in the TS method.

A few examples of the TS method of inference are

$$IF\ x\ is\ small\ THEN\ y = 3x-2$$

$$IF\ x\ is\ large\ THEN\ y = x + y + 5$$

A zero order Sugeno fuzzy model uses the rules of the form,

$$IF\ antecedent_1\ AND\ antecedent_2\ THEN\ \underbrace{output\ =\ k}_{consequent}$$

where k is a constant.

In such as case the output of each fuzzy rule will be a constant. The evaluation of fuzzy rules using Sugeno method is shown in Figure 7.16.

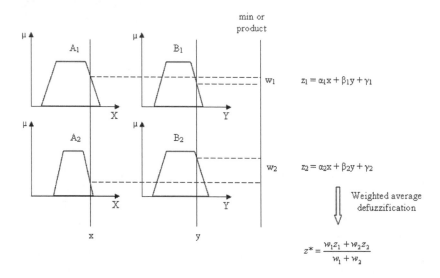

FIGURE 7.16:   The Scheme of Sugeno Inference Method

In Sugeno method the operational steps are similar to the Mamdani inference method. It goes through the same set of steps starting from generation of fuzzy rules to aggregation of rules. The rule evaluation varies when compared with the Mamdani method of inference. Since each rule has a crisp output, the overall output is obtained by weighted average method of defuzzification. Making use of the weighted average method of defuzzification reduces the time consuming process of defuzzification performed in a Mamdani model. Moreover, since the only fuzzy part of the Sugeno model is its antecedent, a clear distinction can be made between the fuzzy and non-fuzzy part.

Sugeno method is computationally effective and is well suited for adaptive and optimization problems, making it more effective in the area of control systems. Sugeno controllers usually have far more adjustable parameters in the rule consequent and the number of the parameters grows exponentially with the increase of the number of input variables.

7.4.3   Tsukamoto Fuzzy Inference Method

In the Tsukamoto fuzzy model, the rule consequent is represented with a monotonical membership function. The general form of a Tsukamoto

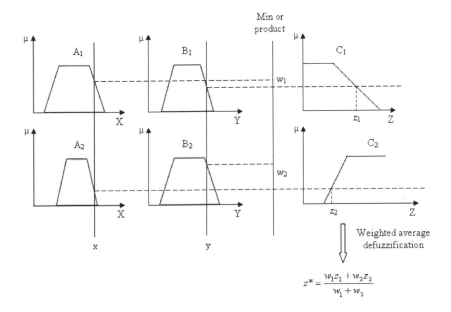

FIGURE 7.17: The Scheme of Tsukamoto Inference Model

rule is

$$IF \; antecedent_1 \; AND \; antecedent_2 \; THEN \; \underbrace{output \; = \; membership \, function}_{consequent}$$

This method also differs from the Mamdani and Sugeno in terms of its rule consequent. The Tsukamoto method of fuzzy inference in shown in Figure 7.17.

The output of each rule is defined as a crisp value induced by the firing strength of the rule. The Tsukamoto model also aggregates each of the rule's output using weighted average method of defuzzification thereby reducing the time consumed for the process of defuzzification.

7.5 Fuzzy Inference Systems in MATLAB

The input crisp values can be mapped to the crisp output by means of using fuzzy logic and this process is referred as fuzzy inference. This mapping offers a foundation from which decisions are made. The fuzzy inference involves all processes membership functions, logical fuzzy operations, and rule base. In that respect there are two types of fuzzy inference systems that could be implemented in the MATLAB's Fuzzy Logic Toolbox: Mamdani-type and Sugeno-type. These inference types have slight differences in the method of determining the output.

Fuzzy inference systems have been successfully applied in fields such as automatic control, data classification, decision analysis, expert systems, and computer vision. Due to its multidisciplinary nature, fuzzy inference systems are related with a number of names, such as fuzzy-rule-based systems, fuzzy expert systems, fuzzy modeling, fuzzy associative memory, fuzzy logic controllers, and just fuzzy systems.

The most commonly used inference type is the Mamdani's fuzzy inference method. Mamdani's technique was among the first control systems built applying fuzzy set theory. This technique was proposed in 1975 by Ebrahim Mamdani to control a steam engine and boiler combination by synthesizing a set of linguistic control rules obtained from experienced human operators. Mamdani's effort was based on Lotfi Zadeh's 1973 paper on fuzzy algorithms for complex systems and decision processes.

Mamdani-type inference, as defined for Fuzzy Logic Toolbox, anticipates the output membership functions to be fuzzy sets. After the aggregation process, there is a fuzzy set for each output variable that needs defuzzification. It is possible, and in many cases much more efficient, to use a single spike as the output membership function rather than a distributed fuzzy set. This type of output is sometimes known as a singleton output membership function, and it can be thought of as a pre-defuzzified fuzzy set. It enhances the efficiency of the defuzzification process because it greatly simplifies the computation required by the more general Mamdani method, which finds the centroid of a 2D function. Rather than integrating across the two-dimensional function to find the centroid, you use the weighted average of a few data points. Sugeno-type systems support this type of model. In general, Sugeno-type systems can be used to model any inference system in which the output membership functions are either linear or constant.

There are five primary GUI tools shown in Figure 7.18 for building, editing, and observing fuzzy inference systems in Fuzzy Logic Toolbox:

- Fuzzy Inference System (FIS) Editor

- Membership Function Editor

- Rule Editor

- Rule Viewer

- Surface Viewer

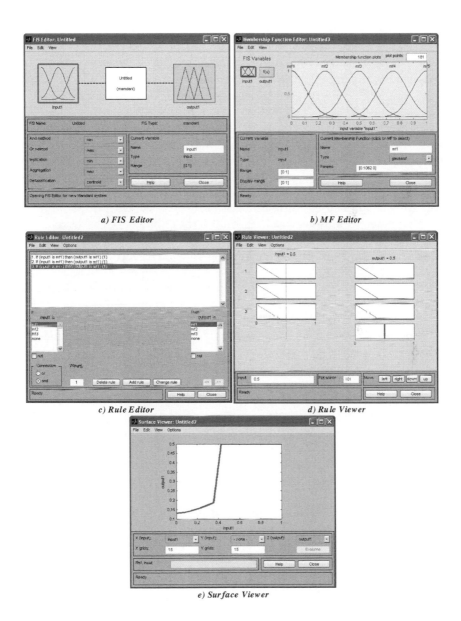

a) FIS Editor

b) MF Editor

c) Rule Editor

d) Rule Viewer

e) Surface Viewer

FIGURE 7.18: GUIs in Fuzzy Logic Toolbox

These GUIs are dynamically linked, in that the changes made to the FIS using one of them, can affect any of the other open GUIs.

The number of inputs, their names etc are handled by the FIS editor. The FIS editor window shown in Figure 7.18a can be opened by entering "fuzzy" in the MATLAB command window. There is no limit on the number of inputs. But, the number of inputs may be limited by the available memory of the machine upon which the inference is implemented.

The Membership Function Editor is used to define the shapes and type of all the membership functions associated with each variable. The MF editor window shown in Figure 7.18b can be opened from the FIS editor window by selecting "Edit" and then choosing "Membership functions" from the drop down menu.

The Rule Editor is used editing the list of rules based on the input and output parameters and this rule base defines the behavior of the system. The Rule editor window shown in Figure 7.18c can be opened from the FIS editor window by selecting "Edit" and then choosing "Rules" from the drop down menu.

A Rule Viewer and a Surface Viewer shown in Figure 7.18d and 7.18e are used to view the status of the FIS editor and can be opened from the Rule editor window by selecting "Edit" and then choosing "Rules" from the drop down menu. The Rule Viewer is a MATLAB based display of the fuzzy inference and is a read only tool. The rule base can be diagnosed and the influence of membership functions on the output can be studied using the rule viewer. The Surface Viewer is used to display the dependency of one of the outputs on any one or two of the inputs-that is, it generates and plots an output surface map for the system.

These five primary editors can interact and exchange information among them. All these editors can exchange inputs and outputs with the MATLAB workspace. Any changes in one specific editor are reflected in the other editors also. For instance, if the names of the membership functions are changed using the Membership Function Editor, those changes are updated in the rules shown in the Rule Editor. The FIS Editor, the Membership Function Editor, and the Rule Editor are all capable of reading and modifying the FIS data, but the Rule Viewer and the Surface Viewer do not modify the FIS data whereas they serve as read only.

7.5.1 Mamdani-Type Fuzzy Inference

Consider a two input one output fuzzy washing machine with input parameters "amount of dirt" and "type of dirt" and the output parameter "washing time". The rules can be formed as follows:

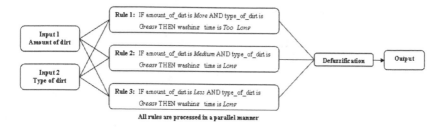

FIGURE 7.19: Basic Structure of the Three-rule Washing Machine

- IF amount_of_dirt is More AND type_of_dirt is Greasy THEN washing _time is Too_Long
  item IF amount_of_dirt is Medium AND type_of_dirt is Greasy THEN washing _time is Long

- IF amount_of_dirt is Less AND type_of_dirt is Greasy THEN washing _time is Long

- IF amount_of_dirt is More AND type_of_dirt is Medium THEN washing _time is Long

- IF amount_of_dirt is Medium AND type_of_dirt is Medium THEN washing _time is Medium

- IF amount_of_dirt is Less AND type_of_dirt is Medium THEN washing _time is Medium

- IF amount_of_dirt is More AND type_of_dirt is Non_Greasy THEN washing _time is Medium

- IF amount_of_dirt is Medium AND type_of_dirt is Non_Greasy THEN washing _time is Short

- IF amount_of_dirt is Less AND type_of_dirt is Non_Greasy THEN washing _time is Too_Short

For simplicity we consider three rules and the basic structure of the three-rule with two-input, one-output, is shown in Figure 7.19.

All the information streams from left to right, that is from two inputs to a single output. The rules of fuzzy logic systems are parallel processed. Rather than sharp flipping between manners based on breakpoints, we will glide smoothly from areas where the system's behavior is dominated by either one rule or another.

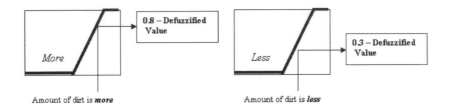

FIGURE 7.20:   Qualifying Inputs through Membership Function

The fuzzy inference process is a five step process consisting of the steps such as fuzzification of the input variables, application of the fuzzy operator (AND or OR) in the antecedent, implication from the antecedent to the consequent, aggregation of the consequents across the rules, and defuzzification. These steps are elaborated more clearly in the following sections.

Fuzzify Inputs

The initial step is to acquire the inputs and find the degree to which they belong. This degree is determined to each of the appropriate fuzzy sets via membership functions. The input is a crisp value that is limited to the universe of discourse and the output is a fuzzy grade of membership in the characterizing linguistic set which is always in the interval [0,1].

The illustration used in this section is built based on three rules, and each of the rules depends on concluding the inputs into a number of different fuzzy linguistic sets like "amount_of_dirt is More", "type_of_dirt is Greasy", etc. Prior to evaluating the rules, the inputs are fuzzified according to each of these linguistic sets. Figure 7.20 shows how well the input is qualified through its membership function, as the linguistic variable "More" and "Less".

In this manner, each input is fuzzified over all the qualifying membership functions required by the rules.

Application of the Fuzzy Operator

After fuzzifying the input values, the degree to which each component of the antecedent has been fulfilled for each rule is known. If the antecedent of a given rule bears more than one part, the fuzzy operator is employed to find one number that represents the result of the antecedent for that rule. This number will then be applied to the output function. The input to the fuzzy operator is two or more membership values from

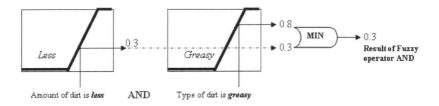

FIGURE 7.21: Fuzzy AND Operator "min"

fuzzified input variables and the output is a single truth-value.

In the MATLAB's Fuzzy Logic Toolbox, two inbuilt AND methods and two inbuilt OR methods are supported. The AND functions are the min (minimum) and prod (product) functions. The OR methods are the max (maximum), and the probabilistic OR method probor. The probabilistic OR method (also known as the algebraic sum) is calculated according to the equation

$$probor(a, b) = a + b - ab$$

Along with these inbuilt methods, the user can create his/her own methods for AND and OR by writing any function and setting that to be the method of choice. Consider the rule of the washing machine example, "IF amount_of_dirt is Less AND type_of_dirt is Greasy THEN washing _time is Too_Long". Here there are two parts in the antecedent section of the rule in terms of amount of dirt and the type of dirt. These two antecedents are connected by the fuzzy AND operator, whose operation is shown in Figure 7.21. The fuzzy AND operator selects the minimum of the two values, 0.3 and 0.8 and returns the minimum value 0.3.

Application of Implication

Prior to employing the implication technique, the rule's weight must be considered. All the rules have a weight (value between 0 and 1), which is applied to the value given by the antecedent. Normally the weight value is 1 and therefore it does not influence the implication process.

As soon as suitable weight has been designated to each rule, the implication method is enforced. A consequent is a fuzzy set constituted by a membership function and reshaped utilizing a function associated with the antecedent as shown in Figure 7.22. The antecedant is a single number which is presented as input for the implication process, and the output of the implication process is a fuzzy set. The implication process is implemented for each and every rule. The MATLAB's fuzzy logic toolbox supports two inbuilt methods, the AND method: min (minimum),

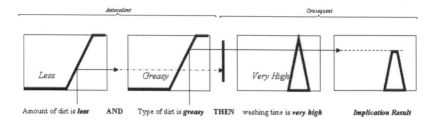

FIGURE 7.22:   Fuzzy AND Operator "min"

which truncates the output fuzzy set, and prod (product), which scales the output fuzzy set produced by the implication process.

Aggregation of the Outputs

The decisions in fuzzy sets are based on testing of all of the rules in an FIS editor, all the rules should be related some way so as to make a decision. All the fuzzy sets that are responsible for generating the output are combined into a single independant fuzzy set in the aggregation process. Aggregation takes place only once for each output variable, exactly before the last step, defuzzification. The input of the aggregation process is the list of truncated output functions returned by the implication process for each rule. The output of the aggregation process is one fuzzy set for each output variable.

As long as the aggregation method is commutative, then the order in which the rules are executed is insignificant. This step uses three inbuilt methods of the MATLAB's fuzzy logic toolbox such as: max (maximum), probor (probabilistic OR), and sum (sum of each rule's output set).

In Figure 7.23, all three rules have been placed collectively to demonstrate the output of each rule is aggregated into a single fuzzy set whose membership function assigns a weight value for every output value.

Defuzzify

The aggregate fuzzy output from the previous step is taken as input to the defuzzification process and the output of the defuzzification process is a single value. Since the aggregate of a fuzzy set from the previous step comprehends a range of output values, it is necessary that the aggregate output must be defuzzified in order to resolve a single output value from the set (Figure 7.24).

The most common defuzzification method is the centroid method, which returns the center of area under the curve. MATLAB's fuzzy logic

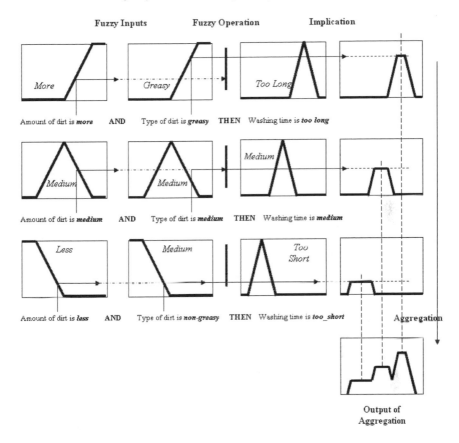

FIGURE 7.23: Aggregation of the Output of Each Rule

toolbox supports five inbuilt methods such as centroid, bisector, middle of maximum, largest of maximum, and smallest of maximum.

The Fuzzy Logic Toolbox aims to provide an open and easily modifiable fuzzy inference system structure. Thus, the Fuzzy Logic Toolbox is designed to be user friendly, to customize the fuzzy inference process for the required application. For instance, the user could replace various MATLAB functions for any of the default functions used in the five steps elaborated above: membership functions, AND methods, OR methods, implication methods, aggregation methods, and defuzzification methods.

Advantages of the Mamdani Method

- It's intuitive.

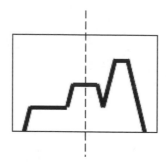

FIGURE 7.24:   Defuzzified Output Using Centroid Method

- It has widespread acceptance.

- It's well-suited to human input.

7.5.2   Sugeno-Type Fuzzy Inference

The entire fuzzy inference process elaborated to in the previous section is called Mamdani's fuzzy inference method, which is one of the primary methodology. In this section, the so-called Sugeno, or Takagi-Sugeno-Kang (TS method) method of fuzzy inference is discussed. The TS method was introduced in 1985, similar to the Mamdani method in several aspects. The first steps of the fuzzy inference process, fuzzifying the inputs and applying the fuzzy operator, are similar to that of Mamdani method. The main difference between Mamdani and Sugeno is that the Sugeno output membership functions are either linear or constant.

The Takagi Sugeno fuzzy model has the rules in the form

If "x is X" and "y is Y", then "z = g(z)"

For a zero-order Sugeno model, the output level z can be a constant or a linear function. In this Sugeno type inference system the consequent part is a crisply defined function. The output level z of each rule is weighted by the firing strength w of the rule. The firing strength for a two input system is given by

w = AndMethod (G1(x), G2(y))

where G1,2 (.) are the membership functions for Inputs 1 and 2. The final output of the system is the weighted average of all rule outputs, computed as

$$\text{Final Output} = \frac{\sum_{i=1}^{N} W_i Z_i}{\sum_{i=1}^{N} W_i}.$$

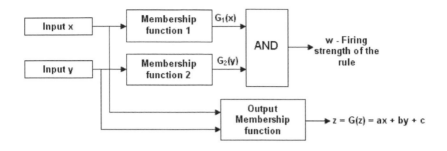

FIGURE 7.25:   Basic Operation of the Sugeno Rule

A Sugeno rule operates as shown in Figure 7.25

Figure 7.26 shows the Sugeno model for a two input one output fuzzy washing machine with input parameters "amount of dirt" and "type of dirt" and the output parameter "washing time". The rules can be formed as follows:

IF amount_of_dirt is More AND type_of dirt is Greasy THEN washing _time is Too_Long

IF amount_of_dirt is Medium AND type_of_dirt is Greasy THEN washing _time is Long

IF amount_of_dirt is Less AND type_of_dirt is Greasy THEN washing _time is Long

IF amount_of_dirt is More AND type_of_dirt is Medium THEN washing _time is Long

IF amount_of_dirt is Medium AND type_of_dirt is Medium THEN washing _time is Medium

IF amount_of_dirt is Less AND type_of_dirt is Medium THEN washing _time is Medium

IF amount_of_dirt is More AND type_of_dirt is Non_Greasy THEN washing _time is Medium

IF amount_of_dirt is Medium AND type_of_dirt is Non_Greasy THEN washing _time is Short

IF amount_of_dirt is Less AND type_of_dirt is Non_Greasy THEN washing _time is Too_Short

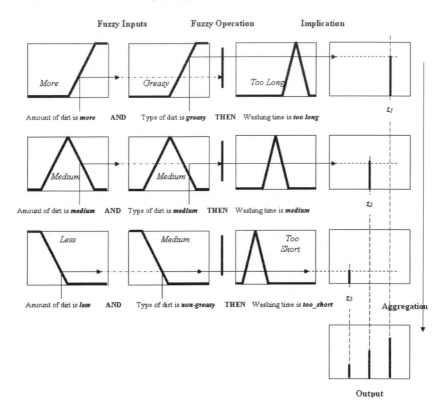

FIGURE 7.26:   Sugeno Fuzzy Model

Three rules are illustrated in Figure 7.26. Each rule is considered as defining the location of a moving singleton. To understand Sugeno method, the singleton output spikes are capable of moving in a linear fashion in the output space, depending on the input. Thus the system notation is very compact and efficient. Higher order Sugeno fuzzy models can be developed, but they introduce significant complexity with little obvious merit. Sugeno fuzzy models whose output membership functions are greater than first order are not supported by the Fuzzy Logic Toolbox.

Due to the linear dependency of rule base on the input variables of a system, the Sugeno method is perfect for serving as an interpolating supervisor of multiple linear controllers to different operating conditions of a dynamic nonlinear system. For instance, the operation of an aircraft may change dramatically with altitude and Mach number. Though linear controllers are easy to compute and well-suited to any given flight condition, it must be updated regularly and smoothly to keep up with the changing state of the flight vehicle. In such cases, a Sugeno fuzzy inference system is extremely well suited to the task of smoothly interpolating the linear gains that would be applied across the input space; it's a natural and efficient gain scheduler. Similarly, a Sugeno system is suited for modeling nonlinear systems by interpolating between multiple linear models.

Advantages of the Sugeno Method

- It's computationally efficient.

- It works well with linear techniques (e.g., PID control).

- It works well with optimization and adaptive techniques.

- It has guaranteed continuity of the output surface.

- It's well-suited to mathematical analysis.

7.5.3   Conversion of Mamdani to Sugeno System

The MATLAB command-line function mam2sug can be used to convert a Mamdani system into a Sugeno system (not necessarily with a single output) with constant output membership functions. It uses the centroid associated with all of the output membership functions of the Mamdani system. The following Mamdani system mam21.fis is taken as the input.

```
% mam21.fis $
  [System]
Name = 'mam21'
Type = 'mamdani'
NumInputs = 2
NumOutputs = 1
NumRules = 4
AndMethod = 'min'
OrMethod = 'max'
ImpMethod = 'min'
AggMethod = 'max'
DefuzzMethod = 'centroid'

  [Input1]
Name = 'angle'
Range = [-5 5]
```

```
NumMFs = 2
MF1='small':'gbellmf',[5 8 -5 0]
MF2='big':'gbellmf',[5 8 5 0]

 [Input2]

Name = 'velocity'
Range = [-5 5]
NumMFs = 2
MF1='small':'gbellmf',[5 2 -5 0]
MF2='big':'gbellmf',[5 2 5 0]

 [Output1]
Name = 'force'
Range = [-5 5]
NumMFs = 4
MF1='negBig':'gbellmf',[1.67 8 -5 0]
MF2='negSmall':'gbellmf',[1.67 8 -1.67 0]
MF3='posSmall':'gbellmf',[1.67 8 1.67 0]
MF4='posBig':'gbellmf',[1.67 8 5 0]

 [Rules]
1 1, 1 (1) : 1
1 2, 2 (1) : 1
2 1, 3 (1) : 1
2 2, 4 (1) : 1
```

The code implementing the mam2sug function is shown below:

```
mam_fismat = readfis('mam21.fis');
sug_fismat = mam2sug(mam_fismat);
subplot(1,2,1);
gensurf(mam_fismat, [1 2], 1);
title('Mamdani system (Output)');
subplot(1,2,2);
gensurf(sug_fismat, [1 2], 1);
title('Sugeno system (Output)');
```

The output in graphical format is shown in Figure 7.27.

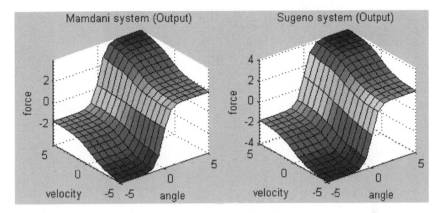

FIGURE 7.27: Illustration of Mamdani to Sugeno Conversion

7.6 Fuzzy Automata and Languages

Due to their diverse behavior application as a discrete model, cellular automata have created more interest in the implementation for several processes. Wolfram's discussion on universality and complexity in cellular automata described four classes of behavior for automata: class 1 for homogeneous stable behavior, class 2 for simple periodic patterns, class 3 for chaotic aperiodic behavior, and class 4 for complex behavior which generates local structures. Recent work by Cattaneo, Flocchini, Mauri, Quaranta Vogliotti, and Santoro introduced an alternate classification scheme that can be obtained by generalizing Boolean cellular automata to a fuzzy automata and then observing the qualitative behavior of a Boolean window embedded in a fuzzy domain. These automata can be classified according to whether the Boolean behavior dominates, a mixture of Boolean and Fuzzy behavior appears, or purely fuzzy behavior appears. The fuzzy behavior can further be divided into subclasses with homogeneous and non-homogeneous behavior. Classical cellular automata consist of an array of states, typically selected from a finite set, along with local rules for updating the array of states. Some of the simplest automata are defined on a one dimensional lattice of cells, are Boolean (two state), and base the future state of a cell upon the states in a 3-cell neighborhood consisting of the cell and its left and right neighbors. Let denote the value, 0 or 1, of the cell at position i at time t.

Table 7.1 shows the behavior of an example automation. It is called Rule 17 since the list of results, , gives the binary digits of 17. Rule 17

TABLE 7.1: The Definition of Rule 17

a_{t-1}^t	a_i^t	a_{t+1}^t	a_i^{t+1}
0	0	0	1
0	0	1	0
0	1	0	0
0	1	1	0
1	0	0	1
1	0	1	0
1	1	0	0
1	1	1	0

can be readily transformed into disjunctive normal form by selecting a variable or its negation for each of the three input states for each possible 1 output. Thus, we get

$$a_i^{t+1} = (a_{i-1}^{-t} \wedge a_i^{-t} \wedge a_{i+1}^{-t}) \vee (a_{i-1}^{-t} \wedge a_i^{-t} \wedge a_{i+1}^{-t})$$

where we denote logical "not" with an overbar. Note that we can capture the essential details of this representation with the matrix

$$\begin{pmatrix} 0 & 0 & 0 \\ 1 & 0 & 0 \end{pmatrix}$$

The rows of the matrix are the two input rows in Table 7.1 that have output value 1.

This definition is then converted into a fuzzy rule by reinterpreting the logical functions in this formula. In fuzzy automata 0 is imagined as false or dead, and 1 is imagined as true or alive and therefore logical "not x" is consistent with 1-x. The logical operation OR between variables x and y (x OR y) can be expressed as min 1, x+y while the logical AND operation (x AND y) is represented by multiplication: x * y. This admits the users to view the disjunctive normal form for Rule 17 as a formula where cells have values in the interval [0,1] that is consistent with the Boolean interpretation on the endpoints 0 and 1. Thus, a fuzzy automation is incurred.

7.7 Fuzzy Control

Most of the intelligent systems are compared with the analogy of biological systems. The intelligent systems perform operations similar to human beings like controlling tasks, recognizing patterns, or making decisions. Though the machines are capable of performing human inelligence, there exists a mismatch between humans and machines. Humans reason facts in an uncertain, imprecise, and fuzzy way, whereas machine reasoning are based on binary reasoning. Fuzzy logic is a means to create machines to be more intelligent by encouraging them to reason in a fuzzy manner like humans.

Controllers that combine intelligent and conventional techniques are generally used in the intelligent control of complex dynamic systems. Therefore, embedded fuzzy controllers automate the traditional human control activity.

Traditional control approach requires modeling of the physical reality which involves three basic models as follows:

Experimental Method

Based on the random experiments conducted on a process, an input-output relation can be characterized in the form of an input-output table. These input-output values can be plotted in a graphical manner also. With a thorough knowledge of the input-output, a user can design a controller. While designing a controller in this method, there are several disadvantages like the process equipment may not be available for experimentation, the procedure would usually be very costly, and for a large number of input values it is impractical to measure the output and interpolation between measured outputs, etc. Users should be aware while determining the expected ranges of inputs and outputs to ensure that they fall within the range of the measuring instruments available.

Mathematical Modeling

Any controlled process can be expressed in terms of difference or differential equations. These equations are solved by applying Laplace transforms and z-transforms. Usually to create simple mathematical models the process is assumed to be linear and time invariant, which implies that the input output relations are linear and the output does not vary with time with respect to the input value. Linear processes are worth while since they render an expert view of the process. Likewise, there is no general theory to obtain analytic solution of nonlinear differential equations and consequently no comprehensive analysis tools are available for nonlinear dynamic systems.

Another assumption is that the process parameters do not change in time (that is, the system is time-invariant) despite system component deterioration and environmental changes. The following problems arise in developing a meaningful and realistic mathematical description of an industrial process:

(1) Poorly understood phenomena

(2) Inaccurate values of various parameters

(3) Model complexity

Heuristics Method

Based on previously obtained knowledge a process is modeled and is referred to as the heuristic method. A heuristic rule is a logical implication of the form: If ¡condition¿ Then ¡consequence¿, or in a typical control situation: If ¡condition¿ Then ¡action¿. Rules associate conclusions with conditions. More or less, the heuristic method is similar to the experimental method of constructing a table of inputs and corresponding output values where instead of having crisp numeric values of input and output variables, one uses fuzzy values:

$$\text{IF input\_voltage} = \text{Large}$$

$$\text{HEN output\_voltage} = \text{Medium.}$$

The advantages of the heuristic method are that:

(1) There is no need to assume the system to be linear.

(2) Heuristic rules can be integrated to the control strategies of human operators.

Fuzzy control schemes are based on experience and experiments rather than from mathematical models and, therefore, linguistic implementations are implemented much faster. Fuzzy control schemes require a large number of inputs, most of which are relevant only for some special conditions. Such inputs are activated only when the related condition prevails. By adding a few computational overheads to the fuzzy rules, the rule base structure remains understandable, leading to efficient coding and system documentation.

7.7.1 Fuzzy Controllers

Almost all commercial fuzzy products such as washing machines, dishwashers, etc. are rule-based systems that get current information in the feedback loop from the device as it operates and control the operation of the device. A fuzzy logic system has four blocks as shown in Figure 7.28. The crisp values are taken as input information from the device and they are are converted into fuzzy values. Each input fuzzy set corresponds to a

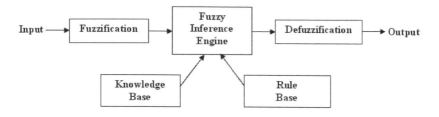

FIGURE 7.28: Fuzzy Controller Block Diagram

fuzzification block. The universe of discourse of the input variables determines the required grading for correct per-unit operation. The grading is really very important since the fuzzy system can be retrofitted with other devices or ranges of operation by just altering the grading of the input and output. The decision-making-logic influences the performance of the fuzzy logic operations, and determines the outputs of each fuzzy rule with reference to the knowledge base. Those are combined and converted to crispy values with the defuzzification block. The output crisp value can be calculated by the center of gravity or the weighted average.

In order to process the input to get the output reasoning there are six steps involved in the creation of a rule based fuzzy system:

1. Identify the inputs and their ranges and name them.

2. Identify the outputs and their ranges and name them.

3. Create the degree of fuzzy membership function for each input and output.

4. Construct the rule base that the system will operate under.

5. Decide how the action will be executed by assigning strengths to the rules.

6. Combine the rules and defuzzify the output.

Fuzzy systems are leading in the areas of consumer products, industrial and commercial systems, and decision support systems. Rather than applying complex mathematical equations, fuzzy logic applies linguistic descriptions to delineate the relationship between the input data and the output action. Fuzzy logic offers an accessible and user-friendly front-end to develop control applications thereby helping designers to focus on the functional objectives, and not on the mathematical theory involved.

7.7.2   A Fuzzy Controller in MATLAB

This section illustrates the steps in implementing a fuzzy controller using MATLAB. The standard Fuzzy Logic Controller routine with the five parts such as membership functions, rule base, fuzzification, fuzzy inference, and defuzzification is as follows:

```
function [u]=Standard_FLC(ke,kec,ku,e,ec)
% Standard FLC function
% Inputs of this function:
%' 1) [ke,kec] are scales of inputs of FLC
% 2) [ku] is output scale of FLC
% 3) e,ec are error and change in error
% 4) Rule_Base is a matrix, it is a m file in
the
% Output of this function:
% 1) u is output of standard FLC
%
%e.g. standard_FLC(1,1,1,3,5,Rule_Base('table'))

% Part_I: Member-ship Functions
%% Creates a new Mamdani-style FIS structure
a=newfis('ST_FLC');

%% to add input parameter of e into FIS
a=addvar(a,'input','e',[-5 5]);

%%% fuzzify e to E
a=addmf(a,'input',1,'NB','trapmf',[-5 -5 -4
-2]);
a=addmf(a,'input',1,'NS','trimf',[-4 -2 0]);
a=addmf(a,'input',1,'ZE','trimf',[-2 0 2]);
a=addmf(a,'input',1,'PS','trimf',[0 2 4]);
a=addmf(a,'input',1,'PB','trapmf',[2 4 5 5]);
figure(1);
plotmf(a,'input',1);

  %%% to add input parameter of ec into FIS
a=addvar(a,'input','ec',[-6.25 6.25]);
%%% fuzzify ec to EC
a=addmf(a,'input',2,'NB','trapmf',[-6.25 -6.25
-5 -2.5]);
a=addmf(a,'input',2,'NS','trimf',[-5 -2.5 0]);
```

```
a=addmf(a,'input',2,'ZE','trimf',[-2.5 0 2.5]);
a=addmf(a,'input',2,'PS','trimf',[0 2.5 5]);
a=addmf(a,'input',2,'PB','trapmf',[2.5 5 6.25
6.25]);
figure(2);
plotmf(a,'input',2);

%%% to add input parameter of u into FIS
a=addvar(a,'output','Fd',[-1.165 1.165]);
%%% fuzzify u to U
a=addmf(a,'output',1,'NB','trapmf',[-2 -2 -1
-0.67]);
a=addmf(a,'output',1,'NM','trimf',[-1 -0.67
-0.33]);
a=addmf(a,'output',1,'NS','trimf',[-0.67 -0.33
0]);
a=addmf(a,'output',1,'ZE','trimf',[-0.33 0
0.33]);
a=addmf(a,'output',1,'PS','trimf',[0 0.33
0.67]);
a=addmf(a,'output',1,'PM','trimf',[0.33 0.67
1]);
a=addmf(a,'output',1,'PB','trapmf',[0.67 1 2
2]);

figure(3);
 plotmf(a,'output',1);
```

% Part II: Rule-bases

```
[Rule_Base]=[ 1 1 7 1 1
             1 2 6 1 1
             1 3 5 1 1
             1 4 4 1 1
             1 5 4 1 1
             2 1 6 1 1
             2 2 5 1 1
             2 3 4 1 1
             2 4 4 1 1
             2 5 4 1 1
             3 1 5 1 1
             3 2 4 1 1
             3 3 4 1 1
```

```
                3 5 3 1 1
                4 1 4 1 1
                4 2 4 1 1
                4 3 4 1 1
                4 4 3 1 1
                4 5 2 1 1
                5 1 4 1 1
                5 2 4 1 1
                5 3 3 1 1
                5 4 2 1 1
                5 5 1 1 1
     %%% add Rule_base into FIS
     a=addrule(a,Rule_Base);
     %
showfis(a);
showrule(a);
figure(8);
gensurf(a);

% Part III Fuzzify

%%% from e to E , ec to EC
E=ke*e;
EC=kec*ec;
%%% confine E
  if E >4
     E=4;
  elseif E<-4
        E=-4;
  end
  %%% confine EC
  if EC >5
    EC=5;
  elseif EC<-5
      EC=-5;
  end

  % Part IV Fuzzy Inference

  FLC_input=[E,EC];

  U=evalfis(FLC_input,a);
    % Part V Defuzzify
```

```
u=ku*U;
end
```

The above code can be used to create any application by modifying the inputs and outputs according to the requirements of the application.

Summary

Fuzzy controllers are implemented in a more specialized way, but they were originally developed from the concepts and definitions presented in this chapter, especially inference and implication. In a fuzzy controller the data passes through a preprocessing block, a controller, and a postprocessing block. Preprocessing consists of a linear or nonlinear scaling as well as a quantization in case the membership functions are discretized (vectors); if not, the membership of the input can just be looked up in an appropriate function. When designing the rule base, the designer needs to consider the number of term sets, their shape, and their overlap. The rules themselves must be determined by the designer, unless more advanced means like self-organization or neural networks are available. There is a choice between multiplication and minimum in the activation. There is also a choice regarding defuzzification; center of gravity is probably most widely used. A fuzzy expert system is an expert system that uses fuzzy logic instead of Boolean logic. In other words, a fuzzy expert system is a collection of membership functions and rules that are used to reason about data. The collective rules are referred to as knowledge base. This knowledge is applied to the input variables and the outputs are determined, the process is known as inferencing. Inference goes through the steps starting from fuzzification to defuzzification as discussed in this chapter. The chapter also described MATLAB implementations based on fuzzy controllers and fuzzy inference.

Review Questions

1. Differentiate Mamdani and Takagi–Sugeno fuzzy rules.

2. Mention the components of a fuzzy expert system.

3. What is defuzzification? Explain the most important defuzzification methods.

4. Define Rule 17. How is it used in Fuzzy Automation?

5. What are the three methods used for a traditional control approach?

6. Describe fuzzy controllers with suitable diagram.

7. Write a MATLAB code to implement the fuzzification process.

8. How will you convert Mamdani rule base into a Sugeno rule base using MATLAB?

9. Mention a few advantages and disadvantages of Mamdani and Sugeno methods.

10. What are fuzzy expert systems?

11. Define knowledge base.

12. Define inference and implication.

Chapter 8

MATLAB Illustrations on Fuzzy Systems

This chapter illustrates some applications of Fuzzy systems such as Fuzzy Washing Machine, Fuzzy Control System, and approximation of sinusoidal functions in MATLAB.

8.1 Illustration 1: Application of Fuzzy Controller Using MATLAB — Fuzzy Washing Machine

Today many household appliances have fuzzy logic built into them to make their use easier. Fuzzy logic is found in shower heads, rice cookers, vacuum cleaners, and just about everywhere. In this section we will look at a simplified model of a fuzzy washing machine.

Like a real washing machine would, the model first tests how dirty the laundry is. Once it knows how dirty the laundry is, it can easily calculate how long it should wash it. To calculate this it uses the graph shown in Figure 8.1:

First it always takes a base of 10 minutes. It does this so that people are happy with its work even if they put completely clean laundry in to wash. It then calculates to what degree it is dirty. If it is 100% dirty it adds two minutes per piece of laundry. Of course a real washing machine would just do these calculations in the end, but this model does it for each individual piece so you can keep track of what is going on easier.

On the graph in Figure 8.1, the base of ten minutes is shown. The point 0,0 is where the laundry is completely clean; non-dirty and non-greasy. The point 0,1 is where the laundry is non-greasy, but dirty. The point 1,0 is where it is greasy but not dirty and 1,1 is greasy and dirty. The washing machine adds 2 minutes per piece for 100% dirty or 100 % greasy and 4 minutes for 100% dirty and greasy.

There has been a boom for fuzzy machines in the last two decades. This is not only because they can do things, which humans had to do

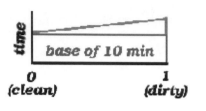 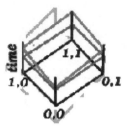

FIGURE 8.1: Graphical Representation of the Time in a Fuzzy Washing Machine

themselves until now, but also because they're much cheaper to build than normal binary machines.

MATLAB Code

```
function varargout = gui1(varargin)
% GUI1, by itself, creates a new GUI1 or raises the
existing
% singleton.
%
% H = GUI1 returns the handle to a new GUI1 or the
handle to
% the existing singleton.
%
% GUI1('CALLBACK',hObject,eventData,handles,...)
calls the local
% function named CALLBACK in GUI1.M with the given
input arguments.
%
% GUI1('Property','Value',...) creates a new GUI1 or
raises the
% existing singleton*. Starting from the left,
property value pairs are
% applied to the GUI before gui1_OpeningFunction
gets called. An
% unrecognized property name or invalid value makes
property application
```

```
% stop. All inputs are passed to gui1_OpeningFcn via
varargin.
%
%
```

Initialization code

```
gui_Singleton= 1;
gui_State = struct('gui_Name', mfilename, ...
    'gui_Singleton', gui_Singleton, ...
        'gui_OpeningFcn', @gui1_OpeningFcn, ...
        'gui_OutputFcn', @gui1_OutputFcn, ...
        'gui_LayoutFcn', [ ] , ...
        'gui_Callback', [ ]);
if nargin & & ischar(varargin1)
gui_State.gui_Callback = str2func(varargin1);
end
if nargout
  [varargout1:nargout] = gui_mainfcn(gui_State,
  varargin:);
else
gui_mainfcn(gui_State, varargin:);
end
```

% End of Initialization code

% - - - Executes just before gui1 is made visible

```
function gui1_OpeningFcn(hObject, eventdata,
handles, varargin)
% This function has no output args, see OutputFcn.
% hObject handle to figure
% eventdata reserved - to be defined in a future
version of MATLAB
% handles structure with handles and user data (see
GUIDATA)
```

```
% varargin command line arguments to gui1 (see
VARARGIN)

% Choose default command line output for gui1
handles.output = hObject;

% Update handles structure
guidata(hObject, handles);

% UIWAIT makes gui1 wait for user response (see
UIRESUME)
% uiwait(handles.figure1);
```

% - - - Outputs from this function are returned to the command line.

```
function varargout = gui1_OutputFcn(hObject,
eventdata, handles)
% varargout cell array for returning output args
(see VARARGOUT);
% hObject handle to figure
% eventdata reserved - to be defined in a future
version of MATLAB
% handles structure with handles and user data (see
GUIDATA)
% Get default command line output from handles
structure
varargout1 = handles.output;
```

% - - - Executes on slider movement.

```
function slider1_Callback(hObject, eventdata,
handles)
% hObject handle to slider1 (see GCBO)
% eventdata reserved - to be defined in a future
version of MATLAB
% handles structure with handles and user data (see
GUIDATA)
```

```
% Hints: get(hObject,'Value') returns position of
slider
% get(hObject,'Min') and get(hObject,'Max') to
determine range of slider
x = get(handles.slider1,'Value')
set(handles.text1,'String',num2str(x))
```

% - - - Executes during object creation, after setting all properties.

```
function slider1_CreateFcn(hObject, eventdata,
handles)
% hObject handle to slider1 (see GCBO)
% eventdata reserved - to be defined in a future
version of MATLAB
% handles empty - handles not created until after
all CreateFcns called
```

```
% Hint:slider controls usuallyhave a light gray background,
  change
% 'usewhitebg' to 0 to use default.
usewhitebg = 1;
if usewhitebg
set(hObject,'BackgroundColor',[.9 .9 .9]);
else
set(hObject,'BackgroundColor',get(0,
  'defaultUicontrolBackgroundColor'));
end
```

% - - - Executes during object creation, after setting all properties.

```
function slider2_CreateFcn(hObject, eventdata,
handles) % hObject handle to slider2 (see GCBO)
% eventdata reserved - to be defined in a future
version of MATLAB
```

```
% handles empty - handles not created until after
all CreateFcns called
% Hint: slider controls usually have a light gray
background,
   change
% 'usewhitebg' to 0 to use default.
usewhitebg = 1;
if usewhitebg

 set(hObject,'BackgroundColor',[.9 .9 .9]);
else
 set(hObject,'BackgroundColor',get(0,'
  defaultUicontrolBackgroundColor'));
end
```

% - - - Executes on button press in pushbutton1.

```
function pushbutton1_Callback(hObject, eventdata,
handles)
% hObject handle to pushbutton1 (see GCBO)
% eventdata reserved - to be defined in a future
version of MATLAB
% handles structure with handles and user data (see
GUIDATA)
p = get(handles.slider1,'Value');
q = get(handles.slider2,'Value');

a=newfis('wash_machine1');
a = addvar(a,'input','x',[0 100]);
a = addmf(a,'input',1,'A1','trimf',[0 0 50]);
a = addmf(a,'input',1,'A2','trimf',[0 50 100]);
a = addmf(a,'input',1,'A3','trimf',[50 100 100]);

a = addvar(a,'input','y',[0 100]);
a = addmf(a,'input',2,'B1','trimf',[0 0 50]);
a = addmf(a,'input',2,'B2','trimf',[0 50 100]);
a = addmf(a,'input',2,'B3','trimf',[50 100 100]);
a = addvar(a,'output','z',[0 60]);
a = addmf(a,'output',1,'C1','trimf',[0 8 12]);
a = addmf(a,'output',1,'C2','trimf',[8 12 20]);
```

```
a = addmf(a,'output',1,'C3','trimf',[12 20 30]);
a = addmf(a,'output',1,'C4','trimf',[30 40 50]);
a = addmf(a,'output',1,'C5','trimf',[50 60 70]);

rulelist=[ ...
1 1 1 1 1
2 2 3 1 1
3 3 5 1 1
1 2 3 1 1
1 3 3 1 1
2 1 2 1 1
2 3 4 1 1
3 1 2 1 1
3 2 4 1 1 ];

a=addrule(a,rulelist);

z = evalfis([p q], a)

set(handles.text3,'String',num2str(z))
set(handles.slider4,'Value',z)
```

% --- Executes on slider movement.

```
function slider3_Callback(hObject, eventdata,
handles)
% hObject handle to slider3 (see GCBO)
% eventdata reserved - to be defined in a future
version of MATLAB
% handles structure with handles and user data (see
GUIDATA)

% Hints: get(hObject,'Value') returns position of
slider
% get(hObject,'Min') and get(hObject,'Max') to
determine range of slider
```

**% - - - Executes during object creation, after
setting all properties.**

```
function slider3_CreateFcn(hObject, eventdata,
handles)
% hObject handle to slider3 (see GCBO)
```

```
% eventdata reserved - to be defined in a future
version of MATLAB
% handles empty - handles not created until after
all CreateFcns called

% Hint: slider controls usually have a light gray
background, change
% 'usewhitebg' to 0 to use default.
usewhitebg = 1;
if usewhitebg

 set(hObject,'BackgroundColor',[.9 .9 .9]);
else
 set(hObject,'BackgroundColor',get(0,
 'defaultUicontrolBackgroundColor'));
end
```

% - - - Executes on slider movement.

```
function slider4_Callback(hObject, eventdata,
handles)
% hObject handle to slider4 (see GCBO)
% eventdata reserved - to be defined in a future
version of MATLAB
% handles structure with handles and user data (see
GUIDATA)

% Hints: get(hObject,'Value') returns position of
slider
% get(hObject,'Min') and get(hObject,'Max') to
determine range of slider
```

 **% - - - Executes during object creation, after
setting all properties.**

```
function slider4_CreateFcn(hObject, eventdata,
handles)
% hObject handle to slider4 (see GCBO)
% eventdata reserved - to be defined in a future
version of MATLAB
```

```
% handles empty - handles not created until after
all CreateFcns called

% Hint: slider controls usually have a light gray
background, change
% 'usewhitebg' to 0 to use default.
usewhitebg = 1;
if usewhitebg

 set(hObject,'BackgroundColor',[.9 .9 .9]);
else
 set(hObject,'BackgroundColor',get(0,
 'defaultUicontrolBackgroundColor'));
end
```

Output

Figures 8.2 and 8.3 show the Graphical User Interface of the Fuzzy Washing Machine when the dirtiness is 70.2041% and 22.449% respectively.

Simulation Results

The inputs variables "x" and "y" denote the type of dirt and the percentage of dirtiness respectively. The output variable "z" denotes the time taken by the washing machine to remove the dirt. Some observations are tabulated in Table 8.1

Thus the above section discussed the operation of a fuzzy washing machine with the GUI and the fuzzy controller designed in MATLAB along with some concluding observations.

TABLE 8.1:   Observations of the Fuzzy Washing Machine

S No	Type of Dirt "x"	Percentage of' Dirtiness "y"	Washing Time "y"
1	72.58	39.59	29.07
2	14.92	03.67	10.39
3	12.50	40.00	17.63
4	26.62	73.06	30.43
5	85.46	87.35	40.27
6	96.37	97.55	49.63

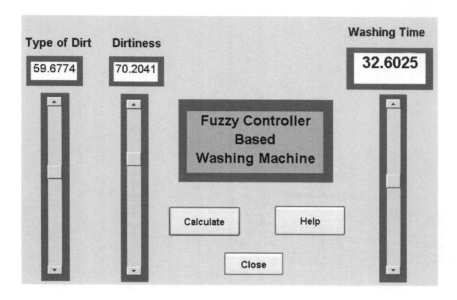

FIGURE 8.2: GUI of the Fuzzy Washing machine when the dirtiness is 70.2041%

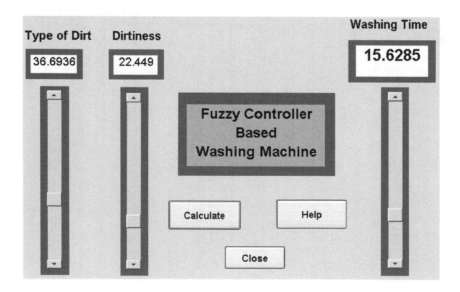

FIGURE 8.3: GUI of the Fuzzy Washing machine when the dirtiness is 22.449%

8.2 Illustration 2 - Fuzzy Control System for a Tanker Ship

This program simulates a fuzzy control system for a tanker ship. It has a fuzzy controller with two inputs, the error in the ship heading (e) and the change in that error (c). The output of the fuzzy controller is the rudder input (delta). We want the tanker ship heading (psi) to track the reference input heading (psi_r). We simulate the tanker as a continuous time system that is controlled by a fuzzy controller that is implemented on a digital computer with a sampling interval of T.

This program can be used to illustrate:

i) Writing code a fuzzy controller (for two inputs and one output, illustrating some approaches to simplify the computations, for triangular membership functions, and either center-of-gravity or center-average defuzzification).

ii) Tuning the input and output gains of a fuzzy controller.

iii) Effect of changes in plant conditions ("ballast" and "full").

iv) How sensor noise (heading sensor noise), plant disturbances (wind hitting the side of the ship), and plant operating conditions (ship speed) can affect performance.

v) How improper choice of the scaling gains can result in oscillations (limit cycles).

vi) How an improper choice of the scaling gains (or rule base) can result in an unstable system.

vii) The shape of the nonlinearity implemented by the fuzzy controller by plotting the input-output map of the fuzzy controller.

MATLAB Code

```
clear% Clear all variables in memory
% Initialize ship parameters
% (can test two conditions, ``ballast" or ``full"):

ell=350;    % Length of the ship (in meters)
```

```
u=5;     % Nominal speed (in meters/sec)
% u=3;   % A lower speed where the ship is more
difficult to control
abar=1;   % Parameters for nonlinearity
bbar=1;

% The parameters for the tanker under ``ballast" conditions
% (a heavy ship) are:

K_0=5.88;
tau_10=-16.91;
tau_20=0.45;
tau_30=1.43;

% The parameters for the tanker under ``full"
 conditions (a ship
% that weighs less than one under ``ballast" conditions)
 are:

% K_0=0.83;
% tau_10=-2.88;
% tau_20=0.38;
% tau_30=1.07;

% Some other parameters are:

K=K_0*(u/ell);
tau_1=tau_10*(ell/u);
tau_2=tau_20*(ell/u);
tau_3=tau_30*(ell/u);
% Initialize parameters for the fuzzy controller

nume=11; % Number of input membership functions for the e
 % universe of discourse (can change this but must
 also
 % change some variables below if you make such
 a change)
numc=11; % Number of input membership functions for the c
 % universe of discourse (can change this but must
 also
 % change some variables below if you make such
 a change)
```

```
% Next, we define the scaling gains for tuning membership
functions for
% universes of discourse for e, change in e (what we
  call c) and
% delta. These are g1, g2, and g0, respectively
% These can be tuned to try to improve the performance.

% First guess:
g1=1/pi;,g2=100;,g0=8*pi/18; % Chosen since:
 % g1: The heading error is at most 180 deg (pi
 rad)
 % g2: Just a guess - that ship heading will
 change at most
 % by 0.01 rad/sec (0.57 deg/sec)
 % g0: Since the rudder is constrained to move
 between +-80 deg
% Tuning:
g1=1/pi;,g2=200;,g0=8*pi/18; % Try to reduce the overshoot
g1=2/pi;,g2=250;,g0=8*pi/18; % Try to speed up the response
% g1=2/pi;,g2=0.000001;,g0=2000*pi/18; % Values tuned to
 get oscillation
% (limit cycle) for COG, ballast, and nominal speed with
no sensor
% noise or rudder disturbance):
% g1: Leave as before
% g2: Essentially turn off the derivative gain
%   since this help induce an oscillation
% g0: Make this big to force the limit cycle
% In this case simulate for 16,000 sec.
% g1=2/pi;,g2=250;,g0=-8*pi/18; % Values tuned to get
 an instability
% g0: Make this negative so that when there
% is an error the rudder will drive the
% heading in the direction to increase the error

% Next, define some parameters for the membership functions

we=0.2*(1/g1);
 % we is half the width of the triangular input
 membership
 % function bases (note that if you change g0, the
 base width
 % will correspondingly change so that we always
```

```
end
% up with uniformly distributed input membership
functions)
% Note that if you change nume you will need to
adjust the
% ``0.2" factor if you want membership functions
that
% overlap in the same way.
wc=0.2*(1/g2);
% Similar to we but for the c universe of
discourse
base=0.4*g0;
% Base width of output membership fuctions of
the fuzzy
% controller

% Place centers of membership functions of the fuzzy
controller:

% Centers of input membership functions for the e
universe of
% discourse of fuzzy controller (a vector of centers)
ce=[-1 -0.8 -0.6 -0.4 -0.2 0 0.2 0.4 0.6 0.8 1]*(1/g1);

% Centers of input membership functions for the c
universe of
% discourse of fuzzy controller (a vector of centers)
cc=[-1 -0.8 -0.6 -0.4 -0.2 0 0.2 0.4 0.6 0.8 1]*(1/g2);

% This next matrix specifies the rules of the fuzzy
controller.
% The entries are the centers of the output membership
functions.
% This choice represents just one guess on how to
synthesize
% the fuzzy controller. Notice the regularity
% of the pattern of rules. Notice that it is scaled by g0,
the
% output scaling factor, since it is a normalized rule
base.
% The rule base can be tuned to try to improve performance.
```

```
rules=
[1      1      1      1      1      1      0.8     0.6     0.3    0.1    0;
 1      1      1      1      1    0.8     0.6      0.3    0.1    0    -0.1;
 1      1      1      1  0.8  0.6      0.3     0.1     0    -0.1   -0.3;
 1      1      1    0.8  0.6  0.3      0.1      0    -0.1   -0.3   -0.6;
 1      1    0.8    0.6  0.3  0.1     0    -0.1   -0.3   -0.6   -0.8;
 1    0.8    0.6    0.3  0.1   0     0.1   -0.3   -0.6   -0.8    -1;
 0.8    0.6  0.3    0.1   0  -0.1   -0.3   -0.6   -0.8   -1     -1;
 0.6    0.3  0.1     0  -0.1  -0.3   -0.6   -0.8   -1     -1     -1;
 0.3    0.1   0   -0.1  -0.3  -0.6   -0.8   -1     -1     -1     -1;
 0.1    0   -0.1  -0.3  -0.6  -0.8   -1     -1     -1     -1     -1;
 0    -0.1   -0.3  -0.8  -1    -1     -1     -1     -1     -1]*g0;


% Now, you can proceed to do the simulation or simply view
 the nonlinear
% surface generated by the fuzzy controller.

flag1=input('\ n Do you want to simulate the \ n fuzzy
 control system \ n for the
tanker? \ n (type 1 for yes and 0 for no) ');

if flag1==1,

% Next, we initialize the simulation:

t=0; % Reset time to zero
index=1; % This is time's index (not time, its index).
tstop=4000;  % Stopping time for the simulation (in
 % seconds)
step=1;  % Integration step size
T=10;% The controller is implemented in discrete
 % time and
   % this is the sampling time for the
   % controller.
   % Note that the integration step size
   % and the sampling
   % time are not the same. In this way
   % we seek to simulate
   % the continuous time system via the
   % Runge-Kutta method and
   % the discrete time fuzzy controller as
   % if it were
```

```
% implemented by a digital computer.
% Hence, we sample
% the plant output every T seconds
% and at that time
% output a new value of the controller
% output.
counter=10;  % This counter will be used to count the
% number of integration
% steps that have been taken in the current
% sampling interval.
% Set it to 10 to begin so that it will
% compute a fuzzy controller
% output at the first step.
% For our example, when 10 integration steps
% have been
% taken we will then we will sample the ship
% heading
% and the reference heading and compute a
% new output
% for the fuzzy controller.
eold=0;  % Initialize the past value of the error (for
% use
% in computing the change of the error, c).
% Notice
% that this is somewhat of an arbitrary
% choice since
% there is no last time step. The same
% problem is
% encountered in implementation.
x=[0;0;0];   % First, set the state to be a vector
x(1)=0;  % Set the initial heading to be zero
x(2)=0;  % Set the initial heading rate to be zero.
% We would also like to set x(3) initially
% but this
% must be done after we have computed the
% output
% of the fuzzy controller. In this case,
% by
% choosing the reference trajectory to be
% zero at the beginning and the other
% initial conditions
% as they are, and the fuzzy controller as
% designed,
```

```
% we will know that the output of the
% fuzzy controller
% will start out at zero so we could have
% set
% x(3)=0 here. To keep things more general,
% however,
% we set the intial condition immediately
% after
% we compute the first controller output
% in the
% loop below.
psi_r_old=0;  % Initialize the reference trajectory

% Next, we start the simulation of the system.
% This is the main loop for the simulation of fuzzy
% control system.

while t <= tstop

% First, we define the reference input psi_r (desired
% heading).

if t<100, psi_r(index)=0; end
 % Request heading of 0 deg
if t>=100, psi_r(index)=45*(pi/180); end
   % Request heading of 45 deg
if t>2000, psi_r(index)=0; end
 % Then request heading of 0 deg
% if t>4000, psi_r(index)=45*(pi/180); end
   % Then request heading of 45 deg
% if t>6000, psi_r(index)=0; end
 % Then request heading of 0 deg
% if t>8000, psi_r(index)=45*(pi/180); end
   % Then request heading of 45 deg
% if t>10000, psi_r(index)=0; end
 % Then request heading of 0 deg
% if t>12000, psi_r(index)=45*(pi/180); end
   % Then request heading of 45 deg

% Next, suppose that there is sensor noise for the
% heading sensor with that is additive, with a uniform
% distribution on [- 0.01,+0.01] deg.
% s(index)=0.01*(pi/180)*(2*rand-1);
```

```
s(index)=0;    % This allows us to remove the noise.

psi(index)=x(1)+s(index);
% Heading of the ship (possibly with sensor noise).

if counter == 10,
    % When the counter reaches 10 then execute the
% fuzzy controller
counter=0;     % First, reset the counter

% Fuzzy controller calculations:
% First, for the given fuzzy controller inputs we
% determine the extent at which the error membership
% functions of the fuzzy controller are on
% (this is the fuzzification part).

c_count=0;,e_count=0;
    % These are used to count the
    % number of non-zero mf
    % certainities of e and c
e(index)=psi_r(index)-psi(index);
% Calculates the error input for
% the fuzzy controller

c(index)=(e(index)-eold)/T; % Sets the value of c

eold=e(index);    % Save the past value of e for
 % use in the above computation the
 % next time around the loop

% The following if-then structure fills the vector mfe
% with the certainty of each membership fucntion of e
% for the current input e. We use triangular
% membership functions.

  if e(index)<=ce(1)
% Takes care of saturation of the left-most
    % membership function
 mfe=[1 0 0 0 0 0 0 0 0 0 0];
% i.e., the only one on is the

 % left-most one
  e_count=e_count+1;,e_int=1;
```

```
% One mf on, it is the
% left-most one.
  elseif e(index) >=ce(nume)
% Takes care of saturation
 % of the right-most mf
  mfe=[0 0 0 0 0 0 0 0 0 0 1];
  e_count=e_count+1;,e_int=nume;
% One mf on, it is the
% right-most one
  else
% In this case the input is on the middle
% part of the universe of discourse for e
% Next, we are going to cycle through the mfs
% to find all that are on
 for i=1:nume
 if e(index)<=ce(i)
mfe(i)=max([0 1+(e(index)-ce(i))/we]);
% In this case the input is to the
% left of the center ce(i) and we compute
% the value of the mf centered at ce(i)
% for this input e
  if mfe(i)~=0
     % If the certainty is not equal to zero then say
     % that have one mf on by incrementing our count
  e_count=e_count+1;
  e_int=i; % This term holds the index last entry
     % with a non-zero term
  end
  else
 mfe(i)=max([0,1+(ce(i)-e(index))/we]);
                    % In this case the input is to the
                    % right of the center ce (i)
  if mfe(i)~=0
  e_count=e_count+1;
  e_int=i; % This term holds the index of the
                    % last entry with a non-zero ter
  end
  end
   end
end

% The following if-then structure fills the vector mfc
% with the certainty of each membership fucntion of the c
```

```
% for its current value (to understand this part of the
% code see the above similar code for computing mfe).
% Clearly, it could be more efficient to make a subroutine
% that performs these computations for each of the fuzzy
% system inputs.

 if c(index)<=cc(1)
     % Takes care of saturation of left-most mf
 mfc=[1 0 0 0 0 0 0 0 0 0 0];
 c_count=c_count+1;
 c_int=1;
 elseif c(index)>=cc(numc)
     % Takes care of saturation of the right-most mf
   mfc=[0 0 0 0 0 0 0 0 0 0 1];
   c_count=c_count+1;
   c_int=numc;
 else
  for i=1:numc
  if c(index)<=cc(i)
  mfc(i)=max([0,1+(c(index)-cc(i))/wc]);
                   if mfc(i)~ =0
                   c_count=c_count+1;
                   c_int=i;
  % This term holds last entry
      % with a non-zero term
  end
 end
 end
 end
```

```
% The next two loops calculate the crisp output using
% only the non- zero premise of error,e, and c. This
% cuts down computation time since we will only compute
% the contribution from the rules that are on (i.e., a
% maximum of four rules for our case). The minimum
% is center-of-gravity used for the premise
% (and implication for the defuzzification case).

num=0;
den=0;
  for k=(e_int-e_count+1):e_int
 % Scan over e indices of mfs that are on
    for l=(c_int-c_count+1):c_int
```

```
% Scan over c indices of mfs that are on
prem=min([mfe(k) mfc(l)]);
   % Value of premise membership function
% This next calculation of num adds up the numerator for
% the center of gravity defuzzification formula.
% rules(k,l) is the output center for the rule.
% base*(prem-(prem)^ 2/2) is the area of a symmetric
% triangle that peaks at one with base width ``base"
% and that is chopped off at a height of prem (since
% we use minimum to represent the implication).
% Computation of den is similar but without rules(k,l).
num=num+rules(k,l)*base*(prem-(prem)^ 2/2);
den=den+base*(prem-(prem)^ 2/2);
% To do the same computations, but for center-average
% defuzzification, use the following lines of code rather
% than the two above (notice that in this case we did not
% use any information about the output membership function
% shapes, just their centers; also, note that the
% computations are slightly simpler for the center-average
% defuzzificaton):
% num=num+rules(k,l)*prem;
% den=den+prem;
   end
    end

 delta(index)=num/den;
    % Crisp output of fuzzy controller that is the input
    % to the plant.
else
% This goes with the ``if" statement to check if the
% counter=10 so the next lines up to the next ``end"
% statement are executed whenever counter is not
% equal to 10

% Now, even though we do not compute the fuzzy
% controller at eachtime instant, we do want to
% save the data at its inputs and output at each time
% instant for the sake of plotting it. Hence, we need to
% compute these here (note that we simply hold the
% values constant):

e(index)=e(index-1);
c(index)=c(index-1);
```

```
delta(index)=delta(index-1);

end
% This is the end statement for the ``if counter=10"
  statement

% Now, for the first step, we set the initial
% condition for the third state x(3).

if t==0, x(3)=-(K*tau_3/(tau_1*tau_2))*delta(index); end

% Next, the Runge-Kutta equations are used to find the
% next state. Clearly, it would be better to use a Matlab
% ``function" for F (but here we do not, so we can have only
% one program).

   time(index)=t;
% First, we define a wind disturbance against the body
% of the ship that has the effect of pressing water
% against the rudder

% w(index)=0.5*(pi/180)*sin(2*pi*0.001*t); % This is an
% additive sine disturbance to the rudder input. It
% is of amplitude of 0.5 deg. and its period is 1000sec.
% delta(index)=delta(index)+w(index);

% Next, implement the nonlinearity where the rudder angle
% is saturated at +-80 degrees

if delta(index) >= 80*(pi/180), delta(index)=80*(pi/180);
end
if delta(index) <= -80*(pi/180), delta(index)=-80*(pi/180);
end
% Next, we use the formulas to implement the Runge-Kutta
% method (note that here only an approximation to the
% method is implemented where we do not compute the
% function at multiple points in the integration step size).

F=[ x(2) ;
  x(3)+ (K*tau_3/(tau_1*tau_2))*delta(index) ;
  -((1/tau_1)+(1/tau_2))*(x(3)+ (K*tau_3/(tau_1*tau_2))
  *delta(index))-...
  (1/(tau_1*tau_2))*(abar*x(2)^ 3 + bbar*x(2)) +
```

```
  (K/(tau_1*tau_2))*delta(index) ];

  k1=step*F;
  xnew=x+k1/2;

F=[ xnew(2) ;
  xnew(3)+ (K*tau_3/(tau_1*tau_2))*delta(index) ;
  -((1/tau_1)+(1/tau_2))*(xnew(3)
     + (K*tau_3/(tau_1*tau_2)) *delta(index))-...
(1/(tau_1*tau_2))*(abar*xnew(2)^ 3 + bbar*xnew(2)) +
  (K/(tau_1*tau_2))*delta(index) ];

  k2=step*F;
  xnew=x+k2/2;

F=[ xnew(2) ;
  xnew(3)+ (K*tau_3/(tau_1*tau_2))*delta(index) ;
  -((1/tau_1)+(1/tau_2))*(xnew(3)+
  (K*tau_3/(tau_1*tau_2))*delta(index))-...
  (1/(tau_1*tau_2))*(abar*xnew(2)^ 3 + bbar*xnew(2))
  + (K/(tau_1*tau_2))*delta(index) ];

k3=step*F;
xnew=x+k3;

F=[ xnew(2) ;
  xnew(3)+ (K*tau_3/(tau_1*tau_2))*delta(index) ;
  -((1/tau_1)+(1/tau_2))*(xnew(3)+
  (K*tau_3/(tau_1*tau_2))*delta(index))-...
  (1/(tau_1*tau_2))*(abar*xnew(2)^ 3 + bbar*xnew(2))
  + (K/(tau_1*tau_2))*delta(index) ];

  k4=step*F;
  x=x+(1/6)*(k1+2*k2+2*k3+k4); % Calculated next state

t=t+step;  % Increments time
index=index+1;   % Increments the indexing term so that
   % index=1 corresponds to time t=0.
counter=counter+1;   % Indicates that we computed one more
%       integration step

end % This end statement goes with the first ``while"
 % statement in the program so when this is
```

```
   complete the simulation is done.

%
% Next, we provide plots of the input and output of the
% ship along with the reference heading that we want
%to track. Also, we plot the two inputs to the
% fuzzy controller.
%

% First, we convert from rad. to degrees
psi_r=psi_r*(180/pi);
psi=psi*(180/pi);
delta=delta*(180/pi);
e=e*(180/pi);
c=c*(180/pi);

% Next, we provide plots of data from the simulation

figure(1)
clf
subplot(211) plot(time,psi,'k-',time,psi_r,'k--')
% grid on
xlabel('Time (sec)')
title('Ship heading (solid) and desired ship heading
(dashed), deg.')
subplot(212)
plot(time,delta,'k-')
% grid on
xlabel('Time (sec)')
title('Rudder angle (δ), deg.')
zoom

figure(2)
clf
subplot(211)
plot(time,e,'k-')
% grid on
xlabel('Time (sec)')
title('Ship heading error between ship heading and
desired heading, deg.')
subplot(212)
plot(time,c,'k-')
% grid on
```

```
xlabel('Time (sec)')
title('Change in ship heading error, deg./sec')
zoom

end % This ends the if statement (on flag1) on whether
 % you want to do a simulation or just
  % see the control surface

% % % % % % % % % % % % % % % % %
% Next, provide a plot of the fuzzy controller surface:
% % % % % % % % % % % % % % % % %

% Request input from the user to see if they want to
% see the controller mapping:

flag2=input('\ n Do you want to see the nonlinear
\ n mapping implemented by the fuzzy \ n
controller? \ n (type 1 for yes and 0 for no) ');

if flag2==1,

% First, compute vectors with points over the whole range
% of the fuzzy controller inputs plus 20% over the end of
% the range and put 100 points in each vector
e_input=(-(1/g1)-0.2*(1/g1)):(1/100)*(((1/g1)+
0.2*(1/g1))-(-(1/g1)-...
  0.2*(1/g1))):((1/g1)+0.2*(1/g1));
ce_input=(-(1/g2)-0.2*(1/g2)):(1/100)*(((1/g2)+
0.2*(1/g2))-(-(1/g2)-...
  0.2*(1/g2))):((1/g2)+0.2*(1/g2));

% Next, compute the fuzzy controller output for all
% these inputs

for jj=1:length(e_input)
  for ii=1:length(ce_input)

c_count=0;,e_count=0; % These are used to count the
  % number of non-zero mf certainities
  of e and c

% The following if-then structure fills the vector mfe
% with the certainty of each membership fucntion of e
```

```
% for the current input e. We use triangular
% membership functions.

  if e_input(jj)<=ce(1)
    % Takes care of saturation of the left-most
                      % membership function
mfe=[1 0 0 0 0 0 0 0 0 0 0]; % i.e.,
the only one on is the
                    % left-most one
  e_count=e_count+1;,e_int=1; % One mf on, it is the
                    % left-most one.
  elseif e_input(jj)>=ce(nume)
                       % Takes care of saturation
                       % of the right-most mf
    mfe=[0 0 0 0 0 0 0 0 0 0 1];
    e_count=e_count+1;,e_int=nume; % One mf on, it is the
                       % right-most one
  else
    % In this case the input is on the
% middle part of the universe of
% discourse for e Next, we are going to
% cycle through the mfs to find all
% that are on
  for i=1:nume
if e_input(jj)<=ce(i)
  mfe(i)=max([0 1+(e_input(jj)-ce(i))/we]);
  % In this case the input is to the
  % left of the center ce(i) and we compute
  % the value of the mf centered at ce(i)
  % for this input e
 if mfe(i)~=0
  % If the certainty is not equal to zero then say
  % that have one mf on by incrementing our count
e_count=e_count+1;
e_int=i;
    % This term holds the index last entry
  % with a non-zero term
end
  else
 mfe(i)=max([0,1+(ce(i)-e_input(jj))/we]);
 % In this case the input is to the
 % right of the center ce(i)
if mfe(i)~=0
```

```
e_count=e_count+1;
e_int=i;
      % This term holds the index of the
      % last entry with a non-zero term
   end
 end
 end
 end


% The following if-then structure fills the vector mfc
% with the certainty of each membership fucntion of the c
% for its current value.

  if ce_input(ii)<=cc(1)
 % Takes care of saturation of left-most mf
 mfc=[1 0 0 0 0 0 0 0 0 0];
  c_count=c_count+1;
  c_int=1;
elseif ce_input(ii)>=cc(numc)
   % Takes care of saturation of the right-most mf
   mfc-[0 0 0 0 0 0 0 0 0 0 1];
   c_count=c_count+1;
   c_int=numc;
else
 for i=1:numc
   if ce_input(ii)<=cc(i)
   mfc(i)=max([0,1+(ce_input(ii)-cc(i))/wc]);
  if mfc(i)~=0
  c_count=c_count+1;
  c_int=i; % This term holds last entry
     % with a non-zero term
   end
   else
mfc(i)=max([0,1+(cc(i)-ce_input(ii))/wc]);
   if mfc(i)~=0
   c_count=c_count+1;
   c_int=i;% This term holds last entry
   % with a non-zero term
   end
  end
 end
 end
```

```
% The next loops calculate the crisp output using only
% the non-zero premise of error,e, and c.

num=0;
den=0;
for k=(e_int-e_count+1):e_int
  % Scan over e indices of mfs that are on
    for l=(c_int-c_count+1):c_int
    % Scan over c indices of mfs that are on
  prem=min([mfe(k) mfc(l)]);
  % Value of premise membership function
% This next calculation of num adds up the numerator
% for the center of gravity defuzzification formula.
  num=num+rules(k,l)*base*(prem-(prem)^ 2/2);
  den=den+base*(prem-(prem)^ 2/2);
% To do the same computations, but for center-average
% defuzzification, use the following lines of code rather
% than the two above:
% num=num+rules(k,l)*prem;
% den=den+prem;
    end
    end

 delta_output(ii,jj)=num/den;
   % Crisp output of fuzzy controller that is the input
   % to the plant.

   end
end

% Convert from radians to degrees:

delta_output=delta_output*(180/pi);
e_input=e_input*(180/pi);
ce_input=ce_input*(180/pi);

% Plot the controller map

figure(3)
clf
surf(e_input,ce_input,delta_output);
view(145,30);
colormap(white);
```

```
xlabel('Heading error (e), deg.');
ylabel('Change in heading error (c), deg.');
zlabel('Fuzzy controller output (δ), deg.');
title('Fuzzy controller mapping between inputs
and output');

end

% % % % % % % % % % % % % % % % % % % % % % % % % % % % %
%
% End of program %
% % % % % % % % % % % % % % % % % % % % % % % % % % % % %
```

Output

Figure 8.4 shows the simulated output of the Fuzzy Control System with the ship heading and the Rudder angle. The error and it derivative

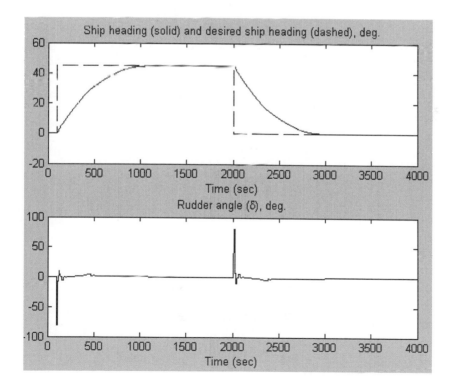

FIGURE 8.4: Simulation of the Fuzzy Control System for the Tanker Showing the Ship Heading, Desired Ship Heading and Rudder Angle

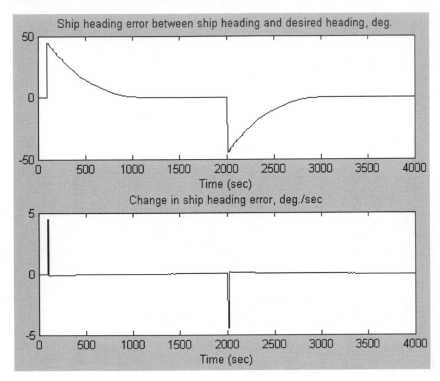

FIGURE 8.5: Error and Change in Ship Heading Error

are plotted in Figure 8.5 and Figure 8.6 shows the nonlinearities between the input and the output.

8.3   Illustration 3 - Approximation of Any Function Using Fuzzy Logic

This section designs a fuzzy system to approximate a function which may not be defined analytically, but the values of the function in n point is defined as for example: $g(x,z)$. The approximation is calculated from a fuzzy set design. The fuzzy function implemented here approximates the given function using Takagi-Sugeno approximation and compares with linear, polynomial, and neural network (single and multi-layer Perceptron) approximation methods.

MATLAB Code

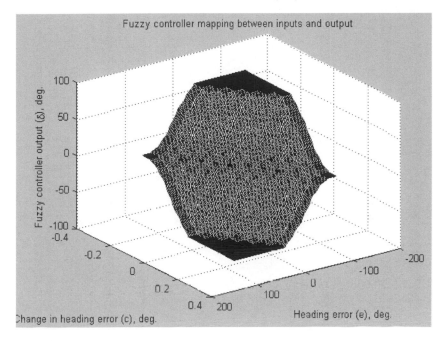

FIGURE 8.6: Nonlinearity between Inputs and Output

```
clear

% For the M=7 case
x7=-6:2:6;

M=length(x7)

for i=1:M,
    G7(i)=exp(-50*(x7(i)-1)^ 2)-0.5*exp(-100*(x7(i)-
1.2)^ 2)+atan(2*x7(i))+2.15+...
    0.2*exp(-10*(x7(i)+1)^ 2)-0.25*
exp(-20*(x7(i)+1.5)^ 2)+0.1*exp(-
10* (x7(i)+2)^ 2) -0.2*exp(-10*(x7(i)+3)^ 2);
    if x7(i) >= 0

G7(i)=G7(i)+0.1*(x7(i)-2)^ 2-0.4;
    end
end
% For the M=121 case
```

```
x121=-6:0.1:6;

M=length(x121)

for i=1:M,
   z(i)=0.15*(rand-0.5)*2;
  % Define the auxiliary variable
   G121(i)=exp(-50*(x121(i)-1)^ 2)-0.5*exp(-100*
(x121(i)
-1.2)^ 2)+atan(2*x121(i))+2.15+...
   0.2*exp(-10*(x121(i)+1)^ 2)-0.25*exp(-20*(x121(i)+
1.5)^ 2)+0.1*exp(-
10*(x121(i)+2)^ 2)-0.2*exp(-10*(x121(i)+3)^ 2);
   if x121(i) >= 0
 G121(i)=G121(i)+0.1*(x121(i)-2)^ 2-0.4;
   end
   G121n(i)=G121(i)+z(i);
  % Adds in the influence of the auxiliary variable
% fpoly(i)=0.6+0.1*x(i);
end

% Next, plot the functions:

figure(1)
plot(x7,G7,'ko')
xlabel('x(i)')
ylabel('y(i)=G(x(i))')
title('M=7')
grid
axis([min(x7) max(x7) 0 max(G7)])

figure(2)
plot(x121,G121,'ko')
xlabel('x(i)')
ylabel('y(i)=G(x(i))')
title('M=121')
grid
axis([min(x121) max(x121) 0 max(G121)])

figure(3)
plot(x121,G121n,'ko')
xlabel('x(i)')
ylabel('y(i)=G(x(i),z(i))')
```

```
title('M=121')
grid
axis([min(x121) max(x121) 0 max(G121)])

% Next, plot some approximator fits to the function
% First, a line

for i=1:M,
   Flinear(i)=((4.5-0.5)/(6-(-6)))*x121(i)+2.5;
end

figure(4)
plot(x121,G121n,'ko',x121,Flinear,'k')
xlabel('x(i)')
ylabel('y(i)=G(x(i),z(i)), and line')
title('Linear approximation')
grid
axis([min(x121) max(x121) 0 max(G121)])

% Next, a polynomial (parabola)

for i=1:M,
   Fpoly(i)=0.6+0.035*(x121(i)+6)^2;
end

figure(5)
plot(x121,G121n,'ko',x121,Fpoly,'k')
xlabel('x(i)')
ylabel('y(i)=G(x(i),z(i)), and parabola')
title('Polynomial approximation')
grid
axis([min(x121) max(x121) 0 max(G121)])

% Next, a single hidden layer perceptron, n1=1

w1=[1.5]';
b1=0;

w=[3]';
b=0.6;

for i=1:M,
   phi1=inv(1+exp(-b1-w1*x121(i)));
```

```
    phi=[phi1]';
    Fmlp(i)=b+w'*phi;
end

figure(6)
plot(x121,G121n,'ko',x121,Fmlp,'k')
xlabel('x(i)')
ylabel('y(i)=G(x(i),z(i)), and perceptron output')
title('Neural network approximation, one neuron')
grid
axis([min(x121) max(x121) 0 max(G121)])

% Next, a single hidden layer perceptron, n1=2
w1=[1.5]';
b1=0;
w2=[1.25]';
b2=-6;
w=[3 1]';
b=0.6;

for i=1:M,
    phi1=inv(1+exp(-b1-w1*x121(i)));
    phi2=inv(1+exp(-b2-w2*x121(i)));
    phi=[phi1 phi2]';
    Fmlp(i)=b+w'*phi;
end
figure(7)
plot(x121,G121n,'ko',x121,Fmlp,'k')
xlabel('x(i)')
ylabel('y(i)=G(x(i),z(i)), and perceptron output')
title('Neural network approximation, two neurons')
grid
axis([min(x121) max(x121) 0 max(G121)])

% Next, we plot the premise membership functions for
% a Takagi-Sugeno fuzzy system approximator

c11=-3.5;
sigma11=0.8;

c12=-0.25;
sigma12=0.6;
```

```
c13=2;
sigma13=0.4;

c14=4.5;
sigma14=0.8;

for i=1:M,
   mu1(i)=exp(-0.5*((x121(i)-c11)/sigma11)^ 2);
   mu2(i)=exp(-0.5*((x121(i)-c12)/sigma12)^ 2);
   mu3(i)=exp(-0.5*((x121(i)-c13)/sigma13)^ 2);
   mu4(i)=exp(-0.5*((x121(i)-c14)/sigma14)^ 2);
   denominator(i)=mu1(i)+mu2(i)+mu3(i)+mu4(i);
end

figure(8)
plot(x121,G121n,'ko',x121,mu1,'k',x121,mu2,'k',
x121,mu3,'k',x121,mu4,'k')
xlabel('x(i)')
ylabel('y(i)=G(x(i),z(i)) and premise membership
function values')
title('Training data and premise membership functions')
grid

% Next, we plot the basis functions for
% a Takagi-Sugeno fuzzy system approximator

for i=1:M,
   xi1(i)=mu1(i)/denominator(i);
   xi2(i)=mu2(i)/denominator(i);
   xi3(i)=mu3(i)/denominator(i);
   xi4(i)=mu4(i)/denominator(i);
end

figure(9)
plot(x121,G121n,'ko',x121,xi1,'k',x121,xi2,'k',x121,xi3,
'k',x121,xi4,'k')
xlabel('x(i)')
ylabel('y(i)=G(x(i),z(i)) and basis function values')
title('Training data and basis functions')
grid

% Next, we plot the Takagi-Sugeno fuzzy
system approximator
```

```
a10=1;
a11=0.5/6;

a20=2.25;
a21=4.4/4;

a30=2.9;
a31=1/12;

a40=1.3;
a41=4.8/8;

  for i=1:M,
    g1(i)=a10+a11*x121(i);
    g2(i)=a20+a21*x121(i);
    g3(i)=a30+a31*x121(i);
    g4(i)=a40+a41*x121(i);
    numerator=g1(i)*mu1(i)+g2(i)*mu2(i)+g3(i)*
    mu3(i)+g4(i)*mu4(i);
    Fts(i)=numerator/(mu1(i)+mu2(i)+mu3(i)+mu4(i));
end

figure(10)
plot(x121,G121n,'ko',x121,Fts,'k')
xlabel('x(i)')
ylabel('y(i)=G(x(i),z(i)), and fuzzy system output')
title('Takagi-Sugeno approximation')
grid
axis([min(x121) max(x121) 0 max(G121)])

% % % % % % % % % % % % % % % % % % % % % % % % % %
% % % %
% End of program %
% % % % % % % % % % % % % % % % % % % % % % % % % %
% % % % %
```

Output

 An unknown function is approximated and its training pattern is shown in Figure 8.7. The function was trained using different approximation techniques such as Linear, Polynomial, Single and Multilayer Neural

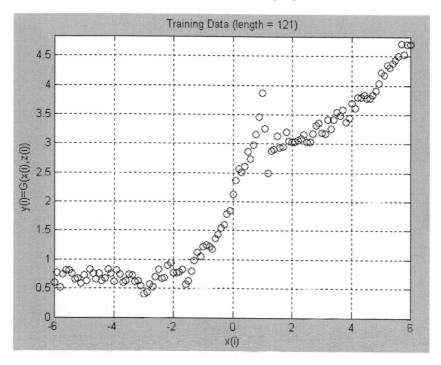

FIGURE 8.7: Training Data with Length = 121

networks. The best performance was obtained with Takagi-Sugeno approximation and the results are plotted in Figure 8.8.

Figure 8.9 shows the training data along with their membership functions and basis functions for a Takagi-Sugeno fuzzy system approximator

8.4   Illustration 4 - Building Fuzzy Simulink Models

To build Simulink systems that use fuzzy logic, simply copy the Fuzzy Logic Controller block out of sltank (or any of the other Simulink demo systems available with the toolbox) and place it in the block diagram. The Fuzzy Logic Controller block in the Fuzzy Logic Toolbox library, can be opened either by selecting **Fuzzy Logic Toolbox** in the Simulink Library Browser, or by typing fuzblock at the MATLAB prompt. The following library (Figure 8.10) appears.

The Fuzzy Logic Toolbox library contains the Fuzzy Logic Controller

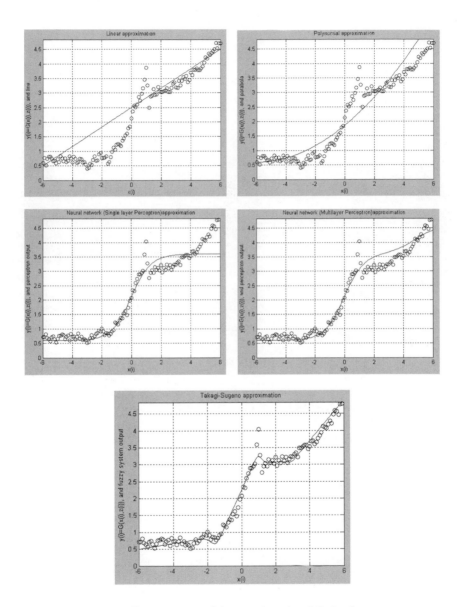

FIGURE 8.8: Comparison of Approximation Methods

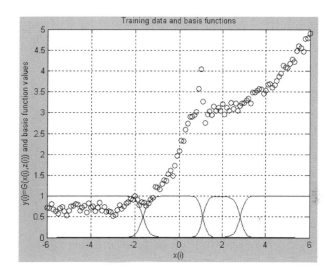

FIGURE 8.9:   Training Data with Membership Functions and Basis Functions for a Takagi-Sugeno Fuzzy System Approximator

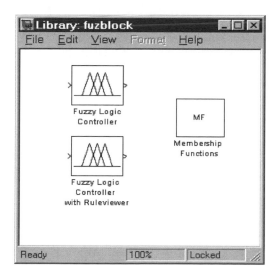

FIGURE 8.10:   Fuzzy Library

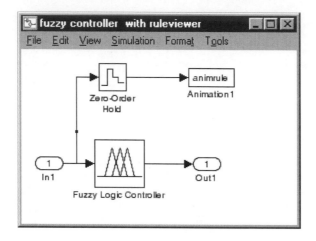

FIGURE 8.11:   Fuzzy Logic Controller Block

and Fuzzy Logic Controller with Rule Viewer blocks. It also includes a Membership Functions sub-library that contains Simulink blocks for the built-in membership functions.

The Fuzzy Logic Controller with Rule Viewer block is an extension of the Fuzzy Logic Controller block. It allows the user to visualize how rules are fired during simulation (Figure 8.11). Double-click the Fuzzy Controller with Rule Viewer block, and the following appears.

To initialize the Fuzzy Logic Controller blocks (with or without the Rule Viewer), double-click on the block and enter the name of the structure variable describing the FIS. This variable must be located in the MATLAB workspace.

Fuzzy Logic Controller Block

For most fuzzy inference systems, the Fuzzy Logic Controller block automatically generates a hierarchical block diagram representation of your FIS. This automatic model generation ability is called the Fuzzy Wizard. The block diagram representation only uses built-in Simulink blocks and therefore allows for efficient code generation.

The Fuzzy Wizard cannot handle FIS with custom membership functions or with AND, OR, IMP, and AGG functions outside of the following list:

    orMethod: max
    andMethod: min,prod
    impMethod: min,prod

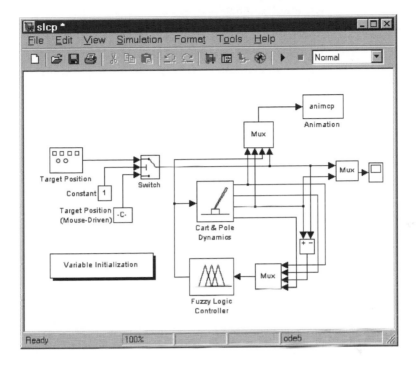

FIGURE 8.12: FIS Model of the Cart and Pole Example

aggMethod: max

In these cases, the Fuzzy Logic Controller block uses the S-function sffis to simulate the FIS.

Cart and Pole Simulation

The cart and pole simulation is an example of an FIS model auto-generated by the Fuzzy Logic Controller block. Type slcp at the MATLAB prompt to open the simulation. The model in Figure 8.12 appears.

Right-click on the Fuzzy Logic Controller block and select Look under mask from the right-click menu. The subsystem shown in Figure 8.13 opens.

Follow the same procedure to look under the mask of the FIS Wizard subsystem to see the implementation of the FIS. The Fuzzy Logic Controller block uses built-in Simulink blocks to implement the FIS. Although the models can grow complex, this representation is better suited than the S-function sffis for efficient code generation.

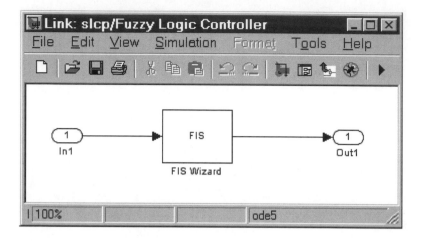

FIGURE 8.13:   Subsystem of the Fuzzy Logic Controller Block

Summary

According to literature the employment of fuzzy logic is not recommendable, if the conventional approach yields a satisfying result, an easily solvable and adequate mathematical model already exists, or the problem is not solvable. MATLAB examples are given for each of the fuzzy concept so that the reader can implement theoretical concepts in practice.

Review Questions

1. Let $y = f(x) = -2x - x^2$.

   (i) Form a fuzzy system using MATLAB, which approximates function f, when $x \in [-10,10]$. Repeat the same by adding random, normally distributed noise with zero mean and unit variance.

   (ii) Simulate the output when the input is sin(t). Observe what happens to the signal shape at the output.

2. Using MATLAB SIMULINK construct a fuzzy controller with three inputs and one output. Consider both continuous and discrete inputs and assume the output as singleton. Analyze the operations of the controller.

3. Write a MATLAB program to construct the nonlinear model of an inverted pendulum.

4. Construct a SIMULINK-based Vehicle control system using a simple generic Mamdani type controller.

5. Conduct a suitable experiment to control the speed of a DC motor. Use SIMULINK to construct the model and plot the torque and speed characteristics.

6. Write a MATLAB code for implementing the PID controller using fuzzy logic and observe the results.

Chapter 9

Neuro-Fuzzy Modeling Using MATLAB

9.1 Introduction

The combination of Artificial Neural Networks (ANN) and Fuzzy Inference Systems (FIS) has drawn the attention of several researchers in various scientific and engineering fields due to the rising needs of intelligent systems to solve the real world complex problems. ANN learns the presented inputs from the base by updating the interconnections between layers. FIS is a most common computation model based on the concept of fuzzy set theory, fuzzy if-then rules, and fuzzy reasoning. There are several advantages in the fusion of ANN and FIS. ANN and FIS can be integrated in several methods depending upon the application. The integration of ANN and FIS can be classified into three categories namely concurrent model, cooperative model, and fully fused model.

This chapter begins with a discussion of the features of each model and generalizes the advantages and deficiencies of each model. The chapter further focuses on the different types of fused neuro-fuzzy systems such as such as FALCON, ANFIS, GARIC, NEFCON, FINEST, FUN, EFuNN, and SONFIN and citing the advantages and disadvantages of each model. A detailed description of ANFIS including its architecture and learning algorithm are discussed. The implementation detail of hybrid neuro-fuzzy model is also delineated. An explanation on Classification and Regression Trees with its computational issues, computational details, computational formulas, advantages, and examples is given in this chapter. The data clustering algorithms such as hard c-means, Fuzzy c-means, and subtractive clustering are also described.

The combination of Neuro and Fuzzy for computing applications is a popular model for solving complex problems. Whenever there is a knowledge expressed in linguistic rules, an FIS can be modelled, and if information is available, or if the parameters can be learned from a simulation (training) then ANNs can be used. While building a FIS, the fuzzy sets, fuzzy operators, and the knowledge base are required to be specified. To implement an ANN for a specific application the architecture and learning algorithm are required. The drawbacks in these approaches appear complementary and consequently it is natural to consider implementing an integrated system combining the neuro-fuzzy concepts.

9.2   Cooperative and Concurrent Neuro-Fuzzy Systems

In the simplest way, a cooperative model can be considered as a preprocessor wherein ANN learning mechanism determines the FIS membership functions or fuzzy rules from the training data. Once the FIS parameters are determined, ANN goes to the background. The rules are formed by a clustering approach or fuzzy clustering algorithms. The fuzzy membership functions are approximated by neural network from the training data. In a concurrent model, ANN assists the FIS continuously to determine the required parameters especially if the input variables of the controller cannot be measured directly. In some cases the FIS outputs might not be directly applicable to the process. In that case ANN can act as a postprocessor of FIS outputs. Figures 9.1 and 9.2 depict the cooperative and concurrent NF models.

9.3   Fused Neuro-Fuzzy Systems

In the merged Neuro-Fuzzy (NF) architecture, the learning algorithms are used to find out the parameters of FIS. The fused NF systems share

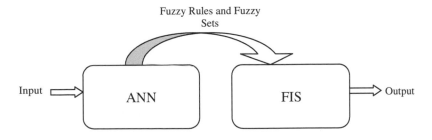

FIGURE 9.1:   Cooperative NF Model

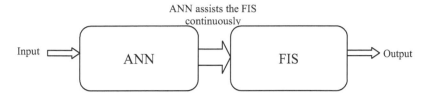

FIGURE 9.2: Concurrent NF Model

information structures and knowledge representations. Learning algorithm can be applied to a fuzzy system by interpreting the fuzzy system in an ANN like architecture. In general the traditional ANN learning algorithms cannot be applied directly to the fuzzy systems since the functions used in the inference process are normally nondifferentiable.

Some of the major works in this area are GARIC, FALCON, ANFIS, NEFCON, FUN, SONFIN, FINEST, EFuNN, dmEFuNN, evolutionary design of neuro fuzzy systems, and many others.

9.3.1 Fuzzy Adaptive Learning Control Network (FALCON)

FALCON has a five-layered architecture as shown in Figure 9.3. There are two linguistic nodes for each output variable among which one node is for storing the training data and a second node for the target output of FALCON. The first hidden layer is accountable for the fuzzification of each input variable. Either a single node or multilayer nodes are used to represent the membership functions. The prerequisites required for the FIS are defined in the second hidden layer and the rule consequents are in the third hidden layer. Unsupervised learning algorithm is applied to the FALCON architecture to locate initial membership functions, rule base, and a gradient descent learning to optimally update the parameters of the MF to create the desired outputs.

9.3.2 Adaptive Neuro-Fuzzy Inference System (ANFIS)

ANFIS implements a Takagi–Sugeno FIS and has a five layered architecture as shown in Figure 9.4. The input variables are fuzzified in the first hidden layer and the fuzzy operators are applied in the second hidden layer to compute the rule antecedent part. The fuzzy rule base is normalized in the third hidden layer and the consequent parameters of the rules are ascertained in the fourth hidden layer. The fifth output layer computes the overall input as the summation of all incoming

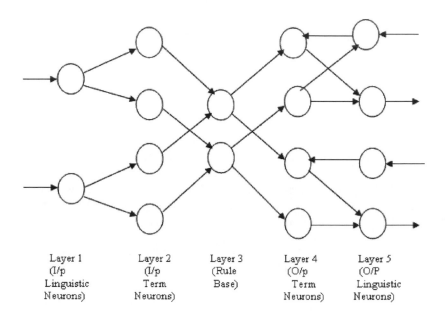

Layer 1   Layer 2   Layer 3   Layer 4   Layer 5
(I/p   (I/p   (Rule   (O/p   (O/P
Linguistic Term   Base)   Term   Linguistic
Neurons)   Neurons)   Neurons)   Neurons)

FIGURE 9.3: Architecture of FALCON

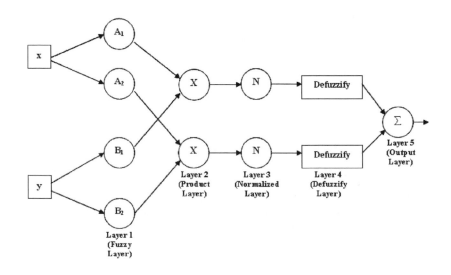

FIGURE 9.4: Structure of ANFIS

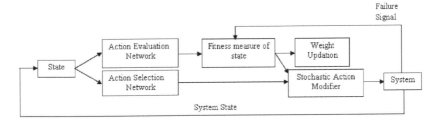

FIGURE 9.5:   Architecture of GARIC

signals. ANFIS uses backpropagation learning to determine the parameters related to membership functions and least mean square estimation to determine the consequent parameters.

A step in the learning procedure has got two parts: In the first part the input patterns are propagated, and the optimal consequent parameters are estimated by an iterative least mean square procedure, while the premise parameters are assumed to be fixed for the current cycle through the training set. In the second part the patterns are propagated again, and in this epoch, backpropagation is used to modify the premise parameters, while the consequent parameters remain fixed. This procedure is then iterated.

9.3.3   Generalized Approximate Reasoning-Based Intelligent Control (GARIC)

GARIC implements a neuro-fuzzy controller by using two neural network modules, the ASN (Action Selection Network) and the AEN (Action State Evaluation Network). The AEN assesses the activities of the ASN. The part ASN of GARIC is a feedforward network with five layers with no connection weights between the layers. Figure 9.5 illustrates the structure of GARIC-ASN. The linguistic values are stored in the first hidden layer and the input unit is connected to this unit. The fuzzy rule base are stored in the second hidden layer. The third hidden layer represents the linguistic values of the control output variable. The defuzzification method used to compute the rule outputs in GARIC is the mean of maximum method. A crisp output is taken from each rule, therefore the outputs must be defuzzified before they are accumulated to the final output value of the controller. The learning method used in GARIC is a combination of gradient descent and reinforcement learning.

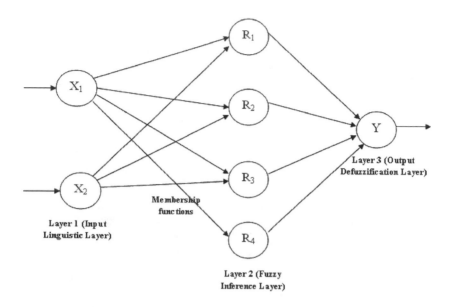

FIGURE 9.6: Architecture of NEFCON

9.3.4 Neuro-Fuzzy Control (NEFCON)

NEFCON is designed to implement Mamdani type FIS and is illustrated in Figure 9.6. The connections between the layers in NEFCON are weighted with fuzzy sets and rules so-called shared weights. These weighted connections assure the unity of the rule base. Among the three layers in the NEFCON structure, the input units present the linguistic values, and the output unit is the defuzzification interface. The intermediate hidden layer serves as the fuzzy inference. NEFCON model uses a combination of reinforcement and backpropagation learning. NEFCON is capable of learning an initial rule base, if no prior knowledge about the system is available or even to optimize a manually defined rule base. NEFCON has two variants: NEFPROX (for function approximation) and NEFCLASS (for classification tasks).

NEFCON Learning Algorithm

The NEFCON rule base is learned using incremental rule learning or decremental rule learning. Incremental rule learning is used in cases if the correct output cannot be predicted and if the rules are created based on the estimated output values. As the learning process proceeds more number of rules are created. In decremental rule learning, initially rules are created due to fuzzy partitions of process variables and unnecessary rules are eliminated in the course of learning. Incremental rule learning is more efficient than the decremental approach. Assume a process with n state variables, consisting of pi fuzzy sets, and a control variable with q linguistic terms, at most N linguistic rules are created, where $N = q\Pi_{i-1}^{n}p_i$. Therefore the rule base is inconsistent. The rules must be consistent with k rules and the condition $k \leq \frac{N}{q}$ must satisfy.

The learning algorithm is performed in stages as follows:

- First stage: Here all rule units are deleted which provide an output value with a sign different from optimal output value.

- Second stage: In this stage, the subsets of rules with similar antecedents taken and the rule which matches the current state is chosen to contribute to the NEFCON output. At the end of this stage, from each rule subset only the rule with the smallest error is kept, and all the other rules are deleted from the NEFCON system.

Decremental Rule Learning

Let R denote the set of all rule units and $Ant(R_r)$ denote the antecedent and $Con(R_r)$ the consequent of a fuzzy rule corresponding to the rule unit R_r. Let S be a process with n state variables $\xi_i \in X_i$ (i $\in$ {1, ... ,n}) which are portioned by p_i fuzzy sets each and one control variable $\eta \in Y$, portioned by q fuzzy sets. Let there also be initial rules with

$$(\forall R, R^\circ \in R((Ant(R) = Ant(R^\circ) \wedge Con(R) = Con(R^\circ) \Rightarrow R = R^\circ))$$

1. For each rule unit R_r a counter C_r (initialized to zero is defined) rϵ {1, ... ,N}. For a fixed number of iterations m_1 the following steps are carried out:

    - Determine the current NEFCON output $O\eta$ using the current state of S.
    - For each rule R_r determine its contribution t_r to the overall output $O\eta$, r ϵ 1, ... ,N.
    - Determine $sgn(\eta_{opt})$ for the current input values.
    - Delete each rule unit R_r with $sgn(t_r) \neq sgn(\eta_{opt})$ and update the value of N.
    - Increment the counter C_r for all R_r with $o_{Rr} > 0$.
    - Apply O_η to S and determine the new input values.

2. For each rule unit R_r a counter Z_r (initialized to zero is defined). For a fixed number of iterations m_2 the following steps are carried out:

   - From all subsets, $R_j = \{R_r \mid \text{Ant}(R_r)\} = \text{Ant}(R_2)(r,s \in \{1, \ldots ,N\})\} \subseteq R$, one rule unit R_{rj} is selected arbitrarily.
   - Determine the current NEFCON output O_η using only the units selected and the current state of S
   - Apply O_η to S and determine the new input values.
   - For each rule R_{rj} determine its contribution t_{rj} to the overall output O_η, $r \in \{1, \ldots ,N\}$.
   - Determine $\text{sgn}(\eta_{opt})$ for the current input values.
   - Add $|O_{R_{rj}}|$ to the counter Z_r of each selected unit R_{rj}.
   - For all selected rule units R_{rj} with $O_{R_{rj}} > 0$, C_{rj} is incremented.

3. Delete all rule units R_{sj} for all subsets R_j from the network for which there is a rule unit $R_{sj} \in R_j$ with $Z_{rj} < Z_{sj}$, and delete all rule units R_r with $C_r < \beta \cdot (m1+m2)$, $0 \leq \beta < 1$, from the network and update the value of N.

4. Apply appropriate FEBP algorithm to the NEFCON system with k=N remaining rule units.

Incremental Rule Learning

In this learning process the rule base is developed from from scratch by adding rule after rule. The rule bases developed by classifying an input vector initially, i.e., finding the membership function corresponding to each variable that generates the highest membership value for the respective input value, thus forming a rule antecedant. The training algorithm attempts to approximate the output value by deducing it from the present fuzzy error. To commence the algorithm, an initial fuzzy partition must be allotted to each variable.

Let S be a process with n state variables $\xi_i \in$ Xi (i $\in$ (1, $\ldots$, n)) which are partitioned by p_i fuzzy sets each and one control variable $\eta \epsilon$ [y_{min}, y_{max}] partitioned by q fuzzy sets. Let there also be initial k predefined rule units, where k may be zero. The following steps give the incremental rule-learning algorithm.

1. For a fixed number m1 of iterations the following steps are carried out.

- For the current input vector $(x_1, \dots, x_n)$ find those fuzzy sets $(\hat{\mu}^{(1)}, \dots \hat{\mu}^{(n)})$ for which $(\forall i, j)(\hat{\mu}^{(1)}(x_i) \geq \hat{\mu}_j^{(i)}(x_i))$ holds.
- If there is no rule with the antecedent.
  If ξ_i is $\hat{\mu}^{(1)}$ and ξ_n is $\hat{\mu}^{(n)}$, **then the fuzzy set $\hat{v}$ such that $(\forall j)(\hat{v}(o) \geq v_j(o))$ holds, where the heuristic output value o is determined by**

$$o = \begin{cases} m+ \mid E \mid .(y_{max} - m) if E \geq 0 \\ m- \mid E \mid .(m - y_{min}) if E < 0 \end{cases}$$

  where m $= \frac{y_{max} + y_{min}}{2}$

- Enter the rule
  If ξ_i is $\hat{\mu}^{(1)}$ and ξ_n is $\hat{\mu}^{(n)}$, **then η is $\hat{v}$; into the NEFCON system**

2. For a fixed number m_2 iterations the following steps are carried out

   - Propagate the current input vector through the NEFCON system and estimate for each rule R_r its contribution tr to the overall output O_η . Compute the desired contribution to the system output value by

$$t_r^* = t_r + \sigma.o_r.E$$

     where F is the extended fuzzy error and $\sigma > 0$ is the learning rate.

   - For each rule unit R_r, determine the new output membership function v_r such that $(\forall i (\hat{v}_r(t_r^*) \geq v_j(t_r^*))$

   - Delete all rules that have not been used in more than in p % of all propagations.

   - Apply appropriate FEBP algorithm to the NEFCON system.

Rule-learning algorithm is mainly applied to examine the present rules and evaluate them. All the rule units undergo the test process and if they do not pass this test they are eliminated from the network. Here the algorithm has to choose the rule with the lowest error value from each subset that consists of rules with identical antecedents, and delete all other rules of each subset. The entire state space must be learned so that a good rule base is produced. Since the incremental rule learning is a complex process, it is optimal for larger systems, while the decremental rule learning can be applied to smaller systems. In order to reduce the cost of the learning procedure for decremental rule learning two criteria are considered:

- If the consequent corresponding to an antecedent is known, then an equivalent rule unit is put into the NEFCON system, and the rule is not deleted.

- For a few states only a subset of all possible rules are to be considered, in that case only these rules are put into the network.

A similar situation is applicable to incremental learning rule, whereby if prior knowledge is available, then rule learning does not need to start from scratch.

9.3.5 Fuzzy Inference and Neural Network in Fuzzy Inference Software (FINEST)

FINEST is capable of two kinds of tuning process, the tuning of fuzzy predicates, combination functions, and the tuning of an implication function. The generalized Modus Ponens is improved in the following four ways

(1) Aggregation operators that have synergy and cancellation nature

(2) A parameterized implication function

(3) A combination function that can reduce fuzziness

(4) Backward chaining based on generalized Modus Ponens

FINEST Architecture and Training Algorithm

The parameter tuning in FINEST is done by the backpropagation algorithm. Figure 9.7 shows the layered architecture of FINEST and the calculation process of the fuzzy inference. FINEST provides a framework to tune any parameter, which appears in the nodes of the network representing the calculation process of the fuzzy data if the derivative function with respect to the parameters is given.

In Figure 9.7, the input values (xi,) are the facts and the output value (y) is the conclusion of the fuzzy inference.

Consider the following fuzzy rules and facts:

Rule i: If x_i is A_{i1} and x_n is A_{in} then y is B_i
Fact j: x_j is A_j° $(j = 1, ..., n, \ i = 1, ..., m)$

The calculation procedure in each layer is as follows:

Layer 1: Converse truth value qualification for condition j of rule i,

$$\tau_{Aij}(a_{ij}) = sup \ \mu_{A_j^\circ}(x_j)$$

$$\mu_{Aij}(x_j) = a_{ij}$$

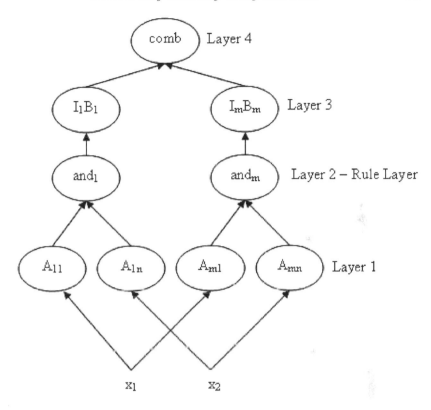

FIGURE 9.7: Architecture of FINEST

Layer 2: Aggregation of the truth values of the conditions of Rule i,

$$\tau_{Ai}(a_i) = sup\{\tau_{Ai}(a_{i1}) \wedge ... \wedge \tau_{Am}(a_{in})\}$$

Layer 3: Deduction of the conclusion from Rule i,

$$\mu_{B_j^\circ}(y) = sup\{\tau_{Ai}(a_i) \wedge I_i(a_i,\ \mu_{Bj}(y))\}$$

Layer 4: Combination of the conclusions derived from all the rules,

$$\mu_{B_j^\circ}(y) = comb(\mu_{B_1^\circ}(y) \wedge ... \wedge\ \mu_{B_m^\circ}(y))$$

In the above equations τ_{Aij} τ_{Ai} B_i°, respectively represent the truth value of the condition "x_i is A_{ij}," of rule i, the truth value of the condition part of rule i, and the conclusion derived from rule i. Besides, the functions I_i and *comb* respectively represent the function characterizing the implication function of rule i, and the global combination function.

Back-propagation method is used to tune the network parameters. All data inside the network are treated as fuzzy sets. The actual output of the network B is given by

$$B^\circ = b^{\circ(1)}/y^{(1)} + ... + b^{\circ(p)}/y^{(p)}$$

The teaching signal T is given by

$$T = t^{(1)}/y^{(1)} + ... + t^{(p)}/y^{(p)}$$

Error function E is expressed by

$$E = \frac{1}{2}\sum_{h=1}^{p}(t^{(h)} - b^{(h)})^2$$

It is possible to tune any parameter, which appears in the nodes of the network representing the calculation process of the fuzzy data if the derivative function with respect to the parameters is given. So it is very important to parameterize the inference procedure.

9.3.6 Fuzzy Net (FUN)

The architecture of FUN consists of three layers, the neurons in the first hidden fuzzify the input variables and also contain the membership functions. The fuzzy AND operations are performed in the second hidden layer. The membership functions of the output variables are stored in the third hidden layer with activation function Fuzzy-OR. The output neuron perfoms defuzzification by using an appropriate defuzzification method. The network is initialized with a fuzzy rule base and the corresponding membership functions. The parameters of the membership functions are altered by using a learning procedure. The learning process is forced by a cost function, which is evaluated after the random alteration of the membership function. Whenever the alteration resulted in an improved performance the alteration is kept, otherwise it is ruined.

FUN Learning Strategies

The FUN learns rules by modifying the connections between the rules and the fuzzy values. The membership functions are learned while the data of the nodes in the first and three hidden layers are changed. FUN can be trained with the standard neural network training strategies such as reinforcement or supervised learning.

Learning of the rules: The rules in FUN are depicted in the net through the connections between the layers. The learning process is carried out as a stochastic search in the rule space: an arbitrarily chosen connection is modified and the new network performance is assessed with a cost function. If the performance is good, the connection is maintained and if the performance is worse the change is ruined. This process continues until the desired output is achieved. Since the learning algorithm ought to maintain the semantic of the rules, it has to be assured that no two values of the same-variable appear in the same rule. This is accomplished by switching connections between the values of the similar variable.

Learning of membership functions: The FUN architecture uses triangular membership functions with three membership function descriptors (MFDs). The membership functions are considered to be normalized, i.e., they always have a constant height of 1. The learning algorithm is based on a mixture of gradient descent and a stochastic search. A maximum change in a random direction is initially assigned to all MFDs. In a stochastic manner one MFD from one linguistic variable is selected, and the network functioning is tested with this MFD which is altered according to the permissible change for this MFD. If the network performs better according to the given cost function, the new value is accepted and next time another change is tried in the same direction. Contrary if the network performs worse, the change is reversed. To guarantee convergence, the alterations are reduced after each training step and shrink asymptotically toward zero according to the learning rate.

9.3.7 Evolving Fuzzy Neural Network (EFuNN)

In EFuNN all nodes are created during learning. The Architecture of EFuNN is shown in Figure 9.8. The information is passed from the input layer to the second layer where the fuzzy membership degrees are calculated based on the input values as belonging to a predefined fuzzy membership function. The fuzzy rulebase is present in the third layer and these rules represent prototypes of input-output data. Connection weights are used to describe rule nodes, and these rules are adjusted through the hybrid learning technique. The fourth layer calculates the degrees to which output membership functions are matched by the input data, and the fifth layer does defuzzification and calculates exact values for the output variables. Dynamic Evolving Fuzzy Neural Network (DmEFuNN) is a expanded version of EFuNN in which the winning rule node's activation is propagated and a group of rule nodes is dynamically selected for every new input vector and their activation values are used to compute the dynamical parameters of the output function. While EFuNN implements fuzzy rules of Mamdani type, DmEFuNN estimates the Takagi-Sugeno fuzzy rules based on a least squares algorithm.

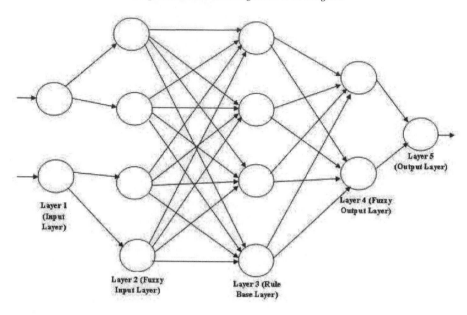

FIGURE 9.8:   Architecture of EFuNN

9.3.8   Self–Constructing Neural Fuzzy Inference Network (SONFIN)

SONFIN implements a modified Takagi-Sugeno FIS and is illustrated in Figure 9.9. In the structure identification of the precondition part, the input space is partitioned in a flexible way according to an aligned clustering based algorithm.

To identify the consequent part, a singleton value is selected by a clustering method and is assigned to each rule initially. Later on, a few supplementary input variables are chosen through a projection-based correlation measure for each rule and are added to the consequent part forming a linear equation of input variables. For parameter identification, the consequent parameters are tuned optimally by either least mean squares or recursive least squares algorithms and the precondition parameters are tuned by backpropagation algorithm.

9.3.9   Evolutionary Design of Neuro-Fuzzy Systems

In the process of evolutionary design of neuro-fuzzy systems, the node parameters, architecture, and learning parameters are adapted according to a five-tier hierarchical evolutionary search procedure. This kind

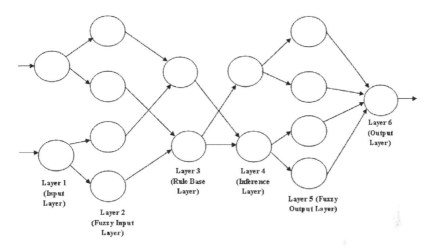

FIGURE 9.9:   Six Layered Architecture of SONFIN

of an evolutionary model can use Mamdani or Takagi Sugeno type fuzzy inference. The layers are depicted in the basic architecture as shown in Figure 9.10. The evolutionary search process will resolve the best quantity of nodes and connections between layers. The fuzzification layer and the rule antecedent layer function similarly as that of other NF models. The consequent part of rule will be found according to the inference system depending on the problem type, which will be adapted accordingly by the evolutionary search mechanism.

Defuzzification operators are adapted according to the FIS chosen by the evolutionary algorithm. For every learning parameter and every inference mechanism, there is the global search of inference mechanisms that continues on a faster time scale in an environment decided by the learning parameters (for inference mechanism only), inference system, and the problem. Likewise, for every architecture, evolution of membership function parameters proceeds at a faster time scale in an environment decided by the architecture, inference mechanism, learning rule, type of inference system, and the problem. Hierarchy of the different adaptation procedures will rely on the prior knowledge. For example, if there is more prior knowledge about the architecture than the inference mechanism then it is better to implement the architecture at a higher level.

The cooperative and concurrent models that were discussed in the beginning sections of this chapter are not fully interpretable due to the presence of the black box concept. But a fused Neuro Fuzzy model is interpretable and capable of learning in a supervised mode. The learning

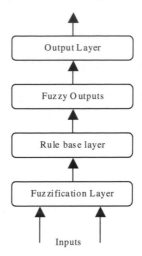

FIGURE 9.10: Architecture Design for Evolutionary Design of Neuro-Fuzzy Systems

TABLE 9.1: Performance of NF systems and ANN

System	Epochs	RMSE
ANFIS	75	0.0017
NEFPROX	216	0.0332
EFuNN	1	0.0140
DmEFuNN	1	0.0042
SONFIN	—	0.0180

process of FALCON, GARIC, ANFIS, NEFCON, SONFIN, FINEST and FUN are only concerned with parameter level adaptation within fixed structures and hence it is suitable only for small problems. The learning process and the determination of the rule base in large scale problems is complicated. In such a case the systems expects the user to provide the architecture details, type of fuzzy operators, etc. An important feature of EFuNN and DmEFuNN is the one pass (epoch) training, which is highly capable for online learning. As the problem becomes more complicated manual definition of NF architecture/parameters becomes complicated.

More specifically, for processes that require an optimal NF system, the evolutionary design approach seems to be the best solution. Table 9.1 provides a comparative performance of some Neuro-Fuzzy systems, where training was performed on 500 data sets and NF models were tested with another 500 data sets.

In this section we have presented the state of art modeling of different neuro-fuzzy systems. Due to the lack of a common framework it remains often difficult to compare the different neuro-fuzzy models conceptually and evaluate their performance comparatively. In terms of RMSE error, NF models using Takagi Sugeno FIS perform better than Mamdani FIS even though it is computationally expensive. The following section will discuss the ANFIS in detail. Some of the major requirements for neuro fuzzy systems to be more intelligent are fast learning, on-line adaptability (accommodating new features like inputs, outputs, nodes, connections, etc.), achieve a global error rate, and computationally inexpensive. Most of the NF systems use gradient descent techniques to learn the membership function parameters. In order to obtain faster learning and convergence, more efficient neural network learning algorithms like conjugate gradient search could be used.

9.4   Hybrid Neuro-Fuzzy Model — ANFIS

Several fuzzy inference systems have been described by different workers but the most commonly-used are Mamdani type and Takagi-Sugeno type, which is also known as Takagi-Sugeno-Kang type. In the case of a Mamdani type fuzzy inference system, both premise (if) and consequent (then) parts of a fuzzy if-then rule are fuzzy propositions. In the case of a Takagi-Sugeno type fuzzy inference system, where the premise part of a fuzzy rule is a fuzzy proposition, the consequent part is a mathematical function, usually a zero or first degree polynomial function. In ANFIS, a Sugeno type model, the parameters associated with a given membership function are chosen so as to tailor the input/output data set.

9.4.1   Architecture of Adaptive Neuro-Fuzzy Inference System

The basic architecture of an ANFIS and its learning algorithm for the Sugeno fuzzy model is illustrated in this section. Assume that the fuzzy inference system has two inputs m and n and one output y. For a first-order Sugeno fuzzy model, a typical rule set with two fuzzy if-then rules can be expressed as:

Rule 1: $If(misA1)and(nisB1)then : y_1 = \alpha_1 m + \beta_1 n + \gamma_1$
Rule 2: $If(misA2)and(nisB2)then : y_2 = \alpha_2 m + \beta_2 n + \gamma_2$

where $\alpha_1, \beta_1, \gamma_1, \alpha_2, \beta_2, \gamma_2$, are linear parameters, and A1 , A2 , B1 and B2 are nonlinear parameters. The corresponding equivalent ANFIS architecture is as shown in Figure 9.11. The entire system architecture consists of five layers, namely, a fuzzy layer, a product layer, a normalized layer, a defuzzy layer, and a total output layer. The functionality of each of these layers is described in the following sections.

Layer 1 - Fuzzy Layer

- Let $O_{1,i}$ be the output of the ith node of the layer 1

- Every node i in this layer is an adaptive node with a node function $O_{1,i} = \mu_{Ai}(x)\, for\, i = 1, 2;\, or\, O_{1,i} = \mu_{Bi}(y)\, for\, i = 1, 2$, where x is the input to node i and Ai is the linguistic label (small, large, etc.) associated with this node function. It other words, $O1, i$ is the membership function of A_i and it specifies the degree to which the given x satisfies the quantifier A_i. The most commonly used membership functions are Bell shaped and Gaussian membership functions.

- The bell shaped membership function is given by

$$f(x,a,b,c) = \frac{1}{1 + \frac{x-c}{a}^{2b}}$$

where the parameter b is usually positive. The parameter c locates the center of the curve.

- The Gaussian membership function is given by

$$A(x) = e^{\frac{(x-c)^2}{2a^2}}, \text{ where } 2\sigma^2 > 0.$$

- The parameters in Layer I are referred to as the Premise parameters.

Layer 2 - Product Layer

- Each node in this layer contains a prod t-norm operator as a node function.

- This layer synthesizes information transmitted by Layer 1 and multiplies all the incoming signals and sends the product out.

- The output of the product layer is given by

$$O_{2,i} = \mu Ai(x)\mu_{Bi}(y) = W_i$$

- Each node in this layer serves as a measure of strength of the rule.

- The output of this layer acts as the weight functions.

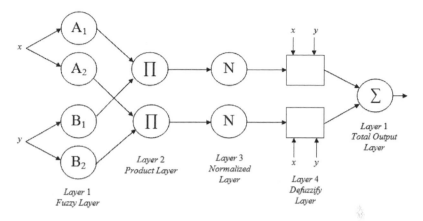

FIGURE 9.11: ANFIS Architecture

Layer 3 - Normalized Layer

- Each node in this layer normalizes the weight functions obtained from the previous product layer.

- The normalized output is computed for the ith node as the ratio of the ith rule's firing strength to the sum of all rule's firing strengths as follows

$$O_{3,i} = \frac{W_i}{w_1 + w_2}$$

Layer 4 - Defuzzify Layer

- The nodes in this layer are adaptive in nature.

- The defuzzified output of this layer is computed according to the formula

$$O_{4,i} = O_{3,i}(\alpha_{4,i} = O3, i(\alpha ix) + \beta_i y + \gamma_i)$$

where α_i, β_i and γ_i are the linear or consequent parameters of the corresponding node i.

Layer 5 - Total Output Layer

- The single node in this layer synthesizes information transmitted by Layer 4 and returns the overall output using the following fixed function

$$O_{5,i} = \frac{\sum w_i y_i}{\sum}$$

9.4.2 Hybrid Learning Algorithm

The hybrid learning algorithm of ANFIS proposed by Jang, Sun, and Mizutani is a combination of Steepest Descent and Least Squares Estimate Learning. Let the total set of parameters be S and let S1 denote the premise parameters and S2 denote the consequent parameters. The premise parameters are known as nonlinear parameters and the consequent parameters are known as linear parameters. The ANFIS uses a two pass learning algorithm: forward pass and backward pass. In forward pass the premise parameters are not modified and the consequent parameters are computed using the Least Squares Estimate Learning algorithm. In backward pass, the consequent parameters are not modified and the premise parameters are computed using the gradient descent algorithm. Based on these two learning algorithms, ANFIS adapts the parameters in the adaptive network. From the architecture, it is clear that the overall output of the ANFIS can be represented as a linear combination of the consequent parameters as

$$f = \frac{w_1}{w_1 + w_2} f_1 + \frac{w_2}{w_1 + w_2} f_2 = \bar{w}_1 f_1 + \bar{w}_2 f_2$$

In forward pass the signals move in forward direction till layer 4 and the consequent parameters are computed while in the backward pass, the error rates are propagated backward and the premise parameters are updated by the gradient descent method.

Forward Pass: Least Squares Estimate Method

The least squares estimate method estimates the unknown values of the parameters, $\alpha_0, \alpha_1 \Lambda$ from the regression function, $f(\bar{x}; \bar{\alpha})$ by finding numerical values for the parameters that minimize the sum of the squared deviations between the observed responses and the functional portion of the model. The mathematical expression to obtain the parameter estimates is

$$Q = \sum_{i=1}^{n} [y_i - f(\bar{x}; \bar{\alpha})]^2$$

Here $\alpha_0, \alpha_1, \ldots$ are the variables to be optimized and $x_1, x_2, \ldots$ are the predictor variable values which are considered as coefficients. For linear models, the least squares minimization is usually done analytically using calculus. For nonlinear models, on the other hand, the minimization must almost always be done using iterative numerical algorithms.

The steps to compute the consequent parameters by maintaining the premise parameters fixed during the forward pass are:

Step 1: Present the input vectors to the ANFIS model.

Step 2: Compute the node outputs of each layer.

Step 3: Repeat Step 2 for all the presented input signal.

Step 4: Compute the consequent parameters using the LSE algorithm.

Step 5: Compute the error measure for each training pair.

Backward Pass: Steepest Descent Method

Though the steepest descent method is one of the oldest methods, it is best suitable for solving constrained as well as unconstrained problems. This method is a most frequently used nonlinear optimization technique.

Let $z = f(\bar{X})$ be a function of $\bar{X}$ such that $\frac{\partial f(\bar{X})}{\partial x_k}$ exists for k = 1,2, ... ,n. The gradient of $f(\bar{X})$, denoted by $\nabla f(\bar{X})$, is the vector $\nabla f(\bar{X}) = \frac{\partial f(\bar{X})}{\partial x_1}, \frac{\partial f(\bar{X})}{\partial x_2}, ... \frac{\partial f(\bar{X})}{\partial x_n}$

The direction of the gradient points to the local steepest downhill direction at the greatest rate of increase of rate of $f(\bar{X})$. Therefore - $\nabla f(\bar{X})$ points in the direction of greatest decrease f($\bar{X}$). An iterative search procedure is performed, starting from point $\bar{P}_0$ in the direction $\bar{S}_0 = \frac{-\nabla f(\bar{P}_0)}{\|-\nabla f(\bar{P}_0)\|}$. Upon this search a point $\bar{P}_1$ is obtained at which the local minimum occurs according to the constraint $\bar{X} = \bar{P} + v\bar{S}_0$. This search and minimization continues to produce a sequence of points with the property $f(\bar{P}_0) > f(\bar{P}_1) > ... > ...if \lim_{k \to \infty} \bar{P}_k = P, then f(\bar{P})$ will be a local minimum.

The basic algorithm of the steepest descent search method is given as follows:

Step 1: Evaluate the gradient vector $\nabla f(\bar{P}_k)$.

Step 2: Calculate the direction of search $\bar{S}_k = \frac{-\nabla f(\bar{P}_k)}{\|-\nabla f(\bar{P}_k)\|}$.

Step 3: Execute a single parameter minimization of $\phi(v) = f(\bar{P}_k + v\bar{S}_k)$ on the interval [0,b]. Upon execution a value $v = h_m in$ is produced at which the local minimum for $\phi(v)$ lies. The relation $\phi(v) = f(\bar{P}_k + h_m in\bar{S}_k)$ shows that this is a minimum for $f(\bar{X})$ along the search line $\bar{X} = \bar{P}_k + v\bar{S}_k$.

Step 4: Search for the next point $\bar{P}_{k+1} = \bar{P}_k + \bar{h}_{min}\bar{S}_k$.

Step 5: Test for termination condition.

Step 6: Repeat the entire process.

The termination condition is obtained when the function values $f(\bar{P}_k)$ and $f(\bar{P}_{k+1})$ are sufficiently close enough to each other and such that the distance $\| P_{k+1} - \bar{P}_k \|$ is a minimum.

9.5   Classification and Regression Trees

9.5.1   CART — Introduction

Classification and Regression Trees also referred to as CART was first proposed by Breiman et al. during 1984. CART are tree-structured, non-parametric computationally intensive techniques that find their application in various areas. Most commonly CART are used by technicians to compute results at a faster rate. These trees are not only used to classify entities into a discrete number of groups, but also as an alternative approach to regression analysis in which the value of a response (dependent) variable is to be estimated, given the value of each variable in a set of explanatory (independent) variables.

In a binary tree classifier a decision is made at each non-terminal node of the tree based upon the value of the available attributes. A threshold value is set and the attribute value is compared against the threshold. If the attribute value is less than the threshold then the left branch of the tree is selected, otherwise the right branch is selected. The classes to which these attributes are to be classified may be the leaves, or terminal nodes of the tree. Based upon these classes, the tree is known as either a classification or regression tree. If the nature of the class is distinct then the tree is known as a classification tree. If the nature of the class is continuous then the tree is known as a regression tree.

Some of the features possesses by CART and made CART to be attractive are

- Intuitive representation, the resulting model is easy to understand and assimilate by humans.

- CART are nonparametric models, therefore they do not require any user intervention and hence are used for exploratory knowledge discovery.

- CART are scalable algorithms - their performance degrades gradually as the size of the training data increases.

- These decision trees are more accurate when compared with other higher models.

While constructing trees the first issue that has to be considered is to determine the binary splits of the parent node. The node splitting criteria is used to determine the binary splits.

9.5.2 Node Splitting Criteria

The algorithm of finding splits of nodes to produce a purer descendant is shown in Figure 9.12:

Let p denote the proportions of the class.

Let t be the root node.

Step 1: The node proportions are defines as p(j—t), where the proportions are belonging to a class j.

Step 2: The node impurity measure of node t is defined as i(t) as a nonnegative function ϕ.

Let Δ denote the candidate split of a node t, such that the split divides node t into t_{left} and t_{right} . The measure of split is now defined as Δi (t) = i(t) - $p_{left}i(t_{left}) - p_{right}i(t_{right})$.

Step 3: Define the set of binary splits such that the candidate split sends all values in t having a "true value" to t_{left} and all values in t having a "false value" to t_{right}.

At each node CART solves the maximization problem given as max[i(t) - $p_{left}i(t_{left}) - p_{right}i(t_{right})$].

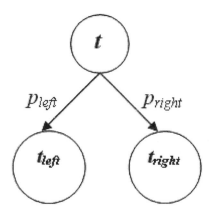

FIGURE 9.12: Node Splitting Algorithm of CART

The node impurity measure is defined using two basic impurity functions Gini splitting rule and Twoing splitting rule.

Gini Splitting Rule: This rule is also known as Gini index and defines the impurity function as $i(t) \sum_{i \neq} p(i|t)p(j|t)$ where i and j denote the class index. $p(i|t)$ denotes the proportion of class i. Gini algorithm is capable of searching the learning sample for the largest class and isolate it from the rest of the data. Gini is best suited for applications with noisy data.

Twoing Splitting rule: The Twoing Rule was first proposed by Breiman during 1984. The value of the impurity measure to be computed is defined as $i(t) = p_{left}p_{right} \sum |p(j|t_{left}) - p(j|t_{right})$ The Twoing Value is actually a goodness measure rather than an impurity measure.

9.5.3   Classification Trees

Construction of a classification tree is a process which requires two phases. During the first phase, the growth phase, a very large classification tree is constructed from the training data. During the second phase, the pruning phase, the ultimate size of the tree T is determined. Most classification tree construction algorithms grow the tree top-down in the greedy way.

The pseudocode of the Classification Tree Algorithm:

Step 1: Split the root node in a binary fashion with response to the query posed. Those attributes of the root node with the answer "true" form the left node and those with the answer "false" form the right node.

Step 2: Apply the node splitting criteria is applied to each of the split point using the equation $\nabla i(t) = i(t) - p_{left}i(t_{l}eft) - p_{right}i(t_{right})$.

Step 3: Determine the best split of the attribute which has high reduction in impurity.

Step 4: Repeat steps 1 to 3 for the remaining attributes at the root node.

Step 5: From the set of chosen best splits, assign rank to the best splits based on the reduction in impurity.

Step 6: Select the attribute and the corresponding split point that have the least impurity of the root node.

Step 7: Assign classes to the nodes based on the fact that minimizes the misclassification costs. The misclassification cost can also be defined by the user.

Step 8: Repeat steps 1 to 7 to each non terminal child node.

Step 9: Continue the splitting process until the stopping criteria is met.

9.5.4 Regression Trees

Breiman proposed the regression tree for exploring multivariate data sets. Initially the regression technique was applied to a data set consisting of a continuous response variable y and a set of predictor variables $x_1, x_2, ..., x_k$ which may be continuous or categorical. Regression trees modeled the response y as a series of 'if-then-else' rules according to the values of the predictors. The underlying concept behind building a regression tree is more or less similar to that of creating or building a classification tree. While creating regression trees, it is not necessary for the user to indicate the misclassification cost. Since the dependant attribute is continuous, the splitting criteria is employed within the node sum of squares of the dependent attribute. The goodness measure of a split is determined based on the reduction in the weighted sum of squares. The pseudocode for constructing a regression tree is:

Step 1: Split the root node and apply a node impurity measure to each split.

Step 2: Determine the reduction in impurity.

Step 3: Determine the best split of the attribute by using node splitting criteria and divide the parent node into right and left child nodes.

Step 4: Repeat steps 1 to 3 until the largest possible tree is obtained.

Step 5: Apply pruning algorithm to the largest tree to produce a set of subtrees and select the right-sized or optimal tree.

9.5.5 Computational Issues of CART

The computation of the best split conditions is quite complex in CART. Tree optimization implies choosing the right size of tree - cutting off insignificant nodes and even subtrees. Almost all the decision trees undergo pruning in order to avoid overfitting of the data. The methodology of removing leaves and branches in order to improve the performance of the decision tree is referred to as pruning. Pruning method allows the tree to grow to a maximum size, and then finally removes all the smaller branches that do not generalize.

Usually when the tree size is defined, a trade-off occurs between the measure of tree impurity and complexity of the tree, which is defined by total number of terminal nodes in the tree. Generally the impurity measure is zero for a maximum tree. During this case, the number of terminal nodes is always a maximum. To predict the optimal size of the tree cross validation technique can be applied.

The procedure involved in building a tree based model can be validated using an intensive method known as cross validation. In cross validation, the tree computed from one set of observations referred to as learning sample is compared with another completely independent set of observations referred to as the testing sample. The learning sample is randomly split into N sections, such that there is even distribution of outcomes in each of the N subsets of data. Among these subsets one is chosen as the testing sample and the remaining N-1 subsets form the learning samples. The model-building process is repeated N times therefore leading to N different models, each one of which can be tested against an independent subset of the data. In CART, the entire tree building and pruning sequence is conducted N times. Thus, there are N sequences of trees produced. If most or all of the splits determined by the analysis of the learning sample are essentially based on "random noise," then the prediction for the testing sample will be very poor. Hence one can infer that the selected tree is not an efficient tree. Using this method, a minimum cost occurs when the tree is complex enough to fit the information in the learning dataset, but not so complex that "noise" in the data is fit.

V-fold cross validation is also used to determine the optimal tree size. V-fold cross validation works by dividing the available data into equal-sized segments and providing one segment at a time for test purposes. If certain classes of the target variable have very small sample sizes it may not be possible to subdivide each class into v subsets. This method has proven to be more accurate since it does not require a separate and independent testing sample. This procedure avoids the problem of over fitting where the generated tree fits the training data well but does not provide accurate predictions of new data.

9.5.6 Computational Steps

The process of computing classification and regression trees can be characterized as involving four basic steps:

- Tree building

- Stopping splitting

- Tree pruning

- Selecting the "right-sized" tree

Tree Building

The initial process of the tree building step is partitioning a sample or the root node into binary nodes based upon the condition "is X ¡ d?" where X is a variable in the data set, and d is a real number. Based on the condition, all observations are placed at the root node. In order to find the best attribute, all possible splitting attributes (called splitters), as well as all possible values of the attribute to be used to split the node are determined. To choose the best splitter, the average "purity" of the two child nodes is maximized. Several measures of purity also known as splitting criteria or splitting functions can be chosen, but the most common functions used in practice are the "Gini measure" and the "Twoing measure." Each node including the root node is assigned to a predefined class. This assignment is based on probability of each class, decision cost matrix, and the fraction of subjects with each outcome in the learning sample. The assignment assures that the tree has a minimum decision cost.

Stopping Splitting

If the splitting is not stopped then the entire process gets very complex. The stopping condition for the tree building process or the splitting process occurs if any of the following condition satisfies:

- If there exists only one observation in each of the child nodes.

- If all observations within each child node have the identical distribution of predictor variables, making splitting impossible.

- If a threshold limit for the number of levels of the maximum tree has been assigned by the user.

Tree Pruning

Tree pruning improve the performance of the decision tree by avoiding overfitting of data. Pruning method allows the tree to grow to a maximum size, and then finally removes all the smaller branches that do not generalize. In minimum cost complexity pruning, a nested sequence of subtrees of the initial large tree is created by weakest-link cutting. With weakest-link cutting (pruning), all of the nodes that arise from a specific non-terminal node are pruned off, and the specific node selected is the one for which the corresponding pruned nodes provide the smallest per node decrease in the resubstitution misclassification rate. If two or more choices for a cut in the pruning process would produce the same per node decrease in the resubstitution misclassification rate, then pruning off the largest number of nodes is favored. A complexity parameter a is used for pruning, as a is increased, more and more nodes are pruned away, resulting in much simpler trees.

Selecting the Right Sized Tree

The maximal tree that has been built, may turn out to be of very high complexity and consist of hundreds of levels. Hence it is required to optimize this tree prior to be applied for the classification process.

The main objective in selecting the optimal tree, is to find the correct complexity parameter a so that the information in the learning samples is fit thereby avoiding overfitting.

The optimized maximal tree should obey a few properties as follows:

- The tree should be sufficiently complex to account for the known facts, while at the same time it should be as simple as possible.

- It should exploit information that increases predictive accuracy and ignore information that does not.

- It should, lead to greater understanding of the phenomena it describes.

Due to overfitting, the tree does not generalize well and when new data are presented, the performance of the tree is sub-optimal. To avoid the overfitting problem, stopping rules can be defined. This is based on threshold, that is when the node purity exceeds a certain threshold, or when the tree reaches a certain depth, the splitting process is stopped. Stopping too early might fail to uncover splits in the child variables with large decrease in node impurity had the tree been allowed to grow further. Breiman pointed out that "looking for the right stopping rule was the wrong way of looking at the problem." They proposed that the

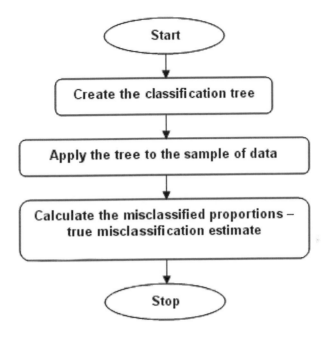

FIGURE 9.13:   Flowchart of Resubstitution Estimate

tree be allowed to be grown to its maximal size first, and then gradually shrunk by pruning away branches that lead to the smallest decrease in accuracy compared to pruning other branches. The search for the "right-sized" tree starts by pruning or collapsing some of the branches of the largest tree from the bottom up, using the cost complexity parameter and cross validation or an independent test sample to measure the predictive accuracy of the pruned tree.

9.5.7   Accuracy Estimation in CART

Almost all classification procedures produce errors, but accuracy is the most important attribute in CART. Breiman proposed three procedures for estimating the accuracy of tree-structured classifiers.

Resubstitution Estimate: The accuracy of the true misclassification rate is estimated as shown in Figure 9.13. The disadvantage of this misclassification estimate is that it is derived from the same data set from which the tree is built; hence, it under estimates the true misclassification rate. The error rate is always low in such cases.

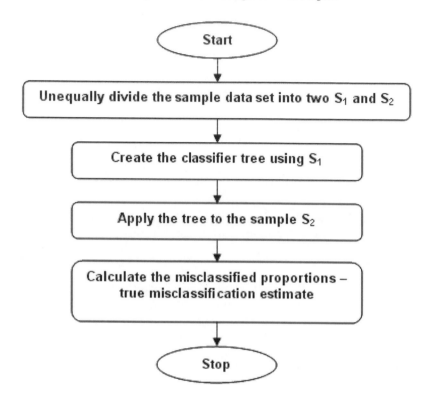

FIGURE 9.14:   Flowchart of Test-Sample Estimate

Test-Sample Estimate: This estimate is used to determine the misclassification rate when the sample size is too large. The methodology to determine the estimate the true misclassification rate is shown in Figure 9.14.

K-Fold Cross-Validation. This estimate is usually best suited for small sample sets. The procedure of computing the true misclassification rate is shown in Figure 9.15. The output obtained from the Figure is a series of test sample resubstitution estimates. These estimates are finally summed up and their average is computed to be the true misclassification rate.

9.5.8   Advantages of Classification and Regression Trees

Tree classification techniques have a number of advantages over many of the alternative classification techniques.

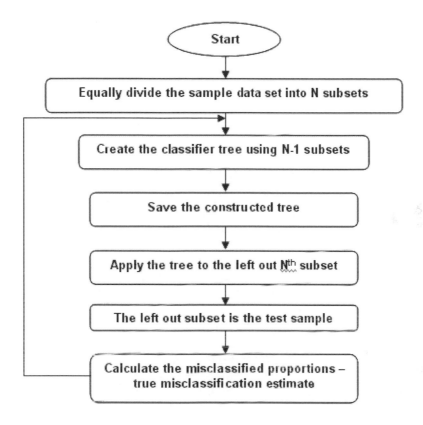

FIGURE 9.15: Flowchart of K-Fold Cross-Validation

- Simplicity of results — The interpretation of results in a tree is really very simple. This simplicity is useful not only for purposes of rapid classification of new observations, but can also frequently generate a much simpler "model" for explaining the reason for classifying or predicting observations in a particular manner.

- Is inherently non-parametric — no assumptions are made regarding the underlying distribution of values of the predictor variables.

- The flexibility to handle a broad range of response types, including numeric, categorical, ratings, and survival data.

- Invariance to monotonic transformations of the explanatory variables.

- It is a relatively automatic machine learning method.

- Ease and robustness of construction.

- Ease of interpretation.

- The ability to handle missing values in both response and explanatory variables.

9.6 Data Clustering Algorithms

One of the basic issues in pattern recognition is clustering which is the process of searching for structures in data. From a finite set of data X, several cluster centers can be determined. Traditional cluster analysis involves the process of forming a partition of X such that the degree of association is strong for data within blocks of the partition and weak for data in different blocks. However, this requirement is too strong in many practical applications and it is thus desirable to replace it with a weaker requirement. When the requirement of a crisp partition of X is replaced with a weaker requirement of a fuzzy partition or a fuzzy pseudo-partition on X, the emerging problem area is referred to as fuzzy clustering. Fuzzy pseudo-partitions are often called fuzzy c-partitions, where c designates the number of fuzzy classes in the partition. Some of the basic methods of fuzzy clustering are fuzzy c-means clustering method [FCM] and subtractive clustering.

9.6.1 System Identification Using Fuzzy Clustering

In general, the identification of the system involves structure identification and parameter identification. The structure is identified from a rule which is formed by removing all the insignificant variables, in the form of IF-THEN rules and their fuzzy sets. Parameter identification includes consequent parameter identification based on certain objective criteria. The model proposed by Takagi and Sugeno, which is called TS fuzzy model, the consequent part is expressed as a linear combination of antecedents. In TS model the system with N rules and m antecedents can be expressed as

$$R^1: \text{IF } x_1 \text{ is } A_{1^1} \text{ AND } x_2 \text{ is } A_{2^1} \text{ AND } \dots \text{ AND } x_m \text{ is } A_{m^1}$$
$$\text{THEN } y^1 = P_{0^1} + P_{1^1} x_1 + \dots + P_{m^1} x_m \dots \tag{9.39}$$

$$R^N: \text{IF } x_1 \text{ is } A_1{}^N \text{ AND } x_2 \text{ is } A_2{}^N \text{ AND } \dots \text{ AND } x_m \text{ is } A_m{}^N$$
$$\text{THEN } y^N = P_0{}^N + P_1{}^N x_1 + \dots + P_m{}^N x_m$$

where x_i is the ith antecedent (i=0,1, ... m)

R^j and y^j respectively represent the jth rule and its consequent (j=0,1, ... N)

P_i^j are the consequent parameters.

When input-output data are available a priori, fuzzy clustering is a technique that can be used for structure identification. Then, the consequent parameters can be optimized by least square estimation (LSE) given by Takagi and Sugeno. The identification of the system using fuzzy clustering involves formation of clusters in the data space and translation of these clusters into TSK rules such that the model obtained is close to the system being identified.

The fuzzy C-means (FCM) clustering algorithm, which has been widely studied and applied, needs a priori knowledge of the number of clusters. If FCM expects a desired number of clusters and if the positions for each cluster center can be guessed, then the output rules depend strongly on the choice of initial values. The FCM algorithm forms a suitable cluster pattern in order to minimize an objective function dependent of cluster locations through iteration. The number and initial location of cluster centers can also be automatically determined through search techniques which is available in the mountain clustering method. This method considers each discrete grid point as a potential cluster center by computing a search measure called the mountain function at each grid point. It is a subtractive clustering method with improved computational effort, in which the data points themselves are considered as candidates for cluster centers instead of grid points. Through application of this method, the computation is simply proportional to the number of data points and independent of the dimension of the problem.

In this method too, a data point with highest potential which is a function of the distance measure, is considered as a cluster center and data points close to new cluster center are penalized in order to control the emergence of new cluster centers.

The different types of clustering algorithms are discussed in the following:

9.6.2 Hard C-Means Clustering

The c-means algorithm tries to locate clusters in the multi-dimensional feature space. The goal is to assign each point in the feature space to a particular cluster. The basic approach is as follows:

Step 1: Seed the algorithm with c cluster centers manually, one for each cluster. (This step requires the prior information regarding the number of clusters into which the points are to be divided.)

Step 2: Assign each data point to the nearest cluster center.

Step 3: Compute a new cluster center for each class by taking the mean values of the coordinates of the points assigned to it.

Step 4: If stopping criteria is not met then go to step 2.

Some additional rules can be added to remove the necessity of knowing precisely how many clusters there are. The rules allow nearby clusters to merge and clusters, which have large standard deviations in coordinate to split. Generally speaking, the c-means algorithm is based on a c-partition of the data space U into a family of clusters $\{C_i\}$, i=1,2 ... c, where the following set-theoretic equations apply,

$$Y_{i=1}^{c} C_i = U \tag{9.40}$$

$$C_i \ I \ C_j = 0, \quad i \neq j \tag{9.41}$$

$$0 \subset C_i \subset U, all \ i \tag{9.42}$$

The set U=$\{u_1, u_2, ...u_K\}$ is a finite set of points in a space spanned by the feature axes, and c is the number of clusters. Here $2 \leq c \leq K$ because c=K clusters just places each data sample into its own cluster, and c=1 places all data samples into the same cluster. Equations (9.40) to (9.42) show that the set of clusters evacuates the whole universe, that there is no overlap between the clusters, and that a cluster can neither be empty nor contain all data samples. The c-means algorithm finds a center in each cluster thereby minimizing an objective function of a distance measure. The objective function depends on the distances between vectors u_k and cluster centers c_i , and when the Euclidean distance is chosen as a distance function, the expression for the objective function is :

$$J = \sum_{i=1}^{c} J_i = \sum_{i=1}^{c} \left(\sum_{k, u_k \epsilon C_j} \|u_k - C_j\|^2 \right) \tag{9.43}$$

where J_i is the objective function within cluster i. The partitioned clusters are typically defined by a cxK binary characteristic matrix M, called the membership matrix, where each element m_{ik} is 1 if the kth data point u_k belongs to cluster i, and 0 otherwise. Since a data point can only belong to one cluster, the membership matrix M has these properties:

- the sum of each column is one

- the sum of all elements is K

If the cluster centers c_i are fixed, the $m_i k$ that minimize J_i can be derived as

$$m_{ik} \left\{ 1 \ if \|u_k - c_i\|^2 \leq \|u_k - c_j\|^2 for \ each \ j \neq i \quad 0 \ otherwise \right. \quad (9.44)$$

That is u_k belongs to cluster i if c_i is the closest center among all centers. If, on the other hand m_{ik} is fixed, then the optimal center c_i that minimizes (5) is the mean of all vectors in cluster i:

$$c_i = \frac{1}{|C_i|} \sum_{k,u_k \epsilon C_i} u_k \quad (9.45)$$

where $|C_i|$ is the number of objects in C_i , and the summation is an element-by-element summation of vectors.

Algorithm

The hard c-means algorithm has five steps.

Step 1: Initialize the cluster centers u_i (i=1,2 ... c) by randomly selecting c points from the data points.

Step 2: Determine the membership matrix M by using equation (9.44).

Step 3: Compute the objective function (9.43). Stop if either it is below a certain threshold value, or its improvement over the previous iteration is below a certain tolerance.

Step 4: Update the cluster centers according to (9.45).

Step 5: Go to step 2.

The algorithm is iterative, and there is no guarantee that it will converge to an optimum solution. The performance depends on the initial positions of the cluster centers, and it is advisable to employ some method to find good initial cluster centers. It is also possible to initialize a random membership matrix M first and then follow the iterative procedure. For example, we have the feature space with two clusters.

The plot of the clusters in Figure 9.16 suggests a relation between the variable x on the horizontal axis and y on the vertical axis. For example, the cluster in the upper right hand corner of the plot indicates, in very loose terms, that whenever x is "high", defined as near the right end of

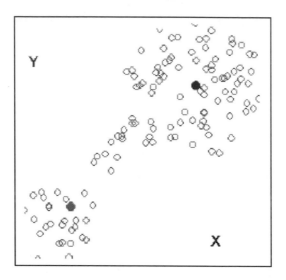

FIGURE 9.16: Example with Two Clusters. Cluster Centers are Marked with Solid Circles

the horizontal axis, then y is also "high", defined as near the top end of the vertical axis. The relation can be described by the rule:

$$\text{IF } x \text{ is high THEN } y \text{ is high} \qquad (9.46)$$

It seems possible to make some intuitive definitions of the two instances of the word 'high' in the rule, based on the location of the cluster center. The cluster in the lower left part of the figure, could be described as

$$\text{IF } x \text{ is low THEN } y \text{ is low} \qquad (9.47)$$

9.6.3   Fuzzy C-Means (FCM) Clustering

Occasionally the points between cluster centers can be assumed to have a gradual membership of both clusters. Naturally this is accommodated by fuzzifying the definitions of "low" and "high" in (9.46) and (9.47). The fuzzified c-means algorithm allows each data point to belong to a cluster to a degree specified by a membership grade, and thus each point may belong to several clusters.

The fuzzy c-means algorithm partitions a collection of K data points specified by m-dimensional vectors uk (k=1,2 ... K) into c fuzzy clusters, and finds a cluster center in each, minimizing an objective function. Fuzzy c-means is different from hard c-means, mainly because it employs fuzzy partitioning, where a point can belong to several clusters with degrees of membership. To accommodate the fuzzy partitioning, the membership matrix M is allowed to have elements in the range [0,1]. A point's total membership of all clusters, however, must always be equal to unity to maintain the properties of the M matrix. The objective function is a generalization of equation (9.43),

$$J(M, c_1, c_2, ...) = \sum_{i=1}^{c} J_i = \sum_{i=1}^{c} \sum_{k=1}^{K} m_{ik}^q d_{ik}^2 \qquad (9.48)$$

where m_{ik} is a membership between 0 and 1, c_i is the center of fuzzy cluster i, $d_{ik} = \|u_k - c_i\|$ is the Euclidean distance between the ith cluster center and k_{th} point, ϵ [1,∞] is a weighting exponent.

There are two necessary conditions for J to reach a minimum:

$$c_i = \frac{\sum_{k=1}^{K} m_{ik}^q u_k}{\sum_{k=1}^{K} m_{ik}^q} \qquad (9.49)$$

and

$$m_{ik} = \frac{1}{\sum_{j=1}^{c} (\frac{d_{ik}}{d_{jk}})^{2/(q-1)}} \qquad (9.50)$$

Algorithm

The algorithm is simply an iteration through the preceding two conditions. In a batch mode operation, the fuzzy c-means algorithm determines the cluster centers c_i and the membership matrix M using the following steps:

Step 1: Initialize the membership matrix M with random values between 0 and 1 within the constraints of (9.44).

Step 2: Calculate c cluster centers c_i (i=1,2 ... c) using (9.49).

Step 3: Compute the objective function according to (9.48). Stop if either it is below a certain threshold level or its improvement over the previous iteration is below a certain tolerance.

Step 4: Compute a new M using (9.50).

Step 5: Go to step 2.

The cluster centers can alternatively be initialised first, before carrying out the iterative procedure. The algorithm may not converge to an optimum solution and the performance depends on the initial cluster centers, just as in the case of the hard c-means algorithm.

9.6.4   Subtractive Clustering

Subtractive clustering is based on a measure of the density of data points in the feature space. The theme behind subtractive clustering is to determine regions in the feature space with high densities of data points. The point with the maximum number of neighbors is selected as center for a cluster. The data points within a prespecified, fuzzy radius are subtracted, and the algorithm looks for a new point with the highest number of neighbors. This process continues until all data points are examined.

Consider a collection of K data points specified by m-dimensional vectors u_k, k=1,2 ... K. Without loss of generality, the data points are assumed normalised. Since each data point is a candidate for a cluster center, a density measure at data point u_k is defined as:

$$D_k = \sum_{j=1}^{K} exp\left(-\frac{\|u_k - u_j\|}{(r_a/2)^2}\right) \qquad (9.51)$$

where r_a is a positive constant. A data point will have a high-density value if it has many neighboring data points. Only the fuzzy neighborhood within the radius r_a contributes to the density measure.

After calculating the density measure for each data point, the point with the highest density is selected as the first cluster center. Let u_{c1} be the point selected and D_{c1} its density measure. Next, the density measure for each data point u_k is revised by the formula:

$$D'_k = D_k - D_{c1} exp\left(-\frac{\|u_k - u_{c1}\|}{(r_b/2)^2}\right) \qquad (9.52)$$

where r_b is a positive constant. Therefore, the data points near the first cluster center u_{c1} will have significantly reduced density measures, making the points unlikely to be selected as the next cluster center. The constant r_b defines a neighborhood to be reduced in density measure. It is normally larger than ra to prevent closely spaced cluster centers; typically $r_b = 1.5* r_a$. Once the density measure for each point is revised, the next cluster center u_{c2} is selected and all the density measures are revised again. The process is repeated until a sufficient number of cluster centers are generated. When applying subtractive clustering to a set of input-output data, each of the cluster centers represents a rule. To generate rules, the cluster centers are used as the centers for the premise sets in a single type of rule base or the radial basis functions in a radial basis function neural network.

TABLE 9.2: Training Set

X1	−1.31	−0.64	0.36	1.69	−0.98	0.02	0.36	−0.31	1.02	−0.31	1.36	-1.31	
X2	−0.63	−0.21	-1.47	0.63	−0.63	1.47	0.21	0.21	−0.63	0.63	-0.63	1.89	-1.47	0.63

9.6.5 Experiments

The experiments illustrate the comparison between the fuzzy C-means clustering algorithm and the subtractive clustering algorithm to determine the most optimal algorithm and apply to the RBF neural networks.

Two training sets have been selected for that purpose:

a) To obtain two clusters for this training set Table 9.2 is used:

By means of the Fuzzy C-Means clustering algorithm the following cluster centers have been derived (Figure 9.17).

The initial cluster centers were generated arbitrarily, whereas the final ones were formed as a result of the FCM algorithm execution. In accordance with the algorithm, objective function values were computed by formula (11) and membership matrix M was calculated by formula (13). Membership function distribution for two clusters is shown in Figure 9.18.

In further experiments an attempt was made to enlarge the number of clusters. The following objective function values were derived:

2 clusters - Objective function = 17.75 (2 iterations)
3 clusters - Objective function = 9.45 (2 iterations)

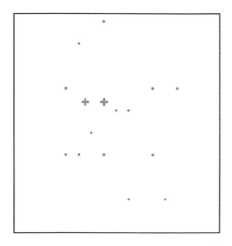 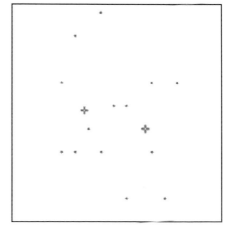

FIGURE 9.17: Initial (a) and Final (b) Cluster Centers

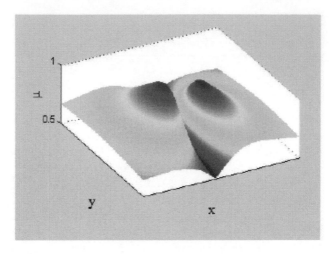

FIGURE 9.18:　Membership Function Distribution

4 clusters - Objective function = 4.84 (2 iterations).
The results are shown in Figures 9.19, 9.20, and 9.21.

　b) With the help of the subtractive algorithm experiments with the experimental dataset have been performed:

Figure 9.20 shows the results of the subtractive algorithm execution with the experimental dataset. The experiments have shown that the subtractive algorithm is hard to apply.

　c) With the help of the FCM algorithm experiments with the SPIRAL database have been performed:

Figures 9.21 and 9.22 shows the result of the FCM algorithm execution with the SPIRAL database.

　The experiments performed with the FCM method have shown that the results of the FCM algorithm execution are close to those of the K-means algorithm operation (in pattern recognition tasks).

　In the given work the possibilities of fuzzy clustering algorithms was described. Fuzzy c-means clustering is a clustering method where each data point may partially belong to more than one cluster with a degree specified by a membership function. FCM starts with an initial guess for the cluster center locations and iteratively updates the cluster centers and the membership grades for each data point based on minimizing a cost function. Subtractive clustering method is a fast, one-pass algorithm for estimating the number of clusters. It partitions the data into clusters

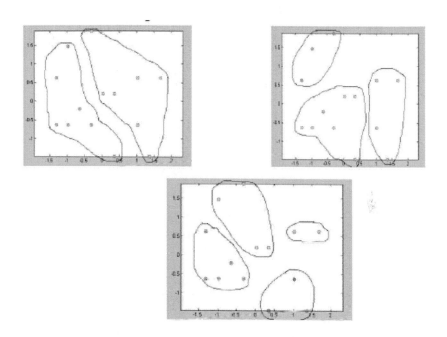

FIGURE 9.19: FCM Algorithm Results: Two, Three and Four Clusters

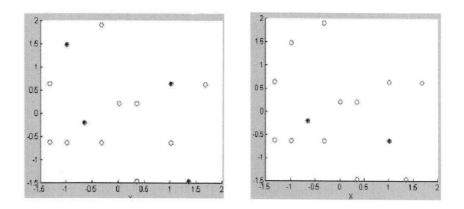

FIGURE 9.20: Subtractive Algorithm Results: Some Cluster Centers

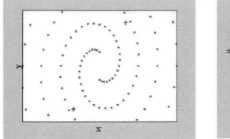

FIGURE 9.21:   Final Cluster Centers (a) and Membership Functions for Two Clusters (b)

and generates a FIS with the minimum numbers of rules required to distinguish the fuzzy qualities associated with each of the clusters. The main difference between fuzzy clustering and other clustering technique is that it generates fuzzy partitions of the data instead of hard partitions. Therefore, data patterns may belong to several clusters, having different membership values in each cluster.

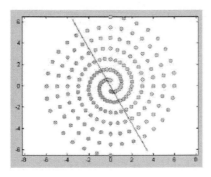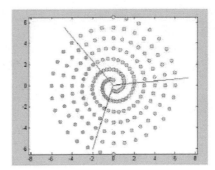

FIGURE 9.22:   FCM Algorithm Results for SPIRAL Data: Two and Three Clusters

Summary

The hybrid neuro-fuzzy approach provides a strong modeling framework for a consistent utilization of both conceptual and empirical components of earth science information for mineral potential mapping. By implementing a fuzzy inference system in the framework of an adaptive neural network, the hybrid neuro-fuzzy approach provides a robust data-based method for estimating the parameters of the fuzzy inference system. The output of a hybrid neuro-fuzzy model is not likely to be affected by the conditional dependence among two or more predictor maps. Moreover, multi-class predictor maps can be conveniently used in a hybrid neuro-fuzzy model. In the hybrid neuro-fuzzy approach, the problems related to dimensionality of input feature vectors can be addressed by using zero-order Takagi-Sugeno type fuzzy inference systems and/or statistical methods like factor analysis. Similar hybrid neuro-fuzzy inference system can be constructed and implemented for modeling larger-scale evidential maps to demarcate specific prospects within the predicted potentially-mineralized zones. The high performance levels of the hybrid neuro-fuzzy and neural network models, described, indicate that machine learning algorithms can efficiently recognize and account for possible conditional dependencies among input predictor patterns.

Review Questions

1. Explain concurrent and cooperative neuro-fuzzy systems.

2. Describe FALCON architecture.

3. Explain briefly on GARIC and NEFCON.

4. Compare FUN and EfuNN.

5. Explain the theoretical background of ANFIS model.

6. In detail, explain the layered architecture and the hybrid learning algorithm of ANFIS model.

7. Describe the different stages in implementing a hybrid neuro fuzzy model.

8. What are classification and regression trees?

9. Delineate on the General Computation Issues and Unique Solutions of C&RT.

10. What are the three estimates of accuracy in classification problems?

11. Mention the advantages of classification and regression trees.

12. What are data clustering algorithms? What is the need for these algorithms?

13. Explain hard c-means clustering algorithm.

14. Describe the Fuzzy C-means clustering algorithm in detail.

15. Differentiate Fuzzy C-means and subtractive clustering algorithm.

16. How is fuzziness incorporated in a neural net framework?

Chapter 10

Neuro-Fuzzy Modeling Using MATLAB

In this chapter, MATLAB illustrations are given on ANFIS, Classification and Regression trees, Fuzzy c-means clustering algorithms, Fuzzy ART Map, and Simulink models on Takagi–Sugeno inference systems.

10.1   Illustration 1 - Fuzzy Art Map

```
   Simplified Fuzzy Art Map
%
% SFAM usage demo.
%
function sfam_demo

% load data
load demodata

% create network
net = create_network(size(data,2))

% change some parameters as you wish
% net.epochs = 1;

% train the network
tnet = train(data, labels, net, 100)

% test the network on the testdata
r = classify(testdata, tnet, testlabels, 10);

% compute classification performance
fprintf(1,'Hit rate: % f \ n', sum(r' ==
```

```
testlabels)*100/size(testdata,1));
% - - - - - - - - - - - - - - - - - - - - - - - - - - - - -
Subprograms used

function results = classify(data, net, labels, debug);
% CLASSIFY Classifies the given data using the given
% trained SFAM.
% RESULTS = CLASSIFY(DATA, NET, LABELS, DEBUG)
% DATA is an M-by-D matrix where M is the number of
% samples and D is the size of the feature
% space. NET is a previously trained SFAM network.
% LABELS is a M-vector containing the correct labels
% for the data.If you don't have them, give it as an
% empty-vector [ ]. DEBUG is a scalar to control
% the verbosity of the program during training.If
% 0, nothing will be printed, otherwise every DEBUG
% iterations an informatory line will be printed.
%

results = [ ];
hits=0;

tic;
for s=1:size (data,1)
   input = data(s,:);
   % Complement code input
   input = [input 1-input];

   % Compute the activation values for each
   % prototype.
   activation = ones(1,length(net.weights));
   for i=1:length(net.weights)
     activation(i)=sum(min(input,net.weights{i}))/...
        (net.alpha + sum(net.weightsi));
   end

   % Sort activation values
   [sortedActivations, sortedIndices] =
                   sort(activation,'descend');
   % Iterate over the prototypes with decreasing
   % activation-value results(s)=-1;
   for p=sortedIndices
      % Compute match of the current candidate
```

```
    % prototype
    match = sum(min(input,net.weightsp))/net.D;

    % Check resonance
    if match>=net.vigilance
        results(s) = net.labels(p);

        if ~ isempty(labels)
            if labels(s)==results(s),
                hits = hits + 1; end;
        end

        break;
    end
end

if mod(s,debug)==0
    elapsed = toc;
    fprintf(1,'Tested % 4dth sample. Hits so far:
    % 3d which is %.3f%%. \ tElapsed
    %.2t seconds.n',s,hits,100*hits/s,elapsed);
    tic;
end
end % samples loop
% - - - - - - - - - - - - - - - - - - - - - - - - -


function net = create_network (num_features, varargin)
% Network can be configured by giving parameters in
% the format: NET = CREATE_NETWORK(NUM_FEATURES,
% 'paramname1', value1,'paramname2', value2, ...)
% Valid parameters are:D, max_categories, vigilance,
% alpha, epochs, beta, epsilon, singlePrecision
%

net = struct ( ...
    'D'               , num_features ...
    ,'max_categories' , 100 ...
    ,'vigilance'      , 0.75 ...
    ,'alpha'          , 0.001 ...
    ,'epochs'         , 10 ...
    ,'beta'           , 1 ...
    ,'weights'        , cell(1) ...
```

```
    ,'labels'          , [ ] ...
    ,'epsilon'         , 0.001 ...
    ,'singlePrecision', false ...
);
for i=1:2:length(varargin)
    net = setfield(net,varargini,varargini+1);
end
% - - - - - - - - - - - - - - - - - - - - - - - - - - - -

function net = train(data, labels, net,debug)

% dbstop in train at 18

for e=1:net.epochs
   network_changed = false;

   tic;
for s=1:size(data,1)

      if mod(s,debug)==0
         elapsed = toc;
         fprintf(1,'Training on % dth sample,
         in % dth epoch. \5# of prototypes=
         % 4d \ tElapsed seconds:
         % f \ n',s,e,length
         (net.weights),elapsed);
         tic;
      end

      input = data(s,:);
      input_label = labels(s);

      % Complement code input
      input = [input 1-input];

      % Set vigilance
      ro = net.vigilance;

      % By default, create_new_prototype=true. Only
      % if 'I' resonates with one of the existing
      % prototypes, a new prot.will not be created
      % create_new_prototype = true;
      % Compute the activation values for each
```

```
% prototype.
activation = ones(1,length(net.weights));
for i=1:length(net.weights)
    activation(i) =
        sum(min(input,net.weights{i}))/...
        (net.alpha + sum(net.weights{i}));
end

    % Sort activation values
    [sortedActivations, sortedIndices] =
            sort(activation,'descend');

    % Iterate over the prototypes with decreasing
activation-value for p=sortedIndices
        % Compute match of the current candidate
        % prototype
        match = sum(min(input,net.weights{p}))
            /net.D; % see note [1]

        % Check resonance
        if match>=ro
                % Check labels
                if input_label==net.labels(p)
                    % update the prototype
                    net.weightsp = net.beta*(min(input,
                        net.weights{p})) + ...
                        (1-net.beta)*net.weightsp;
                    network_changed = true;
                    create_new_prototype = false;
                    break;
                else
                    % Match-tracking begins.
                    % Increase vigilance
                    ro = sum(min(input,net.weights{p}))
                        /net.D + net.epsilon;
                end
            end
    end
    if create_new_prototype
    new_index = length(net.weights)+1;
        if net.singlePrecision
            net.weightsnew_index =
                ones(1,2*net.D,'single');
```

```
        else
           net.weightsnew_index =
                      ones(1,2*net.D);
        end

        net.weightsnew_index =
            net.beta*(min(input,net.weights
               {new_index})) + ...
               (1-net.beta)*net.weightsnew_index;

        net.labels(new_index) = input_label;
        network_changed = true;
      end
    end % samples loop

    if ~ network_changed
       fprintf(1,'Network trained in % d epochs.
                     \ n',e);
       break
    end
  end % epochs loop
```

Observations:

```
tnet =
D: 2
max_categories : 100
vigilance : 0.7500
alpha : 1.0000e-003
epochs : 10
beta : 1
weights : [ ]
labels : [ ]
epsilon : 1.0000e-003
singlePrecision : 0


tnet =
D : 2
max_categories : 100
vigilance : 0.7500
alpha : 1.0000e-003
epochs : 1
```

TABLE 10.1: Training Samples

Training Sample	No. of Prototypes	Elapsed Time (seconds)
100	20	0.047000
200	27	0.047000
300	29	0.062000
400	29	0.047000
500	31	0.063000
600	33	0.062000
700	34	0.063000
800	35	0.062000
900	37	0.063000
1000	38	0.063000

```
beta : 1
weights : 1x38 cell
labels : [2 1 2 2 1 1 2 2 1 2 1 1 1 1 2 2 1 1 1 2
          2 1 1 2 1 2 1 2 1 1 2 1 2 2 2 1 1 2 2 2]
epsilon : 1.0000e-003
singlePrecision : 0
```

Hit rate: 96.000000

TABLE 10.2: Testing Samples

Tested Sample	No. of Hits	Elapsed Time) (seconds)
10	10	0.00
20	20	0.00
30	30	0.02
40	39	0.00
50	49	0.00
60	58	0.02
70	68	0.00
80	77	0.00
90	86	0.02
100	96	0.00

10.2   Illustration 2: Fuzzy C-Means Clustering - Comparative Case Study

Two levels of fuzzy c-means clustering are illustrated in this program. FCM Clustering often works better than Otsu's method which outputs larger or smaller threshold on fluorescence images.

```
- - - - - - - - - - - - - - - - - - - - - - - - - - -
% Main Program
- - - -- - -- -- - -- - - - - - - - - - - - - - -
clear;close all;
im=imread('cell.png');
J = filter2(fspecial('sobel'), im);
fim=mat2gray(J);
level=graythresh(fim);
bwfim=im2bw(fim,level);
 [bwfim0,level0]=fcmthresh(fim,0);
 [bwfim1,level1]=fcmthresh(fim,1);
subplot(2,2,1);
imshow(fim);title('Original');
subplot(2,2,2);
imshow(bwfim);title(sprintf('Otsu,level=% f',
               level));
subplot(2,2,3);
imshow(bwfim0);title(sprintf('FCM0,level=% f',
               level0));
subplot(2,2,4);
imshow(bwfim1);title(sprintf('FCM1,level=% f',
               level1));
- - - - - - - - - - - - - - - - - - - - - - - - - - -
% Subprograms
- - - - - - - - - - - - - - - - - - - - - - - - - - -
function [bw,level]=fcmthresh(IM,sw)
% Thresholding by fuzzy c-means clustering
% [bw,level]=fcmthresh(IM,sw) outputs the binary
      image bw and threshold level of
% image IM using a fuzzy c-means clustering.
% sw is 0 or 1, a switch of cut-off position.
% sw=0, cut between the small and middle class
% sw=1, cut between the middle and large class
```

```
% check the parameters
if (nargin<1)
        error('You must provide an image.');
elseif (nargin==1)
error('You must provide an image.');
elseif (nargin==1)
sw=0;
elseif (sw~=0 && sw~=1)
        error('sw must be 0 or 1.');
end

data=reshape(IM,[],1);
 [center,member]=fcm(data,3);
 [center,cidx]=sort(center);
member=member';
member=member(:,cidx);
 [maxmember,label]=max(member,[],2);
if sw==0
   level=(max(data(label==1))+min(data(label==2)))/2;
else
   level=(max(data(label==2))+min(data(label==3)))/2;
end
bw=im2bw(IM,level);
```

Observations:

Figure 10.1 shows the original image and the clustered images. Fuzzy c means clustering is compared with the Ostu's method and it is seen that FCM has better performance than the Ostu's method. The threshold is 0 or 1, based on the switch cut-off position. If the threshold level is 0 then the cut is between the small and middle class, and if the threshold level is 1 the cut is between the middle and large class.

10.3   Illustration 3 - Kmeans Clustering

This illustration shows the implementation of K-means algorithm using MATLAB to cluster a given image and form a re-clustered image.

```
% MATLAB Code
fprintf('\n Execution Starts...');
```

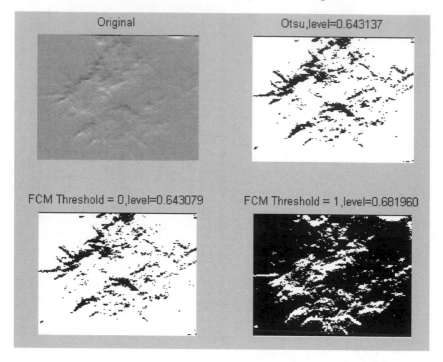

FIGURE 10.1: Plot of clustering using Ostu's method and FCM method

```
clear;
close all;
% Size Of the Image
wd=256;

X=im2double((imread('ALLPICDATA.bmp')));
X=imresize(X,[wd wd]);
X=im2uint8(X);
X=double(X);

mean1 = mean(mean(X(:,:,1)));
mean2 = mean(mean(X(:,:,2)));
mean3 = mean(mean(X(:,:,3)));

mean_mat = [mean1 mean2 mean3];
  [dum,dom]=max(mean_mat);
```

```
switch dom
    case 1
        plane1=2;
        plane2=3;
    case 2
        plane1=1;
        plane2=3;
    case 3
        plane1=1;
        plane2=2;
end

PP=X(:,:,dom);

no_of_cluster = 5;
ipimage=PP;
e=0.03;
thrd=e*mean(mean(ipimage));

% Initial Centroids
miniv = min(min(ipimage));
maxiv = max(max(ipimage));
range = maxiv - miniv;
stepv = range/no_of_cluster;
incrval = stepv;
for i = 1:no_of_cluster
    K(i).centroid = incrval;
    incrval = incrval + stepv;
end

imax=size(ipimage,1);
jmax=size(ipimage,2);

% Initialization Starts...here...
for i=1:imax
    for j=1:jmax
        opimage(i,j).pixel=ipimage(i,j);
    end
end

loop=1;
for ii=1:no_of_cluster
    dist(ii)=K(ii).centroid;
```

```
end
while dist>thrd
temp=K;
% Initial Clustering...
for i=1:imax
    for j=1:jmax
        for k=1:no_of_cluster
        diff(k)=abs(ipimage(i,j)-K(k).centroid);
        end
         [y,ind]=min(diff);
         opimage(i,j).index=ind;
    end
end

% New Centroids...
summ=zeros(no_of_cluster,1);
count=zeros(no_of_cluster,1);

sum1 = zeros(no_of_cluster,1);
sum2 = zeros(no_of_cluster,1);

for i=1:imax
    for j=1:jmax
        for k=1:no_of_cluster
            if(opimage(i,j).index==k)
                summ(k)=summ(k)+opimage(i,j).pixel;
                sum1(k)=sum1(k)+X(i,j,plane1);
                sum2(k)=sum2(k)+X(i,j,plane2);
                count(k)=count(k)+1;

            end
        end
    end
end

% Update Centroids...
for k=1:no_of_cluster
    K(k).centroid=summ(k)/count(k);
end
fprintf('\nNew Loop...');
loop=loop+1
clear dist;
% New distance...
```

```
for ii=1:no_of_cluster
    dist(ii)=abs(temp(ii).centroid-K(ii).centroid);
end
end
% End Of while Loop

for k=1:no_of_cluster
    Other_Plane1(k).centroid=sum1(k)/count(k);
    Other_Plane2(k).centroid=sum2(k)/count(k);
end
% Resultant Image
reimage=zeros(wd,wd,3);
for i=1:imax
    for j=1:jmax
        reimage(i,j,dom) = K(opimage(i,j).index).centroid;
        reimage(i,j,plane1) = Other_Plane2(opimage(i,j).
        index).centroid;
        reimage(i,j,plane1) = Other_Plane2(opimage(i,j).
        index).centroid;
    end
end

% End Of K-Means Algorithm...

% Results...
for k=1:no_of_cluster
    rimage =zeros(imax,jmax,3);
    for i=1:imax
        for j=1:jmax
            if opimage(i,j).index==k
            rimage(i,j,dom) = K(k).centroid;%
            opimage(i,j).pixel;
            rimage(i,j,plane1) = Other_Plane1(k).centroid;
            rimage(i,j,plane2) = Other_Plane2(k).centroid;
            end
        end
    end
    strbuf = sprintf('Cluster No % d ',k);
    figure,imshow(uint8(rimage));
    title(strbuf);
end

figure;
```

FIGURE 10.2:   Input Image

```
imshow(uint8(reimage));title('Re Clustured Image');
fprintf('\n Done...');
```

Observations:

Figures 10.2 to 10.8 show the original image and the clustered images.

FIGURE 10.3:   Clustered Image 1

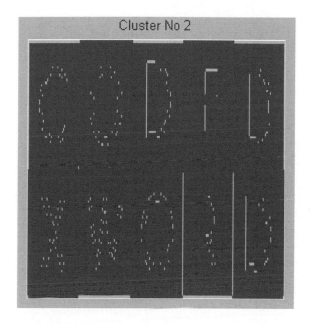

FIGURE 10.4: Clustered Image 2

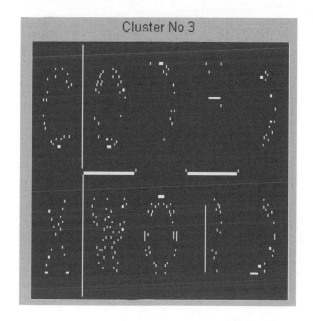

FIGURE 10.5: Clustered Image 3

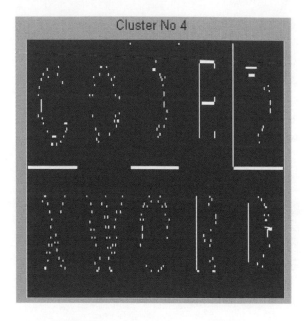

FIGURE 10.6:   Clustered Image 4

FIGURE 10.7:   Clustered Image 5

FIGURE 10.8: Clustered Image 6

10.4 Illustration 4 - Neuro-Fuzzy System Using Simulink

The function $y = f(x) = -2x - x^2$ is used to simulate the neuro fuzzy system using SIMULINK. A Fuzzy System is formed to Approximate Function f, when $x \epsilon$ [-10,10]. The input output data is generated and plotted as shown in Figure 10.9.

```
clc;
x=[-10:.5:10]';
y=(-2*x)-(x.*x);
plot(x,y)
grid
xlabel('x');ylabel('output');title('Nonlinear
characteristics')
    % The data is stored in appropriate form for genfis1
% and anfis and plotted as in Figure 10.10
data=[x y];
trndata=data(1:2:size(x),:);
```

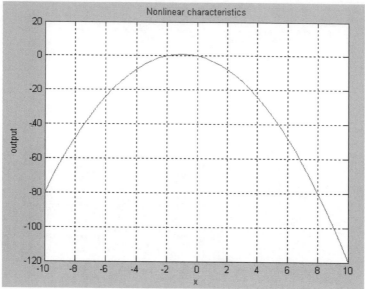

FIGURE 10.9: Plot of the Parabolic Function

```
chkdata=data(2:2:size(x),:);
plot(trndata(:,1),trndata(:,2),'o',chkdata(:,1),
chkdata(:,2),'x')
xlabel('x');ylabel('output');title('Measurement
data');grid

  % The fuzzy system is initialized with the command
genfis1 nu=5; mftype='gbellmf'; fismat=genfis1
(trndata, nu, mftype);

% The initial membership functions produced by genfis1
% are plotted in Figure 10.11

plotmf(fismat,'input',1)
xlabel('x');ylabel('output');title('Initial membership
functions');
grid
  % Next apply anfis-command to find the best FIS system
- max number of
% iterations = 100
```

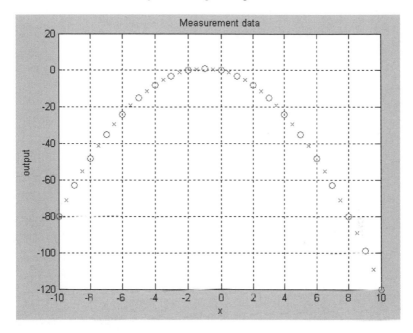

FIGURE 10.10: Training Data (o) and Checking Data (x) Generated from the Parabolic Equation

```
numep=100;
[parab, trnerr, ss, parabcheck, chkerr] = anfis
(trndata, fismat, numep,
[],chkdata);

% The output of FIS system is evaluated using input x
and is plotted
% as shown in Figure 10.12
anfi=evalfis(x,parab);
plot(trndata(:,1),trndata(:,2),'o',chkdata(:,1),
chkdata (:,2), 'x',x,anfi,'-')
grid
xlabel('x');ylabel('output');title('Goodness of fit')
```

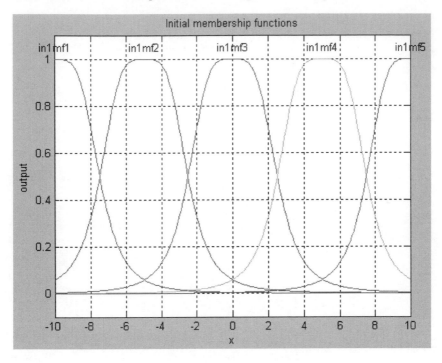

FIGURE 10.11:   Initial Fuzzy System (fismat) for ANFIS

10.5   Illustration 5 - Neuro-Fuzzy System Using Takagi–Sugeno and ANFIS GUI of MATLAB

Step 1: Choose a new Sugeno system from the Fuzzy Toolbox GUI (Figure 10.13)

Step 2: Load the Training data from workspace into the ANFIS editor (Figure 10.14)

Step 3: Generate the membership functions type (Figure 10.15). Here 5 gbellmf is selected with linear type.

Step 4: Train the network to see the performance. The error tolerance is chosen to be 0.001 and number of epochs is limited to 100 (Figure 10.16).

Error converges at the end of 100 epochs. The editor can be used to test the network using a testing pattern.

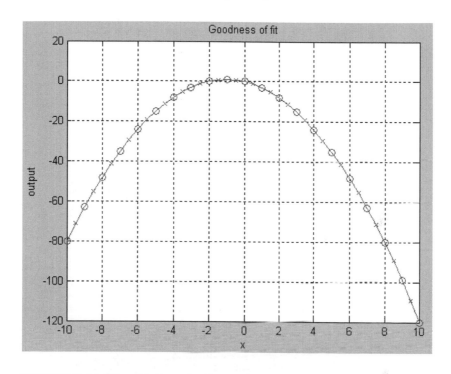

FIGURE 10.12: Fitting the Trained Fuzzy System on Training Data

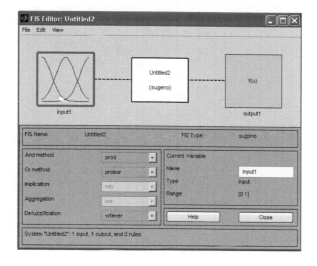

FIGURE 10.13: Display of New Sugeno-Type System

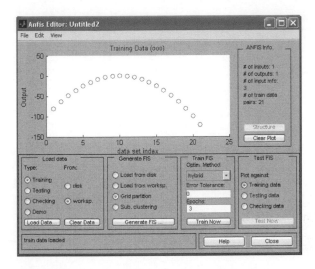

FIGURE 10.14: Training Data from Workspace

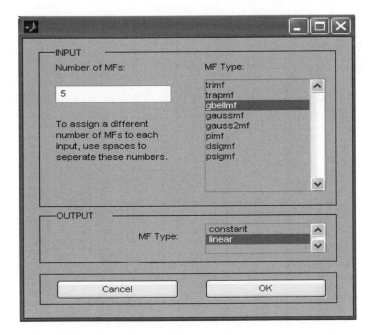

FIGURE 10.15: Generate the Membership Function Types

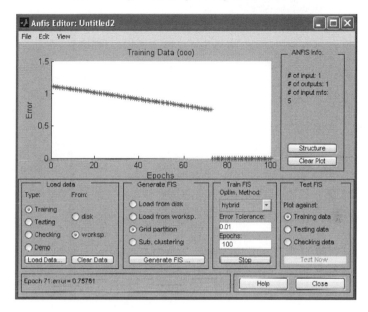

FIGURE 10.16:   Error at the End of 71 Epochs

Summary

By implementing a fuzzy inference system in the framework of an adaptive neural network, the hybrid neuro-fuzzy approach provides a robust data-based method for estimating the parameters of the fuzzy inference system. The output of a hybrid neuro-fuzzy model is not likely to be affected by the conditional dependence among two or more predictor maps. Moreover, multi-class predictor maps can be conveniently used in a hybrid neuro-fuzzy model. In the hybrid neuro-fuzzy approach, the problems related to dimensionality of input feature vectors can be addressed by using zero-order Takagi-Sugeno type fuzzy inference systems and/or statistical methods like factor analysis. For the implementation of Neuro-Fuzzy systems a set of MATLAB illustrations were given in this chapter, which the user can follow to construct their own model.

Review Questions

1. Implement the NEFCON learning rule in MATLAB and simulate the CSTR with PID controller.

2. Compare the maps of Fuzzy c-means and Subtractive clustering using MATLAB.

3. Write a MATLAB program to grow, prune, and plot a classification tree.

4. Implement the ANFIS model in MATLAB with the aid of a suitable example.

5. Using the NEFCON learning rule write a MATLAB program to control the speed of a DC motor.

Chapter 11

Evolutionary Computation Paradigms

11.1 Introduction

Evolutionary computation is a broad subfield of computational intelligence that requires combinatorial optimization problems. The evolutionary computation algorithm is an iterative process like growth or development in a population. The optimal solution is then obtained using parallel processing among the population. These procedures are frequently inspired by biological mechanisms of evolution. A population of individuals is exposed to an environment and responds with a collection of behaviors.

Some of these behaviors are better suited to meet the demands of the environment than are others. Selection tends to eliminate those individuals that demonstrate inappropriate behaviors. The survivors reproduce, and the genetics underlying their behavioral traits are passed on to their offspring. But this replication is never without error, nor can individual genotypes remain free of random mutations. The introduction of random genetic variation in turn leads to novel behavioral characteristics, and the process of evolution iterates. Over successive generations, increasingly appropriate behaviors accumulate within evolving phyletic lines. Evolution optimizes behaviors (i.e., the phenotype), because selection can act only in the face of phenotypic variation. The manner in which functional adaptations are encoded in genetics is transparent to selection; only the realized behaviors resulting from the interaction of the genotype with the environment can be assessed by competitive selection. Useful variations have the best chance of being preserved in the struggle for life, leading to a process of continual improvement. Evolution may in fact create "organs of extreme perfection and complication", but must always act within the constraints of physical development and the historical accidents of life that precede the current population. Evolution is entirely opportunistic, and can only work within the variation present in extant individuals.

The evolutionary computation process can be modeled algorithmically and simulated on a computer. The following difference equation gives a basic overview of an evolutionary model:

$$x[t + 1] = s(v(x[t]))\qquad(11.1)$$

where the population at time t, x[t], is operated on by random variation, v, and selection, s, to give rise to a new population x[t + 1]. The process of natural evolution is continuous while the artificial evolution occurs in discontinuous time intervals. Iterating variation and selection, an evolutionary algorithm is able to reach an optimal population on a response surface that represents the measurable worth of each feasible individual that may live in a population. Evolutionary computation is the field that studies the properties of these algorithms and similar procedures for simulating evolution on a computer.

The history of evolutionary computation spans nearly four decades. Several autonomous attempts were made in order to simulate evolution on a computer during the 1950s and 1960s. There are a few basic sub categories of evolutionary computation such as: evolution strategies, evolutionary programming, and genetic algorithms. All the subcategories begin with a population of contending trial solutions brought to a task at hand. Further new solutions are created by arbitrarily altering the existing solutions. A fitness measure is used to assess the performance of each trial solution and a solution is used to determine the retained parents. The deviations between the procedures are characterized by the typical data representations, the types of variations that are imposed on solutions to create offspring, and the methods employed for selecting new parents.

In this chapter, a brief history of Evolutionary Computation (EC) is discussed. This chapter enlightens the paradigms of EC such as Evolutionary Strategies and Evolutionary Programming. Genetic Algorithms and Genetic Programming will be discussed elaborately in the next chapter. This chapter also describes the advantages and disadvantages of EC.

11.2 Evolutionary Computation

Evolutionary computation is the study of computational methods based on analogy to natural evolution. Evolutionary computation is an example of biologically inspired computation. Other examples of biologically inspired computation include neural networks, which is computation based on analogy to animal nervous systems, artificial life, and artificial intelligence. During the evolutionary computation process a few populations of individuals undergo simulated evolution. While solving certain specific type of problems, these individuals generally comprise potential solutions to the problem. These solutions are presented by a kind of data structure. Selection mechanism is used by simulated evolution in order to find the best fit individuals which survive. The simulated evolution will include some variation-generating methods in which individuals can generate new individuals which are different from existing structures. Mutation and recombination (simulated mating) are two common variation-generating methods. Different evolutionary computation methods will differ in

- The data structure representation of individuals

- Variation producing mechanisms

- Parent selection methods

- Survival selection methods

- Population size

Following is a basic pseudo code for an evolutionary computation algorithm:

```
Generate an initial population in a random manner
While no stopping condition do
 Select parents based on fitness
 Produce offspring with some variants of their
parents
 Select a few individuals to die based on
fitness
End While
```

The fitness function of individuals is usually determined by the problem to be solved. Evolutionary computation is more commonly used in the field of optimization, when a function with several parameters is specified, and the aim is to find the parameter settings that give the largest or optimal value of the function. EC does not assure to find the optimal value of the function, EC merely attempts to find a parameter setting that gives a "large" or "near optimal" function value. Hence, ECs could be represented as "search algorithms" instead of "optimization algorithms". The function that is optimized does not need to be a mathematical function-it may be the result of a computer simulation or a real-world experiment. For example, in genetic programming, the objective can be to find a formula or computer program that accomplishes a certain task. The function that is optimized is the result of applying the formula or program (represented as parse tree) to a number of test cases.

11.3   Brief History of Evolutionary Computation

Several scientists and researchers proved that evolutionary computation can be used as a tool for solving optimization problems in engineering during the 1950s and 1960s. The basic idea was evolving a population of candidate solutions to a given problem, using operators inspired by genetic variation and natural selection. Many other researchers developed evolutionary algorithms for optimization and machine learning problems. History itself is an evolutionary process, not just in the sense that it changes over time but also in the more strict sense that it undergoes mutation and selection.

The selection function is performed under the process of mutation. The history of evolutionary computation is broadly related more with science fiction than a factual record. The idea to use recombination in population-based evolutionary simulations did not arise in a single major innovation but in fact was commonly, if not routinely, applied in multiple independent lines of investigation in the 1950s and 1960s. Population geneticists and evolutionary biologists considered computers to study "life-as-it-could-be" (i.e., artificial life) rather than "life-as-we-know-it".

The idea to use Darwinian principles for automated problem solving originated in the fifties, even before computers were used on a large scale. Later in the sixties, three different interpretations of this idea were developed: Evolutionary programming introduced by Lawrence J. Fogel in the U.S.A., John Henry Holland introduced a method genetic algorithm, and Ingo Rechenberg and Hans-Paul Schwefel of Germany introduced evolution strategies. There was great development in all these areas independantly until the early nincties. Since then these techniques were combined as one technology, called evolutionary computing. During this period, a fourth paradigm following the general ideas of evolutionary programming, genetic algorithm, and evolutionary strategies had emerged and it was named genetic programming. These terminologies denote the whole field by evolutionary computing and consider evolutionary programming, evolution strategies, genetic algorithms, and genetic programming as sub-areas.

11.4   Biological and Artificial Evolution

The basic idea behind Evolutionary Computation (EC) lies under the principles of biological evolution. A set of terminology and expressions used by the EC community is illustrated in this section.

11.4.1   Expressions Used in Evolutionary Computation

The basis of EC technique is very simple, Darwin's theory of evolution, and specifically survival of the fittest. Much of the terminology is borrowed from biology. A set of chromosomes comprise a population with each chromosome representing a possible solution. Every individual chromosome is made up of a collection of genes, these genes are the variables to be optimized.

The process of evolutionary algorithm involves a sequence of operations: creating an initial population (a collection of chromosomes), evaluating this population, then evolving the population through multiple generations. At the end of each generation the fittest chromosomes, i.e., those that represent the best solution, from the population are retained and are allowed to crossover with other fit members. Crossover is performed in order to create chromosomes that are more fit than both the parents by taking the best characteristics from each of the parents. Thus over a number of generations, the fitness of the chromosome population will increase with the genes within the fittest chromosome representing the optimal solution. The entire process of EC is similar to living species, in which they evolve to match their changing environmental conditions. Table 11.1 presents a brief overview of some of the terminology adopted from biology and used in EC.

11.4.2   Biological Evolution Inspired by Nature

The process of evolution is the gradual development of living organisms. Living organisms evolve through the interaction of competition, selection, reproduction, and mutation processes. The evolution of a population of organisms highlights the differences between an organism's "genotype" and "phenotype." The phenotype is the way in which response is contained in the physiology, morphology and behavior of the organism. The genotype is the organism's underlying genetic coding

TABLE 11.1: A Summary of the Basic Expressions Used within EC

Biological Expression	EC Implication
Chromosome	String of symbols
Population	A set of chromosomes
Deme	A local population of closely related chromosomes, a subset of the total population
Gene	A feature, character or detector
Allele	Feature value
Locus	A position in a chromosome
Genotype	Structure
Phenotype	A set of parameters, an alternative solution or a decoded structure

(DNA).

Though the basic principles of evolution (natural selection and mutation) are understood in ample, both population genetics and phylogenetics have been radically transformed by the recent availability of large quantities of molecular data. For example, in population genetics (study of mutations in populations), more molecular variability was found in the 1960s than had been expected. Phylogenetics (study of evolutionary history of life) makes use of a variety of different kinds of data, of which DNA sequences are the most important, as well as whole-genome, metabolic, morphological, geographical, and geological data.

The basic concept of evolutionary biology shows that organisms share a common origin and have diverged through time. The details and timing of these divergences of an evolutionary history are important for both intellectual and practical reasons, and phylogenies are central to virtually all comparisons among species. The area of phylogenetics has helped to trace routes to infectious disease transmission and to identify new pathogens.

For instance, consider the problem of estimating large phylogenesis, which is a central challenge in evolutionary biology. Assume that there are three species X, Y, and Z. Three possible tree structures can be formed from these three species: (X, (Y, Z)); (Y, (X, Z)); and (Z, (X, Y)). Here (X, (Y, Z)) denotes X and Z share a common ancestor, and this in turn shares a different common ancestor with X. Thus even if one picks a tree in random, there is a one in three chance that the tree chosen will be correct. But the number of possible trees grows very rapidly with the number of species involved. For a small phylogenetic problem involving 10 species, there are 34,459,425 possible trees. For a problem involving

22 species, the number of trees exceeds 10^{23}. Today, most phylogenetic problems involve more than 80 species and some datasets contain more than 500 species.

With such large search spaces, it is clear that exhaustive search for the single correct phylogenetic tree is not a feasible strategy, regardless of how fast computers become in the foreseeable future. Researchers and investigators have developed numerous methods for coping with the size of these problems, but most of these methods have severe deficiencies. Thus, the algorithmic of evolutionary biology is a fertile area for research.

A few features across the species can be compared. This comparative method has furnished much of the evidence for natural selection and is probably the most widely used statistical method in evolutionary biology. Always comparative analysis must account for phylogenetic history, since the similarity in features common to multiple species that originate in a common evolutionary history can inappropriately and seriously bias the analyses. Numerous methods have been developed to accommodate phylogenics in comparative analysis, but most of these techniques presume that the phylogeny is known without error. Yet, this is apparently unrealistic, since almost all phylogenetics have a large degree of uncertainity. An important question is therefore to understand how comparative analysis can be performed that accommodates phylogenetic history without depending on any single phylogeny being correct.

The genomic changes occur when an organism adapts to a new set of selection pressures in a new environment. The genomic changes are an interesting problem in evolutionary biology. Since the process of adaptive change is difficult to study directly, there are many important and unanswered questions regarding the genetics of adaptation. For instance, questions regarding the number of mutations involved in a given adaptive change, whether there is any change in this number when different organisms or different environments are involved, whether distribution of fitness changes during adaptation, etc. Though some of the questions remain unanswered evolutionary trends are leading in almost all the research areas.

11.4.3 Evolutionary Biology

In 1789, Thomas Malthus wrote his "Essay on the Principle of Population", in which he recognized that a) population growth rate is a function of population size, and therefore b) left unchecked, a population will grow exponentially. However, since environments have only finite resources, a growing population will eventually reach a point, the Malthusian crunch, at which organisms a) will have to compete for those resources, and b) will produce more young than the environment can support. Crunch time was Darwin's ground-breaking entry point into the discussion.

As many more individuals of each species are born than can possibly survive; and as, consequently, there is a frequently recurring struggle for existence, it follows that any being, if it varies however slightly in any manner profitable to itself, under the complex and sometimes varying conditions of life, will have a better chance of surviving, and thus be naturally selected.

In short, the combination of resource competition and heritable fitness variation leads to evolution by natural selection. When the Malthusian crunch comes, if

a) there is any variation in the population that is significant in the sense that some individuals are better equipped for survival and reproduction than others, and

b)those essential advantages can be passed on to offspring. Then the population as a whole will gradually become better adapted to its environment as more individuals are born with the desirable traits.

In short, the population will evolve. Of course, this assumes that populations can change much faster than geographic factors. In addition to

a) the pressure to favor certain traits over others (known as selection pressure) that the Malthusian crunch creates and,

b) the desired traits enable the features of well-adapted individuals to spread throughout and eventually dominate the population,

The concept of variation is also essential to evolution, not merely as a precondition to population takeover by a dominant set of traits, but as a perpetual process insuring that the population never completely stagnates lest it falls out of step with an environment that, inevitably, does change.

Thus, the three essential ingredients for an evolutionary process are:

1. Selection - some environmental factors must favor certain traits over others.

2. Variation - individuals must consistently arise that are significantly (although not necessarily considerably) different from their ancestors.

3. Heritability - children must, on average, inherit a good many traits from their parents to insure that selected traits survive generational turnover.

These three factors are implicit in the basic evolutionary cycle depicted in Figure 11.1. Beginning in the lower left, a collection of genetic blueprints (more accurately, recipes for growth) known as genotypes are present in an environment. These might be (fertilized) fish eggs on the bottom of a pond, or the collection of all 1-day-old embryos in the wombs of a gazelle population. Each genotype goes through a developmental process that produces a juvenile organism, a young phenotype. At the phenotypic level, traits (such as long legs, coloration patterns, etc.) encoded by genotypes become explicit in the organism. In Figure 11.1, selection pressure is present in all processes surrounded by solid-lined pentagons. These represent the metaphorical sieve of selection through which populations (of genotypes and phenotypes) must pass. Already during development, this sieve may filter out genotypes that encode fatal growth plans, e.g., those that may lead to miscarriages in mammals.

The sieve is persistent, therefore, even the genotypes that encode plans for fit juveniles have little guarantee of proliferation. Juveniles must overcome the the downpours of life while arising to maturity in order to enter a new arena of competition, that is the right to produce offspring. Thus, by the time organisms reach the upper right-hand corner, the mating phenotypes, the genotypic pool has been narrowed considerably. Due to lack of mate, individuals need not perish; a few can be reprocessed and try on during the next mating season, as represented by the aging/death filter above the adult phenotype collection. Heading down from the upper right corner of Figure 11.1, we return to the genotypic level via the production of gametes. This occurs within each organism through mitosis (copying) and meiosis (crossover recombination), resulting in many half-genotypes (in diploid organisms) that normally embody minor variations of the parent's genes. On pairing, these half-genotypes pair up to produce a complete genotype, but slightly different from that of each parent. There is normally an overproduction of gametes; only a chosen few become part of the new genotypes. Although pinpointing the exact locations of variation and heritability are difficult, since genetic mutations can occur to germ cells (i.e., future gametes) at any time during life, it seems fair to say that the major sources of genetic variation are the (imperfect) copying and recombination processes during mitosis and meiosis, respectively. Inheritance is then located along the bottom path from gametes to genotypes, since this is the point at which the parents DNA (and the traits it encodes) officially make it into the next generation. Through repeated cycling through this loop, a population gradually adapts to a (relatively static) environment. In engineering terms, the population evolves such that its individuals become better designs or better solutions to the challenges that the environment poses.

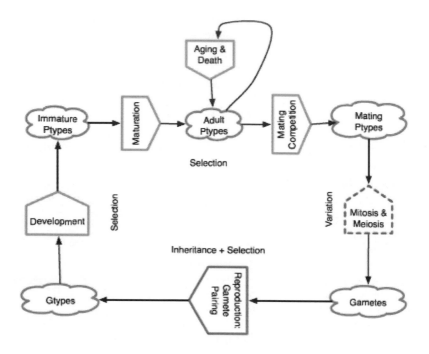

FIGURE 11.1:   The Basic cycle of Evolution. Clouds Represent Populations of Genotypes (gtypes) or Phenotypes (ptypes), While Polygons Denote Processes. Solid Pentagons Represent Processes that Filter the Pool of Individuals, While the Dashed Pentagon for Mitosis and Meiosis Indicates an Increase (in this Case, of Genetic Material).

11.5   Flow Diagram of a Typical Evolutionary Algorithm

A population of structures is maintained in EC to evolve according to rules of selection and other operators, such as recombination and mutation. Each individual in the population receives a measure of its fitness in the environment. During selection more attention is concentrated on high fitness individuals, thus exploiting the available fitness information. The processes of recombination and mutation disturb these individuals, thereby providing general heuristics for exploration. Though simple from a life scientist's viewpoint, these algorithms are sufficiently composite to provide robust and powerful adaptive search mechanisms. Figure 11.2

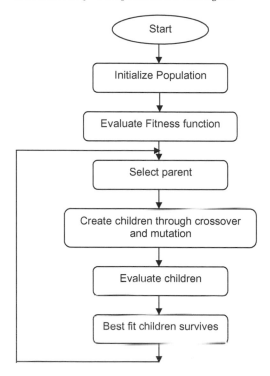

FIGURE 11.2:   Flowchart of a Typical Evolutionary Algorithm

outlines a typical evolutionary algorithm (EA). The population is initialized as the initial step and then evolved from generation to generation by repeated operations like evaluation, selection, recombination, and mutation. The population size N is generally constant in an evolutionary algorithm, although there is no *a priori* reason (other than convenience) to make this assumption. An evolutionary algorithm typically initializes its population randomly, although domain specific knowledge can also be used to bias the search. The evaluation process measures the fitness of each individual and may be as simple as computing a fitness function or as complex as running an elaborate simulation. Selection is often performed in two steps, parent selection and survival. Parent selection decides who becomes parents and how many children the parents have. Children are created via recombination, which exchanges information between parents, and mutation, which further perturbs the children. The children are then evaluated. Finally, the survival step decides who survives in the population.

11.6　Models of Evolutionary Computation

As discussed in the history of evolutionary computation, the origin of evolutionary algorithms can be traced to at least the 1950's (e.g., Fraser, 1957; Box, 1957). The methodologies that have emerged in the last few decades such as: "evolutionary programming (EP)" (L.J. Fogel, A.J. Owens, M.J. Walsh Fogel, 1966), "evolution strategies (ES)" (I. Rechenberg and H.P. Schwefel Rechenberg, 1973), "genetic programming (GP)" (de Garis and John Koza) and "genetic algorithms (GA)" (Holland, 1975) are discussed in this section. Though these techniques have similarities, each of these methods implements the algorithm in its own unique manner. The key differences lie upon almost all aspects of evolutionary algorithms, including the choices of representation for the individual structures, types of selection mechanism used, forms of genetic operators, and measures of performance. The important differences (and similarities) are also illustrated in the following sections, by examining some of the variety represented by the current family of evolutionary algorithms.

These approaches in turn have inspired the development of additional evolutionary algorithms such as "classifier systems (CS)" (Holland, 1986), the LS systems (Smith, 1983), "adaptive operator" systems (Davis, 1989), GENITOR (Whitley, 1989), SAMUEL (Grefenstette, 1989), "genetic programming (GP)" (de Garis, 1990; Koza, 1991), "messy GAs" (Goldberg, 1991), and the CHC approach (Eshelman, 1991). This section will focus on the paradigms such as GA, GP, ES, and EP shown in Table 11.2.

11.6.1　Genetic Algorithms (GA)

Genetic Algorithm (GA) is a basic model of machine learning, which gains its behavior from a group of mechanisms of evolution in nature. The process is done by creating an individual from a population of individuals represented by chromosomes. The individuals in the population then

TABLE 11.2:　Paradigms in Evolutionary Computation

Paradigm	Created by
Genetic Algorithms	J.H. Holland
Genetic Programming	de Garis and John Koza
Evolutionary Programming	L.J. Fogel, A.J. Owens, M.J. Walsh
Evolution Strategies	I. Rechenberg and H.P. Schwefel

go through a process of simulated "evolution". Genetic algorithms are applied in numerous application areas mainly for optimization. The most common example that can be cited is the traveling sales man problem, in which the best shortest route has to be determined by optimizing the path.

Practically, the genetic model of computation can be implemented by having arrays of bits or characters to represent the chromosomes. Simple bit manipulation operations allow the implementation of crossover, mutation and other operations. Though a substantial amount of research has been performed on variable-length strings and other structures, the majority of work with genetic algorithms is focused on fixed-length character strings. The users should focus on both this aspect of fixed-length and the need to encode the representation of the solution being sought as a character string. Since these are crucial aspects that distinguish genetic programming, which does not have a fixed length representation and there is typically no encoding of the problem.

The genetic algorithm is applied by evaluating the fitness of all of the individuals in the population. Once the fitness function is evaluated, a new population is created by performing operations such as crossover, fitness-proportionate reproduction, and mutation on the individuals. Every time the iteration is performed, the old population is rejected and the iteration continues using the new population.

A single iteration of this cycle is referred to as a generation. Indeed, behavior in populations in nature is not found as a whole, but it is a convenient implementation model.

The first generation (generation 0) of this process operates on a population of randomly generated individuals. From there on, the genetic operations, in concert with the fitness measure, operate to improve the population. The flowchart of a typical genetic algorithm is given in Figure 11.2a.

11.6.2   Genetic Programming (GP)

The main idea behind the development of genetic programming (GP) was to allow automatic programming and program induction. GP can be considered as a specialized form of genetic algorithm, which manipulates with variable length chromosomes using modified genetic operators. Genetic programming is not capable of distinguishing between the search space and the representation space. Yet, it is not difficult to introduce genotype/phenotype mapping for any evolutionary algorithm formally whereas it could be a one-to-one mapping instead.

The search space of GP includes the problem space as well as the space of representation of the problem. The technical search space of genetic

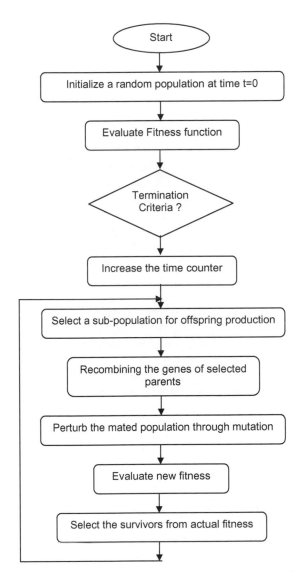

Figure 11.2a : Flowchart of Genetic Algorithm

programming is the search space of all possible recursive compositions over a primitive set of symbols on which the programs are constructed. The genetic programs are represented in a tree form or in a linear form. The major operator for GP is crossover, and the operator interchanges subtrees of the parent trees which are randomly chosen without the syntax of the programs being interrupted. The mutation operator picks a random sub-tree and replaces it by a randomly generated one.

Genetic programming traditionally develops symbolic expressions in a functional language like LISP. An evolved program can contain code segments which when removed from the program would not alter the result produced by the program, i.e., semantically redundant code segments. Such segments are referred to as introns. The size of the evolved program can also grow uncontrollably until it reaches the maximum tree depth allowed while the fitness remains unchanged. This effect is known as bloat. The bloat is a serious problem in genetic programming, since it usually leads to time consuming fitness evaluation and reduction of the effect of search operators. Once it occurs, the fitness nearly stagnates.

These programs are expressed in genetic programming as parse trees, rather than as lines of code. Thus, for example, the simple program "a + b * c" would be represented as follows

or, to be precise, as suitable data structures linked together to achieve this effect. Because this is a very simple thing to do in the programming language Lisp, many GP users tend to use Lisp. However, this is simply an implementation detail. There are straightforward methods to implement GP using a non-Lisp programming environment.

GP operates on programs in the population, which are composed of elements from the function set and the terminal set. These are fixed sets of symbols from which the solution of problems are selected. Mostly GP does not use mutation operator. Genetic programming is capable of using numerous advanced genetic operators. For example, automatically defined functions (ADF) allow the definition of subprograms that can be called from the rest of the program. Then the evolution is to find the solution as well as its decomposition into ADFs together. The fitness function is either application specific for a given environment or it takes the form of symbolic regression. Irrespective of the fitness function, the evolved program must be executed in order to find out what it does. The outputs of the program are usually compared with the desired outputs for

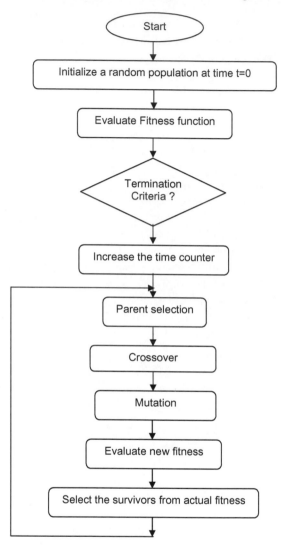

FIGURE 11.3: The GP Algorithm

given inputs. A terminating mechanism of the fitness evaluation process must be introduced to stop the algorithm. Figure 11.3 outlines a typical genetic programming (GP) algorithm.

11.6.3 Evolutionary Programming (EP)

Evolutionary Programming (EP), a stochastic optimization strategy originally developed by Lawrence J. Fogel in 1960. The operation of EP is similar to genetic algorithms with the only difference in that EP places an emphasis on the behavioral linkage between parents and their offspring, rather than seeking to emulate specific genetic operators as observed in nature. Evolutionary programming is similar to evolution strategies, although the two methods were developed uniquely. Similar to both ES and GAs, EP is an efficient method to solve optimization problems compared to techniques such as gradient descent or direct, analytical discovery are not possible. Optimization problems with combinatory and real-valued function optimization in which the optimization surface or fitness landscape is "rugged", are easily and efficiently computed using evolutionary programming.

Like GAs, in EP the fitness landscape can be characterized in terms of variables, and that there exists an optimum solution in terms of these variables. For instance in a traveling sales man problem, each solution is a path. The length of the route or path is a numerical value, which acts as the fitness function. The fitness function is characterized as a hyper surface proportional to the path lengths in a population space. The objective is to find the shortest path in that space.

The basic EP method involves three steps (Figure 11.13a), which are recursive until a threshold for iteration is exceeded or an adequate solution is obtained:

Step 1: Randomly choose an initial population of trial solutions. The number of solutions in a population is closely related to the speed of optimization, but there is no answer to predict the number of appropriate solutions (other than >1) and to predict the discarded solutions.

Step 2: Every solution that is produced is copied into a new population. These offsprings are then mutated according to a distribution of mutation types, ranging from minimum to a maximum with continuous mutation types between. The severity of mutation is guessed on the basis of the functional change enforced on the parents.

Step 3: Every offspring is evaluated by computing its fitness function. Generally, a stochastic tournament selection is applied to determine N solutions, which are held back for the population of solutions. There is no constraint on the population size to be be held constant, or that only a single offspring be generated from each parent. Normally EP does not use any crossover as a genetic operator.

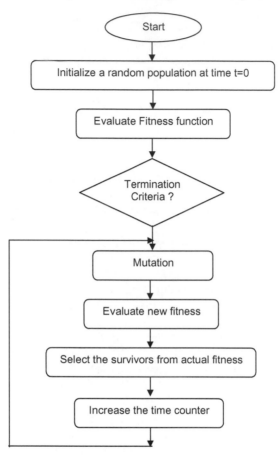

Figure 11.3a :   Flowchart of Evolutionary Programming

11.6.4   Evolutionary Strategies (ESs)

Evolutionary Strategies (ESs) were independently developed with operators such as selection, mutation, and a population of size one. Schwefel brought in recombination (crossover) and populations with more than one individual. Since the initial applications of ES were in hydrodynamic optimization problems, evolution strategies typically use real-valued vector representations.

Evolution Strategies were devised mainly to solve technical optimization problems (TOPs), and until recent years, ES were exclusively recognized by civil engineers. Normally there is no closed form analytical objective function for TOPs and therefore, no practical optimization technique exists, except for the engineer's intuition.

In a two-membered ES, one parent generates one offspring per generation by applying normally distributed mutations, i.e., smaller steps occur more likely than big ones, until a child performs better than its ancestor and takes its place. Due to this simple structure, theoretical results for step size control and convergence velocity could be derived. The ratio between offsprings produced by successful mutations and offsprings produced by all mutations should be 1/5: based on this property the 1/5 success rule was invented. This initial algorithm, that applied only mutation, was further extended to a (m+1) strategy, which incorporated recombination (crossover) due to several, i.e., m parents being available. The mutation scheme and the exogenous step size control were taken across unchanged from two-membered ESs.

Later on Schwefel extrapolated these strategies to the multi-membered ES denoted by (m+l) (plus strategy) and (m,l) (comma strategy) which imitates the following basic principles of organic evolution: a population, leading to the possibility of recombination with random mating, mutation, and selection. In the plus strategy, the parental generation is considered during selection, while in the comma strategy only the offspring undergoes selection, and the parents die off. m (usually a lowercase m, denotes the population size, and l, usually a lowercase lambda denotes the number of offspring generated per generation). The algorithm of an evolutionary strategy is as follows:

```
(define (Evolution-strategy population)
(if (terminate? population)
 population
 (evolution-strategy
   (select
(cond (plus-strategy?
   (union (mutate
     (recombine population))
   population))
   (comma-strategy?
 (mutate
  (recombine population)))))))))
```

Every individual of the ES' population consists of the following genotype representing a point in the search space:

Object Variables

The real-valued x(i) variables should be evolved through recombination (crossover) and mutation such that the objective function reaches its global optimum.

Strategy Variables

The real-valued s(i) variables or mean step sizes determine the mutability of the x(i). They represent the standard deviation (0, ... oi) of s(i)) gaussian distribution (GD) being added to each x(i) as an undirected mutation. If the expectancy value is 0, then the parents will produce offspring similar to themselves. In order to make a doubling and a halving of a step size equally probable, the s(i) mutate log-normally, distributed, i.e., exp(GD), from generation to generation. These step sizes hide the internal model the population has made of its environment, i.e., a self-adaptation of the step sizes has replaced the exogenous control of the (1+1) ES.

This concept works because selection sooner or later prefers those individuals having built a good model of the objective function, thus producing better offspring. Hence, learning takes place on two levels: (1) at the genotypic, i.e., the object and strategy variable level and (2) at the phenotypic level, i.e., the fitness level.

Depending on an individual's x(i), the resulting objective function value f(x), where x denotes the vector of objective variables, serves as the phenotype (fitness) in the selection step. In a plus strategy, the m best of all (m+l) individuals survive to become the parents of the next generation. Using the comma variant, selection takes place only among the l offspring. The next strategy is more naturalistic and hence more eminent, since no individual may survive evermore, which can be done by using the plus variant. The comma strategy performs better vague conventional optimization algorithms. An everlasting adaptation of the step sizes can occur only if the highly fit individuals are forgotten. By this process long stagnation phases due to misadapted sizes can be avoided. These individuals have established an internal model that is no more appropriate for additional progress, and therefore it is better to be discarded.

The convergence of the evolutionary strategy can be obtained by choosing a suitable ratio of m/l. For a faster convergence the preferred ratio of m/l should be very small around (5,100). To find an optimal solution usually the preferred ratio is chosen around (15,100).

The following agents greatly influence self-adaptation within ESs:

Randomness

Mutation cannot be designed as a pure random process since the child would be completely independent of its parents.

Population size

A sufficiently large population is the best choice to obtain optimal solution. Instead of allowing only the current best solutions to reproduce, the set of good individuals are also allowed to reproduce.

Cooperation

When the population m >1, the individuals should recombine their knowledge there with others (cooperate) because one cannot expect the knowledge to accumulate in the best individual only.

Deterioration

To obtain better progress in the future, one should assume deterioration from one generation to the next. A restrained life-span in nature is not a sign of failure, but an important means of forbidding a species from freezing genetically.

ESs have turned out to be successful when compared to other additional iterative methods. ESs are capable of adapting to all sorts of problems in optimization, since they require less data regarding the problem. ESs are also capable of solving high dimensional, multimodal, nonlinear problems subject to linear and/or nonlinear constraints. ESs have been adapted to vector optimization problems, and they can also serve as a heuristic for NP-complete combinatorial problems like the traveling salesman problem or problems with a noisy or changing response surface. A detailed description of Evolutionary Strategies is given in Section 11.9.

11.7 Evolutionary Algorithms

Evolutionary Algorithms (EAs) are stochastic search and optimization heuristics derived from the classic evolution theory, which are implemented on computers in the majority of cases. The basic idea is that if only those individuals of a population reproduce, which meet a certain selection criteria, and the other individuals of the population die, the population will converge to those individuals that best meet the selection criteria. Figure 11.4 shows the structure of a simple evolutionary algorithm.

In the initial stage of computation, a number of individuals (the population) are randomly initialized. The objective function or fitness func-

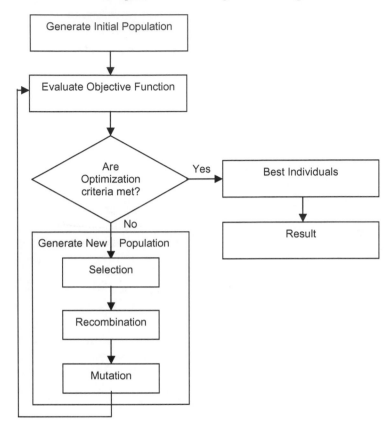

FIGURE 11.4:   Structure of a Single Population Evolutionary Algorithm

tion is then evaluated for these individuals during which the first generation is produced. The individuals of a population have to represent a possible solution of a given problem while solving optimization problems. The generations are created iteratively until the stopping criteria are met. The individuals are selected according to their fitness for the production of offspring. Parents are recombined to produce offspring and all these offspring are mutated with a certain probability. Then the offspring are evaluated for fitness. The offspring are inserted into the population replacing the parents, producing a new generation.

    The generations are produced until the optimization criteria are reached. A single population evolutionary algorithm is more powerful and performs well on a wide variety of optimization problems. But usually, better results can be obtained by introducing multiple subpopula-

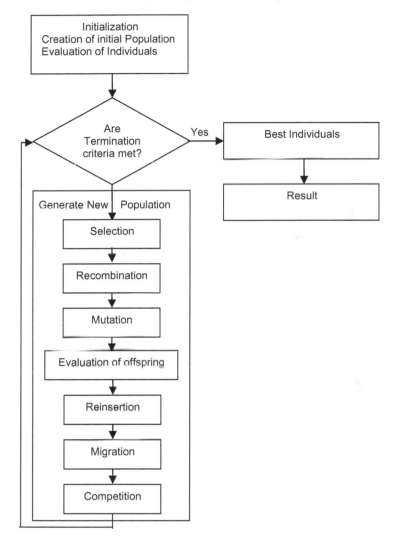

FIGURE 11.5: Structure of an Extended Multi-Population Evolutionary Algorithm

tions. Every subpopulation evolves over a few generations similar to the single population before one or more individuals are exchanged between the subpopulation. The multi-population evolutionary algorithm models the evolution of a species in a way more similar to nature than the single population evolutionary algorithm. Figure 11.5 shows the structure of such an extended multi-population evolutionary algorithm.

11.7.1   Evolutionary Algorithms Parameters

The term evolutionary computation or evolutionary algorithms include the domains of genetic algorithms (GA), evolution strategies, evolutionary programming, and genetic programming. Evolutionary algorithms have a number of parameters, procedures or operators that must be specified in order to define a particular EA. The most important parameters of EA are

- Representation

- Fitness function

- Population

- Parent selection mechanism

- Variation operators, recombination and mutation

- Reproduction operator

The initialization procedure and a termination condition must be defined along with the above parameters.

11.7.2   Solution Representation

The method of representing solutions/individuals in evolutionary computation methods is known as representation. This can encode appearance, behavior, and physical qualities of individuals. An expressive and evolvable design of a representation is a hard problem in evolutionary computation. The conflict in genetic representations in one of the major criteria drawing a line between known classes of evolutionary computation.

Evolutionary algorithm more commonly uses linear binary representations among which the most typical representation is an array of bits. Arrays of other types of data and structures can also be used in a similar manner. The primary attribute that makes these genetic representations favorable is that their sections are well co-ordinated due to their fixed size. Due to this well co-ordinated arrangement simple crossover operation is easily facilitated. Evolutionary algorithms are capable of operating on variable length representations also, but crossover implementation is more complex in this case.

Human-based genetic algorithm (HBGA) provides a mode to avoid solving hard representation problems by outsourcing all genetic operators to outside agents (in this case, humans). In this case, the algorithm need not be aware of a particular genetic representation used for any solution. Some of the common genetic representations are

- binary array

- genetic tree

- parse tree

- binary tree

- natural language

11.7.3   Fitness Function

A fitness function is a special type of objective function that measures or evaluates the optimality of a solution in an evolutionary algorithm so that a particular chromosome may be graded against all the other chromosomes. Optimal chromosomes, or at least chromosomes which are more optimal, are permitted to breed and integrate their datasets by various methods, thereby creating a new generation that will (hopefully) be still better.

Fitness functions can also be considered in terms of a fitness landscape, which shows the fitness for each possible chromosome. An ideal fitness procedure correlates closely with the algorithm's goal, and may be computed quickly. The speed of execution is very important, as a typical genetic algorithm must be iterated several times in order to produce a usable result for a non-trivial problem.

The definition of the fitness function is not straightforward in a few cases and is often performed iteratively if the fittest solutions produced by GA are not what are desired. In a few cases, it is very difficult or impossible to derive or even guess what the fitness function definition might be. Interactive genetic algorithms address this difficulty by outsourcing evaluation to external agents. When the genetic representation has been outlined, the next step is to assign a value to each chromosome corresponding to the fitness function. Generally there is no problem in determining the fitness function. As a matter of fact, most of the time, it is implicitly characterized by the problem that is to be optimized. Specific attention should be taken while selection is performed according to the fitness of individuals. The fitness function not only indicates how good the solution is, but should also correspond to how close the chromosome is to the optimal one.

The fitness function represents the chromosome in terms of physical representation and evaluates its fitness based on traits of being desired in the solution. The fitness function has a greater value when the fitness characteristic of the chromosome is better than others. Additionally, the fitness function inserts a criterion for selection of chromosomes.

11.7.4 Initialization of Population Size

Generally, there are two effects to be considered for population initialization of EA: the initial population size and the procedure to initialize population.

The population size has to increase exponentially with the complexity of the problem (i.e., the length of the chromosome) in order to generate best solutions. New studies have shown that acceptable results can be obtained with a much smaller population size. A large population is quite useful, but it demands excessive costs in terms of both memory and time. There are two methods to generate the initial population: heuristic initialization and random initialization. Heuristic initialization searches a small part of the solution space and never finds global optimal solutions because of the lack of diversity in the population. Therefore, random initialization is applied so that the initial population is generated with the encoding method. The algorithm of population initialization is shown below:

```
Procedure:population initialization
   Begin
   Select the first job into the first position
   randomly;
   For i=0 to code_size(n) //n: length of chromosome
      For j=0 to job_size(m) //m:number of jobs
         Insert job j that has not inserted into
         the chromosome
      temporarily;
         Evaluate the fitness value of this
         complete chromosome so far;
         Keep the fitness value in memory;
      End For
   Evaluate the best fitness value from the memory
End For
```

11.7.5 Selection Mechanisms

Survival of the fittest or natural selection is the process by which favorable genetic traits to a greater extent become common in successive generations of a population of reproducing organisms, and unfavorable heritable traits become less common. Natural selection acts on the phenotype, or the observable characteristics of an organism, such that individuals with favorable phenotypes are more likely to survive and reproduce than those with less favorable phenotypes. The phenotype's genetic basis, genotype associated with the favorable phenotype, will grow in frequency over the following generations. Across time, this process can lead to adaptations that differentiate organisms for specific ecological niches and could finally result in the emergence of new species. More generally, natural selection is the mechanism by which evolution may occur in a population of a particular organism.

Parents are selected according to their fitness by means of one of the following algorithms:

- roulette wheel selection

- stochastic universal sampling

- local selection

- truncation selection

- tournament selection

Roulette Wheel Selection

The fundamental idea behind this selection process is to stochastically select from one generation and to create the basis of the next generation. The fittest individuals have a greater chance of survival than weaker ones. This replicates nature in that fitter individuals will tend to have a better probability of survival and will go forward to form the mating pool for the next generation. Naturally, such individuals may have genetic coding that may prove useful to succeeding generations.

Illustration

The normal method used is the roulette wheel (as shown in Figure 11.6). Table 11.3 lists a sample population of 5 individuals (a typical population of 400 would be difficult to illustrate).

These individuals consist of 10 bit chromosomes and are being used to optimise a simple mathematical function (we can assume from this example we are trying to find the maximum). If the input range for x is between 0 and 10, then we can map the binary chromosomes to base 10 values and then to an input value between 0 and 10. The fitness values are then taken as the function of x. From the table it is inferred that individual No. 3 is the fittest and No. 2 is the weakest. By summing these fitness values the percentage total of fitness can be obtained. This gives the strongest individual a value of 38% and the weakest 5%. These percentage fitness values can then be used to configure the roulette wheel. Table 11.3 highlights that individual No. 3 has a segment equal to 38% of the area. The number of times the roulette wheel is spun is equal to

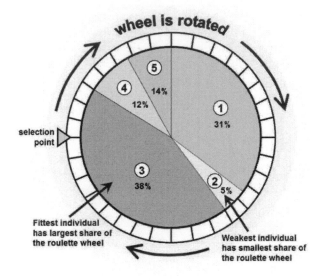

FIGURE 11.6: Roulette Wheel Selection - Operation

size of the population. As can be seen from the way the wheel is now divided, each time the wheel stops this gives the fitter individuals the greatest chance of being selected for the next generation and subsequent mating pool. From this illustration it is known that as the generations progress and the population gets fitter the gene pattern for individual No. 3: 0100000101_2 will become more prevalent in the general population because it is fitter, and more apt to the environment we have put it in - in this case the function we are trying to optimize.

TABLE 11.3: Evaluation of the Fitness Function f(x) $= x^2 + 2x + 5$

No. of Population	Chromosome	Base 10 Values	X	Fitness) (X)	Percentage
1	0001101011	107	1.05	6.82	31
2	1111011000	984	9.62	1.11	5
3	0100000101	261	2.55	8.48	38
4	1110100000	928	9.07	2057	12
5	1110001011	907	8087	3.08	14

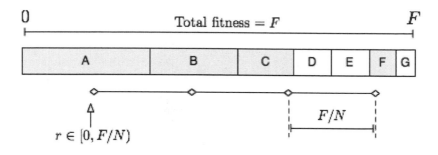

FIGURE 11.7: Stochastic Universal Sampling

Stochastic Universal Sampling

Stochastic universal sampling (SUS) is a single-phase sampling algorithm with minimum spread and zero bias. SUS uses N equally spaced pointers, instead of the single selection pointer as that used in roulette wheel methods. Here N is the number of selections required. The population is shuffled randomly and a single random number in the range [0 Sum/N] is generated, ptr. The N individuals are then chosen by generating the N pointers spaced by 1, [ptr, ptr+1, ... ,ptr+N-1], and selecting the individuals whose fitnesses span the positions of the pointers. An individual is thus guaranteed to be selected a minimum of times and no more, thus achieving minimum spread. In addition, as individuals are selected entirely on their position in the population, SUS has zero bias. The roulette wheel selection methods can all be implemented as O(NlogN) although SUS is a simpler algorithm and has time complexity O(N).

For 7 individuals to be selected, the distance between the pointers is 1/7. Figure 11.7 shows the selection for the above example. For a sample of 1 random number in the range [0, 1/7]:0.1.

Local Selection

Local selection (LS) is defined as a selection scheme that minimizes interactions among individuals. Locality in selection schemes has been a persistent theme in the evolutionary computation community. This kind of a selection process leads to a more realistic reproduction scheme in an evolutionary model of real populations of organisms. In such a model, an agent's fitness must result from individual interactions with the environment, which contain shared resources along with other agents, rather than from global interactions across the population.

In local selection each agent is considered as a candidate solution in the population and is first initialized with some random solution and an initial reservoir of energy. If the algorithm is implemented sequentially, parallel execution of agents can be simulated with randomization of call order. In each iteration of the algorithm, an agent explores a candidate solution (possibly including an action) similar to itself. The agent is taxed with Ecost for this action and collects Δ E from the environment.

While selecting the individuals, an agent compares its current energy level with a threshold θ. If its energy is higher than θ, the agent reproduces. This reproduced individual or the mutated individual is then evaluated for its fitness and becomes a member of the population, with half of its parent's energy. The agent is killed if it runs out of energy.

Since the threshold θ is a constant value independent of the rest of the population, the selection is local. Due to local selection the communication among agent processes is minimized thereby leading to several positive consequences. Initially, two agents compete for shared resources only if they are situated in the same portion of the environment. The environment plays a major role in driving the competition and the selective pressure. The lifetime of an agent is unpredicted, and there is no decision to decide how often it reproduces or when it dies. The search is influenced directly by the environment. The LS mechanism is an implicitly niched scheme and therefore influences the maintenance of population diversity naturally. This process makes the search algorithm more amenable to cover and multi-modal optimization than to standard convergence criteria. The bias is to exploit all resources in the environment, rather than to locate the single best resource.

Instead of determining the population size initially, the size of the population emerges from the carrying capacity of the environment. This is determined by factors like

- the costs incurred by any action

- the replenishment of resources

Both of these factors are independent of the population. Finally, the removal of selection's centralized bottleneck makes the algorithm parallelizable and therefore amenable to distributed implementations.

Though local selection has a number of advantages, it also has a few disadvantages and limitations as well. Local selection can guess a population of agents who can execute code on remote servers in a distributed environment, but have to look up data on a central machine for every action they perform. An exemplary case of such a situation would be a distributed information retrieval task in which agents share a centralized page cache. Due to the communication overhead and synchronization issues, the parallel speedup achievable in this case would be seriously hindered. As this scenario indicates, the feasibility of distributed implementations of evolutionary algorithms based on local selection requires that the environment can be used as a data structure. Similar to natural organisms, agents must be able to mark the environment so that local interactions can take advantage of previous experience.

Truncation Selection

In truncation selection the individuals' quantitative expression of a phenotype is more than or less than a certain value referred to as the truncation point. Truncation selection or block selection is a peculiar breeding technique which ranks all individuals according to their fitness and selects the best ones as parents. In truncation selection a threshold T is defined such that the T% best individuals are selected. Truncation selection has been most commonly used in evolution strategies. It is also often used in quantitative genetics where artificial selection performed by breeders is studied. A specific genetic algorithm, the Breeder Genetic Algorithm incorporates ideas from breeders to perform parameter optimization tasks and the convergence model for the bit counting function. Block selection is equivalent to truncation selection since for a given population size n one simply gives s copies to the n/s best individuals. Both implementations are identical when s $=100/T$.

If the fitness is normally distributed, quantitative genetics defines the selection intensity i that expresses the selection differential S(t)= $\bar{f}(t^*)-\bar{f}(t)$ in the function of the standard deviation : σ (t)

$$S(t)= i \sigma (t)$$

The convergence model can be easily computed using the above definition as:

$$t \mp 1 - \bar{f}(t) = i\sigma(t)$$

Since f(t) = lp(t) and σ^2 = lp(t)(1-p(t)), therefore

$$p(t+1)\text{-}p(t)=\tfrac{i}{\sqrt{l}}\sqrt{p(t)(1-p(t))}$$

Approximating the difference equation with the differential equation

$$\frac{dp(t)}{dt} = \frac{i}{\sqrt{l}}\sqrt{p(t)(1-p(t))}$$

the solution becomes

$$p(t)=0.5\left(1+\sin\left(\frac{i}{\sqrt{l}}t + \arcsin(2p(0)-1)\right)\right)$$

For a randomly initialized population $p(0) = 0.5$

$$p(t) = 0.5 \left(1 + \sin\left(\frac{i}{\sqrt{l}} t \right) \right)$$

Tournament Selection

In tournament selection a set of individuals are chosen in a random manner and the best solution for reproduction is picked out. The number of individuals in the set is mostly equal to two but larger tournament sizes can be used in order to increase the selection pressure. Here we consider the case of optimizing the bit counting function with a tournament size $s = 2$. Under the assumption of a normally distributed function the fitness difference between two randomly sampled individuals in each tournament is also normally distributed with mean $\mu_{\Delta f}(t) = 0$ and variance $\sigma_{\Delta f}^2(t) = 2\sigma^2(t)$. Since the best of the two competing individuals are only selected, only the absolute value of the fitness difference is considered. The average fitness difference between two randomly sampled individuals is thus given by the mean value of those differences that are greater than $\mu_{\Delta f}$, which is equivalent to the mean value of one half of the normal distribution truncated at its mean value 0. The mean value of the right half of a standard normal distribution is given by $\sqrt{2/\pi} = 0.7979$.

Tournament selection selects the best out of every random pair of individuals so the population average fitness increase from one generation to the next is equal to half the mean value of the difference between two randomly sampled individuals:

$$\bar{f}(t+1) - \bar{f}(t) = \frac{1}{2} 0.7979 \sigma_{\Delta f}(t) = \frac{1}{2} 0.7979 \sqrt{2} \sigma(t) = \frac{1}{\sqrt{\pi}} \sigma(t)$$

Since $f(t) = lp(t)$ and $\sigma^2 = lp(t)(1-p(t))$, therefore

$$p(t+1) - p(t) = \sqrt{\frac{p(t)(1 - p(t))}{\pi l}}$$

Approximating the difference equation with the differential equation

$$\frac{dp(t)}{dt} = \sqrt{\frac{p(t)(1 - p(t))}{\pi l}}$$

the solution becomes

$$p(t) = 0.5 \left(1 + \sin\left(\frac{t}{\sqrt{\pi l}} + arcsin(2p(0) - 1) \right) \right)$$

For a randomly initialized population $p(0) = 0.5$, the convergence model is

$$p(t) = 0.5 \left(1 + \sin\left(\frac{t}{\sqrt{\pi l}} \right) \right)$$

MATLAB Code Snippet for Selection

The following subroutine shows the implementation of selection process in MATLAB.

```
function result = select(popWithDistrib)
% Select some genotypes from the population,
% and possibly mutates them.

selector = rand;
total_prob = 0;
% Default to last in case of rounding error
genotype = popWithDistrib(end,2:end);
for i = 1:size(popWithDistrib,1)
total_prob = total_prob + popWithDistrib(i,1);
if total_prob >selector
genotype = popWithDistrib(i,2:end);
break;
end
end
result = mutate(genotype);
```

11.7.6 Crossover Technique

Crossover is a binary variation operator that combines information from two parent genotypes into one or two offspring genotypes. Crossover is also known as recombination. Recombination is a stochastic operator. In recombination by mating two individuals with different but desirable features, mating two individuals with different but desirable features can produce an offspring that combines the features of both parents. Numerous offsprings created by random recombination in the evolutionary algorithms, have more improved characteristics over their parents. Depending on the representation of the variables of the individuals the types of recombination can be applied:

- Discrete recombination

- Real valued recombination

- Binary valued recombination

- Other types of crossover

    o Arithmetic crossover

    o Heuristic crossover

TABLE 11.4:   Discrete Recombination

Individual 1	a	b	c	d	e	f	g	h
Individual 2	A	B	C	D	E	F	G	H

$\downarrow$

Child	a	b	C	d	E	f	g	H

Discrete Recombination

Consider the following two individuals with 8 variables each:

Given two parents, Individual 1 and Individual 2, the child is created by discrete recombination as shown in Table 11.4. Discrete recombination can be used with any kind of variables (binary, integer, real or symbols).

Real Valued Recombination

This recombination technique is applied for the recombination of individuals with real valued variables. Consider the following two individuals with 6 variables each, the child is created from the parents as shown in Table 11.5.

Binary Valued Recombination (Crossover)

This crossover technique is applied for individuals with binary variables. The recombination is done by selecting two parents, then a crossover point is selected, and then the parents mutually swap the information with respect to the crossover point. Depending upon the number of cross points binary valued crossover is divided into single point, two point, and multi point crossover.

Single point crossover - one crossover point is selected, the best among

TABLE 11.5:   Real Valued Recombination

Individual 1	a	b	c	d	e	f
Individual 2	A	B	C	D	E	F

$\downarrow$

Child	(a+A)/2	(b+B)/2	(c+C)/2	(d+D)/2	(e+E)/2	(f+F)/2	(g+G)/2	(h+H)/2

TABLE 11.6: Single Point Crossover

Parent 1	1	1	0	0	1	0	1	1
				+				
Parent 2	1	1	0	1	1	1	1	1
				=				
Offspring	1	1	0	0	1	1	1	1

the binary string of chromosomes along the crossover point between the two parents is used to produce a new offspring as shown in Table 11.6.

Two point crossover - two crossover points are selected, from the binary string between the two crossover points the best chromosomes form the new individual as in Table 11.7.

Uniform crossover - bits are randomly copied from the first or from the second parent as shown in Table 11.8.

Arithmetic Crossover

A crossover operator that linearly combines two parent chromosome vectors to produce two new offspring according to the following equations:

$$Offspring1 = a * Parent1 + (1 - a) * Parent2$$
$$Offspring2 = (1 - a) * Parent1 + a * Parent2$$

where a is a random weighting factor (chosen before each crossover operation). Consider the following 2 parents (each consisting of 4 float genes),

TABLE 11.7: Two Point Crossover

Parent 1	1	1	0	0	1	0	1	1
					+			
Parent 2	1	1	0	1	1	1	1	1
					=			
Offspring	1	1	0	1	1	1	1	1

TABLE 11.8:   Uniform Crossover

Parent 1		1	1	0	0	1	0	1	1	
						+				
Parent 2		1	1	0	1	1	1	1	1	
						=				
Offspring		1	1	0	0	1	1	1	1	

which have been selected for crossover:

$$Parent\ 1:\ (0.3)\ (1.4)\ (0.2)\ (7.4)$$
$$Parent\ 2:\ (0.5)\ (4.5)\ (0.1)\ (11.6)$$

If $a = 0.7$, the following two offspring would be produced:

$$Offspring\ 1:\ (0.36)\ (2.33)\ (0.17)\ (6.86)$$
$$Offspring\ 2:\ (0.402)\ (2.981)\ (0.149)\ (6.842)$$

Heuristic Crossover

A crossover operator that uses the fitness values of the two parent chromosomes to determine the direction of the search. The offspring are created according to the following equations:

$$Offspring1 = BestParent + r * (BestParent - WorstParent)$$
$$Offspring2 = BestParent$$

where r is a random number between 0 and 1.

It is possible that offspring1 will not be feasible. This can happen if r is chosen such that one or more of its genes fall outside of the allowable upper or lower bounds. For this reason, heuristic crossover has a user settable parameter (n) for the number of times to try and find an r that result in a feasible chromosome. If a feasible chromosome is not produced after n tries, the WorstParent is returned as Offspring1.

Crossover Probability

This is simply the chance that two chromosomes will swap their bits. A good value for crossover probability is around 0.7. Crossover is performed by selecting a random gene along the length of the chromosomes and swapping all the genes after that point.

MATLAB Code Snippet for Crossover

```
function [x,y] = crossover(x,y)
% Possibly takes some information from one genotype and
% swaps it with information from another genotype

if rand <0.6
gene_length = size(x,2);
% site is between 2 and gene_length
site = ceil(rand * (gene_length-1)) + 1;
tmp = x(site:gene_length);
x(site:gene_length) = y(site:gene_length);
y(site:gene_length) = tmp;
end
```

11.7.7   Mutation Operator

Mutation is a genetic operator applied among populations, that is they are used to maintain genetic diversity between two succeeding population of chromosomes. During mutation an arbitrary bit in a genetic sequence will be changed from its original state. Mutation is implemented by generating a random variable for each bit in a sequence. This random variable gives information about the particular bit whether it is modified or not.

The purpose of mutation in EAs is to allow the algorithm to avoid local minima by preventing the population of chromosomes from becoming too similar to each other, thus slowing or even stopping evolution. This reasoning also explains the fact that most EA systems avoid only taking the fittest of the population in generating the next but rather a random (or semi-random) selection with a weighting toward those that are fitter.

Real Valued Mutation

In real valued mutation, randomly created values are added to the variables with a low probability. Thus, the probability of mutating a variable (mutation rate) and the size of the changes for each mutated variable (mutation step) must be defined. An example for mutation is given in Table 11.9.

TABLE 11.9:   Individual before and after Real Valued Mutation

Individual before mutation	A	B	C	D	E	F	G	H
Individual after mutation	A	B	D	E	C	F	G	H

Binary Mutation

In binary valued mutation, the variable values of the individuals are flipped randomly, since every variable has only two states. Thus, the size of the mutation step is always 1. Table 11.10 shows an example of a binary mutation for an individual with 11 variables, where variable 6 is mutated.

Advanced Mutation

The methodology of producing better members from the existing members is referred to as mutation. This process concentrates on the similarities between the parents of a given child; the more similarities there are the higher the chance for a mutation. This process cuts down the probability of having premature convergence.

Reproduction, crossover, and mutation operators are the main commonly used operators in genetic algorithms among others. In the reproduction process, only superior strings survive by making several copies of their genetic characteristics. In the crossover process, two cross sections of parent strings with good fitness values are exchanged to form new chromosomes. Conversely, in the mutation process, only parts of a given string are altered in the hope of obtaining genetically a better child. These operators limit the chances of inferior strings to survive in the successive generations. Moreover, if superior strings existed or are created during the process, they will likely make it to the subsequent generations.

Other types of mutation

Flip Bit

This mutation operator, which can be applied only for binary string, merely inverts the value of the chosen gene that is 0 is flipped to 1 and 1 is flipped to 0).

Boundary

TABLE 11.10:   Individual before and after Binary Mutation

Individual before mutation	1	1	0	0	0	1	1	1
Individual after mutation	1	1	0	0	0	0	1	1

Applicable for integer and float type string of chromosomes, this mutation operator replaces the value of the chosen gene with either the upper or lower bound for that gene. The upper bound or lower bound is chosen randomly.

Non-Uniform

As the generations progress the non-uniform mutation operator increases the mutation probability such that the amount of the mutation is close to 0. This type of operator prevents the population from stagnating in the early stages of the evolution and then further allows the genetic algorithm to fine tune the solution in the later stages of evolution. This mutation operator is applicable only for integer and float genes.

Uniform

The uniform mutation operator replaces the value of the chosen gene with a uniform random value selected between the user-specified upper and lower bounds for that gene. This mutation operator can only be used for integer and float genes.

Gaussian

The Gaussian mutation operator applied a unit Gaussian distributed random value to the chosen gene. The new gene value is clipped if it falls outside of the user-specified lower or upper bounds for that gene. This mutation operator can only be used for integer and float genes.

Mutation Probability

The probability of mutating a variable is related with the number of variables. The lesser the number of variables, larger is the mutation probability. Mutation Probability is the chance that a bit within a chromosome will be flipped (0 becomes 1, 1 becomes 0). This is usually a very low value for binary encoded genes, say 0.001.

MATLAB code snippet for mutation

```
function result = mutate(genotype)
% Possibly mutates a genotype
result = abs(genotype - (rand(size(genotype,1),
size(genotype,2))<0.03));
```

11.7.8   Reproduction Operator

Reproduction operator is applied to select the individuals that possess high quality solutions and that are capable of transmitting the information to the next generation. Two basic operators such as crossover and mutation are applied in this stage. The crossover operation is performed on two parents to create a new individual by combining parts of the chromosome from each of the parents. This offspring undergoes mutation, a process which which produces some small changes in the individual solution. Thus crossover exploits predetermined good solutions, while mutation allows the EA to explore different parts of the search space. Once the required number of offspring has been generated, some of these individuals are selected to form the population in the next generation.

Usually the reproduction operator conserves the population size. The most common techniques used to select good solutions for reproduction are rank grouping, proportionate grouping, and tournament grouping. This kind of selection process is performed to assign ranks thereby avoiding faster convergence.

The individuals in the given population are arranged or placed based on their expected value and fitness instead of the absolute fitness. This process of positioning is referred to as rank selection. The fitness is not scaled since very often the absolute fitness is masked. Excluding the absolute fitness avoids convergence problems, which is an advantage in the rank selection method. The difficulty in this method of selection is the critical situation involved in indentifying that an individual is far better than the closest competitor. In rank selection, the population is ranked initially, based on which each of the chromosome will receive the fitness. Each individual is assigned a selection probability based on the individuals' rank which in turn is based on the objective function values.

In fitness proportionate selection, each origination has two elapses. The first elapse occurs when the expected value of the fitness is computed for each individual. In addition, it lapses to perform the mean fitness. The fitness proportionate technique is a time-consuming technique and more similar to ranking. The advantage of using this technique is that it is more effective in parallel implementation.

The tournament selection is more efficient than the ranking methods since it does not consume much time. The pair of individuals for mating are selected from the indiscriminate population. An unbiased number, n is selected from population, such that n is in the range [0,1], and m is a parameter. These two parameters should be chosen such that n ¡ m. If this condition does not agree then the two individuals are taken back to the initial population for reselection. In tournament grouping, two strings compete for a spot in the new generation and the winner string takes the spot. This process is recursive and the superior string enters the competition twice to produce two spots. In the same manner, the inferior string will also have the chance to compete and loose twice. Therefore, every string will either have 0, 1, or 2 representations in the new generation. It has been shown that tournament grouping for replication (reproduction) operator conjoins at a highly rapid pace and needs less processes than the other operator.

11.8 Evolutionary Programming

Evolutionary Programming, originally conceived by Lawrence J. Fogel in 1960, is a stochastic optimization strategy similar to genetic algorithms, but instead places emphasis on the behavioral linkage between parents and their offspring, rather than seeking to emulate specific genetic operators as observed in nature. Evolutionary programming is similar to evolution strategies, although the two approaches developed independently.

Similar to both ES and GAs, EP is a useful method of optimization when other techniques such as gradient descent or direct, analytical discovery are not possible. In combinatoric and real-valued function optimization, the optimization surface or fitness landscape is "rugged", possessing many locally optimal solutions that are well suited for evolutionary programming.

11.8.1 History

The 1966 book, *Artificial Intelligence through Simulated Evolution* by Fogel, Owens and Walsh is the landmark publication for EP applications, although many other papers appear earlier in the literature. In this book, finite state automata were evolved to predict symbol strings generated from Markov processes and non-stationary time series. Such evolutionary prediction was motivated by a recognition that prediction is a keystone to intelligent behavior (defined in terms of adaptive behavior, in that the intelligent organism must anticipate events in order to adapt behavior in light of a goal).

In 1992, the First Annual Conference on Evolutionary Programming was held in La Jolla, California. Later on several other conferences were also held. These conferences encourage the group of academic, commercial, and researchers engaged in both developing the theory of the EP technique and in applying EP to a wide range of optimization problems, both in engineering and biology.

Rather than list and analyze the sources in detail, several fundamental sources are listed below which should serve as good pointers to the entire body of work in the field.

11.8.2   Procedure of Evolutionary Programming

In EP, similar to GAs, there is an underlying assumption that a fitness landscape can be characterized in terms of variables, and that there is an optimum solution (or multiple such optima) in terms of those variables. For instance, while finding the shortest path in a Traveling Salesman Problem, each solution would be a path. The length of the path could be expressed as a number, which would serve as the solution's fitness. The fitness landscape for this problem could be characterized as a hypersurface proportional to the path lengths in a space of possible paths. The goal would be to find the globally shortest path in that space, or more practically, to find very short tours very quickly.

The basic EP method involves 3 steps (Repeat until a threshold for iteration is exceeded or an adequate solution is obtained).

Step 1: Randomly choose an initial population of trial solutions. The number of solutions in a population is closely related to the speed of optimization, but there is no answer to predict the number of appropriate solutions (other than >1) and to predict the discarded solutions.

Step 2: Every solution that is produced is copied into a new population. These offsprings are then mutated according to a distribution of mutation types, ranging from minimum to a maximum with continuous mutation types between. The severity of mutation is guessed on the basis of the functional change enforced on the parents.

Step 3: Every offspring is evaluated by computing its fitness function. Generally, a stochastic tournament selection is applied to determine N solutions, which are held back for the population of solutions. There is no constraint on the population size to be be held constant, nor that only a single offspring be generated from each parent. Normally EP does not use any crossover as a genetic operator.

There is no requirement that the population size be held constant, however, nor that only a single offspring be generated from each parent. It should be pointed out that EP typically does not use any crossover as a genetic operator.

11.8.3 EPs and GAs

EPs and GAs are more or less similar in their functionality. Though they are similar, there a few differences between the two techniques. First, there is no constraint on the representation. In a GA approach the problem solutions are encoded as a string of representative tokens, the genome. In the EP technique, the representation follows from the problem.

Another major difference is, the mutation operation simply changes aspects of the solution according to a statistical distribution which weights minor variations in the behavior of the offspring as highly probable and substantial variations as increasingly unlikely. Further, the severity of mutations is often reduced as the global optimum is approached.

11.8.4 Algorithm of EP

The steps of the Evolutionary Programming Algorithm are

Step 1: Start with an initial time t=0
Step 2: Initialize a random population of individuals at t−0
Step 3: Evaluate fitness of all initial individuals of population
Step 4: Test for termination criterion (time, fitness, etc.). If a termination criterion is reached go to Step 8.
Step 5: Perturb the whole population stochastically
Step 6: Evaluate its new fitness
Step 7: Select the survivors from actual fitness
Step 8: End

11.8.5 Flowchart

The following flowchart in Figure 11.8 outlines the steps involved in using an Evolutionary Programming method to find a globally optimal solution. While compared to the other optimization techniques, EP is considered more as a method rather than an algorithm. There are many parameters that need to be set to use this methodology in a particular computer program. The following section explains the different steps in the flowchart.

Choosing a Solution's Coding Scheme

In all global optimization methods, some scheme needs to be used to store a putative solution. In a Standard Genetic Algorithm, a bit-string is used. Bit-string is not the best method to store a solution, since it flatly does not work for some types of problems. The stored values

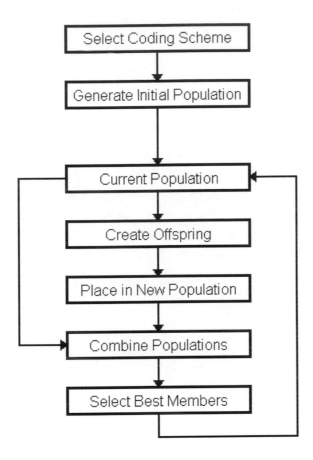

FIGURE 11.8: Flowchart of EP

are quantized, and there is no computational advantage of using bit-manipulation unless the user programs it in an assembler. Depending upon the application the coding scheme can be chosen by the user.

The Traveling Salesman Problem Type

For the Traveling Salesman Problem (TSP) and its variants, a possible solution can be uniquely described by an array of integers of fixed length. For instance, the array (A,B,C,G,H) could mean that the salesman starts at City A and then travels to Cities B, C, G and H before returning to City A.

The TSP differs from the others in the fact that the first city is known ahead of time and is set to City A for convenience. Problems similar to the TSP, have a solution that is an N-element array of integers from 1 to N. Each integer represents a particular city; and each of these integers must be present in the solution array exactly once. The object is to find the best order of these known integers.

The (0,1)-Knapsack Problem Type

The problems in this category also have solutions that consist of an array of integers of known length. In this type of problems, there is no dependency between the integers in each position of the array, meaning that the same integer can appear in more than one location.

The problem is to decide on the items from a large group to be placed into a knapsack such that the knapsack is not over-filled and the value of the items placed into it is a maximum. For instance, the solution (0,1,0,0,1,1,0,1,1,0) states that Items 2, 5, 6, 8 and 9 will be placed into the knapsack while the remaining will not be placed. The quality of this solution is the profit the company receives from these items, minus any penalty that is assessed for over-filling the container. For this problem, all solutions are represented by an array of 10 integers, where the integers are either 0 or 1, but the number of 1's present in the optimal solution is not known ahead of time.

A slightly different situation arises when the user tries to generate Quantitative Structure/Activity Relationships (QSARs) and Quantitative Structure/Property Relationships (QSPRs). In these cases, the user can start with a dataset that contains the values of structure-based descriptors for a set of related compounds and either their biological activity or some physical property (such as their HPLC retention time). The objective is to generate a numerical relationship that uses a small number of the available descriptors and predicts the activity/property as accurately as possible. For example, if 10 descriptors are present in the data set and the goal is to generate a 3-descriptor relationship, the solution array (0,1,0,0,0,1,1,0,0,0) means that the relationship has the form

$$C_0 + C_1 * D_2 + C_2 * D_6 + C_3 * D_7$$

The coefficients, C_i, can be determined by performing least-squares fit to some or all of the data using their known values of the activity/property and Descriptors 2, 6 and 7. The quality of this solution (relationship) can be either the RMS error of the fit or the error observed for other compounds in the dataset that were not used to determine the coefficients. For this type of problem, we not only know the length of the solution array, but we also know the number of 1's and 0's that must be present.

This coding scheme can be extended to use other integers, where integers other than 1 represent certain functions of the descriptors (such as its square or natural logrithm). Here, we still know the total number of nonzero elements in the solution array.

A slightly different coding scheme applies to Ordering Multiple Items from Multiple Possible Suppliers. For example, if it is required to order 10 items, the solution array (1,3,2,2,3,1,1,3,2,1) means that Items 1, 6, 7 and 10 are ordered from Supplier 1, Items 3, 4, and 9 from Supplier 2, and Items 2, 5, and 8 from Supplier 3. The quality of this solution is determined by calculating the total cost of obtaining these items from these suppliers. This cost includes shipping and handling, and may be affected by multiple-item discounts offered by a given supplier.

Though this may appear to be a different type of problem, all of these problems have solution arrays of a known length and, most importantly, the value of the array in one position does not exclude any value for the array at another position (with the caveat that in QSAR/QSPR problems the total number of nonzero array elements is fixed).

The Autonomous Vehicle Routing Problem

The Autonomous Vehicle Routing (AVR) Problem consists of having an unmanned vehicle travel from Point A to Point B using the least amount of energy. This energy depends upon both the total length of the trip and the topology of the ground covered (so a straight line is usually not the best path). The total landscape is usually placed on a grid and the vehicle moves from one grid area to an adjacent area.

The reason why this problem is different from the others mentioned in this section is that the length of the optimal solution array is not known. If the vehicle is not allowed to move diagonally, for example, each element of the solution array can contain the numbers 1 through 4 representing the four Cardinal directions (N, E, S, and W). Therefore, a route that starts with (1,2,2,3, ...) means that the vehicle moves one space to the north, two spaces east, one space south, and so on. The route continues until either Point B is reached, or a maximum number of moves has been taken. The quality of this solution can then depend upon several factors.

- The terrain cost of visiting each region on the route.

- A penalty for traveling outside an allowed region.

- A penalty for not reaching Point B in the maximum number of steps.

Again, the value of an element in the solution array does not exclude any values for other array elements, but the total number of steps in the optimal solution is not known at the start of the search.

Finally, diagonal moves can be allowed by letting each array element be a number between 1 and 8.

The Location/Allocation Problem Type

In contrast to the other problem types described above, a solution vector for the Location/Allocation problem and the Conformation of a Molecule is a fixed length array of floating-point numbers, not integers. In the Location/Allocation problem, a company has many customers spread across a map and the object is to find the optimal location of a fixed number of distribution centers. If the user wants to place two distribution centers, the solution vector contains the following four floating-point numbers (X_1,Y_1,X_2,Y_2), where (X_1,Y_1) is the map coordinates of the first distribution center and $(X2,Y2)$ is the map coordinates of the second. All customers would be serviced by the closest distribution center, and the quality of the solution could be either an unweighted sum of the distances to each customer, or a weighted sum of this distance so that a customer with many deliveries per month should be closer to the distribution center than one with few deliveries.

To determine the optimal conformation of a molecule, the molecule's geometry is uniquely determined by the values of the dihedral angle for all "rotatable bonds". Therefore, if a molecule contains 4 rotatable bonds, a possible solution is described by an array of four floating-point numbers which set the values of these four dihedral angles. The quality of this solution is inversely proportional to the internal energy of this geometry, and the object is to find the values of the dihedral angles that yield the lowest internal energy.

Both of these problems have solutions that are fixed-length arrays of floating-point numbers, and the value of an array element in one position does not exclude any possible values at other positions in the array.

Extending the Coding Scheme

Up to this point, the coding scheme has only been used to store a putative solution to the problem. There is no reason why this coding can't contain additional information.

Evolutionary Programming can use different *mutation operators* for a particular coding scheme. If two such operators are available in a particular algorithm, there is a probability that one or the other will be used. The value of this probability used to generate a putative solution can be stored in the coding scheme of that solution. This means that one or more parameters describing how an offspring will be produced are stored in the coding schemes of the parent(s). These parameters can be mated/mutated to form a parameter set for the offspring, and they will be produced using this set. This means that the optimization will try to optimize both the quality of the solution and the genetic operators used to produce offspring.

This storing of the offspring generation parameters in the coding scheme is one of the ways that Evolutionary Programming and Evolutionary Strategies differ. This overview is designed to be general enough to encompass both methodologies.

Generating an Initial Population

After a coding Scheme for the particular problem is chosen, the user needs to decide the number of putative solutions the program will store. In the lingo of Genetic Methods this is called the *population size*, and will be denoted by the parameter N_{POP}. Large values of N_{POP} are generally good because this allows the population of solutions to span more of the search space. The disadvantage of this is that the resulting program will take significantly longer to converge as N_{POP} increases. Therefore, factors which determine a good choice for N_{POP} include:

1. The complexity of the problem and its landscape.

2. The amount of memory available on the computer.

3. The amount of time available to solve the program.

The first factor is obviously unknown at the start of the solution search, so it is best advised to run the search using different values of N_{POP} and see if there is much variation in the final solution as N_{POP} increases. If a large variation is found, N_{POP} should be further increased until a consistent solution is found.

The second factor is important since the arrays holding the putative solutions for both the current population and all generated offspring should be held in the computer's memory. If the solution array is very large and N_{POP} is set to a very large value, the required storage may exceed the available memory. This will either cause the job to halt or will cause part of memory to be written to the hard disk. If the latter occurs, the job will run very slowly.

The third factor generally asks whether it is better to have a pretty good solution quickly or a very good solution slowly. When Evolutionary Programming methodology was used to solve different problems, it was found that the goal of the search was different. In conformational searches using Genetic Methods, to accurately determine the global energy minimum structure of several flexible polypeptides, N_{POP} is set to several hundred.

Once the value of N_{POP} for a particular simulation is determined, an initial population must be generated that is consistent with the coding scheme of the problem. In general, a random number generator is used to build the initial population.

For example, in a 10-city Traveling Salesman Problem, all solutions must contain the numbers 1 through 10 in different orders with the caveat that 1 is always in the first position. To build the initial population, a random number generator is used to fill an array RAN(I) with 10 floating-point numbers between 0.0 and 1.0. Set the first element of this array to 1.5 since the user always starts at City 1, and find the element of this array with the smallest value. The position of this element in RAN(I) determines the second city to visit, and this element in RAN(I) is changed to 1.5. This process is continued until the order for all 10 cities is determined. Calculate the total route length for this path and store the result and its length in the initial population.

For a (1,0)-Knapsack problem that has a total of 10 items, look at 10 elements of RAN(I) and place a 0 in the solution array if that particular entry is less than a threshold value (say 0.5) and 1 otherwise. In a QSAR/QPSR problem, only a small, fixed number of elements in the solution array should have nonzero values. Here, the random number array can be used to choose positions along the solution array to place these nonzero values. The solution array and its quality are placed in the initial population.

In an Autonomous Vehicle Routing Problem, the maximum possible length of a solution needs to be chosen. For example, if the terrain is divided into a 100x100 grid, the user may want to give the vehicle at most 300 steps to get from Point A to Point B. A random number generator can be used to fill all of these positions with the allowed step directions (either the integers 1 to 4 or 1 to 8). Starting at Point A, these steps are used to move the vehicle across the terrain, and the user can store a running total of the cost of this route as it is traversed. If at any time Point B is reached, the vehicle is stopped and the "cost" of this path is known. If Point B is not found in the 300 steps, a penalty is added that depends on how far from B the final position is. This array of steps and the final cost is stored in the initial population.

Finally, for a Location-Allocation Problem, and Conformational Searches, the solution array consists of floating point numbers. The random number generator can be used to pick random values in the allowed region for each element. The quality of this solution and the solution array are stored in the initial population. The procedure is repeated until

N_{POP} solutions are stored in the initial population.

Instead of randomly generating the initial population, it is also possible to use the results of other optimization methods to generate the initial population. For example in Theoretical Examinations of Polypeptide Folding when searched for the global optimum in the conformation of a flexible polypeptide the user can use the results from an earlier Simulated Annealing investigation as the initial population for a Genetic Algorithm search.

Some care should be taken in doing this since having an initial population too close to suboptimal solutions may make it harder to find the global optimum. Several tests were run using both a Genetic Algorithm and Evolutionary programming to find the global minimum (-16.1 kcal/mol) of a polypeptide. In each case several simulations were run using different seeds to the random number generator to randomly build initial conformations. In the first case, the initial conformation was totally unrestricted and some of the initial energies were greater than 10000 kcal/mol. In the second case, all initial conformations (after local minimization) must have energy below 1000 kcal/mol. In the third and fourth cases, a putative solution is only added to the initial population if, after local minimization, its energy was less than 20.0 and 0.0 kcal/mol, respectively.

Each time these populations were run in a GA or EP simulation, the second case produced the best results (usually the correct global minimum). The first case was so laden with bad solutions that a very good solution never emerged. The third and fourth cases contained initial solutions that were a little too good. If an offspring explored a new area of search space, its energy was almost always larger than any of the conformations in the initial population. Since a "survival of the fittest" strategy is used in these methods, these "unfit" conformations were never able to significantly contribute to the generation of new solutions. Other areas of search space, including the one containing the global minimum, were never sufficiently explored and this solution was never found.

Therefore, it is required that the initial population has solutions that are spread out across as much of the search space as possible, and to not be of such poor quality that the resulting offspring are also very unfit. Conversely, the initial population should not be too good since the solutions may be clustered in regions of search space with suboptimal solutions, but are good enough to have all other solutions die off due to lack of fitness.

The Current Population

The Current Population represents the set of N_{POP} solutions stored by the program at any given time that is used to create new solutions. In a generational algorithm where offspring are placed in a new population (Evolutionary Programming and some Genetic Algorithms), the Current Population remains constant for the entire generation. For *nongenerational* Genetic Algorithms, the Current Population is changed each time an offspring is produced. At the start of the search, the Current Population is just the randomly generated Initial Population.

In a Genetic Algorithm program, there may be the option of choosing parents directly from the Current Population, or use this population to create a Mating Population from which parents are chosen. A Mating Population should only be created if the user wants to use the same probabilistic selection procedure for both parents. For example, if the user decides to use a tournament selection procedure, then he can initially run a large number of these tournaments (at least N_{POP}) and place the tournament winner in a Mating Population. To select two parents for each mating, the user simply needs to randomly select two entries from the Mating Population.

If different selection procedures will be used for each parent, such as a tournament selection for one and a random for the other, there is no advantage to building two Mating Populations (one for each parent), but it can be done.

Finally, if a Mating Population is used in a generational Genetic Algorithm, there is the option of merging either the Current Population or the Mating Population with the new population before using a selection procedure to choose the Current Population for the next generation. Current Population is recommended for the following reason. If the new population is dominated by solutions that do not do a good job at solving the problem (unfit or high-COST solutions), their solution arrays may be missing a key component that is needed to find the global optimum. This would suggest that this component is also either missing or under represented in the Mating Population. If these two populations are combined and a selection procedure is used to create the Current Population for the next generation, the odds are high that this key component will not be present. If this occurs, finding the global optimum will be very unlikely. On the other hand, there is a chance that this key component is present in members of the Current Population, but values of the other array elements caused them to be unfit enough to warrant placement in the mating Population. If the Current Population is merged with the new population, there is a better chance that this key component can find its way into the Current Population for the next generation.

The other situation would be that the new population contains many solutions that have a high fitness (low COST). Merging this population with the Current Population will only increase the chance that the good, new solutions will be transferred to the Current Population for the next generation.

If the Mating Population is merged with the new population, many copies of good solutions will be present in the merged population. This means that the Current Population for the next generation has a high probability of looking quite similar to the Mating Polulation of the previous generation. If this is done for several generations, all members of the Current Population could be the same solution and the search effectively stops.

Offspring Generation in Evolutionary Programming Methods

In all Evolutionary Programming methods, each member of the Current Population is used to generate an offspring. Each offspring is placed into a new population. When all offspring have been generated, the Current Population is merged with the new population of offspring, and a selection procedure is used to generate a Current Population for the next generation.

This section describes some of the *mutation operators* that can be used to create a new offspring from a single member of the Current Population. This list is in no way complete, but gives the reader an idea of some possible mutation operators. The type of operators that can be used obviously depends upon the Coding Scheme, so one or more example operators will be described for each scheme.

In addition, many animal populations nurture their new offspring and allow them to mature before their fitness for survival is tested. Therefore, though it is not used in any of the standard Evolutionary Programming formalisms, it is found useful to employ a *maturation operator* in certain circumstances. This operator performs a local optimization of the initial solution of each offspring before its COST or fitness is determined. Some examples of this operator will also be presented.

The Traveling Salesman Problem Type

All of these problems have a solution that is an array of integers from 1 to N. Each integer represents a particular city, item, task, or game; and each of these integers must be present in the solution array exactly once. Since the object is to find the best order of these known integers, the mutation operator must simply reorder some of the elements. Two possible mutation operators are pair switching and region inversion.

In pair switching, two positions along the solution vector are randomly selected and the cities in these positions are switched. This operation is shown in the following 8-element example

```
(1,4,7,5,2,6,8,3) Parent solution
    |   | Selected solutions
(1,4,8,5,2,6,7,3) offspring solution
```

A region inversion is performed by cutting the solution array at two random points (with more than one element between the cut points) and inverting the array elements between the cuts. This is shown in the following example.

```
(1,4,7,5,2,6,8,3) Parent solution
    |   |      Cut solutions
(1,4,6,2,5,7,8,3) offspring solution
```

Either or both of these mutations may be performed one or more times to generate a new solution.

If a maturation operator is used, it could be a simple pair-wise switching of adjacent elements in the array. In a Traveling Salesman Problem, City 1 is always left in the first position, so this operation would start with switching the cities in the second and third positions of the solution array. If this results in a shorter total path the switch is accepted; otherwise the cities are returned to their original positions. The cities in the third and fourth positions can then be switched and so on until switching the last pair is tried. This procedure can be done as a single pass down the route, or can be repeated a given number of times, or until none of the switches results in a shorter path length.

The (0,1)-Knapsack Problem Type

Of the three types of problems placed into this group, the simple bin packing is the easiest to treat. The coding scheme for a putative solution consists of an array containing only 0's and 1's, with no constraints on the number of each. An offspring can be created by making a copy of the parent's array, randomly selecting one or more elements and changing the values of these elements. For example, if positions 3 and 5 are chosen in an 8-object problem we would have

```
(1,0,0,1,1,0,1,1) Parent solution
    |     |     Selected solutions
(1,0,1,1,0,0,1,1) offspring solution
```

This means that the new solution would try to place Items 1, 3, 4, 7, and 8 into the bin. The quality of this solution is determined by the profit to your company from shipping these items minus any penalty for exceeding the capacity of this shipping bin.

If a maturation operator is used, its goal would be to either remove the penalty or increase the profit of the goods. If the selected items exceed the capacity of the bin, the maturation operator could simply look down the list of items to be shipped in order of increasing profit, and remove the first one that places the user under the limit. If no single item can be found, remove the item that yields the least profit and check the others again. If, on the other hand, the selected items to not completely fill the bin, the maturation operator could look through the list of unselected items in order of decreasing profit and add the first one that does not cause the bin capacity to be exceeded. Once this item is added to the list of packed items (the 0 at its position in the solution array is changed to a 1), the remaining items can be checked to see if another can be added.

For the QSAR/QSPR problem type, it is generally desirable to keep the number of nonzero elements in the solution array fixed to the number of descriptors required in the solution. This means that the user can randomly choose a selected descriptor and a non-selected one and change the values of the elements at these positions. This is shown in the following example.

```
(1,0,0,0,1,0,1,0) Parent solution
   |   |   Selected solutions
(1,0,1,0,0,0,1,0) offspring solution
```

An additional mutation operator can be used if the elements of the solution array can have values different than 0 and 1 to represent different functional forms of the descriptors. For example, if the allowed values range from 0 to 4, this mutation operator can randomly choose a descriptor that is to be included and randomly change the value of its array element to any allowed, nonzero value (1, 2, 3, or 4). Conversely, a maturation operator can be employed that examines each selected descriptor and sequentially uses all allowed, nonzero values while keeping the other elements of the array fixed. The value of this element would then be set to the value that yields the best result. Once this is done for all chosen descriptors, this process can be repeated until no nonzero element is changed.

Finally, in a QSPR examination it was found that if the COST of the solution contained a term that added a penalty for having the wrong number of descriptors, better results were obtained when an extra descriptor is added first and then a descriptor is removed in a later generation. Care had to be taken during this study to ensure that the penalty was not too large to prohibit any extra terms from being added, and not too small to allow the program to use too many descriptors. The population of solutions was examined at the end of each generation, and the value of the penalty was adjusted if either situation occurred. This process may be equivalent to using a mutation operator to randomly add a descriptor and then immediately employ a maturation operator that sequentially removes each of the selected descriptors. In the above example, one of the five positions containing a zero value would be randomly chosen and replaced with a 1. The solution array now contains four nonzero elements and the four possible solutions with three nonzero elements would be examined. The one that produces the best solution would represent this new offpsring.

For the problem of Ordering Multiple Supplies from Multiple Possible Suppliers, the mutation operator would randomly select one or more of the supplies and randomly pick a possible supplier that is different than the one used by the parent solution. An example of this is as follows.

```
(3,2,3,4,1,5,1,2)     Parent Solution
   | Selected Position
   1    Possible Suppliers Exculding Supplier 3
   2
   5
   |
(3,2,5,4,1,5,1,2) offspring Solution
```

No simple maturation operator is available for this process, since it would only result in using the mutation operator again and may result in the offspring looking just like its parent.

The Autonomous Vehicle Routing Problem

For this problem, each solution vector consists of a maximum length array containing integers describing the direction of the next step. The mutation operator would be similar to the one used above in that one or more positions along the array are chosen and the integer describing the next step's direction is changed to another allowed value.

For this problem, it is possible to use a maturation operator that ensures that no grid of the map is visited more than once. A second array can be constructed that contains the grid number that the vehicle is currently occupying. If the same number appears more than once, the path that formed this closed loop can be removed and random directions can be added to the end of the path to fill in the array. Another maturation operator could check the path and make sure that the vehicle does not leave the allowed region. If this occurs, the direction of the next step at the border point can be changed to ensure that the path is not retraced and the vehicle stays in the allowed region. Both of these maturation operators may be of great value in finding good routes for the vehicle.

The Location/Allocation Problem Type

A solution vector for the Location/Allocation problem and the Conformation of a Molecule is a fixed length array of floating-point numbers. Therefore, the mutation operator simply adds or subtracts a random amount to one or more of the elements in the solution array. For the Location/Allocation problem, the program would first have to ensure that the new coordinate lies on the map of the problem. If not, a different random amount should be chosen.

In practice, a maximum step size (MXSTEP) can be set and the following quantity

$$(1.0\text{-}2.0*R)*MXSTEP$$

is added to an array element, where R is a random number in the range (0.0,1.0).

In the Location/Allocation problem no maturation operator exists that is different from a reapplication of the mutation operator. Therefore, it may be advisable to reduce the size of MXSTEP at the end of each generation, so that subsequent generation explores good regions of solution space more carefully.

In the Conformation of a Molecule problem a maturation operator exists, and it is simply finding a local minimum structure. Therefore a mutation operator that completely randomizes one or more of the dihedral angles along with this maturation operator is used, and does not change the mutation operator throughout the search.

In some of the problem types presented above, it is possible to use more than one maturation operator and/or use the same maturation operator more than once. Therefore, an Evolutionary Programming program may want to include probabilities of using a particular mutation operator or using it more than once. These probabilities can be used for all offspring generation and can stay constant or be changed from one generation to the next. Conversely, as stated in Choosing a Solution's Coding Scheme, these probabilities can be stored in the solution array of each putative solution. These particular probabilities can be mutated when they are transferred to the offspring and then used to generate the offspring's solution.

The last point is that all standard applications of Evolutionary Programming have each member of the Current Population generate a single offspring which is added to a new population. It is also just as possible to have each member generate multiple offspring, and only the most-fit (lowest-COST) offspring would be placed in the new population. This may significantly increase the computation time of the algorithm, but may also improve the final results.

Placing Offspring in a New Population

In all Evolutionary Programming algorithms and generational Genetic Algorithms, selected offspring are placed into a new population. In Evolutionary Programming algorithms, each of the NPOP solutions in the Current Population is used to generate one or more offspring by using a mutation operator and optionally a maturation operator. The most-fit (lowest-COST) offspring is placed in the new population. In a generational Genetic Algorithm, pairs of parents are selected from either the Current or Mating Population and use a mating operator to generate either a single offspring, a complimentary pair of offspring, or multiple offspring. The most-fit offspring, and optionally its compliment, are placed in the new population. This continues until NPOP pairs of parents have generated offspring.

The advantage of such a generational algorithm is that the Current/Mating Population remains constant for the entire generation. This means that the production of offspring can easily be distributed across multiple CPUs of a network. If a maturation operator is employed, this offspring generation is by far the most time consuming step, and distributing this workload results in a near-linear speedup.

Once the new population is constructed, it is merged with the Current Population. From this merged population, a new Current Population is selected for the next generation.

Merging the Populations

In all applications of Evolutionary Programming, and selected applications of a Genetic Algorithm, the offspring are placed in a new population. In general, Evolutionary Programming methods have each member of the Current Population generate a single offspring with the aid of one or more mutation operators. This offspring is then placed in the new population. It is also possible for each member of the current population generate multiple offspring. In this case, the offspring with the lowest COST (highest fitness) is placed in the new population. This continues until all members of the Current Population have placed an offspring in the new population.

In Genetic Algorithms, a mating operator is the major mechanism for generating offspring from a pair of parents. Each mating can produce either a single offspring or a complimentary pair of offspring. It is also possible for a selected pair of parent solutions to generate multiple offspring or multiple pairs. In this case, the lowest COST offspring, and optionally its compliment, are placed in the new population. If the number of solutions in the Current Populations is NPOP, this process continues until NPOP pairs of parents have been selected and have placed one or two offspring in the current population.

These procedures are called generational algorithms because the Current Population does not change while the new population of offspring is being generated. The advantage of this procedure is that the generation of offspring can be performed in a distributed fashion across a network of computers. The potential disadvantage of this method is that an offspring that does a very good job at solving the problem cannot be used right away to generate new offspring, and this may slow the convergence of the population to the global optimum. In addition, it is possible for a relatively fit offspring to never be used as a parent in a generational algorithm.

Once the new population of offspring has been generated, it is merged with the Current Population. NPOP members of this merged population are then chosen to become the Current Population of the next generation. This procedure is repeated for either a user-supplied number of times (NGEN), or until a user-supplied convergence criteria has been met. The Current Population at the end of the search represents the final population of solutions.

In applications of a Genetic Algorithm, it is possible for certain members of the Current Population to be selected using a fitness-based probabilistic method and placed in a Mating Population. Though it is only members of the Mating Population that are randomly chosen to become parents and create offspring, it is my opinion that the Current Population should be merged with the new population of offspring and not the Mating Population. There may be values of certain elements of the solution array that are needed to find the global optimum solution, but members of the Current Population that possess these values may not have been selected for the Mating Population. If the Mating Population is merged with the new population, these required values will be lost. In addition, the Mating Population probably has multiple copies of the most-fit solutions from the current population. If they are merged with the new population, the chances are good that most or all of them will be selected for the Current Population of the next generation. Their numbers will again be amplified in the next Mating Population, and after only a few generations the Current Population will be dominated by multiple copies of a few solutions. Though this will increase the convergence rate of the algorithm, the Current Population may become trapped in the region of a suboptimal solution. Therefore, to maintain diversity in the Current Population and increase the chances of finding the global minimum, the Current Population and not the Mating Population should be merged with the new population of offspring.

Choosing a Current Population

In all Evolutionary Programming algorithms, each of the N_{POP} members of the Current Population generate an offspring using a mutation operator, and this offspring is placed in a new population. In a generational Genetic Algorithm, NPOP pairs of parents are chosen from either the Current or Mating Population and use a mating operator to place either a single offspring or a complimentary pair of offspring in the new population.

Once the new population is constructed, it is merged with the Current Population. From this merged population, N_{POP} members are selected to form the Current Population for the next generation (or the final set of solutions). Since similar methods are used to select parents for each mating in a Genetic Algorithm, various selection procedures are presented in a separate section.

Since a "survival of the fittest" criteria must be employed to ensure that the best solutions to date survive to the next generation, a random selection procedure should not be used here (though it can be used to select one of the parents in a Genetic Algorithm). In other words, only fitness-based selection procedures should be used to determine which solutions in the merged population become members of the next generation's Current Population.

A *deterministic selection* procedure is an obvious choice since the N_{POP} most-fit (lowest-COST) members of the merged population survive. This procedure causes the fitness of the best solution to date and the average fitness of the Current Population to never decline from generation to generation. It is also found to increase the convergence of the population. A corollary to this is that the diversity of the solutions is quickly reduced. This occurs because the Current Population soon becomes one or more clusters of solutions around reasonably good solutions. If one of these clusters is around the global optimum solution, it will quickly be found. Conversely, if none of the clusters is around the global optimum, chances increase that it may never be found. Therefore, for certain types of problems, a probabilistic selection procedure may be advisable.

Though several probabilistic selection procedures have been previously presented, the diversity of the next generation's Current Population can be maximized if a Similarity-Based Selection Procedure is used. If each parent or mating pair only places a single offspring in the new population, the size of the merged population is $2*N_{POP}$, and a tournament size (N_{TRN}) of 2 should be used. Conversely, if a Genetic Algorithm places a complimentary pair of offspring into the new population the merged population has a size of $3*N_{POP}$, and $N_{TRN}=3$ should be used.

This procedure simply searches through the list of available solutions in the merged population and finds the member with the lowest COST (highest fitness). This solution is added to the Current Population of the next generation. The algorithm then looks at the other members of the merged population and finds the N_{TRN}-1 solution(s) that are most similar to the selected one, and all N_{TRN} solutions are removed from further consideration. This procedure continues until only N_{TRN} solutions are left in the merged population, and the one with the lowest COST is chosen as the final member of the next generation's Current Population. Basically, this is a tournament selection where all members of the tournament are chosen based on similarity, and all solutions in the tournament are excluded from further consideration. By basing the similarity on the lowest COST solution available, the Current Population in the next generation is guaranteed to have good solutions with high diversity.

The actual complexity of the problem and/or the user's requirement of speed versus diversity will determine if a deterministic selection, one of the probabilistic selections, or a similarity-based selection procedure should be used.

11.9 Evolutionary Strategies

As discussed earlier, Evolution Strategies (ESs) were developed by Rechenberg and Schwefel at the Technical University of Berlin and have been extensively studied in Europe.

While EP has derived for pure scientific interest, motivation of this topic is, from the beginning, to solve engineering design problems: Rechenberg and Schwefel developed ESs in order to conduct successive wing tunnel experiments for aerodynamic shape optimization. Their important features are threefold:

1. ESs use real-coding of design parameters since they model the organic evolution at the level of individual's phenotypes.

2. ESs depend on deterministic selection and mutation for its evolution.

3. ESs use strategic parameters such as on-line self-adaptation of mutability parameters.

The representation used in evolutionary strategies is a fixed-length real-valued vector. As with the bit-strings of genetic algorithms, each position in the vector corresponds to a feature of the individual. However, the features are considered to be behavioral rather than structural.

Evolution Strategies (ESs) are in many ways very similar to Genetic Algorithms (GAs). As their name implies, ESs too simulate natural evolution. The differences between GAs and ESs arise primarily because the original applications for which the algorithms were developed are different. While GAs were designed to solve discrete or integer optimization problems, ESs were applied first to continuous parameter optimization problems associated with laboratory experiments. ESs differ from traditional optimization algorithms in a few aspects:

- ES search between populations, rather than between individuals.

- The objective function information is sufficient, derivatives are not required.

- The transition rules used by ESs are probabilistic, not deterministic.

The reproduction process in an ES is only by mutation. To apply mutation, the Gaussian mutation operator is made use of. Sometimes an intermediate recombination operator is used, in which the vectors of two parents are averaged together, element by element, to form a new offspring (see Figure 11.9).

These operators reflect the behavioral as opposed to structural interpretation of the representation since knowledge of the values of vector elements is used to derive new vector elements.

Compared to GA and GP, the selection of parents to form offspring is less constrained, since the string representation is simple. Due to this nature of the representation, it is easy to average vectors from many

Parent 1	0.2	0.6	0.9	1.5
Parent 2	1.0	0.8	0.7	0.5

Offspring	0.6	0.7	0.8	1.0

FIGURE 11.9: Intermediate Recombination of Parents

individuals to form a single offspring. In a typical evolutionary strategy, N parents are selected uniformly randomly, more than N offspring are generated through the use of recombination, and then N survivors are selected deterministically. The survivors are chosen either from the best N offspring (i.e., no parents survive) or from the best N parents and offspring.

The basic structure of an ES is very similar to that of a basic GA. The standard optimization routine makes use of the word "population" rather than "solution". A more major difference is that the usual operation of generating a new solution has been replaced by three separate activities - population selection, recombination, and mutation. It is in the implementation of these operations that the differences between ESs and GAs lie. ESs, GAs, and associated algorithms are now known collectively as *evolutionary algorithms* and their use as *evolutionary computation*.

11.9.1　Solution Representation

ESs is still primarily used to solve optimization problems with continuous control variables, and for applications of this sort the natural representation of the control variables as an n-dimensional real-valued vector x is entirely appropriate. In addition, the representation of a solution may include (depending on the specific ES implementation being employed) up to n different variances $c_{ii} = \sigma_i^2$ and up to $n(n-1)/2$ covariances c_{ij} of the generalized n-dimensional normal distribution with zero means and a probability density function:

$$p(z) = \sqrt{\frac{|A|}{(2\pi)^n}}\ exp\ \left(-\frac{1}{2}z^T A z\right) \tag{11.20}$$

where A-1 = $\{c_{ij}\}$ is the covariance matrix and z the vector of random variables. To ensure that the matrix A^{-1} is positive-definite, ES algorithm implementations usually work in terms of the equivalent rotation angles:

$$\alpha_{ij} = \frac{1}{2}tan^{-1}\left(\frac{2c_{ij}}{\sigma_i^2 - \sigma_j^2}\right) \tag{11.21}$$

These variances, co-variances, and rotation angles are known as strategy parameters.

11.9.2　Mutation

In GA implementations mutation is usually a background operator, with crossover (recombination) being the primary search mechanism. In

ES implementations mutation takes a much more central role. In its most general form the ES mutation operator works as follows:

- First, if they are used, the standard deviations and rotation angles (strategy parameters) associated with the individual solution are mutated:

$$\sigma'_i = \sigma_i \, exp(\tau' \times N + \tau \times N) \qquad (11.22)$$

$$\alpha'_{ij} = \alpha_{ij} + \beta \times N \qquad (11.23)$$

where the N are (different) random numbers sampled from a normally distributed one-dimensional random variable with zero mean and unity standard deviation, and τ, τ' and β are algorithm control parameters for which Schwefel recommends the following values:

$$\tau = \frac{1}{\sqrt{2\sqrt{n}}}, \tau' = \frac{1}{\sqrt{2n}}, \beta = 0.0873 \qquad (11.24)$$

n being the number of control variables.

- Then the vector of control variables is mutated:

$$x' = x + n \qquad (11.25)$$

where **n** is a vector of random numbers sampled from the **n**-dimensional normal distribution with zero means and the probability density function in equation (11.20).

11.9.3 Recombination

A variety of recombination operators have been used in ESs. Some, like the GA crossover operator, combine components from two randomly selected parents, while others allow components to be taken from any of the solutions in the parent population. Recombination is applied not only to the control variables but also the strategy parameters. Indeed, in some ES implementations different recombination operators are applied to different components of the solution representation. The most commonly used recombination operators are described in the following sections.

Discrete Recombination

In discrete recombination the offspring solution inherits its components such as the control variables, strategy parameters, etc., from two randomly selected parents.

Global Discrete Recombination

In global discrete recombination the offspring solution inherits its components from any member of the population, the parent to contribute each component being chosen by "balanced roulette wheel selection". Thus, if there are μ members of the population, each has a $1/\mu$ chance of being selected to contribute each component of each offspring solution.

Intermediate Recombination

In intermediate recombination the offspring solution inherits components which are a weighted average of the components from two randomly selected parents:

$$c_0 = \omega c_1 + (1 - \omega)c_2$$

where component c could be a control variable component x_i, a standard deviation σ_i, or a rotation angle α_{ij}, and weighting ω traditionally has a value of 0.5, although some ES implementations allow ω to vary.

Global Intermediate Recombination

Global intermediate recombination is similar to intermediate recombination in that the offspring solution inherits components which are a weighted average of the components from two randomly selected parents, but now new parents for each component of the offspring are chosen from the population by "balanced roulette wheel selection".

It has been found that in practice ESs tend to perform best if discrete recombination is performed on the control variables and intermediate recombination on the strategy parameters.

Selection

The best μ individuals from the population of λ offspring or from the combination of the μ previous parents and their λ offspring are chosen as parents for the next generation. The former scheme is known as (μ, λ)-selection, the latter as $(\mu + \lambda)$-selection. Though $(\mu + \lambda)$-selection is elitist, (μ, λ)-selection is, in general, preferred, because it is better able to adapt to changing environments (a common feature of the parameter optimization problems for which ESs were originally developed). Schwefel recommends a ratio of $\mu : \lambda$ as 1:7.

11.9.4   Population Assessment

An ES does not use derivative information, it just needs to be supplied with an objective function value for each member of each population.

Thus, the evaluation of the problem functions is essentially a "black box" operation as far as the ES is concerned. Obviously, in the interests of overall computational efficiency, the problem function evaluations should be performed efficiently. As long as there are no equality constraints and the feasible space is not disjoint, then infeasible solutions can simply be "rejected". In an ES this means ensuring that those particular solutions are not selected as parents in the next generation.

If these conditions on the constraints are not met, then a penalty function method should be used. A suitable form for an ES is:

$$f_A(x) = f(x) + M^k w^T c_V(x) \qquad (11.26)$$

where $\mathbf{w}$ is a vector of nonnegative weighting coefficients, the vector c_V quantifies the magnitudes of any constraint violations, $\mathbf{M}$ is the number of the current generation, and $\mathbf{k}$ is a suitable exponent. The dependence of the penalty on generation number biases the search increasingly heavily toward feasible space as it progresses.

11.9.5 Convergence Criteria

Two standard convergence tests are used to terminate ES searches. One is that the absolute difference in the objective function values of the best and worst members of the post-selection population is less than a user-specified limit, i.e.,

$$f_w - f_b \leq \omega_c \qquad (11.27)$$

The other is that the relative difference in the objective function values of the best and worst members of the post-selection population is less than a user-specified limit, i.e.,

$$f_w - f_b \leq \frac{\varepsilon_d}{\mu} \mid \sum_{i=1}^{\mu} f_i \mid$$

Thus, absolutely or relatively, the objective function values of the parents must lie close together before the algorithm is deemed to have converged.

11.9.6 Computational Considerations

The procedures controlling the generation of new solutions are so simple that the computational cost of implementing an ES is usually dominated by that associated with the evaluation of the problem functions.

It is therefore important that these evaluations should be performed efficiently and is essential if the optimization is to be performed on a serial computer.

Like GAs, ESs are particularly well-suited to implementation on parallel computers. Evaluation of the objective function and constraints can be done simultaneously for a whole population, by mutation and recombination.

If it is possible to parallelize the evaluation of individual problem functions effectively, some thought and, perhaps, experimentation will be needed to determine the level at which multitasking should be performed. This will depend on the number of processors available, the intended population size and the potential speed-ups available. If the number of processors exceeds the population size, multi-level parallelization may be possible.

Pseudocode

The pseudocode of ES is shown below:

```
t=0
initialize(P(t=0)):
evaluate(P(t=0));
while is Not Terminated () do
Pp(t) = selectBest(μ,P(t));
Pc(t) = reproduce(λ,Pp);
mutate(Pc(t));
evaluate(Pc(t));
if (use Plus Strategy) then P(t+1) = Pc(t)∪ P(t);
else P(t+1) = Pc(t):
t = t +1;
end
```

11.9.7 Algorithm Performance

Figure 11.10 shows the progress of an ES on the two-dimensional Rosenbrock function, $f = (1 - x_1)^2 + 100(x_2 - x_1^2)2$. Each member of the 1st, 10th, 20th, and 30th generations is shown (by a symbol), although, in fact, most of the members of the 1st generation lie outside the bounds of the figure. The convergence of the population to the neighborhood of the optimum at $(1,1)$ is readily apparent. Notice how the 20th generation appears to be converging on $(0.75, 0.7)$, but in the

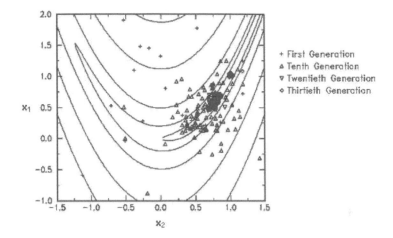

FIGURE 11.10: Progress of ES on Rosenbrock function

next 10 generations the search successfully progresses along the shallow valley to the true optimum.

For this run a (20, 100)-ES was used. The two control variables were subject to discrete recombination (of pairs of parents), while the strategy parameters were subject to global intermediate recombination.

Figure 11.11 shows the progress in reducing the objective function for the same search. Both the objective function of the best individual within each population and the population average objective are shown (note that the scales are different). These are the two standard measures of progress in an ES run. The difference between these two measures is indicative of the degree of convergence in the population.

11.10 Advantages and Disadvantages of Evolutionary Computation

Evolutionary algorithm optimizers are global optimization methods and scale well to higher dimensional problems. They are robust with respect to noisy evaluation functions, and the handling of evaluation functions which do not yield a sensible result in given period of time is straightforward. Evolutionary computation algorithms are applicable to a wide variety of optimization problems. The function to be optimized can have continuous, discrete, or mixed parameters. There are no *a pri-*

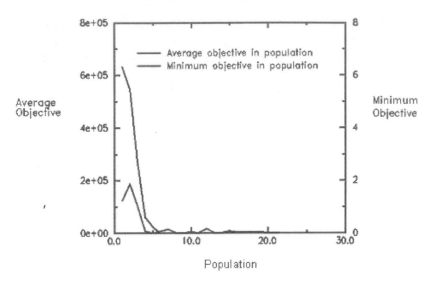

FIGURE 11.11:   Progress in Reducing the Objective Function

ori assumptions about convexity, continuity, or differentiability. Thus, it is relatively easy to apply a general purpose evolutionary computation algorithm to a search or optimization problem. An objective function is noisy if it can give different values for the same parameters. Many real-world problems have this property. Evolutionary computation algorithms are relatively insensitive to noise. Evolutionary algorithms are easy to parallelize. There is a considerable body of research on different methods to parallelize these algorithms. In many cases, they scale too many processors. Evolutionary algorithms can be a natural way to program or to enhance the programming of adaptive and distributed agents.

On the other hand, the best algorithms for solving some particular kind of problem will almost always utilize the characteristics of that problem. In other words, it pays off to understand the problem and to utilize this knowledge in the solution. Usually, the best algorithms to solve specific class problems are not based on evolutionary computation, although in some cases algorithms that combine some other technique such as hill-climbing or local-search with evolutionary computation can do very well. Evolutionary computation algorithms tend to be computation intensive. Further, it is generally impossible to guarantee a particular quality of solution using evolutionary computation. Thus, they are usually not well-suited to real-time applications. There are a great many evolutionary computation algorithms, and these algorithms often have many parameters. It can be difficult to choose an algorithm and once an algorithm is chosen, it can be difficult to tune the parameters of the algorithm.

An advantage as well as disadvantage of EAs lies in the flexibility of design. It allows for adaptation to the problem under study. Algorithmic design of EAs can be achieved in a stepwise manner. This design involves empirical testing of design options and a sound methodological knowledge. Another disadvantage of EAs is that they generally offer no guarantee to identify the a global minimum in a given amount of time. In practical solutions, it is difficult to predict the solution quality attainable within a limited time.

Summary

With a better understanding of the similarities and differences between various implementations of EAs, the community has begun to concentrate on generalizing results initially shown only for specific EAs. For example, Grefenstette and Baker illustrate that many features of EAs do not change when certain properties of selection and scaling are assumed. They also indicate when the features change, if the properties are not met. Although this work is preliminary, it helps explain why a wide variety of EAs have all met with success. As we understand better the strengths and weaknesses of the current evolutionary models, it is also important to revisit the biological and evolutionary literature for new insights and inspirations for enhancements. Booker has recently pointed out the connections with GA recombination theory to the more general theory of population genetics recombination distributions. Muhlenbein has concentrated on EAs that are modeled after breeding practices. In the EP community, Atmar highlights some errors common to evolutionary theory and the EA community. Thus in this chapter, a brief overview of the field of evolutionary computation was provided by describing the different classes of evolutionary algorithms which have served to define and shape the field. By highlighting their similarities and differences, we have identified a number of important issues that suggest directions for future research.

Review Questions

1. Briefly explain the basic template for an evolutionary computation algorithm.

2. Mention the major parameters in which EC methods differ?

3. Define Gene, Allele, Genotype and Phenotype.

4. What are phylogenetic problems? Explain with an illustration.

5. Mention the three essential ingredients for an evolutionary process.

6. Write a brief note on Darwinian Evolution.

7. Explain natural selection.

8. What are the five key observations and inferences of Darwin's theory of evolution?

9. State the paradigms of EC.

10. Explain the algorithm of GA, GP, ES, and EP.

11. What are the components of EC? Explain them in detail.

12. Explain roulette-wheel selection with an example.

13. What are the major differences between truncation and tournament selection?

14. Explain the different types of recombination with suitable examples.

15. What is the difference between shuffle crossover and heuristic crossover?

16. Mention the different types of mutation.

17. Define Flip Bit, Boundary, Non-Uniform, Uniform, and Gaussian type of mutation operators.

18. Differentiate Local and Global reinsertion.

19. Differentiate EPs and GAs.

20. Explain the steps involved in using an Evolutionary Programming with the aid of a flowchart.

21. Briefly explain mutation and recombination in ES.

22. What is the convergence criteria in an ES search?

23. Differentiate EPs and ESs.

24. Mention a few advantages and disadvantages of EC.

Chapter 12

Evolutionary Algorithms Implemented Using MATLAB

Solved MATLAB programs are given in this chapter to illustrate the implementation of Evolutionary Computation in problems such as optimization, proportional-derivative controller, multiobjective optimization, and minimization of functions.

12.1 Illustration 1: Differential Evolution Optimizer

```
%*****************************************************************
% Script file for the initialization and run of the differential
% evolution optimizer
%*****************************************************************

% F_VTR    "Value To Reach" (stop when ofunc <F_VTR)
          F_VTR = -10;

% I_D    number of parameters of the objective
         function
           I_D = 2;

% FVr_minbound,FVr_maxbound        vector of lower
and bounds of initial population
%          the algorithm seems to work especially
well if [FVr_minbound,FVr_maxbound]
% covers the region where the global minimum is
                 expected
%*** note: these are no bound constraints!! ***
        FVr_minbound = -3*ones(1,I_D);
        FVr_maxbound = 3*ones(1,I_D);
        I_bnd_constr = 0; % 1: use bounds as bound
% constraints, 0: no bound constraints
```

```
% I_NP number of population members
        I_NP = 15; % pretty high number -
        needed for demo purposes only


% I_itermax maximum number of iterations(generations)
            I_itermax = 50;


% F_weight DE-stepsize F_weight ex [0, 2]
            F_weight = 0.85;


% F_CR crossover probabililty constant ex [0, 1]
        F_CR = 1;
```

% I_strategy 1 - ->DE/rand/1:
%   the classical version of DE.
%   2 - ->DE/local-to-best/1:
%   a version which has been used by quite a number
%   of scientists. Attempts a balance between robustness
%   and fast convergence.
%   3 - ->DE/best/1 with jitter:
%   taylored for small population sizes and fast convergence.
%   Dimensionality should not be too high.
%   4 - ->DE/rand/1 with per-vector-dither:
%   Classical DE with dither to become even more robust.
%   5 - ->DE/rand/1 with per-generation-dither:
%   Classical DE with dither to become even more robust.
%   Choosing F_weight = 0.3 is a good start here.
%   6 - ->DE/rand/1 either-or-algorithm:
%   Alternates between differential mutation and three-point-
%   recombination.

```
I_strategy = 3

% I_refresh intermediate output will be produced after
%         "I_refresh" iterations. No intermediate
%         output will be produced if I_refresh is <1
I_refresh = 1;


% I_plotting   Will use plotting if set to 1. Will
skip plotting otherwise.
I_plotting = 1;

%--Problem dependent constant values for plotting--
```

```
if (I_plotting == 1)
   FVc_xx = [-3:0.25:3]';
   FVc_yy = [-3:0.25:3]';

   [FVr_x,FM_y]=meshgrid(FVc_xx',FVc_yy') ;
   FM_meshd = peaks(FVr_x,FM_y);

   S_struct.FVc_xx = FVc_xx;
   S_struct.FVc_yy = FVc_yy;
   S_struct.FM_meshd = FM_meshd;
end

S_struct.I_NP           = I_NP;
S_struct.F_weight       = F_weight;
S_struct.F_CR           = F_CR;
S_struct.I_D            = I_D;
S_struct.FVr_minbound   = FVr_minbound;
S_struct.FVr_maxbound   = FVr_maxbound;
S_struct.I_bnd_constr   = I_bnd_constr;
S_struct.I_itermax      - I_itermax;
S_struct.F_VTR          = F_VTR;
S_struct.I_strategy     = I_strategy;
S_struct.I_refresh      = I_refresh;
S_struct.I_plotting     = I_plotting;

%*************************************************************
% Start of optimization
%*************************************************************

[FVr_x,S_y,I_nf] = deopt('objfun',S_struct)
```

Subfunctions Used

```
%%%%%%%%%%%%%%%%%%%%%%%%%%%%%%%%%%%%%%%%%%%%%%%%%%%%%%%%%%%%%%%
% Minimization of a user-supplied function with respect
   to x(1:I_D),
% using the differential evolution (DE) algorithm.
% DE works best if [FVr_minbound,FVr_maxbound] covers
   the region where the
% global minimum is expected. DE is also somewhat
   sensitive to
% the choice of the stepsize F_weight. A good initial
   guess is to
% choose F_weight from interval [0.5, 1], e.g. 0.8.
   F_CR, the crossover
```

```
% probability constant from interval [0, 1] helps to
  maintain
% the diversity of the population but should be close to
  1 for most.
% practical cases.Only separable problems do better with
  CR close to 0
% If the parameters are correlated, high values of F_CR
  work better.
% The reverse is true for no correlation.
% The number of population members I_NP is also not very
  critical. A
% good initial guess is 10*I_D. Depending on the
  difficulty of the
% problem I_NP can be lower than 10*I_D or must be higher
  than
% 10*I_D to achieve convergence.
%
% deopt is a vectorized variant of DE which, however,
  has a
% property which differs from the original version of DE:
% The random selection of vectors is performed by
  shuffling the
% population array. Hence a certain vector can't be
  chosen twice
% in the same term of the perturbation expression.
% Due to the vectorized expressions deopt executes
  fairly fast
% in MATLAB's interpreter environment.
%
% Parameters: fname (I) String naming a function f(x,y)
  to minimize.
% S_struct (I) Problem data vector (must remain
  fixed during the
% minimization ).

I_NP           = S_struct.I_NP;
F_weight       = S_struct.F_weight;
F_CR           = S_struct.F_CR;
I_D            = S_struct.I_D;
FVr_minbound   = S_struct.FVr_minbound;
FVr_maxbound   = S_struct.FVr_maxbound;
I_bnd_constr   = S_struct.I_bnd_constr;
I_itermax      = S_struct.I_itermax;
```

```
F_VTR          = S_struct.F_VTR;
I_strategy     = S_struct.I_strategy;
I_refresh      = S_struct.I_refresh;
I_plotting     = S_struct.I_plotting;

% - - - - - Check input variables - - - - - - - - -


if (I_NP <5)
   I_NP=5;
   fprintf(1,' I_NP increased to minimal value 5\n');
end if ((F_CR <0) | (F_CR >1))
   F_CR=0.5;
   fprintf(1,'F_CR should be from interval [0,1];
      set to default value 0.5\n');
end if (I_itermax <= 0)
   I_itermax = 200;
   fprintf(1,'I_itermax should be >0; set to default
                            value 200\n');
end
I_refresh = floor(I_refresh);

% - -  -  Check input variables - - - - - - - - -
FM_pop = zeros(I_NP,I_D); % initialize FM_pop to gain
speed
% - - FM_pop is a matrix of size I_NPx(I_D+1).It will
be initialized - - -
% - - - with random values between the min and max
values of the - - -
% - - - - - parameters - - - - - - - - - - - - - - - -


fork=1:I_NP
   FM_pop(k,:) = FVr_minbound + rand(1,I_D).* (FVr
   _maxbound - FVr_minbound);
end

FM_popold = zeros (size(FM_pop)); % toggle
population
FVr_bestmem = zeros(1,I_D); % best population member
ever
FVr_bestmemit = zeros (1,I_D); % best population
member in iteration
I_nfeval = 0; % number of function evaluations
```

```
% ---Evaluate the best member after initialization----


I_best_index   = 1;      % start with first
                         % population member
S_val(1)       = feval(fname,FM_pop(I_best_index,:),
S_struct);
S_bestval      = S_val(1);   % best objective
                         % function value so far
I_nfeval = I_nfeval + 1;
for k=2:I_NP              % check the remaining
                         % members
   S_val(k) = feval(fname,FM_pop(k,:),S_struct);
   I_nfeval = I_nfeval + 1;
   if (left_win (S_val(k),S_bestval) == 1)
       I_best_index     = k;      % save its
                         % location
       S_bestval        = S_val(k);
   end
end
FVr_bestmemit = FM_pop(I_best_index,:); % best
                    % member of current iteration
S_bestvalit = S_bestval;              % best
                    % value of current iteration


FVr_bestmem = FVr_bestmemit;            % best
                              % member ever


% ----- DE-Minimization --------------------
% -- FM_popold is the population which has to
   compete. It is -----
% ----- static through one iteration. FM_pop
   is the newly ------
% ----- emerging population. ----------------
FM_pm1   = zeros(I_NP,I_D); % initialize population
                            % matrix 1
FM_pm2   = zeros(I_NP,I_D); % initialize population
                            % matrix 2
FM_pm3   = zeros(I_NP,I_D); % initialize population
                            % matrix 3
FM_pm4   = zeros(I_NP,I_D); % initialize population
                            %  matrix 4
FM_pm5   = zeros(I_NP,I_D); % initialize population
```

```
                                        % matrix 5
FM_bm    = zeros(I_NP,I_D); % initialize FVr_
                              bestmember matrix
FM_ui    = zeros(I_NP,I_D); % intermediate population
                              of perturbed vectors
FM_mui   = zeros(I_NP,I_D); % mask for intermediate
                              population
FM_mpo   = zeros(I_NP,I_D); % mask for old population
FVr_rot  = (0:1:I_NP-1); % rotating index array
                              (size I_NP)
FVr_rotd = (0:1:I_D-1); % rotating index array
                              (size I_D)
FVr_rt   = zeros(I_NP); % another rotating index
                              array
FVr_rtd  = zeros(I_D); % rotating index array for
                  exponential crossover
FVr_a1   = zeros(I_NP);          % index array
FVr_a2   = zeros(I_NP);          % index array
FVr_a3   = zeros(I_NP);          % index array
FVr_a4   = zeros(I_NP);          % index array
FVr_a5   = zeros(I_NP);          % index array

FVr_ind  - zeros(4);

FM_meanv = ones(I_NP,I_D);
I_iter = 1;
while ((I_iter <I_itermax) & (S_bestval.FVr_oa(1)
>F_VTR))
   FM_popold = FM_pop;        % save the old
                              % population
   S_struct.FM_pop = FM_pop;
   S_struct.FVr_bestmem = FVr_bestmem;

   FVr_ind = randperm(4);          % index
                              % pointer array

   FVr_a1   = randperm(I_NP);       % shuffle
                              % locations of vectors
   FVr_rt   = rem(FVr_rot+FVr_ind(1),I_NP); % rotate
                       % indices by ind(1) positions
   FVr_a2   = FVr_a1(FVr_rt+1);       % rotate vector
                              locations
   FVr_rt   = rem(FVr_rot+FVr_ind(2),I_NP);
```

```
FVr_a3   = FVr_a2(FVr_rt+1);
FVr_rt   = rem(FVr_rot+FVr_ind(3),I_NP);
FVr_a4   = FVr_a3(FVr_rt+1);
FVr_rt   = rem(FVr_rot+FVr_ind(4),I_NP);
FVr_a5   = FVr_a4(FVr_rt+1);


FM_pm1   = FM_popold(FVr_a1,:); % shuffled
                               % population 1
FM_pm2   = FM_popold(FVr_a2,:); % shuffled
                               % population 2
FM_pm3   = FM_popold(FVr_a3,:); % shuffled
                               % population 3
FM_pm4   = FM_popold(FVr_a4,:); % shuffled
                               % population 4
FM_pm5   = FM_popold(FVr_a5,:); % shuffled
                               % population 5


for k=1:I_NP    % population filled with
                          % the best member
   FM_bm(k,:) = FVr_bestmemit;     % of the last
                              % iteration
end

   FM_mui = rand(I_NP,I_D) <F_CR;    % all random
                 <numbers F_CR are 1, 0 otherwise

% ---- Insert this if you want exponential
  crossover.--------
% FM_mui = sort(FM_mui'); % transpose, collect
  1's in each column
% for k = 1:I_NP
%    n = floor(rand*I_D);
%    if (n >0)
%        FVr_rtd      = rem(FVr_rotd+n,I_D);
%        FM_mui(:,k)  = FM_mui(FVr_rtd+1,k); % rotate
column k by n
%    end
% end
% FM_mui = FM_mui';     % transpose back
% ---- End: exponential crossover ------------

   FM_mpo  = FM_mui <0.5; % inverse mask to FM_mui
```

```
if (I_strategy == 1)     % DE/rand/1
  FM_ui = FM_pm3 + F_weight*(FM_pm1 - FM_pm2);    % dif
                        % -derential variation
  FM_ui = FM_popold.*FM_mpo + FM_ui.*FM_mui;% crossover
  % FM_origin = FM_pm3;
elseif (I_strategy == 2)   % DE/local-to-best/1

  % FM_ui = FM_popold + F_weight*(FM_bm-FM_popold)
  % +F_weight*(FM_pm1 - FM_pm2);
  FM_ui = FM_popold.*FM_mpo + FM_ui.*FM_mui;
  FM_origin = FM_popold;
elseif (I_strategy == 3)  % DE/best/1 with jitter
  FM_ui = FM_bm + (FM_pm1 - FM_pm2).*((1-0.9999)*
rand (I_NP,I_D)+ F_weight);
  FM_ui = FM_popold.*FM_mpo + FM_ui.*FM_mui;
  FM_origin = FM_bm;
elseif(I_strategy == 4) % DE/rand/1 with per-
                        vector-dither
    f1 = ((1-F_weight)*rand(I_NP,1)+F_weight);
    for k=1:I_D
        FM_pm5(:,k)=f1;
    end
    FM_ui = FM_pm3 + (FM_pm1 - FM_pm2).*FM_pm5;
                % differential variation
    FM_origin = FM_pm3;
    FM_ui = FM_popold.*FM_mpo + FM_ui.*FM_mui;
                % crossover
elseif (I_strategy == 5) % DE/rand/1 with
                % per-vector-dither
    f1 = ((1-F_weight)*rand+F_weight);
    FM_ui = FM_pm3 + (FM_pm1 - FM_pm2)*f1;% dif-
                ferential variation
    FM_origin = FM_pm3;
    FM_ui = FM_popold.*FM_mpo + FM_ui.*FM_mui;
                    % crossover
else       % either-or-algorithm
    if (rand <0.5);   % Pmu = 0.5
    FM_ui = FM_pm3 + F_weight*(FM_pm1 - FM_pm2);
                    % differential variation
    FM_origin = FM_pm3;
    else    % use F-K-Rule: K = 0.5(F+1)
        FM_ui = FM_pm3 + 0.5*(F_weight+1.0)*
        (FM_pm1 + FM_pm2 - 2*FM_pm3);
```

```
        end
      FM_ui = FM_popold.*FM_mpo + FM_ui.*FM_mui;
                        % crossover
   end


% ----- Optional parent+child selection --------
% -------Select which vectors are allowed to enter
the new population---------
   for k=1:I_NP



% ===== Only use this if boundary constraints
are needed =====
    if (I_bnd _constr == 1)
        for j=1:I_D %----boundary constraints via
        bounce back----
          if (FM_ui(k,j) >FVr_maxbound(j))
             FM_ui(k,j) = FVr_maxbound(j) + rand*
             (FM_origin(k,j) FVr_maxbound(j));
          end
          if (FM_ui(k,j) <FVr_minbound(j))
             FM_ui(k,j) = FVr_minbound(j) + rand*
             (FM_origin(k,j) FVr_minbound(j));
          end
      end
    end
    %=========== End boundary constraints ========
    S_tempval = feval(fname,FM_ui(k,:),S_struct);% check
                           cost of competitor
    I_nfeval = I_nfeval + 1;
    if (left_win(S_tempval,S_val(k)) == 1)
        FM_pop(k,:) = FM_ui(k,:);% replace old
          vector with new one (for new iteration)
        S_val(k) = S_tempval; % save value
                        in "cost array"
        % --------we update S_bestval only in case
        of success to save time -------
        if (left_win(S_tempval,S_bestval) == 1)
        S_bestval = S_tempval; % new best value
        FVr_bestmem = FM_ui(k,:); % new best parameter
                        vector ever
        end
    end
```

```
end % for k = 1:NP

        FVr_bestmemit = FVr_bestmem; % freeze the best
     member of this iteration for the coming iteration.
        % This is needed for some of the strategies.

% ---- Output section ----------------------

   if (I_refresh >0)
       if ((rem(I_iter,I_refresh) == 0) | I_iter == 1)
          fprintf(1,'Iteration: % d, Best:
       % f, F_weight: % f, F_CR: % f, I_NP: % d \n',
       I_iter,S_bestval.FVr_oa(1),F_weight,F_CR,I_NP);
          % var(FM_pop)
          format long e;
          for n=1:I_D
               fprintf(1,'best(% d) = % g \n',n,FVr
               _bestmem(n));
          end
          if (I_plotting == 1)
               PlotIt(FVr_bestmem,I_iter,S_struct);
          end
       end
   end

   I_iter = I_iter + 1;
end %---end while ((I_iter <I_itermax) ...
```

Observations

Crossover Probability F_CR = 1.0
Differential evolution step size F_weight = 0.850000
Number of population members I_NP = 15

From Table 12.1, it can be concluded that the Mean squared error is constant from iteration 25 onwards and converges at 49th iteration. The vector distribution values are found to be clustered at the center of the four quadrants as shown in Figure 12.1.

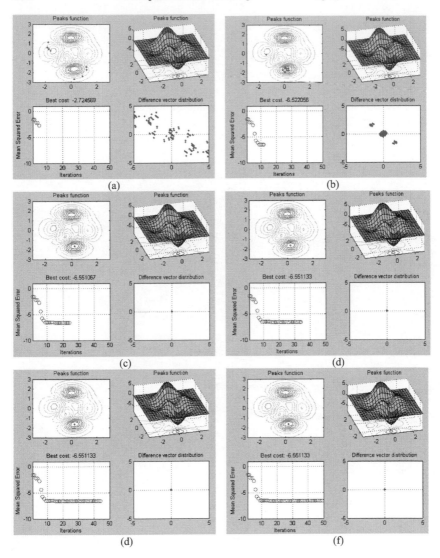

FIGURE 12.1: Differential Evolution Output (a) after 4 Iterations, (b) after 15 Iterations, (c) after 25 Iterations, (d) after 35 Iterations, (e) after 45 Iterations, (f) after 49 Iterations

12.2 Illustration 2: Design of a Proportional-Derivative Controller Using Evolutionary Algorithm for Tanker Ship Heading Regulation

%%
% Stochastic Optimization for Design of a Proportional-Derivative

TABLE 12.1:   MSE and Vector Values for Different Iterations

Iteration	Best Fit Value of MSE	New Best parameter vector $(best(1),best(2))$
4	-2.724569	$best(1) = -1.56597 \; best(2) = 0.247177$
15	-6.550189	$best(1) = 0.236808 \; best(2) = -1.62016$
25	-6.551131	$best(1) = 0.22782 \; best(2) = -1.62551$
35	-6.551131	$best(1) = 0.228336 \; best(2) = -1.62551$
45	-6.551131	$best(1) = 0.228277 \; best(2) = -1.62554$
49	-6.551131	$best(1) = 0.228274 \; best(2) = -1.62555$

```
% Controller for Tanker Ship Heading Regulation
%%%%%%%%%%%%%%%%%%%%%%%%%%%%%%%%%%%%%%%%%%%%


clear          % Clear all variables in memory

% Initialize ship parameters
% (can test two conditions, "ballast" or "full"):
ell=350;         % Length of the ship (in meters)
u=5;             % Nominal speed (in meters/sec)
abar=1;          % Parameters for nonlinearity
bbar=1;

% Define the reference model (we use a first order
  transfer function
% k_r/(s+a_r)):

a_r=1/150;
k_r=1/150;
% Number of evolution steps
Nevsteps=40;

% Size of population
S=4;

% Prop and derivative gains

% Kp=-1.5; Some reasonable size gains - found manually
% Kd=-250;

Kpmin=-5;
Kpmax=0; % Program below assumes this
```

```
Kdmin=-500;
Kdmax=-0; % Program below assumes this

KpKd=0*ones(2,S,Nevsteps);
for ss=1:S
KpKd(1,ss,1)=Kpmin*rand; % Generates a random
            proportional gain in range
KpKd(2,ss,1)=Kdmin*rand; % Generates a random
            proportional gain in range
end

% Store performance measure for evaluating closed
  -loop performance

Jcl=0*ones(S,Nevsteps); % Allocate memory

% Define scale parameters for performance measure
w1=1;
w2=0.01;

% Set parameters that determine size of cloud of
  design points about best one

beta1=0.5;
beta2=50;

% Set the probability that a mutation will occur
(only one can mutate per % generation as is)
pm=1;

%%%%%%%%%%%%%%%%%%%%%%%%%%%%%%%%%%%%%%%%%%%%%%%%%%%%%
% Start evolutionary design loop

for k=1:Nevsteps+1
      % Loop for evolution (note that add one
      % simply to show the best controller up to
      % the last one and so that 0 in the plots
      % corresponds to the initial condition)
    for ss=1:S
%%%%%%%%%%%%%%%%%%%%%%%%%%%%%%%%%%%%%%%%%%%%%%%%%%%%%

% Simulate the controller regulating the ship heading
```

```
% Next, we initialize the simulation:

t=0;            % Reset time to zero
index=1;        % This is time's index (not time, its
                index).
tstop=1200;     % Stopping time for the simulation (in
                seconds)
step=1;         % - normally 20000 Integration step
                size
T=10;           % The controller is implemented in
                % discrete time and this is the
                sampling
                % time for the controller.
                % Note that the integration step size
                and
                % the sampling time are not the same.
                % In this way we seek to simulate
                % the continuous time system via the
                Runge-
                % Kutta method and the discrete time
                % controller as if it were
                % implemented by a digital computer.
                % Hence, we sample the plant output
                every
                % T seconds and at that time output
                a new
                % value of the controller output.
counter=10;     % This counter will be used to count
                the
                % number of integration steps that
                have been
                % taken in the current sampling
                interval.
                % Set it to 10 to begin so that it
                will
                % compute a controller output at the
                first
                % step. For our example, when 10
                integration
                % steps have been taken we will then
                we will
                % sample the ship heading and the
```

```
                reference
                % heading and compute a new output
                for the
                % controller.
eold=0;    % Initialize the past value of the error
           (for use
           % in computing the change of the error, c).
           % Notice
           % that this is somewhat of an arbitrary
           % choice since
           % there is no last time step. The same
           % problem is
           % encountered in implementation.
cold=0;    % Need this to initialize phiold below

psi_r_old=0; % Initialize the reference trajectory
ymold=0; % Initial condition for the first order
                % reference model

x=[0;0;0]; % First, set the state to be a vector
x(1)=0;         % Set the initial heading to be zero
x(2)=0;         % Set the initial heading rate to be
                % zero.
                % We would also like to set x(3)
                % initially
                % but this must be done after we have
                % computed the output of the
                % controller.
                % In this case, bychoosing the
                % reference
                % trajectory to be zero at the
                % beginning
                % and the other initial conditions
                % as they are, and the controller as
                % designed,we will know that the
                % output
                % of the controller will start out
                % at zero
                % so we could have set x(3)=0 here.
                % To keep things more general, however,
                % we set the intial condition
                % immediately
                % after we compute the first
```

```
                   % controller
                   % output in the loop below.


% Next, we start the simulation of the system.
% This is
% the main loop for the simulation of the control
% system.

psi_r=0*ones(1,tstop+1);
psi=0*ones(1,tstop+1);
e=0*ones(1,tstop+1);
c=0*ones(1,tstop+1);
s=0*ones(1,tstop+1);
w=0*ones(1,tstop+1);
delta=0*ones(1,tstop+1);
ym=0*ones(1,tstop+1);


while t <= tstop

% First, we define the reference input psi_r
%(desired heading).
if t>=0, psi_r(index)=0; end
                   % Request heading of 0 deg
if t>=100, psi_r(index)=45*(pi/180); end
                   % Request heading of 45 deg
if t>=1500, psi_r(index)=0; end
                   % Request heading of 0 deg
if t>=3000, psi_r(index)=45*(pi/180); end
                   % Request heading of -45 deg
if t>=4500, psi_r(index)=0; end
                   % Request heading of 0 deg
if t>=6000, psi_r(index)=45*(pi/180); end
                   % Request heading of 45 deg
if t>=7500, psi_r(index)=0; end
                   % Request heading of 0 deg
if t>=9000, psi_r(index)=45*(pi/180); end
                   % Request heading of 45 deg
if t>=10500, psi_r(index)=0; end
                   % Request heading of 0 deg
if t>=12000, psi_r(index)=45*(pi/180); end
                   % Request heading of -45 deg
if t>=13500, psi_r(index)=0; end
```

```
                    % Request heading of 0 deg
if t>=15000, psi_r(index)=45*(pi/180); end
                    % Request heading of 45 deg
if t>=16500, psi_r(index)=0; end
                    % Request heading of 0 deg
if t>=18000, psi_r(index)=45*(pi/180); end
                    Request heading of 45 deg
if t>=19500, psi_r(index)=0; end
                    % Request heading of 0 deg


% Next, suppose that there is sensor noise for the
% heading sensor with that on [- 0.01,+0.01] deg.
% is additive, with a uniform distribution if flag==0,
% s(index)=0.01*(pi/180)*(2*rand-1); else s(index)=0;
% This allows us to remove the noise.
% end

psi(index)=x(1)+s(index); % Heading of the ship
                (possibly with sensor noise).

if counter == 10, % When the counter reaches 10 then
                execute the controller

counter=0;      % First, reset the counter

% Reference model calculations:
% The reference model is part of the controller
% and to simulate it we take the discrete
% equivalent of the reference model to compute
% psi_m from psi_r
%
% For the reference model we use a first order
% transfer function k_r/(s+a_r) but we use the
% bilinear transformation where we replace s
% by (2/step)(z-1)/(z+1), then find the z-domain
% representation of the reference model,
% then convert this to a difference equation:

% ym(index)=(1/(2+a_r*T))*((2-a_r*T)*ymold+...
                k_r*T*(psi_r(index)+psi_r_old));

ymold=ym(index);
psi_r_old=psi_r(index);
```

```
% This saves the past value of the ym and psi_r
% so that we can use it the next time around the loop

% Controller calculations:

e(index)=psi_r(index)-psi(index); % Computes error
                         (first layer of perceptron)
c(index)=(e(index)-eold)/T; % Sets the value of c

eold=e(index);% Save the past value of e for use in
% the above
                % computation the next time around
% the loop

%%%%%%%%%%%%%%%%%%%%%%%%%%%%%%%%%%%%%%%%%%%%%%%%%%%%%%
% A proportional-derivative controller:

delta(index)=KpKd(1,ss,k)*e(index)+KpKd(2,ss,k)*c
(index);
%%%%%%%%%%%%%%%%%%%%%%%%%%%%%%%%%%%%%%%%%%%%%%%%%%%%%%

else % This goes with the "if" statement to check if the
     % counter=10 so the next lines up to the next "end"
     % statement are executed whenever counter is not
     % equal to 10

% Now, even though we do not compute the controller at
% each time instant, we do want to save the data at its
% inputs and output at each time instant for the sake of
% plotting it. Hence, we need to compute these here
% (note that we simply hold the values constant):

e(index)=e(index-1);
c(index)=c(index-1);
delta(index)=delta(index-1);

ym(index)=ym(index-1);

end % This is the end statement for the "if
    % counter=10"
     % statement
```

```
% Next, the Runge-Kutta equations are used to find
% the next state. Clearly, it would be better to
% use a Matlab "function" for F (but here we do
% not, so we can have only one program).

        time(index)=t;

% First, we define a wind disturbance against the
% body of the ship that has the effect of pressing
% water against the rudder

% if flag==0, w(index)=0.5*(pi/180)*sin(2*pi*0.001*t);
% This is an additive sine disturbance to
% the rudder input. It is of amplitude of
% 0.5 deg. and its period is 1000sec.
% delta(index)=delta(index)+w(index);
% end

% Next, implement the nonlinearity where the rudder
% angle is saturated at +-80 degrees

if delta(index) >= 80*(pi/180), delta(index)=80*
                        (pi/180); end
if delta(index) <= -80*(pi/180), delta(index)=-80*
                        (pi/180); end
% The next line is used in place of the line
% following it to change the speed of the ship
% if flag==0,
%% if t>=1000000,
%% if t>=9000, % This switches the ship
            speed (unrealistically fast)
% u=3;        % A lower speed

% else

u=5;
% end

% Next, we change the parameters of the ship to tanker
% to reflect changing loading conditions (note that we
% simulate as if the ship is loaded while moving,but we
% only change the parameters while the heading is zero
% so that it is then similar to re-running the
```

```
% simulation, i.e., starting the tanker operation at
% different times after loading/unloading has occurred).

% The next line is used in place of the line following
% it to keep "ballast" conditions throughout the
% simulation
if flag==0,
% t>=1000000,
% if t>=0,      % This switches the parameters,
                  % possibly in the middle of the
% simulation
K_0=0.83;       % These are the parameters under
                  % "full" conditions
tau_10=-2.88;
tau_20=0.38;
tau_30=1.07;

else

K_0=5.88;       % These are the parameters under
                  % "ballast" conditions
tau_10=-16.91;
tau_20=0.45;
tau_30=1.43;

end

% The following parameters are used in the definition of
% the tanker model:

K=K_0*(u/ell);
tau_1=tau_10*(ell/u);
tau_2=tau_20*(ell/u);
tau_3=tau_30*(ell/u);

% Next, comes the plant:
% Now, for the first step, we set the initial condition
% for the third state x(3).

if t==0, x(3)=-(K*tau_3/(tau_1*tau_2))*delta(index); end

% Next, we use the formulas to implement the Runge-
% Kutta method (note that here only an approximation
```

```
% to the method is implemented where we do not compute
% the function at multiple points in the integration
% step size).

F=[ x(2) ;
    x(3)+ (K*tau_3/(tau_1*tau_2))*delta(index) ;
    -((1/tau_1)+(1/tau_2))*(x(3)+ (K*tau_3/
                (tau_1*tau_2))*delta(index))-
        ...(1/(tau_1*tau_2))*(abar*x(2)^3 + bbar*x(2)) +
                (K/(tau_1*tau_2))*delta(index) ];

    k1=step*F;
    xnew=x+k1/2;

F=[ xnew(2) ;
    xnew(3)+ (K*tau_3/(tau_1*tau_2))*delta(index) ;
    -((1/tau_1)+(1/tau_2))*(xnew(3)+
                (K*tau_3/(tau_1*tau_2))*delta(index))-
      ..(1/(tau_1*tau_2))*(abar*xnew(2)^3 + bbar*xnew(2))
                + (K/(tau_1*tau_2))*delta(index) ];

    k2=step*F;
    xnew=x+k2/2;

F=[ xnew(2) ;
    xnew(3)+ (K*tau_3/(tau_1*tau_2))*delta(index) ;
    -((1/tau_1)+(1/tau_2))*(xnew(3)+
                (K*tau_3/(tau_1*tau_2))*delta(index))-
    ... (1/(tau_1*tau_2))*(abar*xnew(2)^3 + bbar*xnew(2))
                + (K/(tau_1*tau_2))*delta(index) ];

    k3=step*F;
    xnew=x+k3;

F=[ xnew(2) ;
    xnew(3)+ (K*tau_3/(tau_1*tau_2))*delta(index) ;
    -((1/tau_1)+(1/tau_2))*(xnew(3)+
    (K*tau_3/(tau_1*tau_2))*delta(index))-...
    (1/(tau_1*tau_2))*(abar*xnew(2)^3 + bbar*xnew(2))
                + (K/(tau_1*tau_2))*delta(index) ];

    k4=step*F;
    x=x+(1/6)*(k1+2*k2+2*k3+k4); % Calculated next state
```

```
t=t+step;        % Increments time
index=index+1;   % Increments the indexing
                 % term so that index=1
                 % corresponds to time t=0.
counter=counter+1;% Indicates that we computed
                 % one more integration step

end % This end statement goes with the first "while"
    % statement
    % in the program so when this is complete the
    % simulation is done.
%%%%%%%%%%%%%%%%%%%%%%%%%%%%%%%%%%%%%%%%%%%%%%%%%%%%%%%

% Compute how well the controller did in terms of
% deviation
% from the reference model and the control energy

Jcl(ss,k)=w1*(psi-ym)*(psi-ym)'+w2*delta*delta';

end % End evaluation of each controller in the population

[Jbest(k),bestone(k)]=min(Jcl(:,k)); % Store the best
                           controller for plotting
Kpbest(k)=KpKd(1,bestone(k),k);
Kdbest(k)=KpKd(2,bestone(k),k);

% Generate next generation of gains:

KpKd(:,bestone(k),k+1)=KpKd(:,bestone(k),k);
                    % Use elitism - keep the best one

% Create a cloud of points around the best one
for ss=1:S
   if ss ~= bestone(k)
       KpKd(:,ss,k+1)=KpKd(:,bestone(k),k+1)+
              [beta1*randn; beta2*randn];
                           % Perturb random
                           % points about the best one
       if KpKd(1,ss,k+1)<Kpmin,KpKd(1,ss,k+1)=Kpmin; end
              % Fix gains if perturbed out of range
    if KpKd(1,ss,k+1)>Kpmax, KpKd(1,ss,k+1)=Kpmax; end
     if KpKd(2,ss,k+1)<Kdmin, KpKd(2,ss,k+1)=Kdmin; end
```

```
      if KpKd(2,ss,k+1)>Kdmax, KpKd(2,ss,k+1)=Kdmax; end
    end
end

% Next place a mutant, do not mutate the best one

if pm>rand,   % Replace either the first or last
              % member, provided not the best one
        if bestone(k) ~= 1
            KpKd(1,1,k+1)=Kpmin*rand; % Generates
            % a random proportional gain in range
            KpKd(2,1,k+1)=Kdmin*rand; % Generates
            % a random derivative gain in range
        else
            % So, the bestone is the first one so
            % replace the last
            KpKd(1,S,k+1)=Kpmin*rand;
            KpKd(2,S,k+1)=Kdmin*rand;
        end
end


end % End evolutionary design loop (iterate in k)
%%%%%%%%%%%%%%%%%%%%%%%%%%%%%%%%%%%%%%%%%%%%%%%%%%%%%
%
% Next, we provide plots of the input and output
% of the ship along with the reference heading
% that we want to track for the last best
% design.
%
%%%%%%%%%%%%%%%%%%%%%%%%%%%%%%%%%%%%%%%%%%%%%%%%%%%%%


% Take easy approach of simply choosing the best
% gains from the last step of the evolutionary process
% and simulating the control system for those to
% show the final performance

flag=0; % Test under off-nominal conditions
% flag=1;
%%%%%%%%%%%%%%%%%%%%%%%%%%%%%%%%%%%%%%%%%%%%%%%%%%%%%
% Simulate the controller regulating the ship heading

% Next, we initialize the simulation:
```

```
t=0;           % Reset time to zero
index=1;       % This is time's index (not time,
               % its index).
tstop=1200;    % Stopping time for the simulation
               (in seconds)- normally 20000
step=1;        % Integration step size
T=10;          % The controller is implemented in
               % discrete time and this is the
               % sampling
               % ime for the controller. Note
               % that the
               % integration step size and the
               % sampling
               % time are not the same. In this
               % way we
               % seek to simulate the continuous
               % time
               % system via the Runge-Kutta
               % method and
               % the discrete time controller as
               % if it
               % were implemented by a digital
               % computer.
               % Hence, we sample the plant output
               % every
               % T seconds and at that time output
               % a new
               % value of the controller output.
counter=10;    % This counter will be used to
               % count the
               % number of integration steps
               % that have
               % been taken in the current
               % sampling
               % interval. Set it to 10 to
               % begin so
               % that it will compute a
               % controller
               % output at the first step.
               % For our
               % example, when 10 integration
               % steps
               % have been taken we will then we
```

```
                    % will
                    % sample the ship heading and
                    % the
                    % reference heading and compute
                    % a new
                    % output for the controller.
eold=0;             % Initialize the past value of
                    % the error
                    % (for use in computing the change of
                    % the error,
                    % c). Notice that this is somewhat
                    % of an arbitrary
                    % choice since there is no last time
                    % step.
                    % The same
                    % problem is encountered in
                    % implementation.
cold=0;             % Need this to initialize phiold below

psi_r_old=0;        % Initialize the reference trajectory
ymold=0;            % Initial condition for the first
                    % order reference model

x=[0;0;0];          % First, set the state to be a vector
x(1)=0;             % Set the initial heading to be zero
x(2)=0;             % Set the initial heading rate to
                    % be zero.
                    % We would also like to set x(3)
                    % initially
                    % but this must be done after we have
                    % computed
                    % the output of the controller. In
                    % this case, by
                    % choosing the reference trajectory
                    % to be
                    % zero at the beginning and the other
                    % initial
                    % conditions as they are, and the
                    % controller as
                    % designed, we will know that the
                    % output of the
                    % controller will start out at zero
                    % so we could
```

```
                    % have set x(3)=0 here. To keep
                    % things more
                    % general, however, we set the intial
                    % condition
                    % immediately after we compute the
                    % first
                    % controller output in the loop below.

% Next, we start the simulation of the system.
                % This
% is the main loop for the simulation of the
                % control system.

psi_r=0*ones(1,tstop+1);
psi=0*ones(1,tstop+1);
e=0*ones(1,tstop+1);
c=0*ones(1,tstop+1);
s=0*ones(1,tstop+1);
w=0*ones(1,tstop+1);
delta=0*ones(1,tstop+1);
ym=0*ones(1,tstop+1);

while t <= tstop

% First, we define the reference input psi_r
%(desired heading).

if t>=0, psi_r(index)=0; end
                % Request heading of 0 deg
if t>=100, psi_r(index)=45*(pi/180); end
                % Request heading of 45 deg
if t>=1500, psi_r(index)=0; end
                % Request heading of 0 deg
if t>=3000, psi_r(index)=45*(pi/180); end
                % Request heading of -45 deg
if t>=4500, psi_r(index)=0; end
                % Request heading of 0 deg
if t>=6000, psi_r(index)=45*(pi/180); end
                % Request heading of 45 deg
if t>=7500, psi_r(index)=0; end
                % Request heading of 0 deg
if t>=9000, psi_r(index)=45*(pi/180); end
                % Request heading of 45 deg
```

```
if t>=10500,  psi_r(index)=0; end
              % Request heading of 0 deg
if t>=12000,  psi_r(index)=45*(pi/180); end
              % Request heading of -45 deg
if t>=13500,  psi_r(index)=0; end
              % Request heading of 0 deg
if t>=15000,  psi_r(index)=45*(pi/180); end
              % Request heading of 45 deg
if t>=16500,  psi_r(index)=0; end
              % Request heading of 0 deg
if t>=18000,  psi_r(index)=45*(pi/180); end
              % Request heading of 45 deg
if t>=19500,  psi_r(index)=0; end
              % Request heading of 0 deg
% Next, suppose that there is sensor noise for
% the heading
% sensor with that is additive, with a uniform
% distribution
% on [- 0.01,+0.01] deg.
% if flag==0,  s(index)=0.01*(pi/180)*(2*rand-1);
else
s(index)=0;   % This allows us to remove the noise.
% end

psi(index)=x(1)+s(index);% Heading of the ship
                      % (possibly with sensor noise).

if counter == 10,          % When the counter
                 % reaches 10
               % then execute the controller

counter=0;            % First, reset the counter

%%%%%%%%%%%%%%%%%%%%%%%%%%%%%%%%%%%%%%%%%%%%%%%%%%%%%
% Next, we provide plots of data from the simulation
%%%%%%%%%%%%%%%%%%%%%%%%%%%%%%%%%%%%%%%%%%%%%%%%%%%%%

% First, we convert from rad. to degrees
psi_r=psi_r*(180/pi);
psi=psi*(180/pi);
delta=delta*(180/pi);
e=e*(180/pi);
c=c*(180/pi);
```

```
ym=ym*(180/pi);

% Next, we provide plots

figure(1)
clf
subplot(211)
plot(time,psi,'k-',time,ym,'k--',time,psi_r,'k-.')
zoom
grid on
title('Ship heading (solid) and desired ship heading
                        (dashed), deg.')
xlabel('Time in sec');
ylabel('Heading in deg');
subplot(212)
plot(time,delta,'k-')
zoom
grid on
title('Rudder angle, output of controller (input to the
                        ship), deg.')
xlabel('Time in sec');
ylabel('Rudder angle in deg');

figure(2)
clf
plot(time,psi-ym,'k-')
zoom
grid on
title('Ship heading error between ship heading and
                    reference model heading, deg.')
xlabel('Time in sec');
ylabel('Ship Heading Error in deg');

figure(3)
clf
plot(0:Nevsteps,Jbest,'k-')
zoom
grid on
title('Performance measure J_c_l for best controller')
xlabel('Generation');
ylabel('Closed loop performance ');

figure(4)
```

```
clf
plot(0:Nevsteps,Kpbest,'k-')
zoom
grid on
title('K_p gain');
xlabel('Generation');
ylabel('Gain');

figure(5)
clf
plot(0:Nevsteps,Kdbest,'k-')
zoom
grid on
title('K_d gain'); xlabel('Generation');
ylabel('Gain');
%%%%%%%%%%%%%%%%%%%%%%%%%%%%%%%%%%%%%%%%%%%%%%%%%%%
% End of program %
%%%%%%%%%%%%%%%%%%%%%%%%%%%%%%%%%%%%%%%%%%%%%%%%%%%
```

Observations

The output angle is converted from radians to degrees and the graphs are plotted. Figure 12.2 shows the ship heading and the rudder angle. The error between the ship heading and the reference model heading is

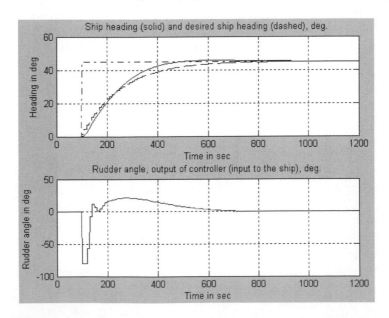

FIGURE 12.2: Plots of Ship Heading and Rudder Angle

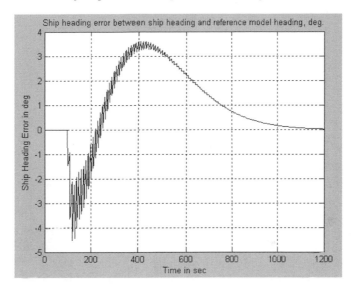

FIGURE 12.3: Plot of the Error between Ship Heading and Reference Model Heading

plotted in Figure 12.3. The closed loop performance measure, proportional gain, and the derivative gain are plotted in Figures 12.4, 12.5 and 12.6 respectively.

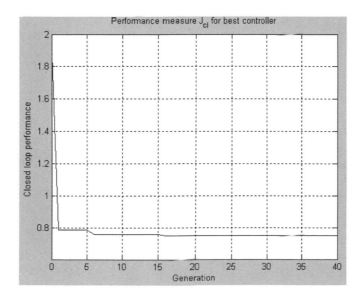

FIGURE 12.4: Performance Measure for the Best Controller

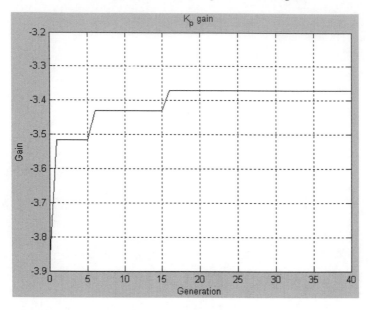

FIGURE 12.5:   Proportional Gain

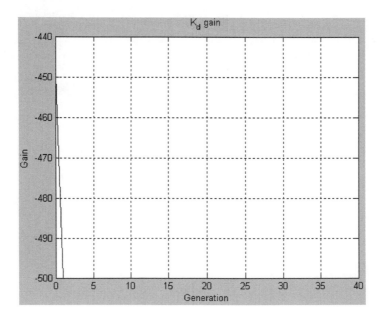

FIGURE 12.6:   Derivative Gain

12.3 Illustration 3: Maximizing the Given One-Dimensional Function with the Boundaries Using Evolutionary Algorithm

```
%%%%%%%%%%%%%%%%%%%%%%%%%%%%%%

% Main Program

%%%%%%%%%%%%%%%%%%%%%%%%%%%%%%

% boundaries
bounds = [-10 10];
% pop size
n = 10;
% number of iterations
numits = 100;
% numer of mutations per it
nummut = 1;

f = @multipeak;

blength = bounds(2)-bounds(1);

% initial population

pop = rand(1,n)* blength + bounds(1);

for it=1:numits
        % fitness eval
        for i=1:n, fpop(i) = feval(f, pop(i)); end
        maxf(it) = max(fpop);
        meanf(it) = mean(fpop);
        % subtract lowest fitness in order to normalize
        m=min(fpop);
        fpop=fpop-m;
        cpop(1) = fpop(1);
        for i=2:n, cpop(i) = cpop(i-1) + fpop(i); end

        % SELECTION
        total_fitness = cpop(n);
```

```
          % use roulette selection (-> need pos. fitness!)
          for i=1:n
            p=rand*total_fitness;
            % now find first index
            j=find(cpop-p>0);
            if isempty(j)
              j=n;
             else
              j=j(1);
            end
            parent(i)=pop(j);
      end
% pop, fpop, parent, pause
      % REPRODUCTION
      % parents 2i-1 and 2i make two new children 2i-1
      % and 2i
      % crossover

      % use arithmetic crossover
      for i=1:2:n
          r=rand;
          pop(i)   = r*parent(i)    + (1-r)*parent(i+1);
          pop(i+1) = (1-r)*parent(i) + r*parent(i+1);
      end

      % mutation
      % use uniform mutation
      for i=1:nummut
          pop(ceil(rand*n)) = bounds(1) + rand*blength;
      end
end
pop
for i=1:n, fpop(i) = feval(f, pop(i)); end
fpop

close all
ezplot(@multipeak,[-10 10])
hold on
 [y,xind]=max(fpop);
plot(pop(xind),y,'ro')
title('Multipeak at (6.4488,44.3352)')
xlabel('Population')
ylabel('Variable y')
```

```
figure, plot(maxf), hold on, plot(meanf,'r-');
title('Maximum and Mean of the 1-D function');
xlabel('Population')
ylabel('Variable y')
%%%%%%%%%%%% End of Main %%%%%%%%%%%%%%%%

%%%%%%%%%%%%%%%%%%%%%%%%%%%%%%%%%%%%%
% Subprogram used
%%%%%%%%%%%%%%%%%%%%%%%%%%%%%%%%%%%%%
function y = multipeak(x)
% Demonstration evaluation function
%  f(x)=x+10sin(5x)+7cos(4x) + 25

y = x + 10*sin(5*x)+7*cos(4*x) + 25;


%%%%%%%%%%%%%%%%%%%%%%%%%%%%%%%%%%%%%
```

Observations

Figure 12.7 shows the multi-peak of the given function which occurs at (6.4488, 44.3352) and Figure 12.8 shows the maximum and mean value of the given function.

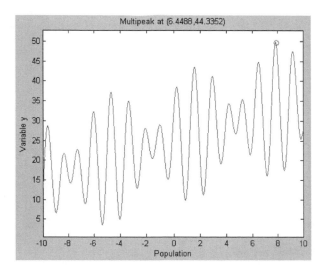

FIGURE 12.7:   Multi-Peak for the Given Function

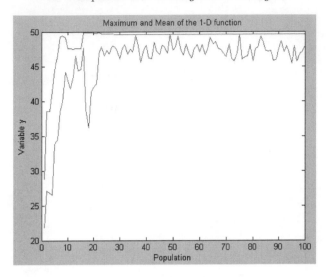

FIGURE 12.8: Maximum and Mean of the Given Function

12.4 Illustration 4: Multiobjective Optimization Using Evolutionary Algorithm (MOEA)

This problem is to find the global optimal space solution for a discontinuous function given by

$$f_1(x) = 1 - e^{-4x_1} \sin^6(6\pi x_1)$$

$$f_2(x) = g(x) \left(1 - \left(\frac{f_1(x)}{g(x)} \right)^2 \right)$$

where $g(x) = 1 + 9 \left(\sum_{i=1}^{6} \frac{x_i}{4} \right)^{0.25}$ subject to $0 \leq x_i \leq 1$

```
% Main program to run the MOEA

    %% Initialize the variables
    % Declare the variables and initialize
     their values
    % pop - population
    % gen - generations
    % pro - problem number
pop = 200;
gen = 1000;
pro = 1;
```

```
switch pro
    case 1
        % M is the number of objectives.
        M = 2;
        % V is the number of decision variables.
        %  In this case it is
        % difficult to visualize the decision
        %  variables space while the
        % objective space is just two dimensional.
        V = 6;
    case 2
        M = 3;
        V = 12;
end


% Initialize the population
chromosome = initialize_variables(pop,pro);
%% Sort the initialized population
% Sort the population using non-domination-sort.
This returns two columns
% for each individual which are the rank and the
crowding distance
% corresponding to their position in the front they
belong.
chromosome = non_domination_sort_mod
%% Start the evolution process
% The following are performed in each generation
% Select the parents
% Perfrom crossover and Mutation operator
% Perform Selection
for i = 1 : gen

    % Select the parents
    % Parents are selected for reproduction to
    generate offspring by
    % tournament selection The arguments are
    pool - size of the mating pool. It is common to
    have this to be half
    % the population size.
    % tour - Tournament size.
    pool = round(pop/2);
    tour = 2;
```

```
parent_chromosome =
tournament_selection(chromosomeP,pool,tour);
% Perform crossover and Mutation operator
% Crossover probability pc = 0.9 and mutation
% probability is pm = 1/n, where n is the number
of decision variables.
% Both real-coded GA and binary-coded GA are
implemented in the original
% algorithm, while in this program only the
real-coded GA is considered.
% The distribution indices for crossover and
mutation operators
% as mu = 20
% and mum = 20 respectively.
mu = 20;
mum = 20;
offspring_ chromosome =
genetic_operator(parent_chromosome,pro,mu,mum);

% Intermediate population
% Intermediate population is the combined
population of parents and
% offsprings of the current generation.
% The population size is almost 1 and
% half times the initial population.
[main_ pop,temp] = size(chromosome);
[offspring_pop,temp] = size(offspring_chromosome);
intermediate_chromosome(main_pop + 1 : main_pop +
offspring_pop,1 : M + V) = ...
offspring_chromosome;

 % Non-domination-sort of intermediate population
 % The intermediate population is sorted again
based on non-domination sort
% before the replacement operator is performed on
the intermediate
% population.
intermediate_chromosome = ...
non_domination_sort_mod(intermediate_chromosome,pro);
% Perform Selection
% Once the intermediate population is sorted only the
best solution is
% selected based on it rank and crowding distance. Each
front is filled in
```

```
% ascending order until the addition of population size
is reached. The
% last front is included in the population based on the
individuals with
  % least crowding distance
        chromosome = replace_ chromosome(intermediate_
        chromosome,pro,pop;)
        if  mod(i,10)
            fprintf('% d\n',i);
        end
    end
```

```
%% Result
% Save the result in ASCII text format.
save solution.txt chromosome -ASCII
```

```
%% Visualize
```

```
% The following is used to visualize the result for the
given problem.
        switch pro case 1
            plot(chromosome(:,V + 1),chromosome(:,V +
            2),'*');
            xlabel('f(x_ 1)');
            ylabel('f(x_2)');
        case 2
            plot3(chromosome(:,V + 1),chromosome(:,V +
            2),chromosome(:,V + 3),'*');
            xlabel('f(x_1)');
            ylabel('f(x_2)');
            zlabel('f(x_3)');
    end
```

```
%%%%%%% End of Main Program %%%%%%%
```

```
%%%%%%%%%%%%%%%%%%%%%%%%%%%%%%
% Subprograms
%%%%%%%%%%%%%%%%%%%%%%%%%%%%%
```

```
function f = initialize_variables(N,problem)
% function f = initialize_variables(N,problem)
% N - Population size
% problem - takes integer values 1 and 2 where,

        % '1' for MOP1
        % '2' for MOP2
```

```
%

% This function initializes the population with N
% individuals and each
% individual having M decision variables based on
%the selected problem.
% M = 6 for problem MOP1 and M = 12 for problem
MOP2.
% The objective space
% for MOP1 is 2 dimensional while for MOP2 is 3
% dimensional.
% Both the MOP's has 0 to 1 as its range for all the
% decision variables.
min = 0;
max = 1;
      switch problem case 1
         M = 6;
         K = 8;
      case 2
         M = 12;
         K = 15;

end
for i = 1 : N

      % Initialize the decision variables
      for j = 1 : M
         f(i,j) = rand(1); % i.e f(i,j) = min + (max -
         min)*rand(1);
      end
      % Evaluate the objective function
      f(i,M + 1: K) = evaluate_objective(f(i,:),problem);

end

%%%%%%%%% End of initialize_ variables %%%%%%%%%
%% Non-Donimation Sort


% This function sort the current popultion based on
 non-domination. All the
% individuals in the first front are given a rank of
 1, the second front
% individuals are assigned rank 2 and so on. After
 assigning the rank the
% crowding in each front is calculated.
```

```
function f = non _domination_sort_mod(x,problem)
[N,M] = size(x);
     switch problem    case 1
        M = 2;
        V = 6;
     case 2
        M = 3;
        V = 12;

end

front = 1;


% There is nothing to this assignment, used only to
 manipulate easily in
% MATLAB. F(front).f = [];
individual = [];
     for i = 1 : N % Number of individuals that
     dominate this individual
     % Individuals which this individual dominate
     individual(i).p = [];
     for j = 1 : N
        dom_less = 0;
        dom_equal = 0;
        dom_more = 0;
        for k = 1 : M
           if (x(i,V + k) < x(j,V + k))
              dom_less = dom_less + 1;
           elseif (x(i,V + k) == x(j,V + k))
              dom_ equal = dom_equal + 1;
           else
              dom_ more = dom_more + 1;
           end
              if dom_ less == 0 & dom_ equal  = M
              individual(i).n = individual(i).n + 1;
           elseif dom_more == 0 & dom_equal  = M
              individual(i).p = [individual(i).p j];
           end
        end
        if individual(i).n == 0
           x(i,M + V + 1) = 1;
           F(front).f = [F(front).f i];
        end
```

```
end
% Find the subsequent fronts
while  isempty(F(front).f)

   Q = [];
   for i = 1 : length(F(front).f)
      if  isempty(individual(F(front).f(i)).p)
          for j = 1 : length(individual(F(front).f(i)).p)
             individual(individual(F(front).f(i)).p(j)).n =
             ...
             individual(individual(F(front).f(i)).p(j)).n -
             1;
             if individual(individual(F(front).f(i)).p(j)).n
             == 0
                x(individual(F(front).f(i)).p(j),M + V + 1)
                = ...
                front + 1;
             Q = [Q individual(F(front).f(i)).p(j)];
          end
      end
   end
end
front = front + 1;
 F(front).f = Q;

   end [temp,index_of_fronts] = sort(x(:,M + V + 1));
   for i = 1 : length(index_of_fronts)

   sorted_based_on_front(i,:) =
   x(index_of_fronts(i),:);

   end
   current_index = 0;

% Find the crowding distance for each individual in
each front
   for front = 1 : (length(F) - 1) objective = [];
   distance = 0;
   y = [];
   previous_ index = current_index + 1;
   for i = 1 : length(F(front).f)
      y(i,:) = sorted_based_on_front(current_index +
      i,:);
    end
   current_index = current_index + i;
   % Sort each individual based on the objective
   sorted_based_on_objective = [];
```

```
for i = 1 : M
  [sorted_based_on_objective, index_of_objectives]
  = ...
    sort(y(:,V + i));
  sorted_based_on_objective = [];
  for j = 1 : length(index_of_objectives)
    sorted_based_on_objective(j,:) =
    y(index_of_objectives(j),:);
  end
  f_max = ...
  sorted_based_on_objective(length(index_of_objectives),
  V + i);
f_min = sorted_based_on_objective(1, V + i);
y(index_of_objectives(length(index_of_objectives)),M
+ V + 1 + i)... = Inf;
for j = 2 : length(index_of_objectives) - 1
  next_obj = sorted_based_n_objective(j + 1,V + i);
  previous_obj = sorted_based_on_objective(j - 1,V + i);
  if (f_max - f_min == 0)
    y(index_of_objectives(j),M + V + 1 + i) = Inf;
  else
    y(index_of_objectives(j),M + V + 1 + i) = ...
    (next_obj  previous_obj)/(f_max - f_min);
  end
  end
end
distance = [];
distance(:,1) = zeros(length(F(front).f),1);
for i = 1 : M
distance(:,1) = distance(:,1) + y(:,M + V + 1 + i); end
y(:,M + V + 2) = distance;
y = y(:,1 : M + V + 2);
z(previous_index:current_index,:) = y;;

  end
  f = z();
  %%%%%%%% End of non_domination_sort_mod %%%%%%%

  function f = selection_individuals(chromosome,
  % pool_size,tour_size)

  % function selection_individuals(chromosome
  %,pool_size,tour_size) is the
  % selection policy for selecting the individuals
  % for the mating pool. The
```

```
% selection is based on tournament selection.
% Argument 'chromosome' is the
% current generation population from which the
% individuals are selected to
% form a mating pool of size 'pool_size'
% after performing tournament
% selection, with size of the tournament being
% 'tour_size'. By varying the
% tournament size the selection pressure
% can be adjusted.

[pop,variables] = size(chromosome);
rank = variables - 1;
distance = variables;

for i = 1 : pool_size

for j = 1 : tour_size
   candidate(j) = round(pop*rand(1));
   if candidate(j) == 0
   candidate(j) = 1; end
   if j > 1
     while  isempty(find(candidate(1 : j - 1) ==
     candidate(j)))
        candidate(j) = round(pop*rand(1));
        if candidate(j) == 0
          candidate(j) = 1;
        end
     end
   end
end
for j = 1 : tour_size
   c_obj_rank(j) = chromosome(candidate(j),rank);
   c_obj_distance(j) =
   chromosome(candidate(j),distance);
end
min_candidate = ...
   find(c_obj_rank == min(c_obj_rank));
if length(min_candidate)  = 1
   max_candidate = ...
   find(c_obj_distance(min_candidate) ==
   max(c_obj_distance(min_candidate)));
   if length(max_candidate)  = 1
```

```
      max_candidate = max_candidate(1);
end
f(i,:) =
chromosome(candidate(min_candidate(max_candidate)),:);
else
  f(i,:) = chromosome(candidate(min_candidate(1)),:);

end
%%%%%%%% End of selection_individuals %%%%%%%%%
function f = genetic_operator(parent_chromosome
,pro,mu,mum);

% This function is utilized to produce offsprings
% from parent chromosomes.
% The genetic operators corssover and mutation which
% are carried out with
% slight modifications from the original design.
% For more information read
% the document enclosed.

[N,M] = size(parent_chromosome);
switch pro

case 1
   M = 2;
   V = 6;
case 2
   M = 3;
   V = 12;

end
p = 1;
was_crossover = 0;
was_mutation = 0;
l_limit = 0;
u_limit = 1;
for i = 1 : N

if rand(1) < 0.9
   child_1 = [];
   child_2 = [];
   parent_1 = round(N*rand(1));
   if parent_1 < 1
      parent_1 = 1;
   end
   parent_2 = round(N*rand(1));
   if parent_2 < 1
      parent_2 = 1;
```

```
end
while
isequal(parent_chromosome(parent_1,:),parent_chromosome
(parent_2,:))

            parent_2 = round(N*rand(1));
            if parent_2 < 1
               parent_2 = 1;
            end
        end
        parent_1 = parent_chromosome(parent_1,:);
        parent_2 = parent_chromosome(parent_2,:);
        for j = 1 : V
           % SBX (Simulated Binary Crossover)
           % Generate a random number
           u(j) = rand(1);
           if u(j) <= 0.5
              bq(j) = (2*u(j))^(1/(mu+1));
           else
              bq(j) = (1/(2*(1 - u(j))))^(1/(mu+1));
           end
           child_1(j) = ...
              0.5*(((1 + bq(j))*parent_1(j)) + (1 -
              bq(j))*parent_2(j));
           child_2(j) = ...
              0.5*(((1 - bq(j))*parent_1(j)) + (1 +
              bq(j))*parent_2(j));
              if child_1(j) > u_limit child_1(j) = u_limit;
           elseif child_1(j) < l_limit
              child_1(j) = l_limit;
           end
           if child_2(j) > u_limit
              child_2(j) = u_limit;
           elseif child_2(j) < l_limit
              child_2(j) = l_limit;
           end
        end
        child_1(:,V + 1: M + V) =
        evaluate_objective(child_1,pro);
        child_2(:,V + 1: M + V) =
        evaluate_objective(child_2,pro);
        was_crossover = 1;
        was_mutation = 0;
     else
        parent_3 = round(N*rand(1));
        if parent_3 < 1
           parent_3 = 1;
```

```
        end
        % Make sure that the mutation does not result in
        variables out of
        % the search space. For both the MOP's the range
        for decision space
        % is [0,1]. In case different variables have
        different decision
        % space each variable can be assigned a range.
        child_3 = parent_chromosome(parent_3,:);
        for j = 1 : V
           r(j) = rand(1);
           if r(j) < 0.5
              delta(j) = (2*r(j))^(1/(mum+1)) - 1;
           else
              delta(j) = 1 - (2*(1 - r(j)))^(1/(mum+1));
           end
           child_3(j) = child_3(j) + delta(j);
           if child_3(j) > u_limit
              child_3(j) = u_limit;
           elseif child_3(j) < l_limit
              child_3(j) = l_limit;
           end
        end
        child_3(:,V + 1: M + V) =
        evaluate_objective(child_3,pro);
        was_mutation = 1;
        was_crossover = 0;
     end
     if was_crossover
        child(p,:) = child_1;
        child(p+1,:) = child_2;
        was_cossover = 0;
        p = p + 2;
     elseif was_mutation
        child(p,:) = child_3(1,1 : M + V);
        was_mutation = 0;
        p = p + 1;
     end

end
f = child;
%%%%%%%%%%%%%%%%%% End of genetic_operator %%%%%%%%%%%%%%%%%
function f = replace_chromosome(intermediate_
% chromosome,pro,pop)
%% replace_chromosome(intermediate_chromosome,pro,pop)
% This function replaces the chromosomes based
% on rank and crowding
```

```
% distance. Initially until the population size is
% reached each front is
% added one by one until addition of a complete
% front which results in
% exceeding the population size. At this point
% the chromosomes in that
% front is added subsequently to the population
% based on crowding distance.

[N,V] = size(intermediate_chromosome);
switch pro

case 1
   M = 2;
   V = 6;
case 2
   M = 3;
   V = 12;

end

% Get the index for the population sort based on the rank
[temp,index] = sort(intermediate_chromosome(:,M + V + 1));

% Now sort the individuals based on the index
for i = 1 : N


sorted_chromosome(i,:) =
intermediate_chromosome(index(i),:);

end

% Find the maximum rank in the current population
max_rank = max(intermediate_chromosome(:,M + V + 1));

% Start adding each front based on rank and crowing
% distance until the
% whole population is filled.
previous_index = 0;
for i = 1 : max_rank

current_index = max(find(sorted_chromosome(:,M + V + 1) ==
i));
if current_index > pop
   remaining = pop - previous_index;
   temp_pop = ...
      sorted_chromosome(previous_index + 1 : current_index,
      :);
```

```
    [temp_sort,temp_sort_index] = ...
       sort(temp_pop(:, M + V + 2),'descend');
    for j = 1 : remaining
       f(previous_index + j,:) =
       temp_pop(temp_sort_index(j),:);
    end
    return;
elseif current_index < pop
    f(previous_index + 1 : current_index, :) = ...
       sorted_chromosome(previous_index + 1 : current_index,
       :);
else
    f(previous_index + 1 : current_index, :) = ...
    sorted_chromosome(previous_index + 1 : current_index,
    :);
    return;
end
previous_index = current_index;

end %%%%%%%%%%%% End of replace_chromosome %%%%%%%%%%%%
function f = evaluate_objective(x,problem)

% Function to evaluate the objective functions for the
% given input vector
% x. x has the decision variables

switch problem

case 1
    f = [];
    %% Objective function one
    f(1) = 1 - exp(-4*x(1))*(sin(6*pi*x(1)))^6;
    sum = 0;
    for i = 2 : 6
       sum = sum + x(i)/4;
    end
    %% Intermediate function
    g_x = 1 + 9*(sum)^(0.25);
    %% Objective function one
    f(2) = g_x*(1 - ((f(1))/(g_x))^2);
case 2
    f = [];
    %% Intermediate function
    g_x = 0;
    for i = 3 : 12
       g_x = g_x + (x(i) - 0.5)^2;
    end
    %% Objective function one
```

TABLE 12.2: Parameters Used to Optimize the Function

Population	Generations	Pool size	Tour size	Distribution index for crossover η_c	Distribution index for mutation η_m
200	1000	100	2	20	20

```
f(1) = (1 + g_x)*cos(0.5*pi*x(1))*cos(0.5*pi*x(2));
%% Objective function two
f(2) = (1 + g_x)*cos(0.5*pi*x(1))*sin(0.5*pi*x(2));
%% Objective function three
f(3) = (1 + g_x)*sin(0.5*pi*x(1));
```

```
end
%%%%%%%%% End of evaluate_objective %%%%%%%%%%%
```

Observations

The parameters use to optimize the function are given in Table 12.2. The set of objective functions resulted in the solution as shown in Figure 12.9.

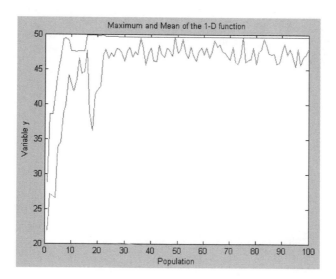

FIGURE 12.9: Optimal Objective Space Solution

12.5   Illustration 5: Evolutionary Strategy for Nonlinear Function Minimization

The function to be minimized is given as a subprogram in this section. The user may change the function according to the requirements.

```
% - - - - - - - - - - - - - - - - - - - - - - - - - - - - - -
% Main program
% - - - - - - - - - - - - - - - - - - - - - - - - - - - - - -
% - - - - - - - - - - - - Initialization - - - - - - - - - -


% User defined input parameters (need to be edited)


% User defined input parameters (need to be edited)
strfitnessfct = 'testfunc'; % name of objective/fitness
                            % function
N = 10; % number of objective variables/problem dimension
xmean = rand(N,1); % objective variables initial point
sigma = 0.5; % coordinate wise standard deviation
             % (step size)
stopfitness = 1e-10; % stop if fitness <stopfitness
             % (minimization)
stopeval = 1e3*N^2; % stop after stopeval number of
             % function evaluations


% Strategy parameter setting: Selection
lambda = 4+floor(3*log(N)); % population size,
             % offspring number
mu = floor(lambda/2); % number of parents/points
             % for recombination
weights = log(mu+1)-log(1:mu)'; % muXone array for
             % weighted recombination
% lambda=12; mu=3; weights = ones(mu,1); % uncomment
             % for (3_I,12)-ES
weights = weights/sum(weights); % normalize recombination
             % weights array
mueff=sum(weights)^2/sum(weights.^2); % variance-effective
             % size of mu
% Strategy parameter setting: Adaptation
cc = 4/(N+4); % time constant for cumulation for
             % covariance matrix
cs = (mueff+2)/(N+mueff+3); % t-const for cumulation
             % for sigma control
```

```
mucov = mueff; % size of mu used for calculating
              % learning rate ccov
ccov = (1/mucov) * 2/(N+1.4)^2 + (1-1/mucov) * ...
% learning rate for ((2*mueff-1)/((N+2)^2+2*mueff));
% covariance matrix
damps = 1 + 2*max(0, sqrt((mueff-1)/(N+1))-1) + cs;
                          % damping for sigma
                          % usually close to
                          % former damp == damps/cs


% Initialize dynamic (internal) strategy parameters and
% constants
pc = zeros(N,1); ps = zeros(N,1);   % evolution paths
% for C and sigma
B = eye(N,N);   % B defines the coordinate system
D = eye(N,N);   % diagonal matrix D defines the scaling
C = B*D*(B*D)';   % covariance matrix
chiN=N^0.5*(1-1/(4*N)+1/(21*N^2));   % expectation of
  % ||N(0,I)|| == norm(randn(N,1))


  % - - - - - - - - - Generation Loop - - - - - - - - - -

counteval = 0; % the next 40 lines contain the 20 lines
of interesting code
while counteval < stopeval

   % Generate and evaluate lambda offspring
   arz = randn(N,lambda); % array of normally distributed
   mutation vectors
   for k=1:lambda,
      arx(:,k) = xmean + sigma * (B*D * arz(:,k)); % add
      mutation % Eq. (1)

      arfitness(k) = feval(strfitnessfct, arx(:,k)); %
      objective function call
      counteval = counteval+1;
   end
   % Sort by fitness and compute weighted mean into xmean
   [arfitness, arindex] = sort(arfitness); % minimization
   xmean = arx(:,arindex(1:mu))*weights; % recombination,
   new mean value
   zmean = arz(:,arindex(1:mu))*weights; % ==
   sigma^-1*D^-1*B'*(xmean-xold)
   % Cumulation: Update evolution paths
   ps = (1-cs)*ps + sqrt(cs*(2-cs)*mueff) * (B * zmean); %
   Eq. (4)
```

```
hsig = norm(ps)/sqrt(1-(1-cs)^(2*counteval/lambda))/chiN
< 1.4 + 2/(N+1);
pc = (1-cc)*pc ...
    + hsig * sqrt(cc*(2-cc)*mueff) * (B * D * zmean); %
    Eq. (2)
    % Adapt covariance matrix C
    C = (1-ccov) * C ...        % regard old matrix % Eq. (3)
    + ccov * (1/mucov) * (pc*pc' ... % plus rank one update
                + (1-hsig) * cc*(2-cc) * C) ...
    + ccov * (1-1/mucov) ...          % plus rank mu update
    * (B*D*arz(:,arindex(1:mu))) ...
    * diag(weights) * (B*D*rz(:,arindex(1:mu)))';
% Adapt step size sigma
sigma = sigma * exp((cs/damps)*(norm(ps)/chiN - 1)); %
Eq. (5)
% Update B and D from C
% This is O(N^3). When strategy internal CPU-time is
critical, the
% next three lines should be executed only every
(alpha/ccov/N)-th
% iteration, where alpha is e.g. between 0.1 and 10
C=triu(C)+triu(C,1)'; % enforce symmetry
[B,D] = eig(C); % eigen decomposition, B==normalized
eigenvectors
D = diag(sqrt(diag(D))); % D contains standard
deviations now
% Break, if fitness is good enough
if arfitness(1) <= stopfitness
    break;
end
disp([num2str(counteval) ': ' num2str(arfitness(1))]);

end % while, end generation loop

    % - - - - - - - - - - Ending Message - - - - - - - -
disp([num2str(counteval) ': ' num2str(arfitness(1))]);
xmin = arx(:, arindex(1));   % Return best point
                    % of last generation.
            % Notice that xmean is expected to be even
            % better.
% - - - - - - - - - - - - - - - - - - - - - - - - - - - -


% - - - - - - - - - - - - - - - - - - - - - - - - - - - -
% Test function used in the main program
% - - - - - - - - - - - - - - - - -
function f=testfunc(x)
    N = size(x,1); if N < 2 error('dimension must
be greater one'); end
```

```
    f=1e4.î((0:N-1)/(N-1)) * x.î2;
% - - - - - - - - - - - - - - - -
```

Observations

The number of generations and the corresponding fitness value is shown in Table 12.3. The best value was reached at the end of 6620 generations.

TABLE 12.3:   Fitness Value Corresponding to the Generations

Generations	Fitness	Generations	Fitness
100	25145.3466	3500	3.5216
200	8264.9339	3600	3.2085
300	11418.9724	3700	2.4869
400	6593.6342	3800	1.6749
500	2795.3487	3900	1.2039
600	2623.5998	4000	0.76946
700	1973.3535	4100	0.41446
800	1412.9365	4200	0.18765
900	1090.126	4300	0.13595
1000	799.1704	4400	0.11234
1100	572.8327	4500	0.10385
1200	428.8273	4600	0.093081
1300	289.5838	4700	0.087091
1400	157.6233	4800	0.081874
1500	45.9125	4900	0.073765
1600	45.5945	5000	0.062782
1700	36.5345	5100	0.048267
1800	30.4814	5200	0.034492
1900	26.6116	5300	0.01582
2000	21.8548	5400	0.0036852
2100	18.9982	5500	0.0029413
2200	17.2331	5600	0.00083211
2300	15.0744	5700	0.00013754
2400	12.4608	5800	5.154e-005
2500	9.9533	5900	1.2286e-005
2600	9.3563	6000	5.2099e-006
2700	6.835	6100	1.157e-006
2800	6.0509	6200	1.7217e-007
2900	4.7504	6300	2.835e-008
3000	4.438	6400	1.3613e-008
3100	4.3084	6500	1.6395e-009
3200	4.2513	6600	3.0056e-010
3300	4.0217	6610	3.2443e-010
3400	3.8919	6620	8.7214e-011

Summary

A set of MATLAB programs were included in this chapter for the reader to implement and simulate EAs. With the rapid growth of the field, there is a particularly pressing need to extend existing and developing new analysis tools which allow us to better understand and evaluate the emerging varieties of EAs and their applications.

Review Questions

1. Write a MATLAB program to tune a PI controller using Evolutionary Strategy.

2. Minimize the Griewank function

   f(x) = sum((x-100).$\hat{2}$,2)./4000 - prod(cos((x-100)./(sqrt(repmat([1:N], length (x(:,1),1)))),2)·| 1 using Evolutionary Algorithm

3. Write a MATLAB code on Stochastic Ranking for Constrained Evolutionary Optimization for the following fitness function and its constraints:

```
% fitness function
f = sum(x(:,1:3),2) ;
% constraints g<=0
g(:,1) = -1+0.0025*(x(:,4)+x(:,6)) ;
g(:,2) = -1+0.0025*(x(:,5)+x(:,7)-x(:,4)) ;
g(:,3) = -1+0.01*(x(:,8)-x(:,5)) ;
g(:,4) =
-x(:,1).*x(:,6)+833.33252*x(:,4)+100*x(:,1)-83333.333
;
g(:,5) =
-x(:,2).*x(:,7)+1250*x(:,5)+x(:,2).*x(:,4)-1250*x(:,4)
;
g(:,6) =
-x(:,3).*x(:,8)+1250000+x(:,3).*x(:,5)-2500*x(:,5) ;
```

4. Write a MATLAB program to design a Proportional Integral-Derivative Controller using Evolutionary Algorithm.

5. Write a MATLAB program to minimize the Rosenbrock function using Evolutionary Algorithm.

Chapter 13

MATLAB-Based Genetic Algorithm

13.1 Introduction

A few researchers and creationists stated that evolution is not useful as a scientific theory since it produces no practical advantages and that it has no significance to daily life. But with the evidence of biology this statement can be claimed false. Various developments in the natural phenomena through evolution give sound theoretical knowledge about several practical applications. For instance, the observed development of resistance — to insecticides in crop pests, to antibiotics in bacteria, to chemotherapy in cancer cells, and to anti-retroviral drugs in viruses such as HIV — these instances are consequences of the laws of mutation and selection. Having a thorough knowledge of evolution has lead researchers to design and develop strategies for dealing with these harmful organisms. The evolutionary postulate of common descent has aided the development of new medical drugs and techniques by giving researchers a good idea of which organisms they should experiment on to obtain results that are most likely to be relevant to humans. Based on the principle of selective breeding, people were able to create customized organisms found in nature for their own benefit. The canonical example, of course, is the many varieties of domesticated dogs, but less well-known examples include cultivated maize (very different from its wild relatives, none of which have the familiar "ears" of human-grown corn), goldfish (like dogs, we have bred varieties that look dramatically different from the wild type), and dairy cows (with immense udders far larger than would be required just for nourishing offspring).

A few critics might charge that creationists can work on these things without recourse to evolution. For instance, creationists often explain the development of resistance to antibiotic agents in bacteria, or the changes wrought in domesticated animals by artificial selection, by presuming that God decided to create organisms in fixed groups, called "kinds" or baramin. Natural evolution or artificial selection can evolve different varieties within the originally created groups such as "dog-kind," or "cow-kind," or "bacteria-kind".

In the last few decades, the continuing advance of modern technology has brought about something new. Evolution is capable of producing practical benefits in almost every field, and at the same time, the creationists cannot claim that their explanation fits the facts just as well. The major field of evolution is computer science, and the benefits come from a programming methodology called genetic algorithms. In this chapter, a detailed description of genetic algorithm, its operators and parameters are discussed. Further, the schema theorem and technical background along with the different types of GA are also elaborated in detail. Finally MATLAB codes are given for applications such as maximization of a given function, traveling sales man problem, and GA based PID parameter tuning.

13.2   Encoding and Optimization Problems

Generally there are two basic functions of genetic algorithm: problem encoding and evaluation function, which are problem dependent. For instance, consider a basic parameter optimization problem where a set of variables should be optimized to either maximum or to a minimum. Such kinds of problems can be viewed as a black box with a series of control dials representing different optimization parameters. The result of the black box is a value returned by an evaluation function that gives a measure about the specific combination of parameter settings, which solves the optimization problem. The goal is to optimize the output by setting various parameters. In more traditional terms, to minimize or maximize some function $F(X_1, X_2, ..., X_M)$.

Generally nonlinear problems are chosen for optimization using genetic algorithms. Therefore each parameter cannot be treated as an independent variable. Usually the combined effects of the parameters must be considered in order to maximize or minimize the output of the black box. This interaction between the variables during combination is known as epistasis among the genetic community. Generally the initial assumption made is that the variables representing parameters can be represented by bit strings. This means that the variables are discretized in an *a priori* fashion, and that the range of the discretization corresponds to some power of 2. For instance, with 12 bits per parameter, we obtain a range with 4096 discrete values.

For continuous valued parameters discretization is not a specific problem. The discretization process provides enough resolution to make it possible to adjust the output with the desired level of precision. It also assumes that the discretization is in some sense representative of the underlying function. If some parameter can only take on an exact finite set of values then the coding issue becomes more difficult. For instance, if there are exactly 1100 discrete values which can be assigned to some variable X_i. We need at least 10 bits to cover this range, but this codes for a total of 1024 discrete values. The 76 unnecessary bit patterns may result in no evaluation, a default worst possible evaluation, or some parameter settings may be represented twice so that all binary strings result in a legal set of parameter values. Solving such coding problems is usually considered to be part of the design of the evaluation function. Aside from the coding issue, the evaluation function is usually given as part of the problem description. Sometimes, deciding and developing an evaluation function can involve developing a simulation. In few other cases, the evaluation may be based on performance and may represent only an approximate or partial evaluation.

The evaluation or the fitness function must also be relatively fast to measure the fitness of the problem. Since a genetic algorithm operates on a population of potential solutions, it incurs the cost of evaluating this population. Furthermore, the population is replaced (all or in part) on a generational basis. The members of the population reproduce, and their offspring must then be evaluated.

13.3   Historical Overview of Genetic Algorithm

The genetic algorithms were first developed during the late 1950s and early 1960s and this technique was programmed on computers by evolutionary engineers and biologists to model the aspects of natural evolution. Evolutionary computation was definitely in the air in the formative days of the electronic computer. During 1962, researchers such as G.E.P. Box, G.J. Friedman, W.W. Bledsoe, and H.J. Bremermann all independently proposed and developed evolution-inspired algorithms for function optimization and machine learning. But the work did not gain much importance. Ingo Rechenberg introduced the evolutionary strategy in 1965, which more resembled hill-climbers than genetic algorithms. In this strategy, there was no population or crossover; one parent was mutated to produce one offspring, and the better of the two was kept and became the parent for the next round of mutation. The versions that were introduced later used the idea of a population.

The evolutionary programming technique came into existence during 1966 and was introduced by L.J. Fogel, A.J. Owens and M.J. Walsh in America. In evolutionary programming the candidate solutions to optimization problems were represented as simple finite-state machines; like Rechenberg's evolution strategy, their algorithm worked by randomly mutating one of these simulated machines and keeping the better of the two. Similar to evolution strategies, the wider formulation of the evolutionary programming technique is still an area of ongoing research.

Based on the properties of natural evolution Genetic Algorithms (GA) were invented to mimic the natural properties. The idea of GA is to use the natural power of evolution to solve optimization problems. The father of the original Genetic Algorithm was John Holland who invented it in the early 1970's. The initial intention of Holland was to was determine exactly how adaptation occurs in nature and then develop ways that natural adaptation might become a part of computer systems instead of creating algorithms.

More interest was created toward evolutionary computation based on the foundational algorithms. Initially genetic algorithms were being applied to a broad range of subjects, from abstract mathematical problems like bin-packing and graph coloring to tangible engineering issues such as pipeline flow control, pattern recognition and classification, and structural optimization. All these applications were mainly theoretical. As research in this field increased, genetic algorithms migrated into the commercial sector, their rise fueled by the exponential growth of computing power and the development of the Internet. Today, evolutionary computation is a thriving field, and genetic algorithms are "solving problems of everyday interest" in areas of study as diverse as stock market prediction and portfolio planning, aerospace engineering, microchip design, biochemistry and molecular biology, and scheduling at airports and assembly lines.

13.4   Genetic Algorithm Description

Genetic algorithms are developed on the basis of Darwin's theory of evolution. To obtain a solution to a problem through genetic algorithms, the algorithm is started with a set of solutions (represented by chromosomes) termed as the population. Choosing random solutions of one population forms a new population. The new population is formed assuming that the new one will be better than the old one. Parent solutions are selected from the population to form new solutions (offspring) based on their fitness measure. This process is repeated over several iterations or until some condition (for example, number of populations or improvement of the best solution) is satisfied.

In Holland's original work, GAs were proposed to understand adaptation phenomena in both natural and artificial systems and they have three key features that distinguish themselves from other computational methods modeled on natural evolution:

- The use of bit string for representation

- The use of crossover as the primary method for producing variants

- The use of proportional selection

Among the evolutionary computation paradigms, genetic algorithms are one of the most common and popular techniques. The traditional genetic algorithm represented the chromosomes using a fixed-length bit string. Each position in the string is assumed to represent a particular feature of an individual, and the value stored in that position represents how that feature is expressed in the solution. Generally, the string is "evaluated as a collection of structural features of a solution that have little or no interactions". The analogy may be drawn directly to genes in biological organisms. Each gene represents an entity that is structurally independent of other genes.

The more classical reproduction operator used is one point crossover, in which two strings are used as parents and swapping a sub-sequence between the two strings forms new individuals. Another popular operator is bit-flipping mutation, in which a single bit in the string is flipped to form a new offspring string. A great variety of other crossover and mutation operators have also been developed. The basic difference between the various operators is whether or not they introduce any new information into the population. For instance the genetic operator crossover does not while mutation does. The strings are manipulated in a constrained manner according to the string interpretation of the genes. Consider for instance, any two genes at the same location on two strings may be swapped between parents, but not combined based on their values.

A few individuals among the population of solutions are selected to be parents probabilistically based upon their fitness values; the newly created offspring replace the parents. This approach is used to solve design optimization problems including discrete design parameters and then real parameter optimization problems. Early applications of GAs are optimization of gas pipeline control, structural design optimization, aircraft landing strut weight optimization, keyboard configuration design, etc. It should be mentioned that GAs have contributed to establish the schema theorem and recognize the role and importance of crossover through attentive theoretical analysis.

13.5   Role of Genetic Algorithms

The genetic algorithm is often cast as a function optimizer, manipulating a population of solutions using recombination and mutation. This formulation omits mechanisms present in many genetic algorithms. Here we will be most concerned with the distinction between the space of genotypes, in which genetic operators apply and the space of phenotypes, in which individuals' fitness are evaluated. Development is the process by which genotypes are transformed to phenotypes, and this term is divided into maturation and learning. Maturation refers to the process by which a genotype is mapped into a phenotype and learning refers to phenotypic plasticity remaining in the mature organism as evidenced by adaptive responses to the environment.

The standard genetic algorithm proceeds as follows: an initial population of individuals is generated at random or heuristically. Every evolutionary step, known as a *generation*, the individuals in the current population are decoded and evaluated according to some predefined quality criterion, referred to as the fitness, or fitness function. To create a new population or the next generation, a few individuals are selected according to their fitness and they are the parent solutions. Several selection methods are currently in use, but one of the simplest is Holland's original *fitness-proportionate selection*, where individuals are selected with a probability proportional to their relative fitness. The expected number of times an individual is chosen is approximately proportional to its relative performance in the population. Therefore best fit or good individuals have a better chance of reproducing, while low-fitness ones are more likely to disappear.

Selection is not the only process of choosing the best solution in the population. Genetically inspired operators, crossover, and mutation create the best-fit solutions. Crossover is performed between two selected individuals (parents) by swapping the genomic parts to form new individuals (offspring) solutions with a crossover probability p_{cross} (the crossover probability or crossover rate); in its simplest form, substrings are exchanged after a randomly selected crossover point. This operator tends to enable the evolutionary process to move toward promising regions of the search space. The mutation operator is introduced to prevent premature convergence to local optima by randomly sampling new points in the search space. It is carried out by flipping bits at random, with some (small) probability p_{mut}. Genetic algorithms are stochastic iterative processes that are not guaranteed to converge; the termination condition may be specified as some fixed maximal number of generations

or as the attainment of an acceptable fitness level. The pseudocode of
the standard genetic algorithm is shown below:

```
Initialize population of chromosomes P(g)
Evaluate the initialized population by computing
its fitness measure
While not termination criteria do
    g:=g+1
    Select P(g+1) from P(g)
    Crossover P(g+1)
    Mutate P(g+1)
    Evaluate P(g+1)
End While
```

13.6   Solution Representation of Genetic Algorithms

Solutions and individuals can be represented using genetic representation to encode appearance, characteristics, behavior, and physical qualities of individuals. The most demanding task in evolutionary computation is the genetic representation. The genetic representation is one of the major criteria drawing a line between known classes of evolutionary computation.

The most commonly used representation form in genetic algorithm is binary representation. For more standard applications an array of bits is used for representation. Due to the fixed size, the formats can be easily aligned which makes GA more convenient for operation, leading to flexible crossover operation. If the strings are of variable length, then crossover operation becomes more complex.

The algorithm need not be aware of a particular genetic representation used for any solution. Some of the common genetic representations are

- binary array
- genetic tree
- parse tree
- binary tree
- natural language

A problem dependant fitness function is defined over the genetic representation to evaluate the quality of the represented solution.

13.7 Parameters of Genetic Algorithm

Parameter tuning in GA involves the tuning of the parameters such as the crossover probability and mutation probability. Several studies are proposed by researchers and scientists in the field of GA to decide the value of these probabilities. Based on various experiments conducted on different experiments and according to DeJong, for a population size of 50, the crossover probability Px was best set to 0.6, and the mutation probability Pm to 0.001, elitism to 2 without any windowing.

13.7.1 Standard Parameter Settings

The efficiency of a GA is greatly dependent on its tuning parameters. Some of the major researchers have worked on the parameters of GA and given a set of specifications as follows:

Dejong Settings

Dejong's settings are the standard for most GAs. Dejong has shown that this combination of parameters works better than many other parameter combinations for function optimization.

- Population size 50

- Number of generations 1,000

- Crossover type= typically two point

- Crossover rate of 0.6

- Mutation types= bit flip

- Mutation rate of 0.001

The crossover method is assumed to be one or two point crossover. For more disruptive methods (such as uniform crossover), use a lower crossover rate (say 0.50). The mutation rate given above is *per bit*, whereas in many public domain codes, the mutation rate is input as a *per chromosome* probability.

Grefenstette Settings

This is generally the second most popular set of parameter settings. It is typically used when the computational expense of figuring the objective function forces you to have a smaller population.

- Population size 30

- Number of generations not specified

- Crossover type= typically two point

- Crossover rate of 0.9

- Mutation types= bit flip

- Mutation rate of 0.01

The crossover method is again assumed to be one or two point crossover and the mutation rate is also *per bit*.

MicroGA Settings

Though these settings are not widely employed presently, those who have used them to optimize functions report as much as four-fold reductions in the number of evaluations required to reach given levels of performance.

- Population Size = 5

- Number of generations = 100

- Crossover type= uniform

- Crossover probability= 0.5

- Mutation types= jump and creep

- Mutation probabilities= 0.02 and 0.04

David's code continuously restarts with random chromosomes when convergence is detected.

13.8   Schema Theorem and Theoretical Background

A schema is a template that identifies a subset of strings with similarities at certain string positions. Schemata are a special case of cylinder sets; and so form a topological space.

Consider a binary string of length 6. The schema 10*0*1 describes the set of all strings of length 6 with 1's at positions 1 and 6 and a 0 at positions 2 and 4. The * is a wildcard symbol, which means that positions

3 and 5 can have a value of either 1 or 0 similar to a tristate value. The number of fixed positions in the schema is known as the order and the distance between the first and last specific positions is the defining length $\delta(H)$. The order of 10*0*1 is 4 and its defining length is 5. The *fitness of a schema* is the average fitness of all strings matching the schema. The fitness of a string is a measure of the value of the encoded problem solution, as computed by a problem-specific evaluation function. With the genetic operators as defined above, the schema theorem states that short, low-order, schemata with above-average fitness increase exponentially in successive generations. Expressed as an equation:

$$m(H, t+1) \geq \frac{m(H,t)f(H)}{a_t}[1-p]$$

Here m(H,t) is the number of strings belonging to schema H at generation t, f(H) is the observed fitness of schema H and at is the observed average fitness at generation t. The probability of disruption p is the probability that crossover or mutation will destroy the schema H. It can be expressed as

$$p = \frac{\delta(H)}{l-1}p_c + o(H)p_m$$

where o(H) is the number of fixed positions, l is the length of the code, p_m is the probability of mutation, and p_c is the probability of crossover. So a schema with a shorter defining length $\delta(H)$ is less likely to be disrupted. The Schema Theorem is an inequality rather than an equality. This is often a big question among investigators. The answer is in fact simple: the Theorem neglects the small, yet non-zero probability, that a string belonging to the schema h will be created "from scratch" by mutation of a single string (or recombination of two strings) that did not belong to h in the previous generation.

13.8.1 Building Block Hypothesis

Though genetic algorithms are relatively simple to implement, their behavior seems to be difficult to understand. In practice it seems very difficult to understand why they are often successful in generating solutions of high fitness. The building block hypothesis (BBH) consists of:

- A description of an abstract adaptive mechanism that performs adaptation by recombining "building blocks", i.e., low order, low defining-length schemata with above average fitness.

- A hypothesis that a genetic algorithm performs adaptation by implicitly and efficiently implementing this abstract adaptive mechanism.

According to Goldberg the abstract adaptive mechanism is defined as: "short, low order, and highly fit schemata are sampled, recombined [crossed over], and resampled to form strings of potentially higher fitness. In a way, by working with these particular schemata [the building blocks], instead of building high-performance strings by trying every conceivable combination, better and better strings can be contstructed from the best partial solutions of past samplings".

Just as a child creates magnificent fortresses through the arrangement of simple blocks of wood (building blocks), so does a genetic algorithm seek near optimal performance through the juxtaposition of short, low-order, high-performance schemata, or building blocks.

13.8.2   The Dynamics of a Schema

The Schema Theorem only shows how schemas dynamically change, and how short, low-order schemas whose fitness remain above the average mean receive exponentially growing increases in the number of samples. It cannot make more direct predictions about the population composition, distribution of fitness, and other statistics more directly related to the genetic algorithm itself. The pseudocode of the general algorithm is as follows:

```
Initialize population
While not end of population do
    Calculate Fitness
    Choose two parents selection methods
    Choose one of the offspring by single
        -point crossover
    Mutate each bit with probability pm,
    Place each bit in population.
End While
```

The method of approximating the dynamics can be calculated through the following method. Let S be a schema with at least one instance in the given population at the time t. Let the function $\hat{u}(S,t)$ be the number of instances of S at time t, and let the function (S,t) be the observed average of S at time t. Now, it is required to calculate the expected number of instances of schema S during the next time interval. Assuming that $E(x)$ is the expected value of x, to determine $E(n(S,t+1))$, the number of instances at t+1 is equivalent to the number of offspring produced. To proceed the calculation, the selection process that is to be used has to be determined. If a fitness-proportionate selection method is to be used, then the expected number of offspring is equal to the fitness of the string over the average fitness of the generation or:

$$\text{fitness-proportionate} = \frac{f(x)}{\hat{f}(t)} \tag{13.3}$$

So, assuming that x ε S (x is a subset, or instance, of S), the equation obtained is:

$$E(n(S,t+1)) = \sum_{x \epsilon S} \frac{f(x)}{\hat{f}(t)} \tag{13.4}$$

Now, using the definition of (S,t) the final equation is:

$$\sum_{x \epsilon S} \frac{f(x)}{n(S,t)} = \hat{u}(S,t)$$

$$\therefore E(n(S,t+1)) = \frac{\hat{u}(S,t)}{\hat{f}(t)} n(S,t) \tag{13.5}$$

Therefore the basic formula for tracking the dynamics of a schema in a genetic algorithm is obtained. Genetic algorithms are simple, since they also simulate the effects of mutation and crossovers. These two effects have both constructive and destructive effects - here the destructive effects are considered.

Figure 13.1 summarizes the schema theorem.

Selection - $fitness - proportionate = \dfrac{f(x)}{\hat{f}(t)}$

Crossover - $p_C(S) \geq 1 - p_c * \dfrac{\delta(S)}{m-1}$

Mutation - $p_m(S) = (1 - p_m)^{o(S)}$

Schema Theorem - $E(n(S,t+1)) \geq \dfrac{\hat{u}(S,t)}{\hat{f}(t)} n(S,t) \left(1 - p_c \dfrac{d(S)}{m-1}\right) \left((1 - p_m)^{o(S)}\right)$

FIGURE 13.1: Derivation of the Schema Theorem

13.8.3 Illustrations Based on Schema Theorem

Illustration 1:

1. Consider the following two schema

$* * * * * 1\ 1\ 1$ and $1 * * * * * 1*$

(a) **What is the defining length? What is the order?**

(b) **Let $p_m = 0.001$ (probability of simple mutation). What is the probability of survival of each schema?**

(c) **Let $p_c = 0.85$ (probability of cross-over). Estimate the probability of survival of each schema?**

Solution:

(a). Order:

$$o(S_1) = 3;\ o(S_2) = 2;$$

Defining Length:

$$\delta(S_1) = 2;\ \delta(S_2) = 6;$$

(b). Given: $p_m = 0.001$
Mutation:

$$p_m(S) = (1 - p_m)^{o(S)}$$
$$p_m(S_1) = (1 - p_m)^{o(S_1)} = (1 - 0.001)^3 = 99.7\%$$
$$p_m(S_2) = (1 - p_m)^{o(S_2)} = (1 - 0.001)^2 = 99.8\%$$

(c). Given: pc = 0.85;
We know that m = 8;
Crossover:

$$p_c(S) \geq 1 - p_c(*)\frac{\delta(S)}{m-1}$$
$$p_c(S_1) = 1 - p_c * \frac{\delta(S_1)}{m-1} = 1 - 0.85 * \frac{2}{8-1} = 76\%$$
$$p_c(S_2) = 1 - p_c * \frac{\delta(S_2)}{m-1} = 1 - 0.85 * \frac{6}{8-1} = 27\%$$

Illustration 2:

A population contains the following strings and fitness values at generation 0 (Table 13.1):

Calculate the expected number of strings covered by the schema $1 * * * *$ in generation 1 if $p_m = 0.01$ and $p_c = 1.0$.

Computational Intelligence Paradigms

TABLE 13.1:   Strings and Fitness Value

S No	String	Fitness
1	10001	20
2	11100	10
3	00011	5
4	01110	15

Solution:

Schema Theorm:

$$E(n(S,t+1)) \geq \frac{\hat{u}(S,t)}{\hat{f}(t)} n(S,t)(1 - p_c \frac{d(S)}{m-1})((1-p_m)^{o(S)})$$

$$\hat{u}(S,t) = \frac{20+10}{2} = 15$$

$$\hat{f}(t) = \frac{20+10+5+15}{4} = 12.5$$

The number of strings covered by the given schema in generation 1 is:

$$E(n(S,t+1)) = 2*\frac{15}{12.5}*\left[1 - 1.0*\frac{0}{4} - 0.01*1\right] = 2*12*(1-0-0.01) = 2.376$$

Illustration 3:
Suppose a schema S which, when present in a particular string, causes the string to have a fitness 25% greater than the average fitness of the current population. If the destruction probabilities for this schema under mutation and crossover are negligible, and if a single representation of the schema is present at generation 0, determine when the schema **H** will overtake the populations of size n = 20; 50; 100; 200.

Solution:

$$\hat{u}(S,t) \geq \hat{u}(S,t-1) * \frac{\overline{\hat{u}(S,t-1)}}{\hat{f}(t-1)} * 1$$

$$hatu(S,t-2)*1.25*1.25$$
$$M$$
$$\hat{u}(S,0)*(1.25)^t$$
$$(1.25)^t$$

Therefore

$$
\begin{aligned}
n &\leq (1.25)^t \\
t &\geq log_{1.25}(n) \\
&\geq \frac{log(n)}{log(1.25)}
\end{aligned}
$$

The schema will overtake the populations when,

$$
\begin{aligned}
n &= 20 \Rightarrow t \geq 14 \\
n &= 50 \Rightarrow t \geq 18 \\
n &= 100 \Rightarrow t \geq 21 \\
n &= 200 \Rightarrow t \geq 24
\end{aligned}
$$

Illustration 4:
Suppose a schema S which, when present in a particular string, causes the string to have fitness 10% less than the average fitness of the current population. Suppose that the schema theorem is equality instead of an inequality and that the destruction probabilities for this schema under mutation and crossover can be ignored. If representatives of the schema are present in 60% of the population at generation 0, calculate when schema H will disappear from the population of size n = 20; 50; 100; 200.

Solution:

$$
\begin{aligned}
\hat{S, t} &= \hat{u}(S, 0) * (0.9)^t \\
&= 0.6 * n * (0.9)^t \\
\frac{1}{2} &> 0.6 * n * (0.9)^t \\
\frac{1}{1.2 * n} &> (0.9)^t \\
t &> log_{0.9}(\frac{1}{1.2 * n})
\end{aligned}
$$

The given schema disappears when,

$$
\begin{aligned}
n &= 20 \Rightarrow t \geq 31 \\
n &= 50 \Rightarrow t \geq 39 \\
n &= 100 \Rightarrow t \geq 46 \\
n &= 200 \Rightarrow t \geq 53
\end{aligned}
$$

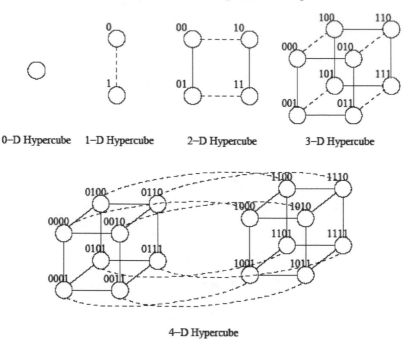

0–D Hypercube   1–D Hypercube   2–D Hypercube   3–D Hypercube

4–D Hypercube

Figure 13.1a: Representation of search space

Illustrations on Search Space, hypercubes, schemata and Gray Codes

Illustration 5:
Suppose we deal with a 'Simple GA' which uses bit-strings of length L for chromosomes.

1. Why is the representation of the search space of size 2L by a hypercube more appropriate than a 'linear representation'.
   Linear Representation

$$00 \ldots 00|00 \ldots 01|00 \ldots 10| \ldots |01 \ldots 11|10 \ldots 00| \ldots$$

Solution:
Properties of hypercube:

  (i) d-dimensional hypercube: 2 (d-1)-dimensional hypercubes

  (ii) every node is connected to or has as neighbors all other nodes for which the bit-representation differs at only 1 position. (Hamming distance 1)

So, neighboring genotypes are represented as neighbors.

13.9   Crossover Operators and Schemata

The basics about crossover and its types were already discussed in Section 11.7.6 in Chapter 11 on evolutionary computation paradigms. Hence this section will delineate the crossover operation with respect to the hyperplanes. The observed representation of hyperplanes in the schemata corresponds to the representation in the intermediate population after selection but before crossover. During recombination, the order-1 hyperplane samples are not affected by recombination, since the single critical bit is always inherited by one of the offspring. The observed distribution of potential samples from hyperplane partitions of order-2 and higher can be affected by crossover. Further, all hyperplanes of the same order are not necessarily affected with the same probability.

13.9.1   1-Point Crossover

Single point crossover, is relatively easy to quantify its effects on different schemata representing hyperplanes. Assume a string encoded with just 12 bits. Now consider the following two schemata.

11********** and 1**********1

The probability that the bits in the first schema will be separated during 1-point crossover is only 1/L-1, since in general there are L-1 crossover points in a string of length L. The probability that the bits in the second rightmost schema are disrupted by 1-point crossover however is (L-1)/(L-1) or 1.0 since each of the L-1 crossover points separates the bits in the schema. This leads to a general observation: when using 1-point crossover the positions of the bits in the schema are important in determining the likelihood that those bits will remain together during crossover.

13.9.2   2-Point Crossover

A 2-point crossover operator chooses two crossover points randomly. The string segment between the two points are swapped. Ken DeJong first observed that 2-point crossover treats strings and schemata as if they form a ring, which can be illustrated as shown in Figure 13.2

```
      b7  b6  b5                 *    *    *
   b8            b4           *              *
   b9            b3        *                    *
   b10           b2           *              *
      b11 b12 b1              *    1    1
```

FIGURE 13.2:   Ring Form

where b1 to b12 represents bits 1 to 12. When viewed in this way, 1-point crossover is a special case of 2-point crossover where one of the crossover points always occurs at the wrap-around position between the first and last bit. Maximum disruptions for order-1 schemata now occur when the 2 bits are at complementary positions on this ring.

From both 1-point and 2-point crossover it is clear that schemata, which have bits that are close together on the string encoding (or ring), are less likely to be disrupted by crossover. More accurately, hyperplanes represented by schemata with more compact representations should be sampled at rates that are closer to those potential sampling distribution targets achieved under selection alone. For current purposes a compact representation with respect to schemata is one that minimizes the probability of disruption during crossover. Note that this definition is operator dependent, since both of the two order-2 schemata are equally and maximally compact with respect to 2-point crossover, but are maximally different with respect to 1-point crossover.

13.9.3   Linkage and Defining Length

In linkage a set of bits act as "coadapted alleles" that tend to be inherited together as a group. In this case an allele would correspond to a particular bit value in a specific position on the chromosome. Of course, linkage can be seen as a generalization of the notion of a compact representation with respect to schema. Linkage under one point crossover is often defined by physical adjacency of bits in a string encoding. Linkage under 2-point crossover is different and must be defined with respect to distance on the chromosome when treated as a ring. Nevertheless, linkage usually is equated with physical adjacency on a string, as measured by defining length.

The defining length of a schemata is based on the distance between the first and last bits in the schema with value either 0 or 1 (i.e., not a * symbol). Given that each position in a schema can be 0, 1 or *, then scanning left to right, if I_x is the index of the position of the rightmost occurrence of either a 0 or 1 and I_y is the index of the leftmost occurrence of either a 0 or 1, then the defining length is merely I_x - I_y. Thus, the defining length of ****1**0**10** is 12-5=6. The defining length of a schema representing a hyperplane H is denoted here by $\triangle$ (H). The defining length is a direct measure of how many possible crossover points fall within the significant portion of a schemata. If 1-point crossover is used, then $\triangle(H)/L$-1 is also a direct measure of how likely crossover is to fall within the significant portion of a schemata during crossover.

13.9.4 Linkage and Inversion

Not only mutation and crossover are genetic operators, inversion is also considered to be a basic genetic operator. The linkage of bits on the chromosomes can be changed such that bits with greater nonlinear interactions can potentially be moved closer together on the chromosome. Generally, inversion is implemented by reversing a random segment of the chromosome. However, before one can start moving bits around on the chromosome to improve linkage, the bits must have a position independent decoding. A common error that some researchers make when first implementing inversion is to reverse bit segments of a directly encoded chromosome. But just reversing some random segment of bits is nothing more than large-scale mutation if the mapping from bits to parameters is position dependent. A position independent encoding requires that each bit be tagged in some way. For example, consider the following encoding composed of pairs where the first number is a bit tag which indexes the bit and the second represents the bit value.

$$((90)(60)(21)(71)(51)(81)(30)(10)(40))$$

Moving around the tag-bit pairs can now change the linkage, but the string remains the same when decoded: 010010110. One must now also consider how recombination is to be implemented.

13.10   Genotype and Fitness

The initial step to solve a problem using GA is to get knowledge about the kind of genotype required by the problem. The parameters of the problem are mapped into a binary string or a fixed length string. The user can decide on the mapping which implies that the user has to choose the number of bits per parameter, the range of the decoded binary-to-decimal parameter, if the strings have constant or changing lengths during the GA operations, etc.

In most of the practical applications the strings use a binary alphabet and the length is maintained constant during the evolution process. All the parameters in GA decode to the same range of values within the length and are allocated the same number of bits for the genes in the string. Recently, this situation is modified in almost practical applications. As the complexity of a problem increases, the applications also require dynamic length strings or else a different non-binary alphabet. Generally integer or float genes are used in several domains such as neural networks training, function optimization with a large number of variables, reordering problems as the Traveling Salesperson Problem, etc.

The binary strings were used in the concept of schema earlier since the number of schemata sampled per string is maximized. Recently, the schema concept is mainly interpreted to explain the usage of non-binary genes. In this new form of interpretation the utilization of higher cardinality alphabets are considered due to their higher expression power. The wildcard symbol (*) acts as an instantiation character for different sets of symbols. More accurately, * is reinterpreted as a family of symbols *x, where x can be any subset of symbols of the alphabet. For binary strings these two interpretations are the same but for higher cardinality alphabets they are not.

As far as the genotypes are considered, several other developments have brought forward the utilization of non-string representations such as trees of symbols which are much more appropriate and flexible for a large number of applications. The genotypes are decoded to yield the phenotype, which is, the string of problem parameters. The decoding process is necessary since the user has to evaluate the string in order to assign it a fitness value. Therefore the input to the fitness function and the fitness value is the phenotype which assists the algorithm to relatively rank the pool of strings in the population.

An appropriate fitness function that is to be created for every problem has to be considered as very highly important for the correct functionality of the GA. The fitness function is used to evaluate the problem environment, and find the best string that solves the problem well. While contrasting the fitness function the following factors should be considered:

- Whether the criteria is to be maximize or minimize.

- Instead of computing the entire solution only approximations can be computed when the fitness function is complex.

- Constraints should also be considered by the fitness function.

- If a fitness function incorporates other different sub-objectives, such as multiobjective function then it presents non-conventional problems.

- The fitness function acts as a black box for the GA. Phenotype is fed as input to the fitness function to obtain the fitness value. This process is achieved by a mathematical function. A complex computer simulator program or a human used to decide how good a string is.

13.11   Advanced Operators in GA

GA has a variety of operators other than the common crossover, mutation, reproduction, etc., discussed in Chapter 5. Some of the advanced operators are discussed in this section.

13.11.1   Inversion and Reordering

The position of the genes in a chromosome can be modified during the GA run by using techniques like inversion and reordering. Inversion is the process in which the order of genes are reversed between two randomly chosen positions within the chromosome. Reordering does nothing to lower epistasis, but greatly expands the search space. An example to illustrate inversion is shown in Figure 13.3.

13.11.2   Epistasis

The independence of bits in a string or a population of strings is known as epistasis. It is defined as the extent to which the "expression" of one gene depends on the values of other genes. Every gene in a chromosome

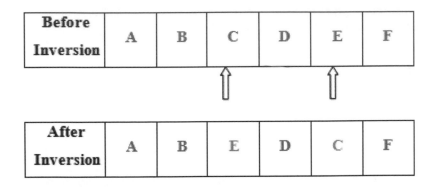

FIGURE 13.3:   Example to Illustrate Inversion

has a different degree of interaction. Any slight modification made in one gene affects the fitness to an extent. This resultant change may vary according to the values of other genes. Certain allele values present in particular bit positions have a strong influence on the fitness of the string. According to Rawlins there exists two extreme cases of epistasis: *maximum epistasis* and the *zero epistasis*. In maximum epistasis, it is possible to find one or more bits whose value influences the contribution of this bit to the fitness of a string. The second case is zero epistasis where all bits are independent.

13.11.3   Deception

   In a deceptive problem, the best low-order schemata contain a deceptive attract or that is maximally distant from the global solution. A deceptive ordering or permutation problem can be defined in the same way by choosing a global and a deceptive attractor that are maximally distant in some sense. Through the process of crossover, these optimal schemata will come together, to form the globally optimum chromosome. Sometimes the schemata which are not present in the global optimum increase in frequency more rapidly than those which are present, thereby misleading the GA away from the global optimum, instead of towards it. This is known as deception. Deception is a special case of epistasis and epistasis is necessary (but not sufficient) for deception. For a very high epistasis GA will not be effective and for a very low epistasis GA will be outperformed by simpler techniques, such as hill climbing.

13.11.4 Mutation and Naïve Evolution

Selection and mutation are combined and termed as naïve evolution. "Naïve evolution" performs a search technique similar to hill-climbing and has been proved to be more powerful even without crossover. Since mutation searches in between populations it is generally capable of finding better solutions when compared to crossover. While the population converges it can be found that mutation is more productive, while crossover is less productive.

13.11.5 Niche and Speciation

Speciation is a useful technique to generate multiple species within the population of evolutionary methods. There are several speciation algorithms which restrict an individual to mate only with similar ones, while others manipulate its fitness using niching pressure to control the selection. Especially the latter ones are naturally based on niching method. In GA, niches are analogous to maxima in the fitness function. If GA is not capable of converging the entire population to a single peak then the GA suffers from genetic drift. The two basic techniques to solve this problem are to maintain diversity, or to share the payoff associated with a niche.

13.11.6 Restricted Mating

Restricted mating or dynamic breeding is performed to prevent crossover between members of different niches. During the restricted mating process, the first parent is selected through tournament selection based upon shared fitness values. The second parent is determined by examining a pool of MF (Mating Factor) members selected randomly without replacement from the population, and selecting the fittest individual coming from the same niche as the first parent. If no such individual is found, then the parent is mated with the most similar individual in the mating pool, which can create lethal offspring. Dynamic inbreeding certainly offers an improvement over unrestricted mating or standard mating. Both standard and dynamic sharing methods have the disadvantage of giving preference to higher peaks because they use tournament selection based on fitness which can cause the extinction of relatively low peaks.

13.11.7 Diploidy and Dominance

Chromosomes that contain two sets of genes are referred as diploid chromosomes. Most of the earlier work on GA on GAs concentrated on haploid chromosomes. Though haploid chromosomes were used due to their simple nature: diploid chromosomes have more benefits. Hollstein modeled diploidy and dominance in genetic algorithms during 1971. In his proposal the genotypes were considered to be diploid. Every individual in the population carried a pair of chromosomes. A diploid chromosome pair can be mapped to a particular phenotype using the dominance map and the phenotype is further used for fitness evaluation. Goldberg and Smith concentrated on the role of dominance and diploidy as abeyance structures (shielding information that may be useful when situations change). In diploid chromosomes the presence of two genes allows two different "solutions" to be remembered, and passed on to offspring. Among the two different solutions one will be dominant, while the other will be recessive. As the environmental conditions are altered, the dominance can shift, thereby making the other gene dominant. This shift can take place much more quickly than would be possible if evolutionary mechanisms had to alter the gene. This mechanism is ideal if the environment regularly switches between two states.

13.12　GA Versus Traditional Search and Optimization Methods

An efficient optimization algorithm must use two techniques to find a global maximum: *exploration* and *exploitation*. Exploration is used to investigate innovative and strange areas in the search space and exploitation is used to make use of the knowledge base from the points found in the past up to that instant to find better points. These two requirements are contradictory, and a good search algorithm must find a tradeoff between the two. This section compares GA with other search methods.

13.12.1　Neural Nets

Genetic Algorithms (GAs) and Neural Networks (NNs) in a broad sense dwell in the class of evolutionary computing algorithms which attempt to imitate natural evolution or data manipulation with respect to day-to-day problems like forecasting the stock market, turnovers, or the identification of credit bonus classes for banks. Both GAs as well as NNs have gained more importance in modern days, particularly in reference to micro-economic questions. In spite of their apparent design disputes, they also share several features in common which are sufficiently interesting for the innovation oriented applications. Due to the dynamics built-in the evolution of both methodologies, they are somewhat isolated scientific professions that interact rarely. Both GAs and NNs are adaptive, they learn, and can deal with highly nonlinear models and noisy data and are robust, "weak" random search methods. They do not require gradient information or smooth functions whereas they mostly concentrate on nonlinear functions. In both cases, the modelling and coding ought to be executed carefully since they are complex. These algorithms work best for real world applications. GAs can optimize parameters of a neural net and GAs can also be used to fine tune the NN parameters.

13.12.2 Random Search

In random search, the search approach for complex functions is a random, or an enumerated search. The points in the search space are selected randomly, and their fitness is evaluated. Since this is not an efficient strategy, it is seldom applied. GAs have proved to be more efficient than random search. In cases where GAs cannot effectively solve problems in which the fitness measure is only specified right/wrong, there is no way for the GA to converge on the solution. In such cases, the random search can be applied to find the solution as quickly as a GA.

13.12.3 Gradient Methods

Several methods have been developed for optimizing well-behaved continuous functions which depend on slope information of the function. The gradient or slope of a function is used to guide the direction of search, known as hill climbing. They perform well on single peak functions or unimodal functions. With multimodal functions, they suffer from the problem of genetic drift. Consequently gradient methods can be coordinated with GAs. Genetic Algorithms are best applied in supervised learning applications when compared to gradient methods. Gradient methods cannot be employed directly in weight optimization domains, in such cases they are combined with GAs. In general gradient methods are capable of finding local optimal and global optimal solutions whereas GAs have proven to find better solutions compared to the gradient methods.

13.12.4 Iterated Search

The combination of random search and gradient search leads to a technique known as the iterated hill climbing search. This method is very simple and performs efficiently if the function does not have too many local maxima. When the random search advances, it proceeds to allocate its trials equally all over the search space. This implies that it will still measure as many as possible points in the regions ascertained to be of low fitness as in regions found to be of high fitness. A GA, commences with an initial random population, and allocates growing trials to regions of the search space found to have high fitness. This is a disadvantage if the maximum is in a small region, surrounded on all sides by regions of low fitness. This kind of function is difficult to optimize by any method, and here the simplicity of the iterated search usually wins.

13.12.5   Simulated Annealing

Since Genetic algorithms (GAs) are adaptive search techniques that are configured to find best-optimal solutions of large scale optimization problems with multiple local maxima, this is basically an altered variant of hill climbing. The terms like solutions, their costs, neighbors, and moves are concerned with simulated annealing whereas terms such as individuals (or chromosomes), their fitness, and selection, crossover and mutation are related with genetic algorithms. The deviation in nomenclature naturally reflects the differences in emphasis, likewise serves to obscure the similarities and the real differences between SA and GA.

Essentially, SA can be thought of as GA where the population size is only one. The current solution is the only individual in the population. As there is only one individual, there is no crossover, but only mutation. As a matter of fact this is the primal difference between SA and GA. As SA produces a new solution by changing only one solution with a local move, GA also produces solutions by fusing two different solutions.

Both SA and GA share the basic assumption that best solutions are more often found "near" already known good solutions than by randomly selecting from the whole solution space. If this were not the case with a particular problem or representation, they would perform no better than random sampling.

Additional positive conclusions bearing on the use of the GA's crossover operation in recombining near-optimal solutions obtained by other methods. Hybrid algorithms in which crossover is used to combine subsections is more effective and efficient than SA or a GA individually.

13.13   Benefits of GA

Some of the advantages of GA are:

- Since the genetic algorithms have multiple offspring for large problems the time taken for evaluation of all the strings is too large, therefore GAs are capable of parallel processing.

- A vast solution set can be scanned by a GA at a very fast rate.

- GAs are well suited for complex, discontinuous, noisy fitness functions.

- GAs do not require computation of partial derivatives and they also do not require a problem to be linearised.

- Since GAs efficiently search the global space, they are capable of converging to the local minima effectively.

13.14   MATLAB Programs on Genetic Algorithm

A set of MATLAB examples are worked out in this section for practical implementation of the GA functions, such as maximization and minimization of given functions, and the traveling salesman problem

13.14.1   Illustration 1: Maximizing the Given One-Dimensional Function within Given Boundaries

Maximize the function $f(x) = x + \sin(x) + \cos(x)$

Solution:
The basic genetic algorithm is used to maximize the given function. To refresh, the algorithm is as follows:

Step 1. Create an initial population (usually a randomly generated string).

Step 2. Evaluate all of the individuals (apply some function or formula to the individuals).

Step 3. Select a new population from the old population based on the fitness of the individuals as given by the evaluation function.

Step 4. Apply genetic operators (mutation and crossover) to members of the population to create new solutions.

Step 5. Evaluate these newly created individuals.

Step 6. Repeat steps 3-6 (one generation) until the termination criteria has been satisfied (usually perform for a certain fixed number of generations.

MATLAB Code

A genetic algorithm is set up to find the maximum of this problem.

```
% boundaries
bounds = [-10 10];
% pop size
n = 10;
% number of iterations
numits = 100;
% numer of mutations per it

nummut = 1;

f = @multipeak;

% ezplot(f,[-10 10])
% pause

blength = bounds(2)-bounds(1);
% initial population
pop = rand(1,n)*blength + bounds(1);

for it=1:numits
  % fitness eval
  for i=1:n, fpop(i) = feval(f, pop(i)); end
  maxf(it) = max(fpop);
  meanf(it) = mean(fpop);
  % subtract lowest fitness in order to normalize
  m=min(fpop);
  fpop=fpop-m;
  cpop(1) = fpop(1);
  for i=2:n, cpop(i) = cpop(i-1) + fpop(i); end

  % SELECTION
  total_fitness = cpop(n);
  % use roulette selection (-> need pos. fitness!)
  for i=1:n
    p=rand*total_fitness;
    % now find first index
    j=find(cpop-p>0);
    if isempty(j)
        j=n;
    else
        j=j(1);
```

```
  end
    parent(i)=pop(j);
  end
  % pop, fpop, parent, pause

  % REPRODUCTION
  % parents 2i-1 and 2i make two new children

  % 2i-1 and 2i crossover
  % use arithmetic crossover
  for i=1:2:n
    r=rand;
    pop(i) = r*parent(i) + (1-r)*parent(i+1);
    pop(i+1) = (1-r)*parent(i) + r*parent(i+1);
  end

  % mutation
  % use uniform mutation
  for i=1:nummut
    pop(ceil(rand*n)) = bounds(1) + rand*blength;
  end
end

pop
for i=1:n, fpop(i) = feval(f, pop(i)); end
fpop

close all
ezplot(@multipeak,[-10 10])
hold on
  [y,xind]=max(fpop);
plot(pop(xind),y,'ro')

figure, plot(maxf), hold on, plot(meanf,'g');
xlabel('Variable x');
ylabel('Max and Mean of the function');

Plotting the given function:
function y = multipeak(x)
% Evaluation function
y = x + sin(x)+cos(x);
```

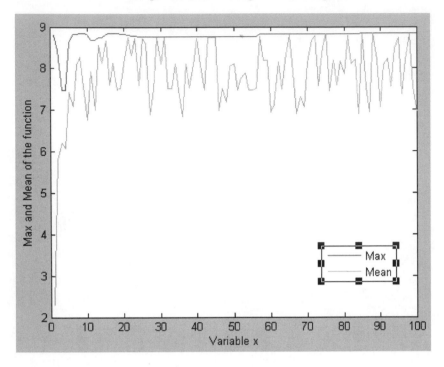

FIGURE 13.4: Maximum and the Mean of the Given Function

Observations:

Figures 13.4 and 13.5 show the maximum and the best fit of the given function.

13.14.2 Illustration 2: Solving Economic Dispatch Problem Using Genetic Algorithm

This program solves the economic dispatch problem by Genetic Algorithm toolbox of MATLAB. The data matrix should have 5 columns of fuel cost coefficients and plant limits.

```
-------------------Main Program-------------------
clear;
clc;
tic;
global data B Pd
data=[0.007 7 240 100 500
```

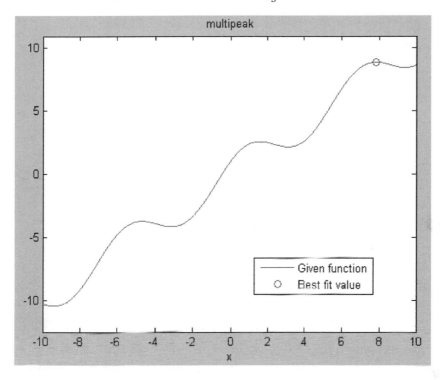

FIGURE 13.5: Plot of the Given Function and the Best Fit

```
0.0095 10 200 50 200
0.009 8.5 220 80 300
0.009 11 200 50 150
0.008 10.5 220 50 200
0.0075 12 120 50 120];
% Loss coefficients it should be square matrix of
% size nXn where n is the no of plants
B=1e-4*[0.14 0.17 0.15 0.19 0.26 0.22
0.17 0.6 0.13 0.16 0.15 0.2
0.15 0.13 0.65 0.17 0.24 0.19
0.19 0.16 0.17 0.71 0.3 0.25
0.26 0.15 0.24 0.3 0.69 0.32
0.22 0.2 0.19 0.25 0.32 0.85
];
% Demand (MW)
Pd=700;
% setting the genetic algorithm parameters.
```

```
options = gaoptimset;
options = gaoptimset('PopulationSize',50,'Generations',
500,'TimeLimit',200,'StallTimeLimit', 100,'PlotFcns',
@gaplotbestf,@gaplotbestindiv);
  [x ff]=ga(@eldga,5,options);
 [ F P1 P1]=eldga(x)
 tic;


----------------------------------------------------
Subprogram
----------------------------------------------------
function[ F P1 P1]=eldga(x)
global data B Pd
x=abs(x);
n=length(data(:,1));
for i=1:n-1
  if x(i)>1;
    x(i)=1;
  else
  end
  P(i)=data(i+1,4)+x(i)*(data(i+1,5)-data(i+1,4));
end

B11=B(1,1);
B1n=B(1,2:n);
Bnn=B(2:n,2:n);
A=B11;
BB1=2*B1n*P';
B1=BB1-1;
C1=P*Bnn*P';
C=Pd-sum(P)+C1;
x1=roots([A B1 C]);
% x=.5*(-B1-sqrt(B1^2-4*A*C))/A
 x=abs(min(x1));
 if x>data(1,5)
  x=data(1,5);
 else
 end
  if x<data(1,4)
x=data(1,4);
  else
  end
 P1=[x P];
```

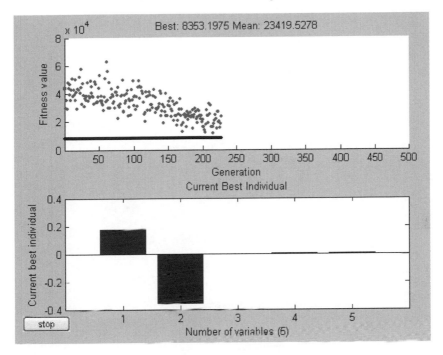

FIGURE 13.6: Best Values of Economic Dispatch Problem

```
for i=1:n
  F1(i)=data(i,1)* P1(i)^2+data(i,2)*P1(i)+data(i,3);
end
P1=P1*B*P1';
 lam=abs(sum(P1)-Pd-P1*B*P1');
 F=sum(F1)+1000*lam;
-----------------End of subprogram------------------
```

Observations:

Figure 13.6 shows the best and mean value of fitness evaluated for solving the economic dispatch problem using genetic algorithms. The total fuel cost at the end of 233 generations was found to be 8.3532x103. The transmission losses involved were 10.7272.

13.14.3   Illustration 3: Traveling Salesman Problem

To find an optimal solution to the Traveling Salesman Problem by using GA to determine the shortest path that is required to travel between cities.

Solution:

The MATLAB function tsp_ga takes input argument as the number of cities. Then the distance matrix is constructed and plotted. The shortest path is then obtained using GA as an optimization tool.

MATLAB Code:

The MATLAB code used to implement the traveling sales man problem is shown below:

```
function varargout = tsp_ga(varargin)
% the function tsp_ga finds a (near) optimal
% solution to the Traveling Salesman Problem
% by setting up a Genetic Algorithm (GA) to
% search for the shortest path (least distance
% needed to travel to each city exactly once)
```

The input argument to the MATLAB function tsp_ga can be any one of the following:

- tsp_ga(NUM_CITIES) where NUM_CITIES is an integer representing the number of cities (default = 50) For example tsp_ga(25) solves the TSP for 25 random cities

- tsp_ga(CITIES) where CITIES is an Nx2 matrix representing the X/Y coordinates of user specified cities For example tsp_ga (10 * RAND (30,2)) solves the TSP for the 30 random cities in the 10*RAND(30,2)) matrix

- tsp_ga(OPTIONS) or tsp_ga(OPTIONS) where OPTIONS include one or more of the following in any order:

- '-NOPLOT' turns off the plot showing the progress of the GA

- -RESULTS' turns on the plot showing the final results as well as the following parameter pairs:

- 'POPSIZE', VAL sets the number of citizens in the GA population VAL should be a positive integer (divisible by 4) – default = 100

- 'MRATE', VAL sets the mutation rate for the GA. VAL should be a float between 0 and 1, inclusive – default = 0.85

- 'NUMITER', VAL sets the number of iterations (generations) for the GA VAL should be a positive integer – default = 500

- For example, the following code solves the TSP for 20 random cities using a population size of 60, a 75% mutation rate, and 250 GA iterations

```
tsp_ga(20, 'popsize', 60, 'mrate', 0.75, 'numiter',
250);
```
The following code solves the TSP for 30 random cities without the progress plot
```
[sorted_cities, best_route, distance] = tsp_ga(30,
'-noplot');
```
The following code solves the TSP for 40 random cities using 1000 GA iterations and plots the results
```
cities = 10*rand(40, 2);
[sorted_cities] = tsp_ga(cities, 'numiter', 1000,
'-results');
```

Declare the parameters used in TSP

```
error(nargchk(0, 9, nargin));
num_cities = 50;
cities = 10*rand(num_cities, 2);
pop_size = 100;
num_iter = 500;
mutate_rate = 0.85;
show_progress = 1;
show_results = 0;

Declarethe Pxrocess Inputs
cities_flag = 0;
option_flag = 0;
for var = varargin
  if option_flag
    if ~isfloat(var1), error(['Invalid value for
      option' upper(option)]);
    end
    switch option
```

```
      case 'popsize',pop_size =
                          4*ceil(real(var1(1))/4);
        option_flag = 0;
      case 'mrate', mutate_rate =
                        min(abs(real(var1(1))), 1);
        option_flag = 0;
      case 'numiter', num_iter =
                          round(real(var1(1)));
        option_flag = 0;
      otherwise, error(['Invalid option '
                            upper(option)])
    end
  elseif ischar(var1)
    switch lower(var1)
      case '-noplot', show_progress = 0;
      case '-results', show_results = 1;
      otherwise, option = lower(var1);
                            option_flag = 1;
    end
  elseif isfloat(var1)
    if cities_flag, error('CITIES or NUM_CITIES
          may be specified, but not both');
    end
    if length(var1) == 1
      num_cities = round(real(var1));
      if num_cities < 2, error('NUM_CITIES must be
          an integer greater than 1');
      end
      cities = 10*rand(num_cities, 2);
                        cities_flag = 1;
    else
      cities = real(var1);
      [num_cities, nc] = size(cities);
                          cities_flag = 1;
      if or(num_cities < 2, nc ^= 2)
      error('CITIES must be an Nx2 matrix of
                  floats, with N > 1')
    end
  end
else
  error('Invalid input argument.')
 end
end
```

```
Construction of the Distance Matrix by using Distance
measurement formula dist_matx = zeros(num_cities);
for ii = 2:num_cities
 for jj = 1:ii-1
   dist_matx(ii, jj) =
       sqrt(sum((cities(ii, :)-cities(jj, :)).^2));
   dist_matx(jj, ii) = dist_matx(ii, jj);
 end
end
The cities and the distance matrix are plotted.
The plot is shown in Figure 13.7.
if show_progress
 figure(1)
 subplot(2, 2, 1)
 plot(cities(:,1), cities(:,2), 'b.')
 if num_cities < 75
   for c = 1:num_cities
     text(cities(c, 1),cities(c, 2),[' ' num2str(c)],
       'Color', 'k', 'FontWeight', 'b')
   end
 end
 title([num2str(num_cities) ' Cities'])
 subplot(2, 2, 2)
 imagesc(dist_matx)
 title('Distance Matrix')
 colormap(flipud(gray))
end

Initialize Population in a random manner
pop = zeros(pop_size, num_cities);
pop(1, :) = (1:num_cities);
for k = 2:pop_size
 pop(k, :) = randperm(num_cities);
end

Calculation of the best route

if num_cities < 25, display_rate = 1;
else
display_rate = 10; end
fitness = zeros(1, pop_size);
best_fitness = zeros(1, num_iter);
```

```
for iter = 1:num_iter
 for p = 1:pop_size
   d = dist_matx(pop(p, 1),pop(p,num_cities));
   for city = 2:num_cities
     d = d + dist_matx(pop(p, city-1), pop(p, city));
   end
   fitness(p) = d;
 end
 [best_fitness(iter) index] = min(fitness);
 best_route = pop(index, :);
```

The best GA route gcalculated from the previous step is computed and plotted.

```
 if and(show_progress, ~mod(iter, display_rate))
    figure(1)
    subplot(2, 2, 3)
    route = cities([best_route best_route(1)], :);
    plot(route(:, 1), route(:, 2)', 'b.-')
    title(['Best GA Route (dist = ' num2str
    (best_fitness(iter)) ')']) subplot(2, 2, 4)
    plot(best_fitness(1:iter), 'r', 'LineWidth', 2)
    axis([1 max(2, iter) 0 max(best_fitness)*1.1])
 end

% Genetic Algorithm Search
 pop = iteretic_algorithm(pop, fitness, mutate_rate);
end

% Plotting the best fitness. The plot is shown in
Figure 13.7 if show_progress
 figure(1)
 subplot(2, 2, 3)
 route = cities([best_route best_route(1)], :);
 plot(route(:, 1), route(:, 2)', 'b.-')
 title(['Best GA Route (dist = ' num2str(best_fitness
                            (iter)) ')'])
 subplot(2, 2, 4)
 plot(best_fitness(1:iter), 'r', 'LineWidth', 2)
 title('Best Fitness')
 xlabel('Generation')
 ylabel('Distance')
 axis([1 max(2, iter) 0 max(best_fitness)*1.1])
end
```

```matlab
if show_results
 figure(2)
 imagesc(dist_matx)
 title('Distance Matrix')
 colormap(flipud(gray))
 figure(3)
 plot(best_fitness(1:iter), 'r', 'LineWidth', 2)
 title('Best Fitness')
 xlabel('Generation')
 ylabel('Distance')
 axis([1 max(2, iter) 0 max(best_fitness)*1.1])
 figure(4)
 route = cities([best_route best_route(1)], :);
 plot(route(:, 1), route(:, 2)', 'b.-')
 for c = 1:num_cities
   text(cities(c, 1),cities(c, 2),[' ' num2str(c)],
        'Color', 'k','FontWeight', 'b')
 end
 title(['Best GA Route (dist =
            ' num2str(best_fitness(iter)) ')'])
end

 [not used indx] = min(best_route);

best_ga_route =
   [best_route(indx:num_cities) best_route(1:indx-1)];
if best_ga_route(2) > best_ga_route(num_cities)
 best_ga_route(2:num_cities) = fliplr(best_ga_route
                               (2:num_cities));
end
varargout1 = cities(best_ga_route, :);
varargout2 = best_ga_route;
varargout3 = best_fitness(iter);
```

The genetic algorithm search function is implemented using the MATLAB Code shown below:

```matlab
function new_pop =
      iteretic_algorithm(pop, fitness, mutate_rate)
 [p, n] = size(pop);

Tournament Selection - Round One
new_pop = zeros(p, n);
```

```
ts_r1 = randperm(p);
winners_r1 = zeros(p/2, n);
tmp_fitness = zeros(1, p/2);
for i = 2:2:p
 if fitness(ts_r1(i-1)) > fitness(ts_r1(i))
   winners_r1(i/2, :) = pop(ts_r1(i), :);
   tmp_fitness(i/2) = fitness(ts_r1(i));
 else
   winners_r1(i/2, :) = pop(ts_r1(i-1), :);
   tmp_fitness(i/2) = fitness(ts_r1(i-1));
   end
end
Tournament Selection - Round Two
ts_r2 = randperm(p/2);
winners = zeros(p/4, n);
for i = 2:2:p/2
 if tmp_fitness(ts_r2(i-1)) > tmp_fitness(ts_r2(i))
   winners(i/2, :) = winners_r1(ts_r2(i), :);
 else
   winners(i/2, :) = winners_r1(ts_r2(i-1), :);
 end
end
new_pop(1:p/4, :) = winners;
new_pop(p/2+1:3*p/4, :) = winners;

Crossover
crossover = randperm(p/2);
children = zeros(p/4, n);
for i = 2:2:p/2
 parent1 = winners_r1(crossover(i-1), :);
 parent2 = winners_r1(crossover(i), :);
 child = parent2;
 ndx = ceil(n*sort(rand(1, 2)));
 while ndx(1) == ndx(2)
   ndx = ceil(n*sort(rand(1, 2)));
 end
 tmp = parent1(ndx(1):ndx(2));
 for j = 1:length(tmp)
   child(find(child == tmp(j))) = 0;
 end
 child = [child(1:ndx(1)) tmp child(ndx(1)+1:n)];
 child = nonzeros(child)';
 children(i/2, :) = child;
```

```
end
new_pop(p/4+1:p/2, :) = children;
new_pop(3*p/4+1:p, :) = children;

Mutate
mutate = randperm(p/2);
num_mutate = round(mutate_rate*p/2);
for i = 1:num_mutate
  ndx = ceil(n*sort(rand(1, 2)));
  while ndx(1) == ndx(2)
    ndx = ceil(n*sort(rand(1, 2)));
  end
  new_pop(p/2+mutate(i), ndx(1):ndx(2)) = ...
    fliplr(new_pop(p/2+mutate(i), ndx(1):ndx(2)));
end
```

Output:

Thus the traveling salesman problem was executed in MATLAB and the results are plotted indicating the number of cities, Distance Matrix, the best GA route, and the best fitness.

Summary

Thus it is observed that genetic algorithms are robust, useful, and are the most powerful apparatus in detecting problems in an array of fields. In addition, genetic algorithms unravel and resolve an assortment of complex problems. Moreover, they are capable of providing motivation for their design and foresee broad propensity of the innate systems. Further, the reason for these ideal representations is to provide thoughts on the exact problem at hand and to examine their plausibility.

Threfore they are employed as a computer encode and identify how the propensities are affected from the transformations with regards to the model. Without genetic algorithms, it is not possible to solve real world issues. The genetic algorithm methods may permit researchers to carry out research, which was not perceivable during this evolutionary technological era. Therefore, one is able to replicate this phenomenon, which would have been virtually impossible to obtain or analyze through

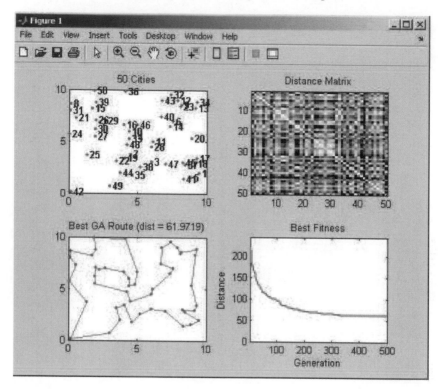

FIGURE 13.7:   Plot of Cities, Distance Matrix, Best GA Route and Best Fitness for Traveling Salesman Problem

traditional methods or through the analysis of certain equations. Because GAs form a subset field of evolutionary computation, optimization algorithms are inspired by biological and evolutionary systems and provide an approach to learning that is based on simulated evolution.

Review Questions

1. What do you mean by Calculus-based schemes and Enumerative-based optimization schemes?

2. Define Genetic Algorithm.

3. Explain the functionality of GA with a suitable flowchart.

4. Mention some of the common genetic representations.

5. State Schema Theorem.

6. Derive schema theorem using the mathematical model.

7. How are the destructive effects of schema theorem compensated? Explain in terms of crossover and mutation.

8. Mention the considerations while constructing a fitness function.

9. Explain inversion and reordering.

10. Write a note on epistatis.

11. Define deception.

12. What do you mean by naïve evolution?

13. Define the process speciation.

14. Explain briefly on restricted mating, diploidy, and dominance.

15. What are the important issues that are considered while implementing a GA?

16. Compare GA with Neural Nets, Random Search, and Simulated Annealing.

17. Mention the advantages of GA.

18. Mention a few application areas of GA.

Chapter 14

Genetic Programming

14.1 Introduction

Genetic programming is a predefined and systematic method to obtain a solution to a problem automatically using computers. Genetic Programming (GP) is a member of evolutionary computation or evolutionary algorithms and this method also follows Darwin's theory of evolution the "survival of the fittest". A set of computer programs are chosen as population of individuals and these individuals reproduce among themselves. As this process increases, the best individuals will survive and seem to be successful in evolving well in the given environment.

GP definition - *"Genetic programming is an automated method for creating a working computer program from a high-level problem statement of a problem. Genetic programming does this by genetically breeding a population of computer programs using the principles of Darwinian natural selection and biologically inspired operations."*

Practically, the genetic programming model can be implemented with arrays of bits or characters to represent the chromosomes. The genetic operations such as crossover, mutation, etc., can be performed using simple bit manipulations. Genetic Programming is the extension of the genetic model of learning into the space of programs. The components that represent the population are not fixed-length character strings and they encode possible solutions to the given problem. These components in the population are programs and when they are executed, they are the candidate solutions to the problem. These programs are expressed in genetic programming as parse trees, rather than as lines of code.

Genetic Programming generates programs by following an evolutionary approach. The process is as follows: The task or the problem that is to be solved should be specified by the user along with the evaluation function. The user has to specify the kind of operation that is to be performed by the programs. Once the specifications are provided, an initial population of programs is randomly generated. Each and every program undergoes a translation, compilation, and execution process.

The performance of these tasks depends on the problem assessed. This process enables the calculation of a fitness value for each of the programs, and the best of them are chosen for reproduction. The chosen programs undergo a mutation operation and the offspring produced are added to the next generation of programs. This process repeats until a termination condition is attained. In this section, a brief history of genetic programming is discussed. To get an idea about programming a basic introduction to Lisp Programming Language is dealt. The basic operations of GP are discussed along with an illustration.

The programs in GP are evolved to solve pre-defined problems. The term evolution refers to an artificial process modeled from natural evolution of living organisms. This process has been abstracted and stripped off of most of its intricate details. It has been transferred to the world of algorithms where it can serve the purpose of approximating solutions to given or even changing problems (machine learning) or for inducing precise solutions in the form of grammatically correct (language) structures (automatic programming).

Genetic programming is a domain-independent method that genetically breeds a population of computer programs to solve a problem. Moreover, genetic programming transforms a population of computer programs into a new generation of programs by applying analogs of naturally occurring genetic operations iteratively. This process is illustrated in Figure 14.1.

The genetic operations include crossover (sexual recombination), mutation, reproduction, gene duplication, and gene deletion. Analogs of developmental processes are sometimes used to transform an embryo into a fully developed structure. Genetic programming is an extension of the genetic algorithm in which the structures in the population are not fixed-length character strings that encode candidate solutions to a problem, but programs that, when executed, are the candidate solutions to the problem.

The programs are represented in a tree form in GP, which is the most common form, and the tree is called program tree (or parse tree or syntax tree). Some alternative program representations include finite automata (evolutionary programming) and grammars (grammatical evolution). For example, the simple expression min(x/y*5, x+y)is represented as shown in Figure 14.2. The tree includes *nodes* (which are also called *points*) and *links*. The nodes indicate the instructions to execute. The links indicate the arguments for each instruction. In the following, the internal nodes in a tree will be called *functions*, while the tree's leaves will be called *terminals*.

The trees and their expressions in genetic programming can be represented using prefix notation (e.g., as Lisp S-expressions). A basic idea of

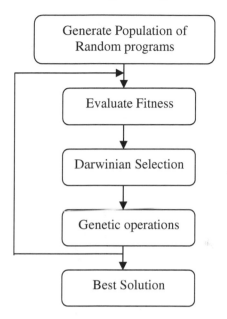

FIGURE 14.1: Main Loop of Genetic Programming

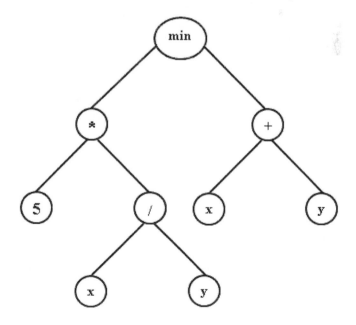

FIGURE 14.2: Basic Tree-Like Program Representation Used in Genetic Programming

lisp programs is required to understand the representations and programming of genetic programming. A brief description about Lisp programming is discussed in Section 14.3. In prefix notation, functions always precede their arguments. In this notation, it is easy to see the correspondence between expressions and their syntax trees. Simple recursive procedures can convert prefix-notation expressions into infix-notation expressions and vice versa.

14.2   Growth of Genetic Programming

The emergence of Genetic Programming started with the utilization of evolutionary algorithms by Nils Aall Barricelli in 1954. The application demanded evolutionary simulations and evolutionary algorithms which were widely accepted as standard optimization methods with reference to the work of Ingo Rechenberg in the 1960s and early 1970s. His team was capable of solving complex engineering problems through evolution strategies. In addition to Rechenberg's work, John Holland's work was the extremely influential one during the early 1970s.

The introductory results on the GP methodology were described by Stephen F. Smith in 1980. Later during 1981 Forsyth reported the evolution of small programs in forensic science for the UK police. The initial statement of modern Genetic Programming stating that, "GP are procedural languages coordinated in tree-based structures and functioned by suitably defined GA-operators", was contributed by Nichael L. Cramer in 1985, and independently by Jrgen Schmidhuber in 1987. Based on these works, John R. Koza, a primary proponent of GP initiated the practical application of genetic programming in several complex optimization and search problems. It ought to be noted that Koza states GP as a generalization of genetic algorithms rather than a specialization.

GP is computationally more intensive and therefore in the 1990s it was primarily employed to work out fundamental elementary problems. Recently due to the exponential growth in CPU power, GP has produced numerous novel and more prominent results in fields such as quantum computing, electronic design, game playing, sorting, searching, and many others. These consequences admit the replication or growth of several post-year-2000 inventions. GP has also been employed to evolvable hardware in addition to computer programs. Initially, formulating a hypothesis for GP has been really difficult and so in the 1990s GP was considered a sort of outcast among search techniques. Just afterwards, a series of breakthroughs occurred in the early 2000s, the theory of GP has had a impressive and fast development, therefore it has been feasible to build precise probabilistic models of GP.

14.3 The Lisp Programming Language

During the late 1950s, a family of programming languages named as Lisp was framed by John McCarthy. The basic computation mechanism of Lisp programs is recursion. The language is implemented to support a wide range of platforms. Based on the computer available the user can execute Lisp in several methods. Likewise there are a few minimal modifications between the LISP interpreters. The two basic types of data structures available in Lisp are the atom and the list. Atoms can be either symbols or numbers. Lists are linked lists where the elements of the list are either atoms or other lists. Lists of atoms are written as follows:

$$(1 \text{ B D } 5)$$

Nested list structures (lists with lists as elements) are written as follows:

$$(A (1\ 2) \text{ D } (4 (6 \text{ G})))$$

This example is a list of four elements. The first is the atom A; the second is the sublist (1 2); the third is the atom D, and the fourth is the sublist (4 (6 G)). Internally, lists are usually represented as single-linked lists. Each node of the list consists of two pointers. The first pointer points either to the corresponding atom or to the corresponding sublist. The second pointer points to the next node in the list.

Pure Lisp is a functional programming language. A functional program is computed by application of functions to arguments. Parameters such as symbols, numbers, or other function calls can be used as arguments. There is no need of assignment statements or variables for computation. Thus, a Lisp program is just a collection of functions that can call each other. Functions in Lisp are represented in the same way as data, namely as nested list structures. Here is a function that squares its argument x:

$$(\text{defun square } (x)(* \text{ x x}))$$

The defun can be thought of as a key word that indicates that a function definition follows. The name of the function is square. Then follows a list of arguments (dummy parameters). Finally is the body of the function. The function returns the value resulting from multiplying x times x.

```
Hereis a slightly more complex example that
   uses the function square:
      (defun hypotenuse(a b))
         (sqrt (+ (square a) (square b))))

   The function to find the length is defined
      as (defun list-length(x)
      (if (null x)
         0
         (+ 1 (length(rest x)))))
```

If the list is null(i. e., empty), then the length is 0. Otherwise, the length is 1 plus the length of the list with the first element removed. Since Lisp functions are stored as data, it is very easy to write functions that define and manipulate other functions. This is exactly what genetic programming does. It is relatively easy to write a simple genetic programming package in Lisp.

14.4   Functionality of Genetic Programming

The operational steps such as creating an individual, creating a random population, fitness test, functions and terminals, the genetic operations, selection functions, crossover operation, mutation and user decisions of a typical genetic programming are described in this section (Figure 14.3).

14.4.1   Generation of an Individual and Population

The basic elements of an individual are its genes. These genes are combined together to form a program. An individual's program is a tree-like structure and as such there are two types of genes: *functions* and *terminals*.

Terminals are defined as leaves or nodes without branches and functions are defined as the nodes with children. The node's children provide the arguments for the function. In the example (Figure 14.4) there are three functions (−, + and /) and four terminals (x, 3, y and 6). The / (division) function requires two arguments: the return value of "−" and the return value of the subtree whose root is "+". The − (subtraction) function takes two arguments which are provided by x and 3. The + (addition) function also takes two arguments which are provided by y and 6.

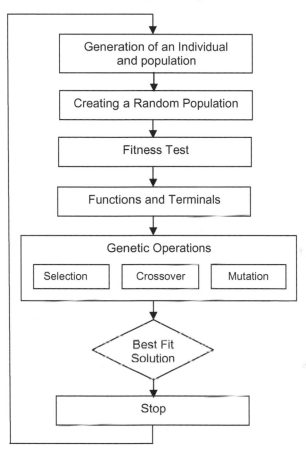

FIGURE 14.3: Functionality of Genetic Programming

The example may be interpreted as $(x-3)/(y+6)$. The genes that will be available to the GP system must be selected or created by the user. This is an important decision as poor selection may well render the system incapable of evolving a solution.

14.4.2 Creating a Random Population

Once the genes are chosen a random population is created. To create random population there are three basic techniques such as grow, full, and ramped-half-and-half:

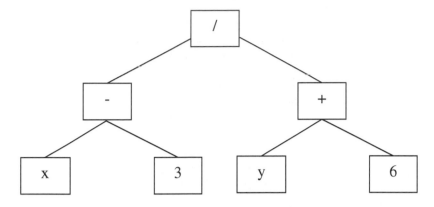

FIGURE 14.4: A Typical Simple Individual That Returns(x-3)/(y+6)

Grow

In the grow technique the whole population is created, and the individuals are created one at a time. The individual created with this technique is a tree of any depth up to a specified maximum, m. The nodes which are created are either terminal or functional nodes. If the node is a terminal, a random terminal is chosen. If the node is a functional node, a random function is chosen, and that node is given a number of children equal to the number of arguments of the function. The algorithm starts for every function nodes' children, and proceeds until the child depth is m. This method does not guarantee individuals of a certain depth (although they will be no deeper than m). Instead it provides a range of structures throughout the population. This technique tends to produce individuals containing only one (terminal) node. Such individuals are quickly bred out if the problem is non-trivial, and therefore not really valuable.

Full

This technique overcomes the limitation of the grow method. In the full method the terminals are guaranteed to be a certain depth. This guarantee does not specify the number of nodes in an individual. This method requires a final depth, d.

1. Every node, starting from the root, with a depth less than d, is made a randomly selected function. If the node has a depth equal to d, the node is made a randomly selected terminal.

2. All functions have a number (equal to the arity of the function) of child nodes appended, and the algorithm starts again. Thus, only if d is specified as one, could this method produce a one-node tree.

Ramped-half-and-half

This technique was used by Koza, To increase the variation in structure both grow and full methods can be used in creating the population-this technique, ramped-half-and-half, is the sole method. Only a maximum depth, md, is specified but the method generates a population with a good range of randomly sized and randomly structured individuals.

1. The population is evenly divided into parts: a total of md-1.

2. One half of the population is produced by the grow method and the other half is produced using the full method. For the first part, the argument for the grow method, m, and the argument for the full method, is d. This continues to part md-1, where the number md is used. Thus a population is created with good variation, utilizing both grow and full methods.

14.4.3   Fitness Test

After the initial random population has been created, the individuals are required to be evaluated for their fitness. Evaluating the fitness function of a program requires how well a program performs into a numerical value. Generally, this requires executing the program several times with different parameters, and evaluating the output each time.

14.4.4   Functions and Terminals

The most important components of genetic programming are the terminal and function sets. All the variables and constants defined by the user are contained in the terminal set while all the functions, subroutines are in the function set. The functions can be any mathematical functions, such as addition, subtraction, division, multiplication, and other more complex functions.

14.4.5   The Genetic Operations

The evolutionary process starts, as soon as the fitness test to all the individuals in the initial random population is completed. The individuals in the new population are formed by genetic operators like *reproduction, crossover,* and *selection.* Whenever a new population is created the old population is destroyed.

14.4.6   Selection Functions

Various selection functions can be used depending upon the application, among which fitness proportionate selection, greedy over-selection, and tournament selection are the most common.

Fitness-Proportionate Selection

In this method, the individuals are selected depending on their ability by comparing them with the entire population. Due to this comparison, the best individual of a population is likely to be selected more frequently than the worst. The selection probability is computed with the following algorithm:

1. The raw fitness is restated in terms of standardized fitness. A lower standardized fitness value implies a better individual. If the raw fitness increases as an individual improves then an individual's standardized fitness is the maximum raw fitness (i.e., the fitness of the best individual in the population) minus the individual's raw fitness. If the raw fitness decreases as an individual improves, standardized fitness for an individual is equal to the individual's raw fitness.

2. Standardized fitness is then restated as adjusted fitness, where a higher value implies better fitness. The formula used for this is:

$$\text{adj(i)} = \frac{1}{1 + std(i)} \qquad (14.1)$$

where adj(i) is the adjusted fitness and std(i) is the standardized fitness for individual i. The use of this adjustment is beneficial for separation of individuals with standardized fitness values that approach zero.

3. Normalized fitness is the form used by both selection methods. It is calculated from adjusted fitness in the following manner:

$$\text{norm(i)} = \frac{adj(i)}{\sum_{k=1}^{M} adj(k)} \qquad (14.2)$$

where norm(i) is the normalized fitness for individual i, and M is the number of individuals in the population.

4. The probability of selection (sp) is:

$$\text{sp(i)} = \frac{norm(i)}{\sum_{k=1}^{M} norm(k)} \qquad (14.3)$$

This can be implemented by:

(a) Order the individuals in a population by their normalized fitness (best at the top of the list).

(b) Chose a random number, r, from zero to one.

(c) From the top of the list, loop through every individual keeping a total of their normalized fitness values. As soon as this total exceeds r stop the loop and select the current individual.

Greedy Over-Selection

Greedy over-selection is mainly applied to reduce the number of generations required for a GP run. The individuals are selected based on their performance but this method biases selection toward the highest performers. The normalized fitness is calculated for every individual.

1. Using the normalized fitness values, the population is divided into two groups. Group I includes the top 20% of individuals while Group II contains the remaining 80%.

2. Individuals are selected from Group I 50% of the time. The selection method inside a group is fitness-proportionate.

Tournament Selection

In this method, pairs of individuals are chosen at random and the most fit one of the two is chosen for reproduction. If two unfit individuals are paired against each other, one of them is guaranteed to reproduce. Sometimes individuals are chosen by assigning a rank using the evaluation function and the ones at the top of the ranking are chosen, which implies that only the best will be chosen for reproduction. As with most AI applications, it's a question of trying out different approaches to see which works for a particular problem.

14.4.7 MATLAB Routine for Selection

```
function selm = gpols_selection(popu,gap,pc,pm,tsels);
% Selection operator, mix the new generation
% selm = gpols_selection(popu,ggap,pc,pm,tsles)
% selm <- matrix (n x 3)
% popu - >population
% ggap - >generation gap (0-1)
% pc - >crossover probability (0-1)
```

```
% pm - >mutation probability (0-1)
% tsels - >selection (integer)
%
% Remark:
% if tsels = 0 - >roulette wheel selection
% if tsels = 1 - >total random selection
% if tsels >= 2 - >tournament selection, tsels = tournament
% size Example values: ggap = 0.8, pc = 0.7, pm = 0.3,
% tsels = 2 Columns of selm:
% 1: index of individual (first parent)
% 2: index of second ind. (second parent) (if crossover)
% 3: 0: direct rep., 1: crossover, 2: mutation
%

% Begin
popun = popu.size;
selm = zeros(popun,3);

% Fitness values and sort fit = zeros(1,popun);
for i = 1:popun,

    fit(i) = popu.chromi.fitness;
end
    if ~isempty(find(fit<0)),s
end
 [fitsort,sortix] = sort(-fit);
fitsort = -fitsort./sum(-fitsort);
fitsum = cumsum(fitsort);
fitsum(end) = 1; % avoid ~1E-16 error from representation

% Copy elite indv.s
i = floor((1-gap)*popun);
if i>=1 & i<=popun,
    selm(1:i,1) = sortix(1:i)';
    i = i+1;
else
    i = 1;
end
nn = i-1;

% New individuals
while nn<popun,
    if tsels > 0,
```

```
      j1 = tournament(fit,tsels);
    else
      j1 = roulette(fitsum,sortix);
    end
    % Select a method
    r = rand;
    if r<pc,
      % Crossover
      if tsels > 0,
        j2 = tournament(fit,tsels);
      else
        j2 = roulette(fitsum,sortix)
      end
      selm(i,1) = j1;
      selm(i,2) = j2;
      selm(i,3) = 1;
      i = i+1;
      nn = nn+2;
    elseif r<pc+pm,
      % Mutation
      selm(i,1) = j1;
      selm(i,3) = 2;
      i = i+1;
      nn = nn+1;
    else
      % Direct copy
      selm(i,1) = j1;
      selm(i,3) = 0;
      i = i+1;
      nn = nn+1;
    end
end
selm = selm(1:i-1,:);
%-----------------------------------------------
function j = tournament(fit,tsels);

n = length(fit);
jj = floor(rand(tsels,1)*n)+1;
 [fitmax,maxix] = max(fit(jj));
j = jj(maxix);
%-----------------------------------------------
function j = roulette(fitsum,sortix);
```

```
v = find(fitsum >= rand(1,1));
j = sortix(v(1));
```

14.4.8   Crossover Operation

The structures in genetic programming are altered using two primary genetic operators, crossover and mutation. In the crossover operation, two parent solutions are sexually combined to form two new offspring solutions. The fitness function is evaluated and the parents are chosen. There are three methods for selecting the solutions for the crossover operation. The first method is based on the fitness probability of the solution. If $f(s_j(t))$ is the fitness of the solution s_i and $\sum_{j=1}^{M} f(s_j(t))$ is the total sum of all the members of the population, then the probability that the solution si will be copied to the next generation is:

$$\frac{f(s_i(t))}{\sum_{j=1}^{M} f(s_j(t))} \tag{14.4}$$

The second method for selecting the solution is tournament selection. The genetic program selects two random solutions among which the solution with the higher fitness wins. This technique simulates biological mating patterns where two members of the same sex compete to mate with a third one of a different sex.

The last method is done by ranking the fitness function of the solution of the population.

Offspring are created in crossover operation by deleting the crossover fragment of the first parent and then inserting the crossover fragment of the second parent. Similarly a second offspring is also created. Consider the two expressions in Figure 14.5, where the expressions are represented in a tree form.

Given a tree based GP the crossover operator works in the following manner:

1. Select two individuals, based on the selection mechanism.

2. Select a random subtree in each parent.

3. Swap the two subtrees or sequences of instructions.

The crossover operator closely mimics the process of biological sexual reproduction. With this biological process as a base, the crossover operator has been used to claim that GP search is more efficient than methods based solely on mutation.

Figure 14.5 illustrates one of the main advantages of genetic programming over genetic algorithms. In genetic programming identical parents

Crossover Operation with different parents

Parents

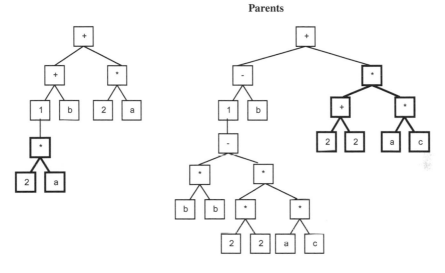

Children

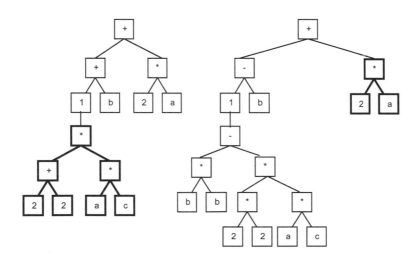

FIGURE 14.5: Crossover operation

Computational Intelligence Paradigms

Crossover Operation with identical parents

Parents

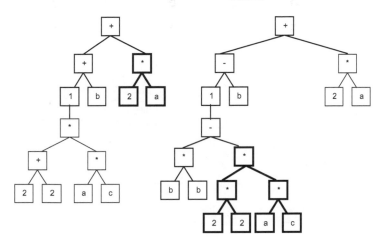

Children

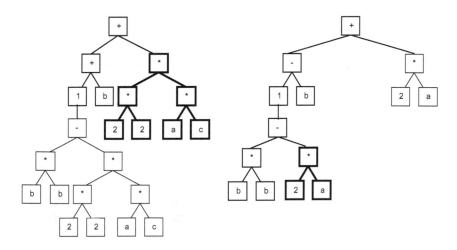

FIGURE 14.5: (*Continued*)

can yield different offspring, while in genetic algorithms identical parents would yield identical offspring. The bold selections indicate the swapped subtrees.

14.4.9 MATLAB Routine for Crossover

```
function [tree1,tree2] = recombinate_trees
(treein1,treein2
% ,mode,symbols);
% Recombinates two trees
% [tree1,tree2] = recombinate_trees(treein1,treein2,
mode,
% symbols)
% tree1,tree2 <- two childs
% treein1,treein2 -> two parents
% mode -> 1: one-point- 2: two-point crossover
% symbols -> cell arrays of operator and terminator
node
% strings
%

% Begin
tree1 = treein1;
tree2 = treein2;
nn = [length(symbols1), length(symbols2)];
% Calculate indexes
switch mode,
    case 1,
            [n,v1] = tree_size(tree1);
            [n,v2] = tree_size(tree2);
            n = max([tree1.maxsize, tree2.maxsize]);
            dummy1 = zeros(n,1);
            dummy2 = zeros(n,1);
            dummy1(v1) = 1;
            dummy2(v2) = 1;
            v = find((dummy1+dummy2)==2);
            ix1 = v(floor(rand*(length(v))+1));
            ix2 = ix1;
    case 2,
            [n,v] = tree_size(tree1);
            ix1 = v(floor(rand*(length(v))+1));
            [n,v] = tree_size(tree2);
            ix2 = v(floor(rand*(length(v))+1));
```

```
    otherwise,
        return;
end
% Repleace subtrees (recombinate)
sub1 = tree_getsubtree(treein1,ix1);
sub2 = tree_getsubtree(treein2,ix2);
tree1 = tree_inserttree(sub2,tree1,ix1,nn(2));
tree2 = tree_inserttree(sub1,tree2,ix2,nn(2));
```

14.4.10   Mutation Operation

In genetic programming two kinds of mutations are possible. In the first technique a function always replaces a function and a terminal replaces a terminal only. Whereas in the second technique, the entire subtree can replace another subtree. Figure 14.6 explains the concept of mutation:

14.4.11   MATLAB Routine for Mutation

```
function tree = tree_mutate(treein,symbols);
% Mutates a tree (mutates one randomly selected node)
% tree = tree_mutate(treein,symbols)
%  tree <- the output tree
%  treein ->the input tree
%  symbols ->cell arrays of operator and terminator
%  node strings %
% Begin
tree = treein;
nn = [length(symbols1), length(symbols2)];
% Mutate one node
 [n,v] = tree_size(tree);
i = v(floor(rand*(length(v))+1));
if i<(tree.maxsize+1)/2 & rand<0.5,
   [tree.nodetyp(i) tree.node(i)] = tree_genrndsymb
                   ((tree.nodetyp (i)==1),nn); else
   while tree.node(i)==treein.node(i) & tree.nodetyp
                   (i)==treein.nodetyp(i),
   [tree.nodetyp(i) tree.node(i)] =
           tree_genrndsymb((tree.nodetyp(i)~=1),nn);
   end
end
%-------------------------------------------------------
function [nodetyp,node] = tree_genrndsymb(p0,nn)
```

Mutation Operation

Original Individual

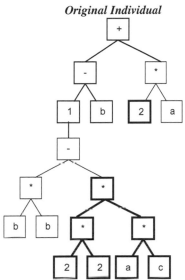

Mutated Individual

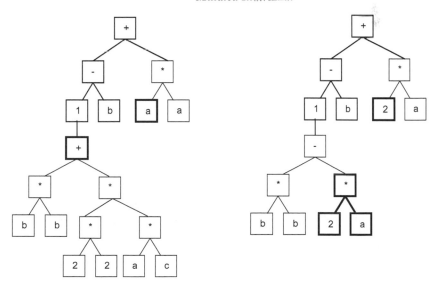

FIGURE 14.6: Mutation Operation

```
% Generate a random symbol (terminate or operate)
% [nodetyp,node] = tree_genrndsymb(p0,nn)
% nodetyp,node <- results
% p0 ->probability of terminate node
% nn ->vector [number of operators, variables]
%

if rand<p0,
   nodetyp = 2;
else
   nodetyp = 1;
end
node = floor(nn(nodetyp)*rand)+1;
```

14.5   Genetic Programming in Machine Learning

In several fields like artificial intelligence, machine learning, or symbolic processing, several problems originate whose resolution can be considered as the search of a computer program, within a space of feasible programs that produce a few desired outputs from the inputs. This search should be executed in such a way that the searched program is the more capable for the problem that is considered. The genetic programming (GP) paradigm furnishes the appropriate framework to implement this type of search in an efficient and flexible mode, because it can adapt to any problem type.

The paradigm of genetic programming is based on the principle of survival of fittest (C. Darwin). Beginning from a randomly-generated initial population, it acquires populations adopting this principle. The new individuals are a product of genetic operations on the current population's better individuals. In association with the genetic algorithms (GAs), GP contributes the philosophy and the characteristics of being heuristic and stochastic.

Within the area of the machine learning, various paradigms are concentrated toward the resolution of problems. In each paradigm the applied structures are different.

Connectionist Model

Here the value of the signal is amplified or diminished with strengths known as weights which are real valued. The solution to the problem is given as a group of real-valued weights.

Application field: NEURAL NETWORKS

Evolutionary Model

In this model, the solutions are fixed-length strings. Each chromosome represents a possible solution to the problem. A conventional genetic algorithm is applied to obtain the best solution (or a good enough solution) among all possible solutions.

Application field: GENETIC ALGORITHMS

Inductive Model

According to this paradigm the solutions to a certain problem are given by decision trees. Each one of these trees classifies each instance of the problem in classes, for which a possible solution exists.

Application field: CLASSIFIER SYSTEMS

All the above mentioned models can be more or less effective in solving a certain problem type. The approach that is used to determine the efficiency of a method is, in the first place, the flexibility to adapt to several types of problems, and in second its easiness to represent the solutions to this problem in a natural and comprehensible way for the user. Computer programs offer flexibility:

- To perform operations with variables of different types.

- To carry out operations conditioned to the results obtained in intermediate points.

- To carry out iterations and recursions.

- To define routines that can be used later on.

Flexibility of a solution includes the concepts of flexibleness in the size, the form, and the structural complexity of the solution. The user should avoid expressing any kind of explanation or previous imposition on the size or form of the result. The true ability of GP resides in its capability of adaptating to the problem type, for what the conditions on the size, the complexity, or the form of the result should come out during its own resolution process. The importance of the representation of the solutions dwells in that the genetic algorithms keep in line the structure of the representation directly and not its own solution.

String-representations do not instantly provide the hierarchical data structure of programs. Likewise, the adaptation of the form, size, and complexity of individuals gets really hard. Consequently, GP has extended toward more complex representations that contribute the necessitous flexibility.

GP uses programs similar to individuals, but the variance of opinions appears with its implementation. Cramer uses a parse-tree-like representation of the code, and specifies suitable operations (for example, exchange of subtrees for recombination). Fujiki and Dickinson have developed a system that is based on the generation of programs that use simple conditional sentences of LISP (COND).

The resolution of numerous problems could be patterned as an evolutionary process in which the fittest individual survives. The simulation of this evolutionary process begins with the generation of an initial population, composed by computer programs that represent the individuals. These programs are generated starting from the group of functions and terminal elements that adapt better to the problem to solve. In most cases, the election of the group of terminal and nonterminals (functions) is critical to make the algorithm work properly.

For instance, to render complex mathematical functions it is interesting to introduce in the group of non terminals such functions as sines, cosines, and so on. For graphic applications, it is usually quite normal to introduce primitive of the type Line, Circle, Torus that graphically represent figures in diverse ways.

The appropriateness of each program is quantified in terms of how of well it performs in the surroundings of the particular problem. This measure is designated as the fitness function. A program is generally measured in different representative cases, and the final measure of its fitness will be an average of all the measures. Generally, the result of the genetic algorithm is the best individual generated in generation n, where n is the maximum number of generations.

14.6 Elementary Steps of Genetic Programming

The task to be solved is identified and a high-level statement of the problem is formed. This statement is the input to the genetic programming system. To map the high-level statement with the GP system, several preparatory steps are required to be specified by the user. The steps are:

(1) Set of terminals (e.g., the independent variables of the problem, zero-argument functions, and random constants) for each branch

of the to-be-evolved program

(2) Set of primitive functions for each branch of the to-be-evolved program

(3) Fitness measure (for explicitly or implicitly measuring the fitness of individuals in the population)

(4) Certain parameters for controlling the run

(5) Termination criterion and method for designating the result of the run

14.6.1　The Terminal Set

All the variables and constants of the evolved programs are stored in the terminal set. Whereas the function set consists of all the problem specific details. Consider a scenario in which a robot is being controlled, it may be that movement functions are parameterized by directions such as left, right, forward, backward, which would form part of the terminal set for that GP application. Likewise constants like pi etc., can be in the terminal set. The terminal set is the terminal node, since in the tree representations, the constants and variables are found at the end of the branches.

14.6.2　The Function Set

Function set is a set of domain related functions used along with the terminal set to construct effective solutions to a given problem. As with the evaluation function, the set of functions will be hand-carved for the particular task. For instance, while evolving a program to control the movement of a washer it tries to use the functions will include things like time taken to wash, dirtiness in the clothes, etc., The function set also includes the set of programmatic functions such as if-then-else and for-loops similar to the traditional "c" programming language.

14.6.3　The Fitness Function

The numeric value assigned to each member of a population to provide a measure of the appropriateness of a solution to the problem in question is known as fitness function.

14.6.4   The Algorithm Control Parameters

The algorithm control parametric quantities include population size, crossover, and mutation probabilities. There are a lot of possible actions to predict the proceedings of the search, and therefore the user ought to fine-tune several parameters to optimize the performance of the GP system. The main consideration will be the size of the population, as this will effect the performance of GP. Larger populations imply less number of generations within the time available, but leads to larger diversity within the population of programs. Another important parameter is the length of the programs that are produced as the programs grow. A major critique of GP approach is that the programs produced are too large and elaborated to be understood, therefore, the length of the programs should be kept relatively small. Other parameters will control various probabilities, including the probability that each genetic operator is employed.

14.6.5   The Termination Criterion

The stopping conditions for the GP engine to stop are very similar to that of Genetic Algorithms. The process is allowed to run for a certain specified amount of time and then determine the best individual and terminate the process. Another approach is to allow the process until it has produced a certain number of generations, then the best individual produced in any generation is chosen. Many GP implementations enable the user to monitor the process and click on the stop button when it appears that the fitness of the individuals has reached a plateau.

This is generally a predefined number of generations or an error tolerance on the fitness. It should be noted that the first 3 components determine the algorithm search space, while the final 2 components affect the quality and speed of search. In order to further illustrate the coding procedure and the genetic operators used for GP, a symbolic regression example will be used. Consider the problem of predicting the numeric value of an output variable, y, from two input variables a and b. One possible symbolic representation for y in terms of a and b would be,

$$y = (a - b)/3$$

Figure 14.7 demonstrates how this expression may be represented as a tree structure.

With this tree representation, the genetic operators of crossover and mutation must be posed in a fashion that allows the syntax of resulting expressions to be preserved. A valid crossover operation is shown where the two parent expressions are given by:

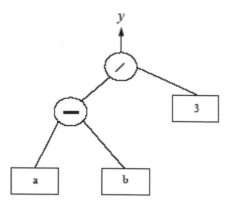

FIGURE 14.7: Representation of a Numeric Expression Using a Tree Structure.

Parent 1: y = (a − b) / 3 (2)

Parent 2: y = (c − b) * (a + c) (3)

Parent 1 has input variables "a" and "b" and a constant "3" while parent 2 has three input variables "a", "b" and "c".

Both expressions attempt to predict the process output, "y".

If the "/" from parent 1 and the "*" from parent 2 are chosen as the crossover points, then the two offspring are given by:

Offspring 1: y = (a − b) / (a + c) (4)

Offspring 2: y = (c − b)* 3 (5)

14.7 Flowchart of Genetic Programming

The most important dissimilarity between genetic programming and genetic algorithms is based on the solution representation. Genetic programming creates computer programs in the Lisp or scheme computer languages as the solution. Genetic algorithms create a string of numbers that represent the solution. Genetic programming uses the basic executional steps as follows to solve problems:

- Step 1: Generate a random initial population

- Step 2: Assign a fitness value to each individual in the population

- Step 3: Create a new population

- Step 4: Choose the best existing solutions

- Step 5: Create new solutions by crossover and mutation

- Step 6: The best solution is chosen as the fittest solution.

The process of genetic programming starts with an initial population of computer programs composed of functions and terminals appropriate to the problem. These individual programs are usually generated by recursively generating a rooted labeled program tree made up of random choices of the primitive functions and terminals. The individuals are generated with a maximum size, which is predefined. Each program in the population has different size and is of different shape (shape of the tree).

Every individual program present in the population is executed separately. After which, each individual program in the population is evaluated in order to understand the performance. This evaluation produces a single explicit numerical value known as the fitness. In most of the practical applications, the measure of fitness is multiobjective which means that it combines two or more different elements. The different elements of the fitness measure are often in competition with one another to some degree. For several practical problems, every individual in the population is executed over a representative sample of different fitness cases. These fitness cases may represent different values of the program's input(s), different initial conditions of a system, different sensor inputs, or different environments. Merely the fitness cases are implemented probabilistically. The initial population is chosen in random, in the sense that the search is a blind random search of the search space of the problem. Generally the individuals in the first generation, generation 0, have very poor fitness. The differences in fitness are then exploited by genetic programming. Genetic programming applies Darwinian selection and the genetic operations to create a new population of offspring programs from the current population.

The genetic operations include crossover (sexual recombination), mutation, reproduction, and the structure-altering operations. Once the individuals are selected from the population by the fitness measure, the individuals undergo genetic operations such as crossover, mutation, etc. During the probabilistic selection process of the individuals, better individuals are preferred over inferior individuals. This implies that the best individual in the population is not necessarily selected and the worst individual in the population is not necessarily passed over.

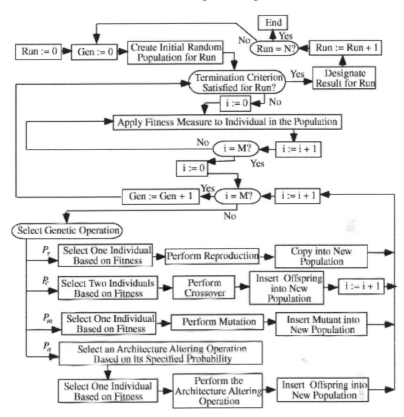

FIGURE 14.8: Genetic Programming Flowchart

The genetic operations produce a population of new offspring from the individuals of the current population. The fitness evaluation and the genetic operations are performed iteratively over several generations. This iterative process of measuring fitness and performing the genetic operations is repeated over many generations. The iterative process stops when the termination criterion is satisfied. The best individual encountered during the iterative process is chosen as the best-fit solution.

There are numerous alternative implementations of genetic programming that vary from the foregoing brief description. Figure 14.8 is a flowchart showing the executional steps of a run of genetic programming. The flowchart shows the genetic operations of crossover, reproduction, and mutation as well as the structure-modifying operations.

14.8   Benefits of Genetic Programming

A few advantages of genetic programming are:

1. Without any analytical knowledge accurate results are obtained.

2. If fuzzy sets are encoded in the genotype, new and more suited-fuzzy sets are generated to describe precise and individual membership functions. This can be done by means of the intersection and/or union of the existing fuzzy sets.

3. Every component of the resulting GP rule-base is relevant in some way for the solution of the problem. Thus null operations that will expend computational resources at runtime are not encoded.

4. This approach does scale with the problem size. Some other approaches to the cart-centering problem use a GA that encodes NxN matrices of parameters. These solutions work badly as the problem grows in size (i.e., as N, increases).

5. With GP no restrictions are imposed on how the structure of solutions should be. Also the complexity or the number of rules of the computed solution is not bounded.

14.9   MATLAB Examples Using Genetic Programming

During the operation of GP, it generates a lot of potential solutions in the form of a tree-structure. These trees may have better and worse terms (subtrees) that contribute more or less to the accuracy of the model represented by the tree. This section focuses on the MATLAB coding to apply Orthogonal Least Squares (OLP) Method to estimate the contribution of the branches of the tree to the accuracy of the model. Illustration 1 discusses the GP-OLS model for a static function and Illustration 2 discusses a dynamical model.

14.9.1   Illustration 1: Static Function Identification

This Genetic Programming code applies Orthogonal Least Squares algorithm (OLS) to improve the search efficiency of GP. It can be used for static equation discovery or structure identification of simple dynamical linear-in-parameters models. This section illustrates the implementation of a static function identification of models.

```
% Static function identification

clear all

% Regression matrix
ndata = 100;
nvar = 3;
X = rand(ndata,nvar);

% Output vector (y = 10*x1*x2+5*x3)
Y = 10*X(:,1).*X(:,2) + 5*X(:,3);
Y = Y + randn(size(Y))*0.01;
% some 'measurement' noise
% GP equation symbols
symbols1 = '+','*';
symbols2 = 'x1','x2','x3';
% length(symbols2) = size(X,2) !

% Initial population
popusize = 40;
maxtreedepth = 5;
popu = gpols_init(popusize,maxtreedepth,symbols);

% first evaluation
opt = [0.8 0.5 0.3 2 1 0.2 30 0.05 0 0];
popu = gpols_evaluate(popu,[1:popusize],
        X,Y,[],opt(6:9));
% info
disp(gpols_result([],0));
disp(gpols_result(popu,1));
% GP loops
for c = 2:20,
  % iterate
  popu = gpols_mainloop(popu,X,Y,[],opt);
  % info
  disp(gpols_result(popu,1));
end

% Result
 [s,tree] = gpols_result(popu,2);
disp(s);
```

```
% ------------------------------------------------
% Subprograms
% ------------------------------------------------
function popu = gpols_init(popusize,maxtreedepth
,symbols);
% Initializes population variable
% popu = gpols_init(popusize,maxtreedepth,symbols)
%    popu <- generated individuals (population
%    variable) popusize ->number of
%    individuals (size of population) maxtreedepth
%    ->maximum tree depth symbols ->
%    cell arrays of operator and terminator node
%    strings. %
% E.g.
%    symbols1 = ('+','*');
%    symbols2 = {'x1','x2','x3'};
%    popu = gpols_init(20,5,symbols);
%
popu.generation = 1;
popu.symbols = symbols;
popu.size = popusize;
for i = 1:popusize,
  popu.chromi.fitness = 0;
  popu.chromi.mse = 0;
  popu.chromi.tree = tree_genrnd(maxtreedepth,symbols);
end
% -------------- End of gpols_init --------------------

function popu = gpn_evaluate(popuin,ixs,X,Y,Q,optv);
% Evaluates individuals and identificates their linear
% parameters popu=gpols_evaluate(popuin,ixs,X,Y,Q,optv)
% popu <- result (population)
% popuin ->input population
% ixs ->vector of indexes of individuals to evaluate
% X ->Regression matrix without bias (!)
% Y ->Output vector
% Q ->Weighting vector (set empty if not used)
% optv ->evaluation parameters (set empty for default)
%   [optv(1)optv(2)]:a1,a2 tree-size penalty parameters
%   (default: 0,0)
%   optv(3): OLS treshold value,range: 0-1 (default:0)
%   optv(4): if == 1 then polynomial evaluation else
%   normal (default: 0)
%
% Output
popu = popuin;
```

```
% Options and parameters
if isempty(optv),
  optv = zeros(1,4);
end
a1 = optv(1);
a2 = optv(2);
olslimit = optv(3);
polye = optv(4);

% WLS matrices
if isempty(Q),
  Q = ones(size(Y,1),1);
end
Q = diag(Q);
X = sqrt(Q)*X;
Y = sqrt(Q)*Y;

% Symbolum list
for i = 1:length(popu.symbols1),
  s = popu.symbols1i(1);
  if s=='*' | s=='/' | s=='^' | s=='\',
    symbols1i = strcat('.',popu.symbols1i);
  else
    symbols1i = popu.symbols1i;
  end
end
for i = 1:size(X,2),
  symbols2i = sprintf('X(:,end

% MAIN loop
for j = ixs,

  % Get the tree
  tree = popu.chromj.tree;

  % Exhange '+' ->'*'under non-'+'
  % (polynom-operation)    if (polye == 1),

    tree = polytree(tree);
  end

  % Collect the '+ parts'
 [vv,fs,vvdel] = fsgen(tree,symbols);

  % Prune redundant parts
  tree = prunetree(vvdel,tree,symbols);
```

```
% Collect the '+ parts'
[vv,fs,vvdel] = fsgen(tree,symbols);
if ~isempty(vvdel),
 error('Fatal error:redundant strings after
                        deleting');
end

% OLSQ
[vfsdel,err] = gpols_olsq(fs,X,Y,olslimit);
tree.err = err;
vvdel = vv(vfsdel);

% Prune redundant parts
tree = prunetree(vvdel,tree,symbols);

% Collect the '+ parts'
[vv,fs,vvdel] = fsgen(tree,symbols);
if ~isempty(vvdel),
 error('Fatal error:redundant strings after
                        deleting');
end

% LSQ
[mse,cfsq,theta] = gpols_lsq(fs,X,Y);
fit = cfsq;

% Tree-size penalty
if a1~=0 & a2~=0,
 Sl = tree_size(tree);
 fit = fit / (1+exp(a1*(Sl-a2)));
end

% Chrom
popu.chromj.tree = tree; % write back the tree
popu.chromj.mse = mse;
popu.chromj.fitness = fit;
popu.chromj.tree.param(1:length(theta)) = theta;
popu.chromj.tree.paramn = length(theta);

end

%----------------------------------------------------
function [tree] = polytree(treein);
tree = treein;
v = [1];
vv = [];
i = 1;
```

```
while i <= length(v),
   ii = v(i);
   if tree.nodetyp(ii)==1 & tree.node(ii)==1,
    v = [v, ii*2, ii*2+1];
   else
    vv = [vv, ii];
   end


   i = i+1;
end
for ii = [vv],
   v = [ii];
   i = 1;
   while i <= length(v),
      if tree.nodetyp(v(i))==1,
         if tree.node(v(i))==1,
            tree.node(v(i)) = 2;
         end
         if v(i)*2+1 <= tree.maxsize,
            v = [v, v(i)*2, v(i)*2+1];
         end
      end
      i = i+1;
   end
end
%- - - - - - - - - - - - - - - - - - - - - - - - - - - - - - - - - - - - - - - -
function [vv,fs,vvdel] = fsgen(tree,symbols);
% Search the '+ parts'
v = [1];
vv = [];
i = 1;
while i <= length(v),
   ii = v(i);
   if tree.nodetyp(ii)==1 & tree.node(ii)==1,
      v = [v, ii*2, ii*2+1];
   else
      vv = [vv, ii];
   end
   i = i+1;
end
fs = [];
i = 1;
for ii = [vv],
   fsi = strcat('(',tree_stringrc(tree,ii,
                            symbols),')');
   i = i+1;
```

```
end
% Search the redundant '+ parts'
vvdel = [];
vvv = [];
i = 1;
while i <= length(fs),
    ok = 0;
    ii = 1;
    while ii<i & ok==0,
        ok = strcmp(fsi,fsii);
        ii = ii+1;
    end
    if ok==1,
        vvdel = [vvdel, vv(i)];
    else
        vvv = [vvv, i];
    end
    i = i+1;
end
%- - - - - - - - - - - - - - - - - - - - - - - - - - - - - - - - - - - - - - - - -
function tree = prunetree(vvdel,treein,symbols);
% Delete subtrees
nn = [length(symbols1), length(symbols2)];
tree = treein;
n = floor(tree.maxsize/2);
tree.nodetyp(vvdel) = 0;
ok = 1;
while ok,
    ok = 0;
    i = 1;
    while i<=n & ok==0,
        if (tree.nodetyp(i)==1)&(tree.nodetyp(i*2)==0|
        tree.nodetyp(i*2+1)==0),
            ok = 1;
            if tree.nodetyp(i*2)==0&tree.nodetyp(i*2+1)==0,
                tree.nodetyp(i*2) = treein.nodetyp(i*2);
                tree.nodetyp(i*2+1) = treein.nodetyp(i*2+1);
                tree.nodetyp(i) = 0;
            elseif tree.nodetyp(i*2)==0,
                tree.nodetyp(i*2) = treein.nodetyp(i*2);
                subtree = tree_getsubtree(tree,i*2+1);
                tree = tree_inserttree(subtree,tree,i,nn(2));
            else
                tree.nodetyp(i*2+1) = treein.nodetyp(i*2+1);
                subtree = tree_getsubtree(tree,i*2);
                tree = tree_inserttree(subtree,tree,i,nn(2));
            end
```

```
    elseif
      (tree.nodetyp(i*2)==0 | tree.nodetyp(i*2+1)==0),
      ok = 1;
      if tree.nodetyp(i*2)==0,
        tree.nodetyp(i*2) = treein.nodetyp(i*2);
      end
      if tree.nodetyp(i*2+1)==0,
        tree.nodetyp(i*2+1) = treein.nodetyp(i*2+1);
      end
    end
    i = i+1;
  end
end
----
function [popu,evnum]=gpols_mainloop(popuin,X,Y,Q,opt);
% Run one evolutionary loop, makes the next generation
% [popu,evnum] = gpols_mainloop(popuin,X,Y,Q,opt)
%    popu <- next generation of the population
%    evnum <- number of fun.evaualtion (usually number
%    of new individuals) popuin ->the population
%    opt ->options vector, GPOLS-parameters
%    X,Y,Q ->input, output and weighting matrices
(see gpols_evaluate)
%
% Remark:
%    opt(1): ggap, generation gap (0-1)
%    opt(2): pc, probability of crossover (0-1)
%    opt(3): pm, probability of mutation (0-1)
%    opt(4): selection type(integer,see gpols_selection)
%    opt(5): rmode, mode of tree-recombination (1 or 2)
%    opt(6): a1, first penalty parameter
%    opt(7): a2, second penalty parameter (0 if there
%               is not penalty)
%    opt(8): OLS treshhold real 0-1 or integer >= 2
%    opt(9): if == 1 ->polynomial evaluation
%    opt(10):if == 1 ->evaluate all indv.s not only
%               new offsprings
%

popun = popuin.size;
ggap = opt(1);
pc = opt(2);
pm = opt(3);
tsels = opt(4);
rmode = opt(5);

% Selection
```

```
selm = gpols_selection(popuin,ggap,pc,pm,tsels);

% New generation
popu = popuin;
newix = [];
nn = 1;
for i=1:size(selm,1),
  m = selm(i,3);
  %*** Crossover ***
  if m==1,
    p1 = selm(i,1);
    p2 = selm(i,2);
    popu.chrom{nn} = popuin.chromp1;
    popu.chrom{nn}.fitness = -1;
    if nn+1<popu.size,
      popu.chrom{nn+1} = popuin.chrom{p2};
      popu.chrom{nn+1}.fitness = -1;
    end
    % recombinate trees
    tree1 = popuin.chrom{p1}.tree;
    tree2 = popuin.chrom{p2}.tree;
    [tree1,tree2] = tree_crossover(tree1,tree2,rmode,
                        popu.symbols);
    popu2.chrom{nn}.tree = tree1;
    if nn+1<=popu.size,
      popu.chrom{nn+1}.tree = tree2;
    end
    % remember the new individuals
    newix = [newix nn];
    nn = nn+1;
    if nn<=popu.size,
      newix = [newix nn];
      nn = nn+1;
    end
  %*** Mutation ***
  elseif m==2,
    p1 = selm(i,1);
    popu.chrom{nn} = popuin.chrom{p1};
    popu.chrom{nn}.fitness = -1;
    % muatate tree
    tree1 = popu.chrom{p1}.tree;
    tree1 = tree_mutate(tree1,popu.symbols);
    popu.chrom{nn}.tree = tree1;
    % remember the new individual
    newix = [newix nn];
    nn = nn+1;
  %*** Direct copy ***
```

```
    else
      p1 = selm(i,1);
      popu.chrom{nn} = popuin.chrom{p1};
      nn = nn+1;
    end
  end

% if opt(10)==1 ->evaluate all indv.s
if length(opt)>9 & opt(10)==1,
  newix = [1:popu.size];
end

function [sout,tree] = gpols_result(popu,info)
% Gets information string about the best solution
% of a population
% [sout,tree] = gpols_result(popu,info);
% sout <- text (string)
% tree <- the best solution
% popu ->population structure
% info ->info mode (1,2)
%
if info == 0,
  sout = sprintf('Iter \t Fitness \t Solution');
  return;
end
best = popu.chrom{1}.fitness;
bestix = 1;
for i = 1:popu.size,
  if popu.chrom{i}.fitness >best,
    best = popu.chrom{i}.fitness;
    bestix = i;
  end
end
tree = popu.chrom{bestix}.tree;
if info == 1,
  sout=sprintf('% 3i.\t% f',popu.generation,best);
  s = tree_stringrc(tree,1,popu.symbols);
  sout = sprintf('% s \t % s',sout,s);
  return;
  end
if info == 2,
  sout = sprintf('fitness:% f, mse:
        % f',best,popu.chrombestix.mse);
  [vv,fs] = fsgen(tree,popu.symbols);
  for i = 1:length(fs),
    sout = sprintf('% s \n % f * % s +',sout,
          tree.param(i),fsi);
```

```
    end
    sout = sprintf('% s \n % f',sout,tree.param(i+1));
    return;
  end
  sout = '???';
%--------------------------------------------------
function [vv,fs,vvdel] = fsgen(tree,symbols);
% Search the '+ parts'
v = [1];
vv = [];
i = 1;
while i <= length(v),
  ii = v(i);
  if tree.nodetyp(ii)==1 & tree.node(ii)==1,
    v = [v, ii*2, ii*2+1];
  else
    vv = [vv, ii];
  end
  i = i+1;
end
fs = [];
i = 1;
for ii = [vv],
  fsi = strcat('(',tree_stringrc(tree,ii,
    symbols),')');
  i = i+1;
end
% Search the redundant '+ parts'
vvdel = [];
vvv = [];
i = 1;
while i <= length(fs),
  ok = 0;
  ii = 1;
  while ii<i & ok==0,
    ok = strcmp(fs{i},fs{ii});
    ii = ii+1;
  end
  if ok==1,
    vvdel = [vvdel, vv(i)];
  else
    vvv = [vvv, i];
  end
  i = i+1;
end
```

Observations

```
Iter  Fitness   Solution
1.    0.900239  (((x1)*(x2))+(x3))+(((x3)+((x2)*(x3)))*(((x1)*(x2))+
                ((x1)+(x1))))
2.    0.916748  ((x1)*(x2))+(((x3)+((x2)*(x3)))*((x3)+((x1)+(x1))))
3.    0.993295  ((x1)*(x2))+(x3)
4.    0.993295  ((x1)*(x2))+(x3)
5.    0.993295  ((x1)*(x2))+(x3)
6.    0.993295  ((x1)*(x2))+(x3)
7.    0.993295  ((x1)*(x2))+(x3)
8.    0.993295  ((x1)*(x2))+(x3)
9.    0.993295  ((x1)*(x2))+(x3)
10.   0.993295  ((x1)*(x2))+(x3)
11.   0.993295  ((x1)*(x2))+(x3)
12.   0.993295  ((x1)*(x2))+(x3)
13.   0.993295  ((x1)*(x2))+(x3)
14.   0.993295  ((x1)*(x2))+(x3)
15.   0.993295  ((x1)*(x2))+(x3)
16.   0.993295  ((x1)*(x2))+(x3)
17.   0.993295  ((x1)*(x2))+(x3)
18.   0.993295  ((x1)*(x2))+(x3)
19.   0.993295  ((x1)*(x2))+(x3)
20.   0.993295  ((x1)*(x2))+(x3)
          fitness: 0.993295, mse: 0.000087
          10.000433 * ((x1)*(x2)) +
          4.996494 * (x3) +
          0.000405
```

14.9.2   Illustration 2: Dynamical Input-Output Model Identification

The method illustrated in this code uses genetic programming to generate nonlinear input-output models of dynamical systems that are represented in a tree structure. The main idea of the tree is to apply the orthogonal least squares (OLS) algorithm to estimate the contribution of the branches of the tree to the accuracy of the model.

```
%----------------------------------------------
% Main Program , For Sub programs follow Illustration 1
% Dynamical input-output model identification
clear all

% Simulation of a dynamic system and generates input
% /output data
t = [0:0.2:20]';
```

```
u = sin(t/2)-0.5;
u = u + randn(size(u))*0.1;
y = zeros(size(u));
y(1) = 0;
y(2) = 0;
for k = 3:length(t),
   dy = 0.7*u(k-1)*u(k-1) - 0.6*y(k-1) - 0.3*y(k-2) - 0.1;
   y(k) = y(k-1) + dy;
end
% Adds some simulated 'measurement noise' to the output
y = y + randn(size(y))*0.02;

% Select the maximum input and output order for
% identification
uorder = 2;
yorder = 2;

% Regressors and outputs for identification
tofs = max(uorder,yorder)+1;
Y = y(tofs:end) - y(tofs-1:end-1); % dy
X = [];
for i=1:yorder,
X = [X, y(tofs-i:end-i)];
end
for i=1:uorder,
X = [X, u(tofs-i:end-i)];
end

% GP equation symbols
symbols{1} = {'+','*'};
for i = 1:yorder,
   symbols{2}{i} = sprintf('y(k-% i)',i);
end
for j = 1:uorder,
   symbols{2}{i+j} = sprintf('u(k-% i)',j);
end

% Initial population
popusize = 40;
maxtreedepth = 5;
popu = gpols_init(popusize,maxtreedepth,symbols);

% first evaluation
opt = [0.8 0.7 0.3 2 2 0.2 25 0.01 1 0];
popu = gpols_evaluate(popu,[1:popusize],X,Y,[],
       opt(6:9));
% info
```

```
disp(gpols_result([],0));
disp(gpols_result(popu,1));
% GP loops
for c = 2:20,
  % iterate
  popu = gpols_mainloop(popu,X,Y,[],opt);
  % info
  disp(gpols_result(popu,1));
end

% Result
[s,tree] = gpols_result(popu,2);
disp(s);
```

Iter	Fitness	Solution
1.	0.284622	(y(k-2))+((u(k-2))+(u(k-1)))
2.	0.284622	(y(k-2))+((u(k-2))+(u(k-1)))
3.	0.284622	(y(k-2))+((u(k-2))+(u(k-1)))
4.	0.683562	(y(k-2))+((u(k-1))*(u(k-1)))
5.	0.683562	(y(k-2))+((u(k-1))*(u(k-1)))
6.	0.894455	(y(k-1))+((u(k-1))*(u(k-1)))
7.	0.894455	(y(k-1))+((u(k-1))*(u(k-1)))
8.	0.894455	(y(k-1))+((u(k-1))*(u(k-1)))
9.	0.903012	((y(k-2))+((u(k-1))*(u(k-1))))+((u(k-1))*(u(k-2)))
10.	0.903012	((y(k-2))+((u(k-1))*(u(k-1))))+((u(k-1))*(u(k-2)))
11.	0.903012	((y(k-2))+((u(k-1))*(u(k-1))))+((u(k-1))*(u(k-2)))
12.	0.903012	((y(k-2))+((u(k-1))*(u(k-1))))+((u(k-1))*(u(k-2)))
13.	0.903012	((y(k-2))+((u(k-1))*(u(k-1))))+((u(k-1))*(u(k-2)))
14.	0.903012	((y(k-2))+((u(k-1))*(u(k-1))))+((u(k-1))*(u(k-2)))
15.	0.903012	((y(k-2))+((u(k-1))*(u(k-1))))+((u(k-1))*(u(k-2)))
16.	0.903012	((y(k-2))+((u(k-1))*(u(k-1))))+((u(k-1))*(u(k-2)))
17.	0.903012	((y(k-2))+((u(k-1))*(u(k-1))))+((u(k-1))*(u(k-2)))
18.	0.903012	((y(k-2))+((u(k-1))*(u(k-1))))+((u(k-1))*(u(k-2)))
19.	0.903012	((y(k-2))+((u(k-1))*(u(k-1))))+((u(k-1))*(u(k-2)))
20.	0.903012	((y(k-2))+((u(k-1))*(u(k-1))))+((u(k-1))*(u(k-2)))

```
fitness: 0.903012, mse: 0.001285
-0.777391 * ((u(k-1))*(u(k-2))) +
-0.370513 * (y(k-2)) +
1.059709 * ((u(k-1))*(u(k-1))) +
-0.047268
```

The GP-OLS algorithm generates linear in parameter models or polynomials models, and the simulation results show that the proposed tool provides an efficient and fast method for selecting input-output model structure for nonlinear processes.

14.9.3 Illustration 3 - Symbolic Regression Problem Using Genetic Programming Toolbox

The following program runs a symbolic regression problem (the quadratic polynomial) with 100 individuals for 25 generations, with automatic adaptation of operator probabilities, drawing several plots in runtime, and finishing with two additional post-run plots.

```
- - - - - - - - - - - - - - - - - - - - - - - - - - - - - - - - - - - - - - - -
% Main Program
- - - - - - - - - - - - - - - - - - - - - - - - - - - - - - - - - - - - - - - -
fprintf('Running symbolic regression demo...');
p=resetparams;

p=setoperators(p,'crossover',2,2,'mutation',1,1);
p.operatorprobstype='variable';
p.minprob=0;

p.datafilex='quartic_x.txt';
p.datafiley='quartic_y.txt';

p.usetestdata=1;
p.testdatafilex='exp_x.txt';
p.testdatafiley='exp_y.txt';

p.calcdiversity={'uniquegen'};
p.calccomplexity=1;
p.graphics={'plotfitness','plotdiversity','plotcomplexity',
'plotoperators'};
p.depthnodes='2';

[v,b]=gplab(25,50,p);

desired_obtained(v,[],1,0,[]);
accuracy_complexity(v,[],0,[]);

figure
plotpareto(v);

drawtree(b.tree);
- - - - - - - - - - - - - - - - - - - - - - - - - - - - - - - - - - - - - - - -
% Subprograms
- - - - - - - - - - - - - - - - - - - - - - - - - - - - - - - - - - - - - - - -
function params=setoperators(params,varargin)
%
%   Input arguments:
%     PARAMS - the algorithm running parameters (struct)
```

```
%      OPNAME - the name of the operator to use (string)
%      NPARENTS - the number of parents required by the
%      operator
%      NCHILDREN - the number of children produced by
%      the operator ...
%   Output arguments:
%      PARAMS - the updated algorithm running parameters
%               (struct) %

params.operatornames={};
params.operatornparents=[];
params.operatornchildren=[];

params.initialfixedprobs=[];
params.initialvarprobs=[];

params=addoperators(params,varargin);

% - - - - - - - - - - End of setoperators - - - - - - - - - -

function [vars,best]=gplab(g,varargin) %
%   Input arguments:
%      NGENS - the number of generations to run
%              the algorithm (integer)
%      POPSIZE - the number of individuals in the
%                population (integer)
%      PARAMS - the algorithm running parameters (struct)
%      VARS - the algorithm variables (struct)
%       VARS.POP - the current population
%       VARS.PARAMS - the algorithm running parameters =
%                     PARAMS
%       VARS.STATE - the current state of the algorithm
%   Output arguments:
%      VARS - the algorithm variables (struct) - see Input
%                     arguments
%      BEST - the best individual found in the run
%             (struct)
if (nargin<2) || (nargin>3)

error('GPLAB: Wrong number of input arguments.
Use either gplab (ngens, vars) to continue a run,
or gplab (ngens, popsize, [optional params]) to
start a run')


elseif isstruct(varargin1)
  % argument 1:the number of additional generations
```

```
        to run
  % argument 2:the algorithm variables
  if ~(isvalid(g,'posint'))
    error('GPLAB: The first argument must be an
    integer greater than 0.')
  end
  end
  start=0;
  continuing=1;
  vars=varargin1;
  n=vars.state.popsize;
  level=vars.state.maxlevel;
  ginic=vars.state.generation+1;
  % start generation number
  gend=ginic-1+g; % end generation number
else
% argument 1:the number of generations to run
% argument 2:the number of individuals in the population
% argument 3:(optional) the parameters of the algorithm

    if ~(isvalid(g,'special_posint') && isvalid(varargin1,
'posint') && varargin1>=2)

    error('GPLAB:The first two arguments must be integers, and
the second >1')

  end
  start=1;
  continuing=0;
  n=varargin1;
  if nargin==3
    vars.params=varargin2;
  else
    vars.params=[];
  end
  vars.state=[];
  vars.data=[];
  ginic=1; % start generation number
  gend=g; % end generation number
end


% check parameter variables:
vars.params=checkvarsparams(start,continuing,
vars.params,n);

% check data variables:
[vars.data,vars.params]=checkvarsdata(start,
```

```
continuing,vars .data,vars.params);

% check state variables:
[vars.state,vars.params]=checkvarsstate(start,
continuing,vars
.data,vars.params,vars.state,n,g);

% initialize random number generator (see help on
RAND):
rand('state',sum(100*clock));
fprintf('\n Running algorithm...\n');

% initiate graphics:
% (if we're not going to run generations or draw
history,
% don't initiate the graphics)
if ~ isempty(vars.params.graphics) &&
(ginic<=gend ||
continuing) gfxState=graphicsinit(vars.params);
end

% initial generation:

if start
  [vars.pop,vars.state]=genpop(vars.params,vars.
  state,vars.data,n);
  if strcmp(vars.params.savetofile,'firstlast') ||
  strcmp(vars.params.savetofile,'every10') ||
  strcmp(vars.params.savetofile,'every100') ||
  strcmp(vars.params.savetofile,'always')
    saveall(vars);
  end
  if ~ strcmp(vars.params.output,'silent')
    fprintf(' # Individuals:
    % d\n',vars.state.popsize);
    if strcmp(vars.params.survival,'resources')
    fprintf('MaxResources:
    % d\n',vars.state.maxresources);
    end
    fprintf('UsedResources:
    % d\n',vars.state.usedresources);
    fprintf('Best so far:
    % d\n',vars.state.bestsofar.id);
    fprintf('Fitness:
    % f\n',vars.state.bestsofar.fitness);
    if vars.params.usetestdata
    fprintf('   Test fitness:
```

```
     % f \n',vars.state.bestsofar.testfitness);
     end
fprintf('  Depth:
     % d \n',vars.state.bestsofar.level);
fprintf('  Nodes:
  % d \n\n',vars.state.bestsofar.nodes);
  end
  % (if we're not going to run generations, don't
  start
  % the graphics:)
  if ~ isempty(vars.params.graphics) && ginic<=gend
    gfxState=graphicsstart(vars.params,vars.state,
           gfxState);
  end
end

if continuing
  if ~ isempty(vars.params.graphics)
    gfxState=graphicscontinue(vars.params,vars.state,
           gfxState);
  end
end

sc=0;

% generations:

for i=ginic:gend

  % stop condition?
  sc=stopcondition(vars.params,vars.state,vars.data);
  if sc
    % unless the option is to never save,save
    % the algorithm
variables now:
    if (~ strcmp(vars.params.savetofile,'never'))
      saveall(vars);
    end
    break % if a stop condition has been reached,
          % skip the for cycle
end

% new generation:

[vars.pop,vars.state]=generation(vars.pop,vars.params,
vars.state, vars.data);
```

```
% save to file?
if (strcmp(vars.params.savetofile,'firstlast')
&& i==g) ||
(strcmp(vars.params.savetofile,'every10')
&& rem(i,10)==0) ||
(strcmp(vars.params.savetofile,'every100')
&& rem(i,100)==0) ||
strcmp(vars.params.savetofile,'always')
  saveall(vars);
end

% textual output:
if ~ strcmp(vars.params.output,'silent')
  fprintf('# Individuals:
  % d\n',vars.state.popsize);
  if strcmp(vars.params.survival,'resources')
  fprintf(' MaxResources:
  % d\n',vars.state.maxresources);
  end
  fprintf('UsedResources:
  % d\n',vars.state.usedresources);
  fprintf('Best so far:
  % d\n',vars.state.bestsofar.id);
  fprintf('Fitness:
  % f\n',vars.state.bestsofar.fitness);
  if vars.params.usetestdata
  fprintf('  Test fitness:
  % f \n',vars.state.bestsofar.testfitness);
  end
fprintf('  Depth:
       % d \n',vars.state.bestsofar.level);
fprintf('  Nodes:
       % d \n\n',vars.state.bestsofar.nodes);
  end
  % plots:
  if ~ isempty(vars.params.graphics)
    gfxState=graphicsgenerations(vars.params,
           vars.state,gfxState);
  end

end % for i=ginic:gend

% messages regarding the stop condition reached:

if sc
  if vars.state.generation==0
    fprintf('\n Stop condition #
```

```
    % d was reached after initial generation.\n',sc);
  else
fprintf('\n Stop condition #
 % d was reached after generation
 % d.\n',sc,vars.state.generation);
  end
else
  fprintf('\n Maximum generation
  % d was reached.\n',vars.state.generation);
end
best=vars.state.bestsofar;
vars.state.keepevals=[];
% clear memory, we don't want to save all this!

fprintf('\nDone!\n \n');

%- - - - - - - - - - End of gplab- - - - - - - - - -

function accuracy_complexity(vars,offsets,bw,sizexy)
%
% Input arguments:
% VARS - all the variables of the algorithm (struct)
% OFFSETS - the offsets for each line,[] for no offset
% BLACKWHITE - the flag to draw a b&w or color plot
%             (boolean)
% SIZEPLOT - the x and y size of plot,[] for default
%             (1x2 matrix)
%
h=vars.state.bestsofarhistory;

if isempty(offsets)
  fitoffset=0;
  leveloffset=0;
  nodesoffset=0;
else
  fitoffset=offsets(1);
  leveloffset=offsets(2);
  nodesoffset=offsets(3);
end

if isempty(sizexy)
  sizexy(1)=0;
  sizexy(2)=0;
end

for i=1:size(h,1)
  g(i)=hi,1;
```

```
  f(i)=hi,2.fitness;
  l(i)=hi,2.level;
  n(i)=hi,2.nodes;
  % noruegueses fizeram
  % n(i)=nodes(hi,2.tree);
end
g=g';
f=f'+fitoffset;
l=l'+leveloffset;
n=n'+nodesoffset;

ff=figure;
set(ff,'Color',[1 1 1]);
if sizexy(1)<=0   sizexy(1)=400; end if sizexy(2)<=0
  sizexy(2)=350;
end

set(ff,'Position',[200 250 sizexy(1) sizexy(2)])
hold on
title('Accuracy versus Complexity');
xlabel('generation');

if fitoffset~=0
  if fitoffset<0
    ylab1=strcat('fitness',int2str(fitoffset));
  else
    ylab1=strcat('fitness+',int2str(fitoffset));
  end
else
  ylab1='fitness';
end

if leveloffset~=0
  if leveloffset<0
    ylab2=strcat('level',int2str(leveloffset));
  else
    ylab2=strcat('level+',int2str(leveloffset));
  end
else
  ylab2='level';
end

if nodesoffset~=0
  if nodesoffset<0
    ylab3=strcat('nodes',int2str(nodesoffset));
  else
    ylab3=strcat('nodes+',int2str(nodesoffset));
```

```
   end
else
  ylab3='nodes';
end
ylaball=[ylab1 ', ' ylab2 ', ' ylab3];
ylabel(ylaball);

if bw
  plot(g,f,'k.-',g,l,'k*-',g,n,'k+-');
else
  plot(g,[f,l,n],'.-');
end legend(ylab1,ylab2,ylab3);

%- - - - - - End of accuracy_complexity - - - - - -

function plotpareto(vars) %
%   Input arguments:
%    VARS - all the variables of the algorithm (struct)
%   Output arguments:
%    none
%

for i=1:length(vars.pop)
  if isempty(vars.pop(i).nodes)
    vars.pop(i).nodes=nodes(vars.pop(i).tree);
  end
end

% build pareto front variable:
paretofront=[];
paretofront(1).fitness=[];
for solution=vars.pop
  while length(paretofront)<solution.nodes
    paretofront(end+1).ind=[];
  end
if isempty(paretofront(solution.nodes).fitness)||
((vars.params.lowerisbetter&&
solution.fitness<paretofront(solution.nodes).fitness)||
(~ vars.params.lowerisbetter&&
solution.fitness>paretofront(solution.nodes).fitness))
    % cross validation
    if vars.params.usetestdata
testindividual=calcfitness(solution,vars.params,vars
.data.test,vars.state,1);
% (1 = test data)
      solution.testfitness=testindividual.fitness;
    end
```

```
      paretofront(solution.nodes).fitness=solution.
                                        fitness;
      paretofront(solution.nodes).ind=solution;
      if vars.params.usetestdata
        paretofront(solution.nodes).testfitness=
        solution.
                                        testfitness;
      end
    end
end
% collect fitness and #nodes:
y=[paretofront.fitness];
x=[];
for i=1:size(paretofront,2)
  if ~ isempty(paretofront(i).ind)
    x=[x i];
  end
end
if length(y) =length(x)
  error('','internal error');
end
% compute pareto front:
best=[];
bestind=[];
sofar=[];
for i=1:length(y)
  if i==1 || y(i) >sofar
    sofar=y(i);
    best=[best sofar];
    bestind=[bestind x(i)];
  end
end

hold on
title('Pareto front');
xlabel('nodes');
ylabel('fitness');
% plot:
plot(x,y,'o-',bestind,best,'ro-');
plot([vars.pop.nodes],[vars.pop.fitness],'g*');
if vars.params.usetestdata
  plot(x,[paretofront.testfitness],'m-');
  legend('best for #nodes','pareto front','current
  population','test fitness');
else
legend('best for #nodes','pareto front','current
 population');
```

```
end

hold off

%- - - - - - - - - End of plotpareto - - - - - - - -

function drawtree(tree,titletext)
%
%  Input arguments:
%    TREE - a GPLAB tree (struct)
%    TITLE - title of the tree figure (optional,string)
%  Output arguments:
%    none
%

% Set tree titles
if nargin <2
  titletext = 'Displaying GPLAB-tree found';
end
% Using new figure to display tree
h=figure; % new figure for this particular tree
set(h,'name',titletext);

% First, count nodes
[tree, count] = walkTreeDepthFirst(tree, 'countLeaves',
                                    [], 0, 0);
state.nodeCount = count;
state.yDist = -1;
% Position leaves (equally spaced)
[tree, state]=walkTreeDepthFirst(tree,'positionLeaves',
                                    [], 0, state);
% Position parents (midway between kids)
[tree, state] = walkTreeDepthFirst(tree,[],
                            'positionParents',0,state);
% Draw tree
[tree,state]=walkTreeDepthFirst(tree,[],'drawNode',0,0 );

axis('off')

function [tree, state] = walkTreeDepthFirst(tree, preDive,
postDive, initialDepth, state )
% Calls preDive(tree, depth, state), enters subnodes,
% calls postDive(tree,depth,state).Useful for walking
 the tree.
if ~ isempty(preDive)
  [tree, state] = feval(preDive, tree, initialDepth,
  state); end
```

```
for i = 1:length(tree.kids)
    [tree.kidsi, state] = walkTreeDepthFirst(tree.
    kidsi, preDive,
    postDive, initialDepth + 1, state);
end
if ~ isempty(postDive)
[tree, state] = feval(postDive, tree, initialDepth,
 state);
end

function [tree,count]=countLeaves(tree,depth,count)
if isempty(tree.kids)
  tree.index = count;
  count = count + 1;
end

function [tree,state]=positionLeaves(tree,depth,state)
  if isempty(tree.kids)
  if state.nodeCount <= 1
    tree.X = 0;
  else
    tree.X = tree.index / (state.nodeCount - 1);
  end
  tree.Y = depth * state.yDist;
end

function [tree,state]=positionParents(tree,depth,state)
if ~ isempty(tree.kids)
  x = [];
  for i = 1:length(tree.kids)
    kid = tree.kids{i};     x = [x kid.X];   end
  tree.X = mean(x);
  tree.Y = depth * state.yDist;
end

function [tree, state] = drawNode(tree, depth, state)
if ~ isempty(tree.kids)
  for i = 1:length(tree.kids)
    kid = tree.kids{i};
    line([tree.X kid.X], [tree.Y kid.Y]);
  end
  line(tree.X, tree.Y, 'marker', 'ƒ', 'markersize', 8)
  opText = tree.op;
  text(tree.X, tree.Y, [' ' opText], 'Horizontal
  Alignment',
  'left','interpreter', 'none')
else
```

```
opText = tree.op;
line(tree.X,tree.Y,'marker','.','markersize', 8)
text(tree.X,tree.Y,opText,'HorizontalAlignment',
'center','VerticalAlignment','top','interpreter',
'none')
end
%- - - - - - - - - - End of drawtree.m- - - - - - - - -
```

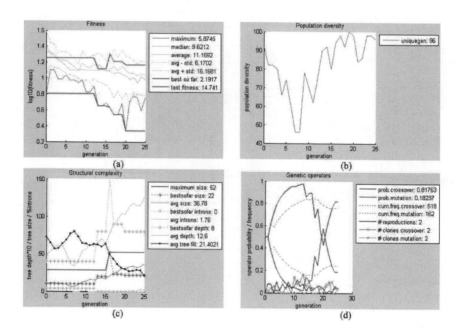

(a)   (b)

(c)   (d)

FIGURE 14.9: Observations of GP after 25 Generations. (a) Plot of the Fitness (b) Plot of the Population Diversity (c) Structural Complexity (d) Genetic Operators

Observations

The parameters after 25 generations are shown below

#Individuals:	50
UsedResources:	821
Best so far:	1019
Fitness:	6.057157

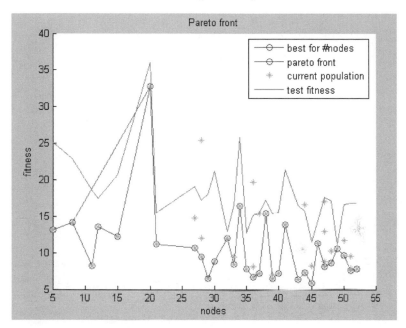

FIGURE 14.10: Pareto Plot Showing the Best Nodes, Current Population and the Test Fitness

Test fitness:	15.218551
Depth:	8
Nodes:	12

Figure 14.9 shows the fitness, population diversity, Structural Complexity, and the Genetic Operators of the symbolic regression problem after 25 generations. The Pareto plot is also shown in Figure 14.10 indicating the best nodes, current population, and the test fitness. The tree found using GP is plotted and is shown in Figure 14.11.

Summary

From the sections discussed in this chapter it is inferred that Genetic Programming (GP) is an automated method for creating a working computer program from a high-level problem statement of a problem. Genetic programming starts from a high-level statement of "what needs to

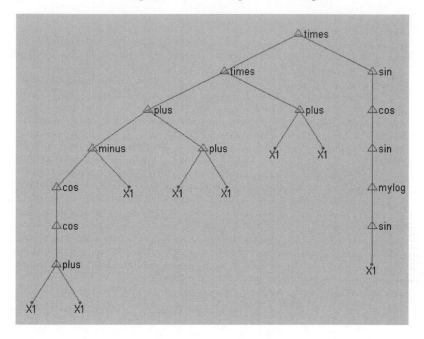

FIGURE 14.11:   Plot of the Tree Found Using GP

be done" and automatically creates a computer program to solve the problem. The chapter provided a set of MATLAB illustrations on GA and GP to provide a practical knowledge to the user.

The fact that genetic programming can evolve entities that are competitive with human-produced results suggests that genetic programming can be used as an automated invention machine to create new and useful patentable inventions. In acting as an invention machine, evolutionary methods, such as genetic programming, have the advantage of not being encumbered by preconceptions that limit human problem-solving to well-troden paths.

Review Questions

1. Define Genetic Programming.

2. Write a short note on Lisp programming language.

3. Explain the basic operations of Genetic Programming.

4. Explain selection in Genetic Programming.

5. How does crossover and mutation occur in Genetic Programming?

6. Mention the paradigms of GA in machine learning and give one example for each.

7. What are the five basic preparatory steps that should be specified by the user in Genetic Programming?

8. Draw and explain the flowchart of Genetic Programming.

9. Mention a few advantages of Genetic Programming.

10. Write a MATLAB program to maximize one dimensional function within the given boundaries using genetic algorithm. Consider the function to be y = x + 10*sin(5*x)+7*cos(4*x) + 25; within the bounds x ϵ[-10,10].

11. Using genetic algorithm, write a MATLAB program to tune a PI controller.

12. Write a MATLAB program to find the shortest path using genetic algorithm.

13. Write a MATLAB program to draw a tree.

14. Obtain the genetic programming selection, mutation, crossover, and insertion in MATLAB.

15. Determine the size of a tree by writing a suitable MATLAB program.

Chapter 15

MATLAB-Based Swarm Intelligence

15.1 Introduction to Swarms

The behaviors of a flock of birds, a group of ants, a school of fish, etc., were the field of study during earlier days. Such collective motion of insects and birds is known to as "swarm behavior." Later on biologists and computer scientists in the field of artificial life studied the modeling of biological swarms to analyze the interaction among the social animals, to achieve goals, and to evolve them. Recently the interest of engineers is increasing rapidly since the resulting swarm intelligence (SI) is applicable in optimization problems in various fields like telecommunicate systems, robotics, electrical power systems, consumer appliances, traffic patterns in transportation systems, military applications, and many more. In swarm intelligence, N agents in the swarm or a social group are coordinating to achieve a specific goal by their behavior. This kind of collective intelligence arises from large groups of relatively simple agents. The actions of the agents are governed by simple local rules. The intelligent agent group achieves the goal through interactions of the entire group. A type of "self-organization" emerges from the collection of actions of the group.

Swarm intelligence is the collective intelligence of groups of simple autonomous agents. The autonomous agent is a subsystem that interacts with its environment, which probably consists of other agents, but acts relatively independently from all other agents. There is no global plan or leader to control the entire group of autonomous agents. Consider for example, the movement of a bird in a flock, the bird adjusts its movements such that it coordinates with the movements of its neighboring flock mates. The bird tries to move along with its flock maintaining its movement along with the others and moves in such a way to avoid collisions among them. There is no leader to assign the movements therefore the birds try to coordinate and move among themselves. Any bird can fly in the front, center, and back of the swarm. Swarm behavior helps birds take advantage of several things including protection from predators (especially for birds in the middle of the flock), and searching for food (essentially each bird is exploiting the eyes of every other bird).

This chapter gives the basic definition of swarms, followed by a description on Swarm Robots. The Biological Models, Characterizations of Stability, and Overview of Stability Analysis of Swarms are also elaborated in this chapter. The chapter deals with the taxonomy of Swarm Intelligence, properties of the Swarm Intelligence system, studies and applications of swarm intelligence. The variants of SI such as Particle Swarm Optimization (PSO) and Ant Colony Algorithms for Optimization Problems are discussed. A few applications of Particle Swarm Optimization such as Job Scheduling on Computational Grids and Data Mining and a few applications of Ant Colony Optimization such as Traveling Salesman Problem (TSP), Quadratic Assignment Problem (QAP), and Data Mining and their implementation in MATLAB are explained in this chapter.

15.2   Biological Background

The collections of birds and other biological species, such as bird flocks, sheep herds, and fish schools, move in an orchestrated manner. The movement of a flock of birds resembles a well-choreographed dance troupe. Similarly a swarm of ants or a school of fish follows the rest of the group even if they are in the opposite direction to search for a desired path. The movement of the birds appears the same either left in uniform or suddenly they may all dart to the right and swoop down toward the ground. This is really a very surprising feature as to know how the birds can coordinate their actions so well. In 1987, a "boid" model was created by Reynold. This boid is a distributed behavioral model, which is used to simulate the motion of a flock of birds on a personal computer. Each boid serves as an independent actor that navigates based on its own perception of the dynamic environment. There are a certain set of rules that are to be observed by the boid.

- The avoidance rule states that an individual boid must move away from boids that are too close, so as to reduce the chance of in-air collisions.

- The copy rule states that a boid must fly in the general direction along with the flock by considering the other boids' average velocities and directions.

- The center rule states that a boid should minimize exposure to the flock's exterior by moving toward the perceived center of the flock.

- The view rule indicates that a boid should move laterally away from any boid the blocks its view. This boid model seems reasonable if we consider it from another point of view, that of it acting according to attraction and repulsion between neighbors in a flock.

The repelling behavior of the flock leads to collision avoidance and helps the flock to maintain its shape. The center rule plays a role in both attraction and repulsion. The swarm behavior of the simulated flock is the result of the dense interaction of the relatively simple behaviors of the individual boids.

If social insects like ants find a prey, and if it cannot move that food particle all alone, then it informs its nest mate by trail laying. After which the group of insects collectively carry the large particle to their nest. Though this concept appears simple, the underlying mechanisms of such a cooperative transport are unclear. Inorder to model such a kind of co-operative transport, Kube and Zhang introduced a simulation model to recover stagnancy using the method of task modeling. Resnick designed StarLogo (an object-oriented programming language based on Logo, to do a series of microworld simulations). He successfully illustrated different self-organization and decentralization patterns in the slime mold, artificial ants, traffic jams, termites, turtle and frogs and so on.

Terzopooulos developed artificial fishes in a 3D virtual physical world. They emulate the individual fish's appearance, locomotion, and behavior as an autonomous agent situated in its simulated physical domain. The simulated fish can learn to control internal muscles to locomote hydrodynamically. They also emulated the complex group behaviors in a certain physical domain. Millonas proposed a spatially extended model of swarms in which organisms move probabilistically between local cells in space, but with weights dependent on local morphgenetic substances, or morphogens. The morphogens are in turn affected by the paths of movements of an organism. The evolution of morphogens and the corresponding flow of the organisms constitute the collective behavior of the group. All living beings learn and evolve in their life span. In the field of artificial life, a variety of species adaptation genetic algorithms are proposed. Sims describes a lifelike system for the evolution and co-evolution of virtual creatures. These virtual creatures perform a competition in physically simulated 3D environments to seize a common resource. During this process, the winners are only capable of surviving and reproducing. Their behavior is limited to physically plausible actions by realistic dynamics, like gravity, friction and collisions. He structures the genotype by the directed graphs of nodes and connections. These genotypes can determine the neural systems for controlling muscle forces and the morphology of these creatures. They simulate co-evolution by adapting the morphology and behavior mutually during the evolution process. They found interesting and diverse strategies and counter-strategies emerge during the simulation with populations of competing creatures.

15.3 Swarm Robots

The major application area of swarm intelligence is swarm robotics. Swarms are capable of providing enhanced task performance, high reliability, low unit complexity and decreased cost over the existing traditional robotic systems. They are also capable of performing tasks that are not possible for a single robot. These swarm robots find their application in areas such as flexible manufacturing systems, spacecraft, inspection/maintenance, construction, agriculture, and medicine work. Several swarm models have been proposed and implemented. Cellular robotic systems were introduced by Beni, consisting of a group of autonomous, non-synchronized, non-intelligent robots coordinating on a finite n-dimensional cellular space under distributed control. These robots operate independently and coordinate their behaviors according to the other robots to attain the predefined global objective of the given task. Hackwood and Beni proposed a model in which the robots operate on the basis of signpost concept. These signposts are capable of modifying the internal state of the swarm units as they move around. Due to these signposts, the entire swarm acts as a collective unit to execute complex behaviors. The self-organization behavior was realized in a general model which had the cyclic boundary condition as the most restrictive assumption. The model requires that sensing swarm "circulate" in a loop during its sensing operation.

Based on the behavior, Brooks proposed a control strategy which was more popular and gained its application to collections of simple independent robots, usually for simple tasks. Ueyama proposed a technique in which the complex robots are arranged in tree-like hierarchies with communication between robots limited to the structure of the hierarchy. Mataric describes a set of experiments with a homogeneous population of robots acting under different communication conditions. As the communication between the robots increases, the swarm robots are capable of solving a wide range of complex behaviors. Swarm robots are more similar to networks of autonomous agents, they are potentially reconfigurable networks of communicating agents capable of coordinated sensing and interaction with the environment. Considering the variety of possible design of groups such as mobile robots, Dudek presented a swarm-robot taxonomy of the different ways in which such swarm robots can be characterized. This proposal was very useful to clarify the strengths, constraints and tradeoffs of various designs. The dimensions of the taxonomic axes are swarm size, communication range, topology, bandwidth, swarm reconfigurability, unit processing ability, and composition. For instance, swarm size includes the cases of single agent, pairs, finite sets, and infinite numbers. Communication ranges include none, close by neighbors, and complete where every agent communicates with every other agent.

Swarm composition can be homogeneous or heterogeneous (i.e., with all the same agents or a mix of different agents). For instance, Hackwood and Beni's model has multiple agents in its swarm, nearby communication range, broadcast communication topology, free communication bandwidth, dynamic swarm reconfigurability, heterogeneous composition, and its agent processing is Turing machine equivalent. Due to the research on decentralized autonomous robotics systems, various fields have received great attention including modeling of swarms, agent planning or decision making and resulting group behavior, and the evolution of group behavior. The decision-making is part of artificial intelligence since several agents coordinate or cooperate to make decisions. Fukuda introduced a distributed genetic algorithm to optimize the distributed planning in a cellular robotics system. The concept of self-recognition for the decision making was also proposed to prove the learning and adaptation capability.

15.4 Stability of Swarms

The stability of a swarm is generally considered as cohesiveness. There are several basic principles for swarm intelligence, such as the proximity, quality, response diversity, adaptability, and stability. The stability of swarms is based upon the relative velocity and distance of adjacent members in the group. To obtain stability, attractant and repellant profiles should exist in the environment such that the group moves toward attractants avoiding repellants. Jin proposed the stability analysis of synchronized distributed control of 1-D and 2-D swarm structures. He proves that synchronized swarm structures are stable in the sense of Lyapunov with appropriate weights in the sum of adjacent errors if the vertical disturbances vary sufficiently more slowly than the response time of the servo systems of the agents.

Li W et al. proved that, if the topology of the underlying swarm is strongly connected, the swarm is then stable in the sense that all agents will globally and exponentially converge to a hyperellipsoid in finite time, both in open space and profiles, whether the center of the hyperellipsoid is moving or not. The swarm boundary and convergence rate are characterized by the eigenparameters of the swarm, which reveals the quantitative relationship between the swarming behavior and characteristics of the coupling topology.

Beni proposed a sufficient condition for the asynchronous convergence of a linear swarm to a synchronously achievable configuration since a large class of distributed robotic systems self-organizing tasks can be mapped into reconfigurations of patterns in swarms. The model and stability analysis is, however, quite similar to the model and proof of stability for the load balancing problem in computer networks.

15.5 Swarm Intelligence

Gerardo Beni and Jing Wang introduced the term swarm intelligence in a 1989 article. Swarm intelligence techniques are population-based stochastic methods used in combinatorial optimization problems in which the collective behavior of relatively simple individuals arises from their local interactions with their environment to produce functional global patterns. Swarm intelligence represents a metaheuristic approach to solving a variety of problems. Although there is typically no centralized control dictating the behavior of the agents, local interactions among the agents often cause a global pattern to emerge. Examples of systems like this can be found in nature, including ant colonies, bird flocking, animal herding, honey bees, bacteria, and many more. Swarm-like algorithms, such as Particle Swarm Optimization (PSO) and Ant Colony Optimization (ACO), have already been applied successfully to solve real world optimization problems in various engineering applications.

Particle Swarm Optimization (PSO) was proposed by James Kennedy and R. C. Eberhart in 1995, inspired by social behavior of organisms such as bird flocking and fish schooling. Beyond its application to solving optimization problems, PSO algorithm also serves as a tool for representing sociocognition of human and artificial agents, based on principles of social psychology. PSO as an optimization tool, provides a population-based search procedure in which individuals called particles change their position (state) with time. In a PSO system, particles fly around in a multidimensional search space. During flight, each particle adjusts its position according to its own experience, and according to the experience of a neighboring particle, making use of the best position encountered by itself and its neighbor. Thus, as in modern GAs and memetic algorithms, a PSO system combines local search methods with global search methods, attempting to balance exploration and exploitation.

Ant Colony Optimization (ACO) is a class of algorithms, whose first member, called Ant System, was initially proposed by Colorni, Dorigo and Maniezzo. ACO algorithms inspired by the foraging behavior of natural ant colonies, are applied to solve optimization problems. The essential trait of ACO algorithms is the combination of a prior information about the structure of a promising solution with a posteriori information about the structure of previously obtained good solutions. A colony of ants moves through states of the problem corresponding to partial solutions of the problem to solve. They move by applying a stochastic local decision policy based on two parameters, called trails and attractiveness. By moving, each ant incrementally constructs a solution to the problem. When an ant completes a solution, or during the construction phase, the ant evaluates the solution and modifies the trail value on the components used in its solution. This pheromone information will direct the search of the future ants.

15.5.1 Properties of a Swarm Intelligence System

The swarm as a whole is capable of presenting an intelligent behavior as a result of the interaction of neighbor individuals, based on simple rules. A typical swarm intelligence system possess the following properties:

- Unity: A swarm is a combination of several individuals.

- Fault tolerance: Swarm intelligent processes do not rely on a centralized control mechanism. Therefore the loss of a few nodes or links does not result in catastrophic failure, but rather leads to graceful, scalable degradation.

- Rule-based behavior: A certain set of rules are observed by the individuals that exploit only local information that the individuals exchange directly or through the environment.

- Autonomy: The overall behavior of the swarm system is self-organized, it does not depend on orders external to the system itself. No human supervision is required.

- Scalability: Population of the agents can be adapted according to the network size. Scalability is also promoted by local and distributed agent interactions.

- Adaptation: The individuals present in the ant system change, die or reproduce, according to the entire network changes.

- Speed: The individuals in the group change their behavior rapidly, according to the neighbors. The propagation is very fast.

- Modularity: The behavior of agents is independent of the others in the group.

- Parallelism: The operations of the individuals are inherently parallel.

15.6 Particle Swarm Optimization (PSO)

In 1995, Dr. Eberhart and Dr. Kennedy developed PSO, a population based on stochastic optimization strategy, inspired by social behavior of flock of birds, school of fish, swarm of bees and even sometimes social behaviour of human. Though PSO is similar to Genetic Algorithms (GA) interms of population initialization with random solutions and searching for global optima in successive generations, PSO does not undergo crossover and mutation, whereas the particles move through the problem space following the current optimum particles. The underlying concept is that, for every time instant, the velocity of each particle also known as the potential solution, changes between its pbest and lbest locations. The particle associated with the best solution (fitness value) seems to be the leader and each particle keeps track of its coordinates in the problem space. This fitness value is stored which is referred to as pbest. Another "best" value that is tracked by the particle swarm optimizer is the best value, obtained so far by any particle in the neighbors of the particle. This location is called lbest. when a particle takes all the population as its topological neighbors, the best value is a global best and is called gbest.

The Canonical Model, Parameters of PSO, Performance Comparison with Some Global Optimization Algorithms, Extended Models of PSO for Discrete Problems and Binary PSO are discussed in this section.

15.6.1 Mathematical Model of PSO

The swarm of particles initialized with a population of random candidate solutions move move through the d-dimension problem space to search the new solutions. The fitness, f, can be calculated as the certain qualities measure. Each particle has a position represented by a position-vector presenti (i is the index of the particle), and a velocity represented by a velocity-vector velocityi. After every iteration the best position-vector among the swarm so is stored in a vector. The update of the velocity from the previous velocity to the new velocity is determined by Equation (15.1). The new position is then determined by the sum of the previous position and the new velocity by Equation (15.2).

$$velocity_{ij}(new) = w * velocity_{ij}(old) + c_1 rand_1(pbest_{ij}(old))$$
$$- present_{ij}(old) + c_2 rand_2(gbest_j(old))$$
$$- present_{ij}(old)) \tag{15.1}$$

$$present_{ij}(new) = present_{ij}(old) + velocity_{ij}(new) \tag{15.2}$$

Here w is the inertia weight, $rand_1$ and $rand_2$ are the random numbers usually chosen between [0,1]. c_1 is a positive constant, called as coefficient of the self-recognition component, c_2 is a positive constant, called as coefficient of the social component and the choice of value is $c_1 = c_2 = 2$ generally referred to as learning factors. From Equation (15.1), a particle decides where to move next, considering its own experience, which is the memory of its best past position, and the experience of its most successful particle in the swarm. In the particle swarm model, the particle searches the solutions in the problem space with a range [-s, s] (If the range is not symmetrical, it can be translated to the corresponding symmetrical range.) In order to guide the particles effectively in the search space, the maximum moving distance during one iteration must be clamped in between the maximum velocity $[-velocity_m ax, velocity_m ax]$ given in Equation (15.3):

$$velocity_{ij} = sign(velocity_{ij}) min(velocity_{ij} | velocity_{max})$$

The value of vmax is p s, with $0.1 = p = 1.0$ and is usually chosen to be s, i.e. p = 1. The pseudo-code for particle swarm optimization algorithm is illustrated as follows:

15.6.2 Algorithm of PSO

The pseudocode of the Particle Swarm Optimization Algorithm is shown below:

```
Initialize the size of the particle swarm n
Initialize the positions and the velocities for
all the particles randomly
While end criterion false do
    t=t+1
    Compute fitness value of each particle
```

$$x\star = \textbf{arg min}_{t-1}^n(f(x*(t-1)), f(x_1(t)), f(x_2(t)),$$
$$.......f(x_t(t)), ...f(x_n(t)));$$

```
    For i=1 to n
```

$$x_t^\#(t) = argmin_{t-1}^n(f(x_t^\#(t-1)), f(x_t(t)))$$

```
    textbfFor j=1 to Dimension
```

Update the j-th dimension value of x_t *and* v_t

$$v_{ij}(t+1) = wvi_j(t) + c_1r_1(x_{ij}^*(t) - x_{ij}(t)) + c_2r_2(x_j^*(t) - x_{ij}(t))$$
$$xi_j(t+1) = x_{ij}(t) + v_{ij}(t+1)$$
$$v_{ij} = sign(v_{ij})min(|v_{ij}|, v_{max})$$

```
    End For
      End For
End While
```

15.6.3 Parameters and Tuning of Parameters in PSO

PSO does not require a large number of parameters to be initialized. The initialization process is quite simple. This section gives a list of parameters and initialization process of PSO.

Number of particles: The number of particles is a very important factor to be considered. For most of the practical applications a best choice of the number of particles is typically in the range [20,40]. Usually 10 particles is a large number which is sufficient enough to get best results. In case of difficult problems the choice can be 100 or 200 particles also.

Inertia Weight: The inertia weight plays a very important role in the convergence behavior of the PSO algorithm. The inertia weight is employed to control the impact of the previous history of velocities on the current one. Accordingly, the parameter w regulates the trade-off between the global (wide-ranging) and local (nearby) exploration abilities of the swarm. Too large an inertia weight aids in global exploration (searching wide ranging areas), while too small an inertia weight aids in local exploration (searching within the nearby areas). To obtain a balance between the global and local exploration the number of iterations required to locate the optimum solution are reduced. The inertia weight is set as a constant initially and in order to promote global exploration of the search space, the parameter is gradually decreased to get more optimal solutions. Usually the best choice of the inertia weight is around 1.2, and as the algorithm progresses this value is gradually decreased to 0.

Learning factors: The parameters c1 and c2, coefficient of self recognition and social components, are not much critical for the convergence of PSO. Fine-tuning of these learning vectors aids in faster convergence and alleviation of local minima. Usually the choice for these parameters is, $c_1 = c_2 = 2$, but some experiment results indicate that $c_1 = c_2$ = #1.49 might provide even better results. Recent papers report that it might be even better to choose a larger self recognition component, c1, than the social component, c_2, such that it satisfies the condition c1+c2 = 4.

Range and dimension of particles: The particle dimension and range is determined based on the problem to be optimized. Various ranges can be chosen for different dimension of particles.

$Velocity_{max}$: The maximum change one particle can take during one iteration is defined as the maximum velocity and denoted as $Velocity_{max}$. Usually the range of particles is set as the maximum velocity. For instance, if a particle belongs to the range [-5, 5], then the maximum velocity is 10.

Stopping condition: The stopping condition may be any one of the following criteria:

- The process can be terminated after a fixed number of iterations like 500, 1000 iterations etc.

- The process may be terminated when the error between the obtained objective function value and the best fitness value is less than a pre-fixed anticipated threshold.

15.6.4 Neighborhood Topologies

Various types of neighbourhood topologies are investigated based on the behavioral movement of swarms The most common neighborhood topologies are:

a. Star (or wheel) topology.

b. Ring (or circle) topology.

c. Von Neumann (or Square) topology.

Star Topology

Similar to computer network topology, star Topology, which is also known as gbest model, is a fully connected neighborhood relation. The star topology is one of the most common network setups where all the particles in the workspace connect to a specific particle selected as a central hub. With this topology, the propagation is very fast, resulting in premature convergence problem. Figure 15.1(a) illustrates the star neighborhood topologies.

Ring Topology

In ring topology (Figure 15.1b), which is also known as lbest model, each particle is connected on a single circle to its immediate neighbors. Since there are no terminated ends, compared to the star topology the flow of information in ring topology is very slow. However, using the ring topology will slow down the convergence rate because the best solution found has to propagate through several neighborhoods before affecting all particles in the swarm. This slow propagation will enable the particles to explore more areas in the search space and thus decrease the chance of premature convergence.

Von Neumann Topology

Kennedy and Mendes studied the various population topologies on the PSO performance and proposed the Von Neumann topology. Their experiments showed that the topology can be observed as a spatial neighborhood when it is determined by the Euclidean distance between the positions of two particles, or as a sociometric neighborhood. Though Von Neumann is a type of lbest model, in Von Neumann topology, particles are connected using a grid network like a lattice structure where each

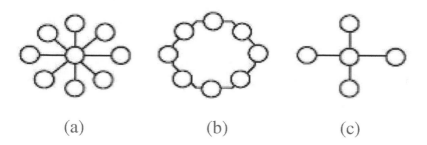

(a)   (b)   (c)

FIGURE 15.1: Neighborhood Topologies a) Star, b) Ring, and c) Von Neumann

particle is connected to its four neighbor particles as shown in Figure 15.1(c). Similar to ring topology, using Von Newmann topology will slow down the convergence rate. Slow propagation will enable the particles to explore more areas in the search space and thus decrease the chance of premature convergence.

15.7   Extended Models of PSO

15.7.1   PSO-I

The local is kept in the standard PSO, as the next position of each particle is sampled from the global search range, sampling continues until the next position of each particle is in the local range. The sampling range is as follows:

If a ≥ 0 then R(global) = $[0,2d_{max}]$ where $d_{max} = max(d_1, d_2)$

If a ≤ 0 and b ≥ 0 then R(global) = $[d_{min}, 0] \bigcup [0, 2d_{max}]$

where $d_{max} = max(d_1, d_2)$ and $d_{min} = min(d_1, d_2)$

If b < 0 then R(global) = $[2d_{min}, 0]$ where $d_{min} = min(d_1, d_2)$

15.7.2   PSO-II

In PSO-2, a particle close to the current best particle is randomly generated in every generation and each dimension with a threshold p which is drawn in the interval $[0,1]$. The velocities of the particles are used in the next generation. "s" is randomly drawn from an uniform distribution over the interval $[-r,r]$, and r is the radius of the area around the global best position x_{gB} in the whole population. r is determined according to the attributes of the test functions. If $l < p$, the actual position is calculated and if $l \geq p$, $x(t+1)$ is computed which is the same as that obtained from SPSO. The actual position is calculated as

$$\bar{x}(t+1) = \begin{cases} x_{gB} + s, & if\ l < p \\ x(t+1), & if\ l \geq p \end{cases}$$

15.7.3 PSO-III

The objective of PSO-III is to adjust the particles to search the solution space more appropriately. The probability of particles landing into the range around x_{gB} is increased, maintaining the probability of particles flying away from x_{gB}, and the probability of particles wandering around the small range containing x_{gB} is increased. Thus the particles are given only two choices, either landing very near x_{gB} to enhance the local search ability or landing far from xgB to keep the global search capability. Therefore PSO-III simply makes the particles search the space more pertinently and efficiently.

In this PSO variant velocity affects the behaviors of the PSO. Here a range $[-r,r]$ is set around the best individual in each dimension for every generation. As the particles in the swarm pass through the range in the next generation, a magnifier operator is used to enlarge the range without changing the velocity of particles. Thus, the particles would get a better chance to land into the range, which is able to check the area around the current best individual more precisely. On the other hand, the velocity of the particles is maintained so that they are able to fly out of the range in certain generations to maintain the global search ability. The transition process of a particle x from the tth to $t+1^{th}$ generation in each dimension can be schematically expressed as four situations as the following respectively, and in each situation, the position after using the magnifier operator in PSO-III will be calculated:

$$
\bar{x}(t+1) = \begin{cases} x(t+1) - (2*r/s - 2*r) & if\ x(t) < L\ and\ x(t+1) > R, \\ x(t+1) + (2*r/s - 2*r) & if\ x(t) > R\ and\ x(t+1) < L, \\ x(t+1) - [(R - x(t))/s - (R - x(t))] & if L < x(t) < R\ and\ x(t+1) > R, \\ x(t+1) + [(x - (t) - L)/s - (x(t) - L)] & if L < x(t) < R\ and\ x(t+1) < R, \end{cases}
$$

r is the radius of the interval with L and R be the left and right boundary, and s is the scale which decides the multiple that enlarges the range. s should not be too small, it should be decided such that the particles do not fall into the range too easily and hard to fly out, which lead to a prematurity. On the other hand, r should reduce along with the growth of generations, because the best individual should converge inside the range. So, an initial value should be fixed for r, from which r reduces linearly to zero. The iterative equation of r is expressed by

$$
r = r * (1 - k/M)
$$

where k is the current iteration number and M is the maximum iteration number.

15.7.4 PSO-IV

In PSO-4, two sub-swarms are used for the evolution. One sub-swarm use the standard formula. The other sub-swarm with n particles using the following formula

$$V_{id}(t+1) = wV_{id}(t) + (rc_1(r_1 - 0.5) + r/2)(P_{iBD}(t) - X_{id}(t))$$
$$+ (rc_2(r_2 - 0.5) + r/2)(P_{gBd}(t) - X_{id}(t))$$

where r is used to tune the sampling range of X and Y. The two sub-swarms use different sampling range to evolve. So the performance of the PSO should be changed.

15.8 Ant Colony Optimization

In nature, ants usually wander randomly, and upon finding food return to their nest while laying down pheromone trails. The other ants find the path (pheromone trail), and follow the trail, returning and reinforcing it if they eventually find food. The pheromone starts to evaporate as time passes. If the time taken for an ant to travel down the path and back again to the nest, the pheromone evaporates thereby making the path less prominent. A shorter path, in comparison will be visited by more ants (can be described as a loop of positive feedback) and thus the pheromone density remains high for a longer time. ACO is implemented as a collective group of intelligent agents, which simulate the ants behavior, walking around the graph representing the problem to solve using mechanisms of cooperation and adaptation. ACO algorithm requires the following definitions:

- The problem needs to be represented appropriately, which would allow the ants to incrementally update the solutions through the use of a probabilistic transition rules, based on the amount of pheromone in the trail and other problem specific knowledge. It is also important to enforce a strategy to construct only valid solutions corresponding to the problem definition

- A problem-dependent heuristic function η that measures the quality of components that can be added to the current partial solution

- A rule set for pheromone updating, which specifies how to modify the pheromone value τ

- A probabilistic transition rule based on the value of the heuristic function η and the pheromone value τ that is used to iteratively construct a solution

15.8.1 Mathematical Model of ACO

ACO was first introduced using the Traveling Salesman Problem (TSP). Starting from its start node, an ant iteratively moves from one node to another. When being at a node, an ant chooses to go to a unvisited node at time t with a probability given by

$$P_{i,j}^k(t) = \frac{[\tau_{i,j(t)}]^\alpha [\eta_{i,j}(t)]^\beta}{\sum_{t\epsilon N_i^k}[\tau_{i,j}(t)]^\alpha [\eta_{i,j}(t)]^\beta} \quad j\epsilon N_i^k \tag{15.16}$$

where N_i^k is the feasible neighborhood of the ant$_k$, that is, the set of cities which ant$_k$ has not yet visited; $\tau_{i,j}(t)$ is the pheromone value on the edge (i,j) at the time t, α is the weight of the pheromone; $\eta_{i,j}(t)$ is *a priori* available heuristic information on the edge (i,j) at the time t, β is the weight of heuristic information. Two parameters α and β determine the relative influence of pheromone trail and heuristic information. $\tau_{i,j}(t)$ is determined by

$$tau_{i,j}(t) = \rho\tau_{i,j}(t-1) + \sum_{k=1}^{n} \triangle\tau_{i,j}{}^k \forall(i,j) \tag{15.17}$$

$$\triangle\tau_{i,j}{}^k(t) = \begin{cases} \frac{Q}{L_k(t)} & \textit{if the edge } (i,j) \textit{ chosen by ant}_k \\ 0 & \textit{otherwise} \end{cases} \tag{15.18}$$

where ρ is the pheromone trail evaporation rate ($0<\rho<1$), n is the number of ants, Q is a constant for pheromone updating. A generalized version of the pseudo-code for the ACO algorithm is illustrated in the following section.

15.8.2 Ant Colony Optimization Algorithm

```
Initialize the number of ants n
While end criterion false do
    t=t+1
    For k=1 to n
    antk is positioned on a starting node;
  For m=2 to problem_size
    Choose the state according to the probabilistic
    transition rules
    Append the chosen move into tabuk(t) for the antk
```

End For
Update the trail pheromone intensity for every edge (i,j)

$$\tau_{i,j}(t) = \rho\tau_{i,j}(t-1) + \sum_{k=1}^{n} \triangle\tau_{i,j}{}^{k}(t)\forall(i,j)$$

$$\triangle\tau_{i,j}{}^{k}(t) = \begin{cases} \frac{Q}{L_k(t)} & if \ the \ edge \ (i,j) chosen \ by \ ant_k \\ 0 & otherwise \end{cases}$$

Compare and update the best solution
  End For
End While

15.9   Studies and Applications of Swarm Intelligence

In this section a few illustrations based on scientific and engineering swarm intelligence studies are given.

15.9.1   Ant-based Routing

Dorigo et al. and Hewlett Packard during the mid-1990's applied Swarm Intelligence to the field of Telecommunication Networks and referred this as Ant Based Routing. The application involves a probabilistic routing table rewarding or reinforcing the route successfully traversed by each "ant" (a small control packet). Reinforcement of the route in the forward, reverse direction and both simultaneously have been studied and analyzed in their research. A symmetric network is used for backward reinforcement, which couples the network bidirectionally. The forward reinforcement rewards a route before the outcome is known.

15.9.2   Clustering Behavior of Ants

The main idea of the clustering algorithm based on swarm intelligence was proposed by Deneubourg. The process of clustering is defined in three stages. In the first stage, the data objects are chosen at random and are projected onto a low dimensional space. During the second phase, simple agents perceive the swarm similarity of the current object with the local region, and compute the probability of pick-up and drop. And finally during the third phase, the cluster centers are formed by the simple agents based on collective behavior.

15.9.3 Graph Coloring

Graph coloring was used as a benchmark for solving distributed constraint satisfaction problems is offered. All the agents in the space correspond one-to-one with the nodes of the proposed graph. The agents whose nodes are directly connected are enemies; agents whose nodes are at shortest graph distance are friends. Usually the agents are attracted to friends while moving from one place to the other such as to avoid enemies. The colors for the graph coloring correspond to specific attraction spatial regions and the agents are attracted toward these regions. The graph coloring solution is given by the distribution of the agents over the color attraction regions. If all the agents are placed in one of the color attraction space regions, the system configuration can be interpreted as defining a complete coloration of the graph. If some agents are outside these regions, the system configuration corresponds to a partial solution to the coloration problem.

15.9.4 Machine Scheduling

Machine scheduling application is used to assign a set of machines $M_1, M_2, ... M_n$ to a set of jobs $J_1, J_2, ..., J_m$. Each job consists of a set of operations $Oj_1, Oj_2, ... Oj_p$ and the machines are assigned jobs based on certain constraints like

- A machine can perform only one operation at a time
- The operations on different jobs do not have any priority
- A running operation cannot be interrupted

Let T represent the time taken for a machine Mi to complete an operation O_{jk}. Let ΣT_i denote the time taken by a machine Mi to complete all the assigned jobs. Thus the objective of a machine scheduling problem is to minimize the maximum completion time defined as max $\{\sum T_i\}$ and to minimize the sum of completion times defined as $\sum_{i=1}^{n} \{\Sigma T_i \}$

For machine scheduling problems the parameters of PSO algorithm are initialized as follows:

Learning vectors c1 = c2 = 1.49
Inertia weight w = 0.9 initially and gradually decreased to 0.1
Particle size = 20

The parameters used for machine scheduling problem by applying the ACO algorithm are as follows:

Number of ants = 5
Weight of pheromone trail $\alpha = 1$
Weight of heuristic information $\beta = 5$
Pheromone evaporation parameter $\rho = 0.8$
Constant for pheromone updating Q = 10

15.9.5 Quadratic Knapsack Problem

The 0-1 quadratic knapsack problem (QKP) is a hard computational problem, which is a generalization of the knapsack problem (KP). This problem was introduced by Gallo, which was used to choose elements from n items for maximizing a quadratic profit objective function subject to a linear capacity constraint. The problem is defined as follows:

$$\text{Objective: To maximize f(x)} = \sum_{i=1}^{n} p_i x_i + \sum_{i=1}^{n-1} \sum_{j=i+1}^{n} p_{ij} x_i x_j$$

$$\text{Subject to the constraint: g(x)} - \sum_{i=1}^{n} w_i x_i - C \leq 0$$

where p_i and p_{ij} are profit coefficients, wi is weight coefficient and c denotes the capacity of the knapsack. All these coefficients are non-negative integers. x_i denotes the item that is to be selected.

The best choice of parameters used by PSO algorithm for solving the QKP problem can be initialized as follows:

Learning vectors c1 = c2 = 1.49445
Inertia weight w = 0.729 initially and gradually decreased to 0.1
Particle size = 20
Range and dimension of particles = 2.8 x 109

The best choice of parameters used by ACO algorithm for solving the QKP problem can be initialized as follows:

Number of ants = 30
Weight of pheromone trail $\alpha = 1$
Weight of heuristic information $\beta = 5$
Pheromone evaporation parameter $\rho = 0.01$
Constant for pheromone updating Q = 10

15.9.6 Traveling Salesman Problem

The objective of the traveling salesman problem is defined to visit a number of cities, plan the trip such that every city is visited at-least once and that the whole trip is as short as possible. The Traveling Salesman Problem (TSP) is a simple combinatorial problem in computational mathematics. The salesman starts from a city and travels through n number of cities cyclically. Once he has visited all the cities he finishes up where he started. The constraint is to minimize the distance traveled and such as to find a proper order of the cities to be traveled. More formally, the TSP can be represented by a complete weighted graph G = (N,A) where N denotes the set of nodes, representing the cities, and A the set of arcs fully connecting the nodes N. Each arc is assigned a value d_{ij} , which is the length of arc (i, j) ϵ A, that is, the distance between cities i and j, with i, j ϵ N.

The best choice of parameters used by PSO algorithm for solving the TSP problem can be initialized as follows:

> Learning vectors c1 = c2 = 1
> Inertia weight w = initially and gradually decreased to 0.1
> Particle size = 20

The best choice of parameters used by ACO algorithm for solving the TSP problem can be initialized as follows:

> Number of ants = 25
> Weight of pheromone trail $\alpha = 1$
> Weight of heuristic information $\beta = 2$
> Pheromone evaporation parameter $\rho = 0.2$
> Constant for pheromone updating Q = 20

15.10 MATLAB Examples of Swarm Intelligence

A few applications of Particle Swarm Optimization and Ant Colony Optimization such as Traveling Salesman Problem (TSP), Quadratic Assignment Problem (QAP), behavior and simulation of swarms in MATLAB are explained in this section.

15.10.1 Illustration 1: Simulation of the Movement of Swarm to Minimize the Objective Function

```
%% Particle Swarm Optimization Simulation
% Simulates the movements of a swarm to minimize
% the objective function
% (x - 15)^ 2 + (y - 20)^ 2

% The swarm matrix is
%
% swarm(index, [location, velocity, best position,
best
% value], [x, y components or the value component])
%

%% Initialization
% Parameters
clear
clc
iterations = 30;
inertia = 1.0;
correction factor = 2.0;
swarm_size = 49;
% - - - - initial swarm position - - - -
index = 1;
for i = 1 : 7
   for j = 1 : 7
 swarm(index, 1, 1) = i;
 swarm(index, 1, 2) = j;
 index = index + 1;
   end
end

swarm(:, 4, 1) = 1000; % best value so far
swarm(:, 2, :) = 0;      % initial velocity
%% Iterations
for iter = 1 : iterations

   %- - evaluating position & quality - - -
   for i = 1 : swarm_size
     swarm(i, 1, 1) = swarm(i, 1, 1) + swarm(i, 2, 1)/
1.3;
% update x position
```

```
    swarm(i, 1, 2) = swarm(i, 1, 2) + swarm(i, 2, 2)/
1.3;
% update y position
x = swarm(i, 1, 1);
y = swarm(i, 1, 2);

val = (x - 15)^ 2 + (y - 20)^ 2; % fitness evaluation

if val <swarm(i, 4, 1) % if new position is better
  swarm(i, 3, 1) = swarm(i, 1, 1); % update best x,
  swarm(i, 3, 2) = swarm(i, 1, 2);% best y postions


  swarm(i, 4, 1) = val;   % and best value
    end
  end

 [temp, gbest] = min(swarm(:, 4, 1)); % global best
position

  %--- updating velocity vectors
  for i = 1 : swarm_size
    swarm(i, 2, 1) = rand*inertia*swarm(i, 2, 1) +
correction_factor*rand*(swarm(i, 3, 1)
  - swarm(i, 1, 1)) +
correction_factor*rand*(swarm(gbest, 3, 1)
  - swarm(i, 1, 1));
% x velocity component
    swarm(i, 2, 2) = rand*inertia*swarm(i, 2, 2) +
correction_factor*rand*(swarm(i, 3, 2)
  - swarm(i, 1, 2)) +
correction_factor*rand*(swarm(gbest, 3, 2)
  - swarm(i, 1, 2));
% y velocity component
  end

  %% Plotting the swarm
  clf
  plot(swarm(:, 1, 1), swarm(:, 1, 2), 'x') % drawing
% swarm
  movements axis([-2 30 -2 30]);
  title('Swarm movements')
  xlabel('Variable x')
```

```
    ylabel('Variable y')
pause(5)
end
```

Observations

Figure 15.2 shows the movement of the swarms for the given objective function. The objective function can be changed according to the requirement.

15.10.2   Illustration 2: Behavior of Particle Swarm Optimization

This section illustrates the MATLAB program used to minimize or maximize a set of functions such as Ackley, Alpine, DeJong, Foxhole, Rosenbrock, etc. The observations are obtained by minimizing the Rosenbrock function.

```
% The PSO tries to find the minimum of the
% Rosenbrock function,
% a standard benchmark
clear all
close all
clc
help demopsobehavior
warning off
functnames = {'ackley','alpine','DeJong_f2','DeJong_f3',
    'DeJong_f4',...
    'Foxhole','Griewank','NDparabola',...
    'Rastrigin','Rosenbrock','f6','f6mod','tripod',...
    'f6_bubbles_dyn','f6_linear_dyn','f6_spiral_dyn'};

    disp('Static test functions, minima don''t change
w.r.t. time/iteration:');
disp('  1) Ackley');
disp('  2) Alpine');
disp('  3) DeJong_f2');
disp('  4) DeJong_f3');
disp('  5) DeJong_f4');
disp('  6) Foxhole');
disp('  7) Griewank');
disp('  8) NDparabola (for this demo N = 2)');
disp('  9) Rastrigin');
disp(' 10) Rosenbrock');
disp(' 11) Schaffer f6');
disp(' 12) Schaffer f6 modified (5 f6 functions translated
from each other)');
```

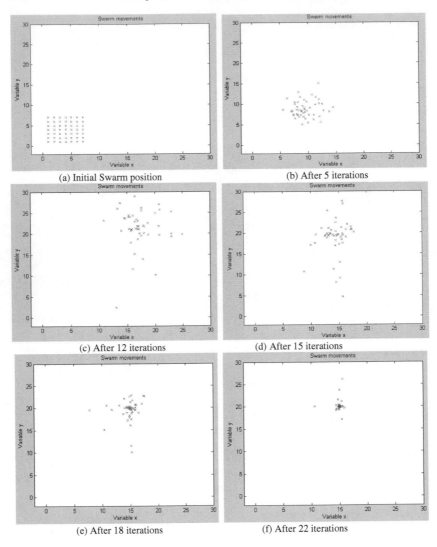

(a) Initial Swarm position

(b) After 5 iterations

(c) After 12 iterations

(d) After 15 iterations

(e) After 18 iterations

(f) After 22 iterations

FIGURE 15.2: Movement of the Swarms - The Given Function Reached Optimum Value at the End of 30 Iterations

```
disp('13) Tripod');
disp(' ');
```

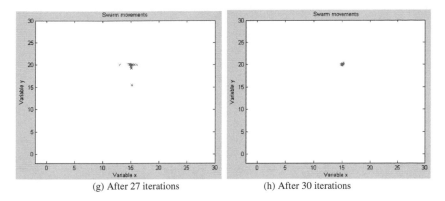

(g) After 27 iterations (h) After 30 iterations

FIGURE 15.2: (*Continued*)

```
disp('Dynamic test functions, minima/environment evolves
over time/iteration:');
disp('14) f6_bubbles_dyn');
disp('15) f6_linear_dyn');
disp('16) f6_spiral_dyn');

functchc-input('Choose test function ? ');
functname = functnames{functchc};


disp(' ');
disp('1) Intense graphics, shows error topology and
surfing particles');
disp('2) Default PSO graphing, shows error trend and
particle dynamics');
disp('3) no plot, only final output shown, fastest');
plotfcn=input('Choose plotting function ? ');
if plotfcn == 1
  plotfcn = 'goplotpso4demo';
  shw = 1; % how often to update display
elseif plotfcn == 2
  plotfcn = 'goplotpso';
  shw = 1; % how often to update display
else
  plotfcn = 'goplotpso';
  shw = 0; % how often to update display
end

% set flag for 'dynamic function on', only used at very
end for tracking plots
dyn_on = 0;
```

```
if functchc==15 | functchc == 16 | functchc == 17
  dyn_on = 1;
end

% xrng=input('Input search range for X, e.g.
% [-10,10] ? ');
% yrng=input('Input search range for Y ? ');
xrng=[-30,30];
yrng=[-40,40];
disp(' ');
% if =0 then we look for minimum, =1 then max
  disp('0) Minimize')
  disp('1) Maximize')
  minmax=input('Choose search goal ?');
  % minmax=0;
  disp(' ');
  mvden = input('Max velocity divisor (2 is
a good choice) ? ');
  disp(' ');
  ps = input('How many particles (24 - 30 is common)? ');
  disp(' ');
  disp('0) Common PSO - with inertia');
  disp('1) Trelea model 1');
  disp('2) Trelea model 2');
  disp('3) Clerc Type 1" - with constriction');
  modl = input('Choose PSO model ? ');
% note: if errgoal=NaN then unconstrained min or max
% is performed
  if minmax==1
    % errgoal=0.97643183; % max for f6 function (close
    % enough for
termination)
    errgoal=NaN;
  else
    % errgoal=0;    errgoal=NaN;
  end
  minx = xrng(1);
  maxx = xrng(2);
  miny = yrng(1);
  maxy = yrng(2);
%--------------------------------------------------
  dims=2;
  varrange=[];
  mv=[];
  for i=1:dims
    varrange=[varrange;minx maxx];
    mv=[mv;(varrange(i,2)-varrange(i,1))/mvden];
```

```
end

ac = [2.1,2.1];% acceleration constants, only
used for modl=0
Iwt = [0.9,0.6]; % intertia weights, only used for
modl=0
epoch = 400; % max iterations
wt_end = 100; % iterations it takes to go from Iwt(1)
to Iwt(2),
 only for modl=0
errgrad = 1e-99; % lowest error gradient tolerance
errgraditer=100; % max # of epochs without error
change >= errgrad
PSOseed = 0; % if-1 then can input particle starting
positions,
if= 0 then all random
% starting particle positions (first 20 at zero, just
for an example)
PSOseedValue = repmat([0],ps-10,1);

psoparams=...
[shw epoch ps ac(1) ac(2) Iwt(1) Iwt(2) ...
wt_end errgrad errgraditer errgoal modl PSOseed];
% run pso
% vectorized version
[pso out,tr,te]=pso_Trelea_vectorized(functname, dims,...
  mv, varrange, minmax, psoparams,plotfcn,PSOseedValue);

%-------------------------------------------------
% display best params, this only makes sense for static
% functions, for dynamic you'd want to see a time history
% of expected versus optimized global best values.
disp(' ');
disp(' ');
disp(['Best fit parameters: ']);
disp([' cost = ',functname,'( [ input1, input2 ] )']);
disp(['--------------------------------']);
disp([' input1 = ',num2str(pso_out(1))]);
disp([' input2 = ',num2str(pso_out(2))]);
disp([' cost = ',num2str(pso_out(3))]);
disp(['   mean cost = ',num2str(mean(te))]);
disp([' # of epochs = ',num2str(tr(end))]);

%% optional, save picture
% set(gcf,'InvertHardcopy','off');
% print -dmeta
% print('-djpeg',['demoPSOBehavior.jpg']);
```

```
-----------------------------------------------
% Sub Functions
-----------------------------------------------
% goplotpso.m
clf
 set(gcf,'Position',[651 31 626 474]); % this is the
     computer dependent part
 % set(gcf,'Position',[743 33 853 492]);
 set(gcf,'Doublebuffer','on');

 plot3(pos(:,1),pos(:,D),out,'b.','Markersize',7)
 hold on
 plot3(pbest(:,1),pbest(:,D),pbestval,'g.'
 ,'Markersize',7);
 plot3(gbest(1),gbest(D),gbestval,'r.','Markersize',25);

 % crosshairs
 offx = max(abs(min(min(pbest(:,1)),min(pos(:,1)))),...
 abs(max(max(pbest(:,1)),max(pos(:,1)))));

 offy = max(abs(min(min(pbest(:,D)),min(pos(:,D)))),...
 abs(min(max(pbest(:,D)),max(pos(:,D)))));
 plot3([gbest(1)-offx;gbest(1)+offx],...
     [gbest(D);gbest(D)],...
     [gbestval;gbestval],...
     'r-.');
 plot3([gbest(1);gbest(1)],...
     [gbest(D)-offy;gbest(D)+offy],...
     [gbestval;gbestval],...
     'r-.');

 hold off

 xlabel('Dimension 1','color','y')
 ylabel(['Dimension ',num2str(D)],'color','y')
 zlabel('Cost','color','y')

 title('Particle Dynamics','color','w','fontweight',
 'bold')

 set(gca,'Xcolor','y')
 set(gca,'Ycolor','y')
 set(gca,'Zcolor','y')
 set(gca,'color','k')
 % camera control
 view(2)
 try
```

```
  axis([gbest(1)-offx,gbest(1)+offx,gbest(D)-offy,
    gbest(D)+offy]);
 catch
    axis([VR(1,1),VR(1,2),VR(D,1),VR(D,2)]);
 end

% error plot, left side
 subplot('position',[0.1,0.1,.475,.825]);
   semilogy(tr(find(~isnan(tr))),'color','m','linewidth',2)
   % plot(tr(find(~isnan(tr))),'color','m','linewidth',2)
   xlabel('epoch','color','y')
   ylabel('gbest val.','color','y')

   if D==1
     titstr1=sprintf(['% 11.6g = % s( [ % 9.6g ] )'],...
 gbestval,strrep(functname,'_','\_'),gbest(1));
   elseif D==2
     titstr1=sprintf(['% 11.6g = % s( [ % 9.6g, % 9.6g ]
 )'],... gbestval,strrep(functname,'_','\_'),
 gbest(1),gbest(2));
   elseif D==3
     titstr1=sprintf(['% 11.6g = % s( [ % 9.6g, % 9.6g,
      % 9.6g ] )'],...

 gbestval,strrep(functname,'_','\_'),gbest(1),gbest(2),
 gbest(3));
   else
     titstr1=sprintf(['% 11.6g = % s( [ % g inputs ] )'],
 ... gbestval,strrep(functname,'_','\_'),D);
   end
   title(titstr1,'color','m','fontweight','bold');

   grid on
% axis tight

   set(gca,'Xcolor','y')
   set(gca,'Ycolor','y')
   set(gca,'Zcolor','y')
   set(gca,'color','k')
   set(gca,'YMinorGrid','off')
% text box in lower right
% doing it this way so I can format each line any
% way I want
subplot('position',[.62,.1,.29,.4]);
   clear titstr
   if trelea==0
     PSOtype = 'Common PSO';
```

```
    xtraname = 'Inertia Weight : ';
    xtraval = num2str(iwt(length(iwt)));

  elseif trelea==2 |trelea==1

    PSOtype = (['Trelea Type ',num2str(trelea)]);
    xtraname = ' ';
    xtraval = ' ';
  elseif trelea==3
    PSOtype = (['Clerc Type 1"']);
    xtraname = '\chi value : ';
    xtraval = num2str(chi);

end
if isnan(errgoal)
 errgoalstr='Unconstrained';
else
 errgoalstr=num2str(errgoal);
end
if minmax==1
 minmaxstr = ['Maximize to : '];
elseif minmax==0
 minmaxstr = ['Minimize to : '];
else
 minmaxstr = ['Target to : '];
end
if rstflg==1
 rststat1 = 'Environment Change';
 rststat2 = ' ';
else
 rststat1 = ' ';
 rststat2 = ' ';
end

titstr={'PSO Model: ' ,PSOtype;...
 'Dimensions : ' ,num2str(D);...
 '# of particles : ',num2str(ps);...
 minmaxstr ,errgoalstr;...
 'Function : ' ,strrep(functname,'_','\_');...
 xtraname ,xtraval;...
 rststat1 ,rststat2};

text(.1,1,[titstr{1,1},titstr{1,2}],'color','g',
'fontweight','bold');
hold on
text(.1,.9,[titstr{2,1},titstr{2,2}],'color','m');
text(.1,.8,[titstr{3,1},titstr{3,2}],'color','m');
```

```
text(.1,.7,[titstr{4,1}],'color','w');
text(.55,.7,[titstr{4,2}],'color','m');
text(.1,.6,[titstr{5,1},titstr{5,2}],'color','m');
text(.1,.5,[titstr{6,1},titstr{6,2}],'color','w'
,'fontweight','bold');
text(.1,.4,[titstr{7,1},titstr{7,2}],'color','r'
,'fontweight','bold');

% if we are training a neural net, show a few more
% parameters
if strcmp('pso_neteval',functname)
  % net is passed from trainpso to pso_Trelea
  % _vectorized incase you are wondering where that
  % structure comes from
  hiddlyrstr = [];
  for lyrcnt=1:length(net.layers)
TF{lyrcnt} = net.layers{lyrcnt}.transferFcn;
Sn(lyrcnt) = net.layers{lyrcnt}.dimensions;
hiddlyrstr = [hiddlyrstr,', ',TF{lyrcnt}];
  end
  hiddlyrstr = hiddlyrstr(3:end);

  text(0.1,.35,['#neur/lyr = [ ',num2str(net.inputs1.
  size),' ',... num2str(Sn),' ]'],'color','c','
  fontweight','normal',... 'fontsize',10);
  text(0.1,.275,['Lyr Fcn: ',hiddlyrstr],...
'color','c','fontweight','normal','fontsize',9);

end

legstr = {'Green = Personal Bests';...
'Blue = Current Positions';...
'Red = Global Best'};
text(.1,0.025,legstr{1},'color','g');
text(.1,-.05,legstr{2},'color','b');
text(.1,-.125,legstr{3},'color','r');

hold off

set(gca,'color','k');
set(gca,'visible','off');

drawnow
---------------------End of goplotpso.m-----------
% function to force a vector to be a single column
function[out]=forcecol(in)
len=prod(size(in));
```

```
out=reshape(in,[len,1]);
----------------------End of forcecol.m-------------
% function to force a vector to be a single row
function[out]=forcerow(in)
len=prod(size(in));
out=reshape(in,[1,len]);
----------------------End of forcerow.m-------------
% returns an offset that can be added to data that
% increases linearly with
% time, based on cputime, first time it is called
% is start time
% equation is: offset = (cputime - tnot)*scalefactor
% where tnot = cputime at the first call
% scalefactor = value that slows or speeds up
% linear movement
%
% usage: [offset] = linear_dyn(scalefactor)

function out = linear_dyn(sf)
 % this keeps the same start time for each run of the
% calling function
 % this will reset when any calling prog is re-saved
% or workspace is
 % cleared
  persistent tnot
 % find starting time
 if ~exist('tnot') | length(tnot)==0
  tnot = cputime;
 end
 out = (cputime-tnot)*sf;
 return
----------------------End of linear_dyn.m----------
% spiral_dyn.m
% returns x,y position along an archimedean spiral of
% degree n
% based on cputime, first time it is called is start time
%
% based on: r = a*(theta^n)
%
% usage: [x_cnt,y_cnt] = spiral_dyn(n,a)
% i.e.,
% n = 2 (Fermat)
% = 1 (Archimedes)
% = -1 (Hyberbolic)
% = -2 (Lituus)

function [x_cnt,y_cnt] = spiral_dyn(n,a)
```

```
% this keeps the same start time for each run of the
% calling function
% this will reset when any calling prog is re-saved or
% workspace is
% cleared
persistent tnot iter
% find starting time
 if ~exist('tnot') |length(tnot)==0
   tnot = cputime;
 % iter = 0;
end
% iter = iter+10 ;

theta = cputime-tnot;
% theta = iter/10000;

r = a*(theta.^n);

x_cnt = r*cos(theta);
y_cnt - r*sin(theta);
return

---------------------End of spiral dyn m-----     --
% Rosenbrock function
%
% used to test optimization/global minimization problems
%
% f(x) = sum([ 100*(x(i+1) - x(i)^2)^2 + (x(i) -1)^2])
%
% x = N element row vector containing [x0, x1, ..., xN]
% each row is processed independently,
% you can feed in matrices of timeXN no prob
%
% example: cost = Rosenbrock([1,2;5,6;0,-50])
% note: known minimum =0 @ all x = 1

function [out]=Rosenbrock(in)

 x0=in(:,1:end-1);
 x1=in(:,2:end);

 out = sum( (100*(x1-x0.^2).^2 + (x0-1).^2) , 2);
---------------------End of Rosenbrock.m-------------
```

Observations

The observations were obtained by minimizing the Rosenbrock function. The parameters such as the dimensions, no. of particles, global best, and the current positions are plotted during different iterations in Figure 15.3.

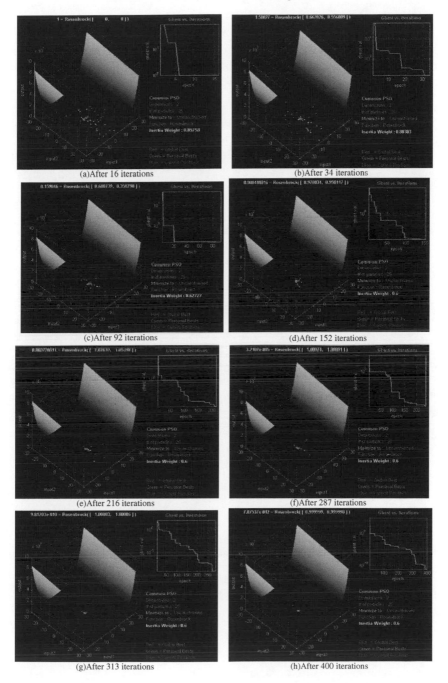

(a)After 16 iterations

(b)After 34 iterations

(c)After 92 iterations

(d)After 152 iterations

(e)After 216 iterations

(f)After 287 iterations

(g)After 313 iterations

(h)After 400 iterations

FIGURE 15.3: Performance of the Rosenbrock Function

```
Best fit parameters:
 cost = Rosenbrock( [ input1, input2 ] )
-------------------------------
    input1      = 1
    input2      = 1
    cost        = 7.0754e-012
    mean cost   = 348.636
    # of epochs = 400
```

15.10.3   Illustration 3: Ant Colony Optimization to Determine the Shortest Path

```
------------------------------------------------------------
%%%%%%%%% Main Program %%%%%%%%%%
------------------------------------------------------------
% getting information
 [x,y,d,t,h,iter,alpha,beta,e,m,n,el]=ants_information;
for i=1:iter
    [app]=ants_primaryplacing(m,n);
    [at]=ants_cycle(app,m,n,h,t,alpha,beta);
    at=horzcat(at,at(:,1));
    [cost,f]=ants_cost(m,n,d,at,el);
    [t]=ants_traceupdating(m,n,t,at,f,e);
    costoa(i)=mean(cost);
    [mincost(i),number]=min(cost);besttour(i,:)=at
    (number,:);
    iteration(i)=i;
end
subplot(2,1,1);plot(iteration,costoa);
title('average of cost (distance) versus number of
cycles');
xlabel('iteration');
ylabel('distance');
 [k,l]=min(mincost);
for i=1:n+1
    X(i)=x(besttour(l,i));
    Y(i)=y(besttour(l,i));
end
subplot(2,1,2);plot(X,Y,'--rs','LineWidth',2,...
        'MarkerEdgeColor','k',...
        'MarkerFaceColor','g',...
        'MarkerSize',10)
xlabel('X');ylabel('y');axis('equal');
```

```
for i=1:n
    text(X(i)+.5,Y(i),['\leftarrow node ',num2str
    (besttour(1,i))]);
end
title(['optimum course by the length of ',num2str(k)]);
------------------------------------------------------
% End of Main
------------------------------------------------------
```

Subprograms

```
% Ant information
function [x,y,d,t,h,iter,alpha,beta,e,m,n,el]=ants_
information;
iter=100;% number of cycles.
m=200;% number of ants.
x=[8 0 -1 2 4 6 3 10 2.5 -5 7 9 11 13];
y=[2 4 6 -1 -2 0.5 0 3.7 1.8 1 0 4 3 2];% take care not
to enter iterative points.
n=length(x);% number of nodes.
for i=1:n % generating link length matrix.
    for j-1:n
        d(i,j)=sqrt((x(i)-x(j))^2+(y(i)-y(j))^2);
    end
end
e=.1;% evaporation coefficient.
alpha-1;% order of effect of ants' sight.
beta=5;% order of trace's effect.
for i=1:n% generating sight matrix.
    for j=1:n
        if d(i,j)==0
            h(i,j)=0;
        else
            h(i,j)=1/d(i,j);
        end
    end
end
t=0.0001*ones(n);% primary tracing.
el=.96;% coefficient of common cost elimination.
-----------------End of ant information-----------
function [cost,f]=ants_cost(m,n,d,at,el);
for i=1:m
    s=0;
    for j=1:n
        s=s+d(at(i,j),at(i,j+1));
    end
    f(i)=s;
end
```

```
cost=f;
f=f-el*min(f);% elimination of common cost.
----------------- End of ants_cost----------------
function [at]=ants_cycle(app,m,n,h,t,alpha,beta);
for i=1:m
    mh=h;
    for j=1:n-1
        c=app(i,j);
        mh(:,c)=0;
        temp=(t(c,:).^beta).*(mh(c,:).^alpha);
        s=(sum(temp));
        p=(1/s).*temp;
        r=rand;
        s=0;
        for k=1:n
            s=s+p(k);
            if r<=s
                app(i,j+1)=k;
                break
            end
        end
    end
end
at=app;% generation of ants tour matrix during a cycle.
---------------- End of ants_cycle -----------------
function [app]=ants_primaryplacing(m,n);
rand('state',sum(100*clock));
for i=1:m
    app(i,1)=fix(1+rand*(n-1));% ants primary placing.
End
-------------- End of ants_primaryplacing ------------
function [t]=ants_traceupdating(m,n,t,at,f,e);
for i=1:m
    for j=1:n
        dt=1/f(i);
    t(at(i,j),at(i,j+1))=(1-e)*t(at(i,j),at(i,j+1))
    +dt;% updating traces.
    end
end
-------------- End of ants_traceupdating --------------
```

Observations:

The average distance is plotted against the number of ant cycles. The optimum distance was found to be 45.562 as shown in Figure 15.4.

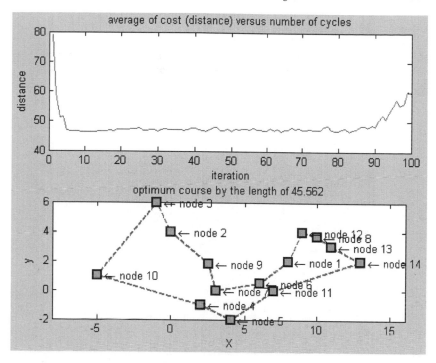

FIGURE 15.4: Plot of Distance and Optimum Length

15.10.4 Illustration 4: Ant Algorithm for the Quadratic Assignment Problem (QAP)

The following MATLAB code implements the QAP to assign n departments to n unique sites.

```
clc;
clear;

% THE DISTANCE/FLOW MATRIX
% UPPER HALF = DISTANCE, LOWER HALF = FLOW
DF =  [NaN  1   2   3   1   2   3   4;
       5   NaN  1   2   2   1   2   3;
       2   3   NaN  1   3   2   1   2;
       4   0   0   NaN  4   3   2   1;
       1   2   0   5   NaN  1   2   3;
       0   2   0   2   10  NaN  1   2;
       0   2   0   2   0   5   NaN  1;
       6   0   5   10  0   1   10  NaN];
```

```
% PROBLEM SIZE (NUMBER OF SITES OR DEPTS)
pr_size = size(DF, 1);
ants = 5; % NUMBER OF ANTS
max_assigns = 100000; % NUMBER OF ASSIGNMENTS
% (WHEN TO STOP)
optimal = 107; % OPTIMAL SOLUTION
a = 1; % WEIGHT OF PHEROMONE
b = 5; % WEIGHT OF HEURISTIC INFO
lamda = 0.8; % EVAPORATION PARAMETER
Q = 10; % CONSTANT FOR PHEROMONE UPDATING
AM = ones(ants,pr_size); % ASSIGNMENTS OF EACH ANT
min_cost = -1;

% HEURISTIC INFO - SUM OF DISTANCES BETWEEN SITES
for i=1:pr_size
     D(i) =sum( DF(1:i-1,i))+sum(DF(i,i+1:pr_size));
end

% START THE ALGORITHM
assign = 1;
while (assign <= max_assigns) & ( (min_cost > optimal) |
        (min_cost == -1) )
     % =============== FIND PHEROMONE ===============

     % AT FIRST LOOP, INITIALIZE PHEROMONE
     if assign==1
         % SET 1 AS INITIAL PHEROMONE
         pher = ones(8);

     % IN THE REST OF LOOPS, COMPUTE PHEROMONE
     else
         for i=1:pr_size
             for j=1:pr_size
                 tmp = zeros(ants,pr_size);
                 tmp( find(AM==j)) = 1;
                 tmp = tmp(:,i);
                 tmp = tmp .* costs';
                 tmp( find(tmp==0) ) = [];
                 tmp = Q ./ tmp;
                 delta(i,j) = sum(tmp);
             end
         end
         pher = lamda * pher + delta;
     end
     % ============ ASSIGN DEPTS TO SITES ============

     % EACH ANT MAKES ASSIGNMENTS
```

```
for ant=1:ants
     % GET RANDOM DEPT ORDER
     depts = rand(pr_size, 2);
     for i=1:pr_size
          depts(i,1) = i;
     end
     depts = sortrows(depts,2);

     % KEEP AVAILABLE SITES IN A VECTOR
     for i=1:pr_size
          free_sites(i) = i;
     end
pref = ones(pr_size,1); % PREFERENCE FOR EACH SITE
prob=ones(pr_size,1);% PROBABILITIES FOR EACH DEPT
for dept_index=1:pr_size
     % GET SUM OF THE PREFERENCES
     % AND THE PREFERENCE FOR EACH SITE
     pref_sum = 0;
     for site_index=1:size(free_sites,2)
          tmp_pher=pher(depts(dept_index),free_sites
               (site_index));
          pref(site_index) = tmp_pher^a *
               ( 1/D(free_sites(site_index)) )^b;
          pref_sum = pref_sum + pref(site_index);
     end

     % GET PROBABILITIES OF ASSIGNING THE DEPT
     % TO EACH FREE SITE
     prob = free_sites';
     prob(:,2) = pref / pref_sum;

     % GET THE SITE WHERE THE DEPT WILL BE ASSIGNED
     prob = sortrows(prob,2);
     AM(ant,dept_index) = prob(1);

     % ELIMINATE THE SELECTED SITE FROM THE
     % FREE SITES
     index = find(free_sites==prob(1));
     prob(1,:) = [];
     free_sites(index) = [];
     pref(index) = [];
end

% GET THE COST OF THE ANT'S ASSIGNMENT
costs(ant) = 0;
for i=1:pr_size
     for j=1:i-1
```

```
                dept_flow = DF(i,j);
                site1 = AM(ant,i);
                site2 = AM(ant,j);
                if site1 \textless site2
                    sites_distance = DF(site1, site2);
                else
                    sites_distance = DF(site2, site1);
                end
                costs(ant) = costs(ant) + dept_flow *
                sites_distance;
            end
        end
        if (costs(ant) \textless min_cost)|(min_cost==-1)
            min_cost = costs(ant);
            ch_assign = AM(ant,:);
        end
        if mod(assign,100) == 0
            disp( sprintf('Assignments so far : % d
                    Cheapest cost : % d', assign,
                    min_cost));
        end
        assign = assign + 1;
        end

end
disp( sprintf('Cheapest Cost : % d', min_cost));
disp( sprintf('Assignments : % d', assign-1));
disp(' ');
disp('Assignment');
disp('----------');
ant_index = find(costs==min(costs));
for i=1:pr_size
    disp( sprintf('Dept % d to Site % d', i,
    % ch_assign(i)));
end
```

Output: Cheapest Cost : 107
Assignments : 4945

```
Assignment
----------
Dept 1 to Site 2
Dept 2 to Site 1
Dept 3 to Site 5
Dept 4 to Site 3
Dept 5 to Site 4
Dept 6 to Site 8
```

```
Dept 7 to Site 7
Dept 8 to Site 6
```

Summary

This chapter introduced the theoretical foundations of swarm intelligence with a focus on the implementation and illustration of particle swarm optimization and ant colony optimization algorithms. We provided a few applications on Particle optimization problems and any colony optimization for easy understanding of the optimization problems. A set of MATLAB programs were also given for implementation of PSO and ACO. Results were analyzed, discussed, and their potentials were illustrated.

Review Questions

1. Define swarm and swarm behavior.

2. Explain briefly on boid.

3. What are swarm robots?

4. Write a note on stability analysis of swarms.

5. Define Swarm Intelligence.

6. Explain the taxonomy of Swarm Intelligence.

7. Mention a few properties of swarm intelligence.

8. Explain briefly the applications of swarm intelligence.

9. Mention the usage of Particle Swarm Optimization and Ant Colony Optimization.

10. Represent PSO in a canonical form.

11. Write the algorithm of PSO and explain the operational steps briefly.

12. Mention a few neighborhood topologies that influence the performance of PSO.

13. Compare PSO with a few global optimization techniques.

14. State the functions used to test the performance of PSO.

15. Write short notes on Fuzzy PSO and Binary PSO.

16. Explain the Job Scheduling application using PSO.

17. Write a note on PSO for data mining.

18. Write the algorithm of ACO and explain the operational steps briefly.

19. How the traveling sales man problem is optimized using ACO?

20. What do you mean by Quadratic assignment problems?

21. Write a note on ACO used in data mining.

22. Write a MATLAB program to implement the PSO algorithm and analyze its performance on Job shop scheduling problem.

23. Optimize the path of a Traveling Salesman Problem using PSO algorithm in MATLAB. Compare the results with ACO.

24. Find the minimum of a function using Hybrid PSO in MATLAB.

25. Write a MATLAB program using PSO to train a feed forward neural network. Assume that the Neural network is approximating a noisy sine function.

26. Implement the Quadratic Assignment Problem in MATLAB using PSO and compare the results with QAP using ACO algorithm.

Appendix A

Glossary of Terms

A

Action: Any process based on satisfied conditions or occurrence of a situation.

Activation: The time-varying value that is the output of a neuron.

Activation Function: A function that translates a neuron's net input to an activation value. A mathematical function that maps the net input of a neuron to its output. Commonly used activation functions are: step, sign, linear, and sigmoid, also referred to as Transfer function.

ADALINE: Acronym for a linear neuron: ADAptive LINear Element.

Adaptability: The ability of an organism to learn in response to changes in its environment over the course of its lifetime. This allows it to improve its fitness over that available from its initial phenotype.

Adaptive: Subject to adaptation; can change over time to improve fitness or accuracy.

Adaptive Behavior: Underlying mechanisms that allow animals, and potentially, robots to adapt and survive in uncertain environments.

Adaptive Learning Rate: A learning rate adjusted according to the change of error during training. If the error at the current epoch exceeds the previous value by more than a predefined ratio, the learning rate is decreased. However, if the error is less than the previous one, the learning rate is increased. The use of an adaptive learning rate accelerates learning in a multilayer perceptron.

Adaptive Neuro-Fuzzy Inference System (ANFIS): A technique for automatically tuning Sugeno-type inference systems based on training data.

Aggregate Set: A fuzzy set obtained through aggregation.

Aggregation: The combination of the consequents of each rule in a Mamdani fuzzy inference system in preparation for defuzzification.

Algorithm: A detailed and unambiguous sequence of instructions that describes how a computation is to proceed and can be implemented as a program.

Allele: Alternative form of a gene; one of the different forms of a gene that can exist at a single locus

Allele Loss: Allele loss is the natural loss of traits in the gene pool over the generations of a run. Another term for allele loss is convergence. Severe allele loss results in a population incapable of solving the problem with the available gene pool.

Ant Colony Optimization (ACO): Heuristic search algorithm for NP-hard optimization, inspired by the foraging behavior of real ant colonies, experimentally showed that the foraging behavior can give rise to the emergence of shortest path when employed by a colony of ants

Antecedent: A conditional statement in the IF part of a rule, also referred to as Premise.

Ants: Ants are social, colony-building insects who live in self-organized groups. Their perceptive spectrum is limited to various smells and a good sense of feel. The individual ants themselves are not capable of complex thinking. There are also no superants who delegate knowledge. The special role of the queen is limited to laying eggs. Why do ant colonies act complexly and intelligently as a whole then? "Single ants aren't smart. Ant colonies are." This effect is called "swarm intelligence": individuals follow simple rules to which complex behavior adds up. For example: there are two routes from the colony to food, a long one and a short one. None of the ants know which way is better and why, they choose the route by chance at first. The ants who choose the shorter route, however, walk back and forth between the colony and the food more often in the same period of time and thus leave more pheromone traces on the shorter path. This scent message signals to the other ants that this is the optimum route.

Approximate Reasoning: Reasoning that does not require a precise matching between the IF part of a production rule with the data in the database.

Artificial Neural Network (ANN): An information-processing paradigm inspired by the structure and functions of the human brain. An ANN consists of a number of simple and highly interconnected processors, called neurons, which are analogous to the biological neurons in the brain. The neurons are connected by weighted links that pass signals from one neuron to another. While in a biological neural network, learning involves adjustments to the synapses, ANNs learn through repeated adjustments of the weights. These weights store the knowledge needed to solve specific problems.

Assertion: A fact derived during reasoning.

Associative Memories: Associative memories work by recalling information in response to an information cue. Associative memories can be auto-associative or hetero-associative. Auto-associative memories recall the same information that is used as a cue, which can be useful to complete a partial pattern. Hetero-associative memories are useful as a memory. Human long-term memory is thought to be associative because of the way in which one thought retrieved from it leads to another. When we want to store a new item of information in long term memory it typically takes us 8 seconds to store an item that can't be associated with a pre-stored item, but only one or two seconds, if there is an existing information structure with which to associate the new item.

Attribute: A property of an object. For example, the object "computer" might have such attributes as "model", "processor", "memory", and "cost".

Automatically Defined Function (ADF): Concept of modularization aiming at an efficiency increase in GP. ADFs are sub-trees, which can be used as functions in main trees. ADFs are varied in the same manner as the main trees.

Axon: A single long branch of a biological neuron that carries the output signal (action potential) from the cell. An axon may be as long as a meter. In an ANN, an axon is modeled by the neuron's output.

B

Backpropagation: An algorithm for efficiently calculating the error gradient of a neural network, which can then be used as the basis of learning. Backpropagation is equivalent to the delta rule for perceptrons, but can also calculate appropriate weight changes for the hidden layer weights of a multilayer perceptron by generalizing the notion of an error correction term. In the simplest case, backpropagation is a type of steepest descent in the search space of the network weights, and it will usually converge to a local minimum.

Backpropagation Learning Rule: Learning rule in which weights and biases are adjusted by error-derivative (delta) vectors backpropagated through the network. Backpropagation is commonly applied to feedforward multilayer networks. Sometimes this rule is called the generalized delta rule.

Backtracking Search: Linear search routine that begins with a step multiplier of 1 and then backtracks until an acceptable reduction in performance is obtained.

Baldwin Effect: If the ability to learn increases the fitness, survival, of an individual, then its offspring will have a high probability of having that ability to learn.

Batch: Matrix of input (or target) vectors applied to the network simultaneously. Changes to the network weights and biases are made just once for the entire set of vectors in the input matrix. (The term batch is being replaced by the more descriptive expression "concurrent vectors.")

Bias: Neuron parameter that is summed with the neuron's weighted inputs and passed through the neuron's transfer function to generate the neuron's output.

Bias Vector: Column vector of bias values for a layer of neurons.

Bi-directional Associative Memory (BAM): A class of neural networks that emulates characteristics of associative memory; proposed by Bart Kosko in the 1980s. The BAM associates patterns from one set to patterns from another set, and vice versa. Its basic architecture consists of two fully connected layers an input layer and an output layer.

Bloat/Code Bloat: Phenomenon of uncontrolled genome growth in GP individuals.

Block: Element of a tree-shaped GP individual. The block set consists of the terminal set and the function set.

Boid: An autonomous agent that behaves like a simplified bird but will display flocking patterns in the presence of other boids.

Boltzmann Selection: In Boltzmann selection, a method inspired by the technique of simulated annealing, selection pressure is slowly increased over evolutionary time to gradually focus the search. Given a fitness of f, Boltzmann selection assigns a new fitness, f0, according to a differentiable function.

Branch: A connection between nodes in a decision tree.

Building Block: A pattern of genes in a contiguous section of a chromosome, which, if present, confers a high fitness to the individual. According to the building block hypothesis, a complete solution can be constructed by crossover joining together in a single individual with many building blocks, which were originally spread throughout the population.

Building Block Hypothesis (BBH): It attempts to explain the functioning of a (binary) GA based on the schema theorem. The BBH is based on the assumption that beneficial properties of a parent are aggregated in (relatively) small code blocks at several locations within the genome. In the offspring they can be merged by crossover. Thus, the BBH suggests that the improved solution is assembled from "partial solutions," the so-called building blocks.

C

Case-Based Reasoning: Case-based reasoning (CBR) solves a current problem by retrieving the solution to previous similar problems and altering those solutions to meet the current needs. It is based upon previous experiences and patterns of previous experiences. Humans with years of experience in a particular job and activity (e.g., a skilled paramedic arriving on an accident scene can often automatically know the best procedure to deal with a patient) use this technique to solve many of their problems. One advantage of CBR is that inexperienced people can draw on the knowledge of experienced colleagues, including ones who aren't in the organization, to solve their problems. Synonym: Reasoning by analogy.

Canonical GA: Causality property of a system, that small variations in the cause only provoke small variations in the effect. In EAs variations of the cause correspond to changes of the genotype and variations of the effect to changes of the phenotype or of the respective objective function value (fitness). Although strong causality remarkably increases the efficiency of EAs, it is not always a prerequisite for their successful application.

CART (Classification and Regression Trees): A tool for data mining that uses decision trees. CART provides a set of rules that can be applied to a new data set for predicting outcomes. CART segments data records by creating binary splits.

Cascade-Forward Network: Layered network in which each layer only receives inputs from previous layers.

Cellular Automata: A regular array of identical finite state automata whose next state is determined solely by their current state and the state of their neighbors. The most widely seen is the game of life in which complex patterns emerge from a (supposedly infinite) square lattice of simple two state (living and dead) automata whose next state is determined solely by the current states of its four closest neighbors and itself.

Certainty Factor: A number assigned to a fact or a rule to indicate the certainty or confidence one has that this fact or rule is valid, also referred to as Confidence factor.

Certainty Theory: A theory for managing uncertainties in expert systems based on inexact reasoning. It uses certainty factors to represent the level of belief in a hypothesis given that a particular event has been observed.

Child: In a decision tree, a child is a node produced by splitting the data of a node located at the preceding hierarchical level of the tree. A child node holds a subset of the data contained in its parent.

Chromosome: Normally, in genetic algorithms chromosome is the bit string, which represents the individual. In genetic programming the individual and its representation are usually the same, both being the program parse tree. In nature many species store their genetic information on more than one chromosome.

Chromosome Mutation: Any type of change in the chromosome structure or number

Class: A group of objects with common attributes. Animal, person, car, and computer are all classes.

Class-Frame: A frame that represents a class.

Classification: Automated classification tools such as decision trees have been shown to be very effective for distinguishing and characterizing very large volumes of data. They assign items to one of a set of predefined classes of objects based on a set of observed features. For example, one might determine whether a particular mushroom is "poisonous" or "edible" based on its color, size, and gill size. Classifiers can be learned automatically from a set of examples through supervised learning. Classification rules are rules that discriminate between different partitions of a database based on various attributes within the database. The partitions of the database are based on an attribute called the classification label (e.g., "faulty" and "good").

Classifier: A rule that is part of a classifier system and has a condition that must be matched before its message (or action) can be posted (or effected). The strength of a classifier determines the likelihood that it can outbid other classifiers if more than one condition is matched.

Classifier System: An adaptive system similar to a post production system that contains many "if ... then" rules called classifiers. The state of the environment is encoded as a message by a detector and placed on the message list from which the condition portion of the classifiers can be matched. "Winning" classifiers can then post their own messages to the message list, ultimately forming a type of computation that may result in a message being translated into an action by an effector. The strengths of the classifiers are modified by the bucket brigade algorithm, and new rules can be introduced via a genetic algorithm.

Cluster Analysis: A method of data reduction that tries to group given data into clusters; data of the same cluster should be similar or homogenous, data of disjunct clusters should be maximally different; assigning each data point to exactly one cluster often causes problems, because in real world problems a crisp separation of clusters is rarely possible due to overlapping of classes; also there are usually exceptions which cannot be suitably assigned to any cluster.

Clustering: Clustering is an approach to learning that seeks to place objects into meaningful groups automatically based on their similarity. Clustering, unlike classification, does not require the groups to be predefined with the hope that the algorithm will determine useful but hidden groupings of data points. The hope in applying clustering algorithms is that they will discover useful but unknown classes of items. A well-publicized success of a clustering system was NASA's discovery of a new class of stellar spectra.

Co-evolution: Evolution of species, not only with respect to their environment, but also as to how they relate to other species. This is a more potent form of evolution to that normally considered, changing the shape of the fitness landscape dynamically.

Combinatorial Optimization: Some tasks involve combining a set of entities in a specific way (e.g., the task of building a house). A general combinatorial task involves deciding the specifications of those entities (e.g., what size, shape, material to make the bricks from), and the way in which those entities are brought together (e.g., the number of bricks, and their relative positions). If the resulting combination of entities can in some way be given a fitness score, then combinatorial optimization is the task of designing a set of entities, and deciding how they must be configured, so as to give maximum fitness.

Competitive Learning: Unsupervised learning in which neurons compete among themselves such that only one neuron will respond to a particular input pattern. The neuron that wins the "competition" is called the winner-takes-all neuron. Kohonen self-organizing feature maps are an example of an ANN with competitive learning.

Computation: The realization of a program in a computer.

Computational Intelligence: An area in which the systems are dealing only with numeric data, have pattern recognition capabilities, do not use knowledge in the artificial sense.

Concurrent Input Vectors: Name given to a matrix of input vectors that are to be presented to a network simultaneously. All the vectors in the matrix are used in making just one set of changes in the weights and biases.

Conditional Independence: Two propositions are independent if they do not affect each other's chance of being true. They are conditionally independent if they are independent given certain conditions. For example, high humidity in the air and a damp sidewalk are not independent, but they are conditionally independent if we know that it has just rained.

Conditional Probability: The probability of a proposition, A, being true given that all we know is some evidence, B. This is expressed as P (A — B).

Conditional Probability Table (CPT): A table of probability values for a node in the network. Each value corresponds to one conditioning case.

Conditioning Case: A permutation of truth-values of all the parents of a given node, listed in a CPT.

Connection: One-way link between neurons in a network.

Connection Strength: Strength of a link between two neurons in a network. The strength, often called weight, determines the effect that one neuron has on another.

Connectionist System : A system characterized by explicit connections between the components resulting in a distributed data structure (as used in neural networks).

Connectivity: The relation of an agent to its neighbors, it can be sparsely connected (only affected by a few neighbors), fully connected (interfacing with every other agent in the system) or some intermediate arrangement. This parameter critically affects the dynamics of the system.

Consequent: The Then part of a If-Then rule, or one clause or expression in this part of the rule.

Constraint: A force of some sort restricting the movement of a system. In life studies the variations of form do not allow infinite variation, something constrains the options available. Complexity studies seek the laws that apply, if any, in these cases and similar areas.

Convergence: For computers, halting with an answer; for dynamical systems, falling into an attractor; for searches (e.g., backpropagation and genetic algorithms), finding a location that cannot be improved upon; for infinite summations, approaching a definite value.

Cooperation: The behavior of two or more individuals acting to increase the gains of all participating individuals.

Crossover: A reproduction operator which forms a new chromosome by combining parts of each of the two "parent" chromosomes. The simplest form is single-point crossover, in which an arbitrary point in the chromosome is picked. All the information from parent A is copied from the start up to the crossover point, then all the information from parent B is copied from the crossover point to the end of the chromosome. The new chromosome thus gets the head of one parent's chromosome combined with the tail of the other. Variations use more than one crossover point, or combine information from parents in other ways.

Crossover Point: Crossover point of a fuzzy set is the element in U at which its membership function is 0.5.

Crossover Probability: A number between zero and one that indicates the probability of two chromosomes crossing over.

Cycle: Single presentation of an input vector, calculation of output, and new weights and biases.

D

Darwinism: Theory of evolution, proposed by Darwin, that evolution comes through random variation of heritable characteristics, coupled with natural selection (survival of the fittest). A physical mechanism for this, in terms of genes and chromosomes, was discovered many years later. Darwinism was combined with the selectionism of Weismann and the genetics of Mendel to form the Neo-Darwinian Synthesis during the 1930s-1950s by T. Dobzhansky, E. Mayr, G. Simpson, R. Fisher, S. Wright, and others.

Dead Neuron: Competitive layer neuron that never won any competition during training and so has not become a useful feature detector. Dead neurons do not respond to any of the training vectors.

Decision Boundary: Line, determined by the weight and bias vectors, for which the net input n is zero.

Decision Making: A process of deriving solution of a complex problem using knowledge from the given domain and data relevant to the problem

Decision Tree: A graphical representation of a data set that describes the data by tree-like structures. A decision tree consists of nodes, branches, and leaves. The tree always starts from the root node and grows down by splitting the data at each level into new nodes. Decision trees are particularly good at solving classification problems. Their main advantage is data visualization.

Defuzzification: The last step in fuzzy inference; the process of converting a combined output of fuzzy rules into a crisp (numerical) value. The input for the defuzzification process is the aggregate set and the output is a single number.

Degree of Membership: A numerical value between 0 and 1 that represents the degree to which an element belongs to a particular set, also referred to as Membership value.

Delta Rule: A procedure for updating weights in a perceptron during training. The delta rule determines the weight correction by multiplying the neuron's input with the error and the learning rate.

Dendrite: A branch of a biological neuron that transfers information from one part of a cell to another. Dendrites typically serve an input function for the cell, although many dendrites also have output functions. In an ANN, dendrites are modeled by inputs to a neuron.

Directed Acyclic Graph (DAG): A graph in which the nodes are connected by arrows (directed), and in which there is no directed path from a given node back to itself (acyclic).

Distance Function: A particular way of calculating distance, such as the Euclidean distance between two vectors.

DNA: Deoxyribonucleic Acid, a double stranded macromolecule of helical structure (comparable to a spiral staircase). Both single strands are linear, unbranched nucleic acid molecules build up from alternating deoxyribose (sugar) and phosphate molecules. Each deoxyribose part is coupled to a nucleotide base, which is responsible for establishing the connection to the other strand of the DNA. The four nucleotide bases Adenine (A), Thymine (T), Cytosine (C) and Guanine (G) are the alphabet of the genetic information. The sequences of these bases in the DNA molecule determines the building plan of any organism.

E

Early Stopping: Technique based on dividing the data into three subsets. The first subset is the training set, used for computing the gradient and updating the network weights and biases. The second subset is the validation set. When the validation error increases for a specified number of iterations, the training is stopped, and the weights and biases at the minimum of the validation error are returned. The third subset is the test set. It is used to verify the network design.

Ecology: The study of the relationships and interactions between organisms and environments.

Ecosystem: A biological system consisting of many organisms from different species.

Ecosystem: The relatively stable balance of different species within a particular area. A food chain, usually cyclic and self-sustaining.

Eigenvalue: The change in length that occurs when the corresponding eigenvector is multiplied by its matrix.

Eigenvector: A unit length vector that retains its direction when multiplied to the corresponding matrix. An (n * n) matrix can have as many as n unique eigenvectors, each of which will have its own eigenvalue.

Elitism: Elitism (or an elitist strategy) is a mechanism which is employed in some EAs which ensures that the chromosomes of the most highly fit member(s) of the population are passed on to the next generation without being altered by Genetic Operators. Using elitism ensures that the minimum fitness of the population can never reduce from one generation to the next. Elitism usually brings about a more rapid convergence of the population. In some applications elitism improves the chances of locating an optimal individual, while in others it reduces it.

Embedding: A method of taking a scalar time series and using delayed snapshots of the values at fixed time intervals in the past so that the dynamics of the underlying system can be observed as a function of the previously observed states.

Emergence: Global behavior of a system is not evident from the local behavior of its elements; a defining characteristic of a complex dynamical system.

Entropy: A measure of a system's degree of randomness or disorder.

Environment: Environment surrounds an organism. Can be "physical" (abiotic), or biotic. In both, the organism occupies a niche which influences its fitness within the total environment. A biotic environment may present frequency-dependent fitness functions within a population, that is, the fitness of an organism's behavior may depend upon how many others are also doing it. Over several generations, biotic environments may foster co-evolution, in which fitness is determined with selection partly by other species.

Epigenesis: Lifetime learning.

Epistasis: A "masking" or "switching" effect among genes. A gene is said to be epistatic when its presence suppresses the effect of a gene at another locus. Epistatic genes are sometimes called inhibiting genes because of their effect on other genes which are described as hypostatic. Epistasis is referred to any kind of strong interaction among genes, not just masking effects. A possible definition is: "Epistasis is the interaction between different genes in a chromosome. It is the extent to which the contribution to fitness of one gene depends on the values of other genes." Problems with little or no epistasis are trivial to solve (hillclimbing is sufficient). But highly epistatic problems are difficult to solve, even for GAs. High epistasis means that building blocks cannot form, and there will be deception.

Epoch: Presentation of the set of training (input and/or target) vectors to a network and the calculation of new weights and biases. Note that training vectors can be presented one at a time or all together in a batch.

Equilibrium: The tendency of a system to settle down to a steady state that isn't easily disturbed, an attractor. Traditionally, equilibrium systems in physics have no energy input and maximise entropy, usually involving an ergodic attractor, but dissipative systems maintain steady states far-from-equilibrium (also non-equilibrium).

Error Jumping: Sudden increase in a network's sum-squared error during training. This is often due to too a large learning rate.

Error Ratio: Training parameter used with adaptive learning rate and momentum training of backpropagation networks.

Error: The difference between the actual and desired outputs in an **ANN** with **supervised learning.**

Error Vector: Difference between a network's output vector in response to an input vector and an associated target output vector.

Euclidean: Pertaining to standard geometry, i.e., points, lines, planes, volumes, squares, cubes, triangles, etc.

Euler's Method: The simplest method of obtaining a numerical solution of a differential equation. There are many other numerical techniques that are more accurate; however, an analytical solution (i.e., a closed form of an integral) is always preferred but not always possible.

Evolution: A process operating on populations that involves variation among individuals, traits being inheritable, and a level of fitness for individuals that is a function of the possessed traits. Over relatively long periods of time, the distribution of inheritable traits will tend to reflect the fitness that the traits convey to the individual; thus, evolution acts as a filter that selects fitness-yielding traits over other traits.

Evolution: This is a universal idea, generalized as "general selection theory" to be the process of "variation, selection, retention" underlying all systemic improvement over time (including "trial and error" learning). The term is often specifically applied however to genetic evolution where some changes, by being more efficient in functional ways, are preferred by natural selection.

Evolution Strategy: A numerical optimization procedure similar to a focused Monte Carlo search. Unlike genetic algorithms, evolution strategies use only a mutation operator, and do not require a problem to be represented in a coded form. Evolution strategies are used for solving technical optimization problems when no analytical objective function is available, and no conventional optimization method exists.

Evolutionarily Stable Strategy: A strategy that performs well in a population dominated by the same strategy. Or, in other words, an "ESS" is a strategy such that, if all the members of a population adopt it, no mutant strategy can invade.

Evolutionary Algorithm (EA): A collective term for all variants of (probabilistic) optimization and approximation algorithms that are inspired by Darwinian evolution. Optimal states are approximated by successive improvements based on the variation-selection-paradigm. Thereby, the variation operators produce genetic diversity and the selection directs the evolutionary search.

Evolutionary Computation (EC): Computation based on evolutionary algorithms. EC encompasses methods of simulating evolution on a computer. The term is relatively new and represents an effort to bring together researchers who have been working in closely related fields but following different paradigms. The field is now seen as including research in genetic algorithms, evolution strategies, evolutionary programming, artificial life.

Evolutionary Computing: Using algorithms that mimic the genetic and evolutionary processes of natural selection to produce software programs or solve problems. The process of natural selection uses many copies of a piece of code (which corresponds to a biological species' DNA), introduces random changes (mutations), allows the best results to continue (survival of the fittest), mixes their traits (reproduction), and repeats the cycle many times (evolution).

Evolutionary Programming (EP): It is a stochastic optimization strategy, which is similar to Genetic Algorithms, but dispenses with both "genomic" representations and with crossover as a reproduction operator. It is a variant of EA, which, like ES, operates on the "natural" problem representation. Only mutation is used as the variation operator together with tournament selection; recombination is not employed.

Evolutionary Systems: A process or system which employs the evolutionary dynamics of reproduction, mutation, competition, and selection. The specific forms of these processes are irrelevant to a system being described as evolutionary.

Evolutionary Theory: The study of evolution based upon neo-Darwinian ideas. Modern complexity science adds additional self-organizational concepts to this theory to better explain organizational emergence.

Evolvable Hardware: Evolvable hardware includes special devices, circuits, or machines, which allow an implementation of the Darwinian evolution paradigm at the material level.

Excitatory: Refers to a neural synapse or weight that is positive such that activity in the source neuron encourages activity in the connected neuron; the opposite of inhibitory.

Exhaustive Search: A problem-solving technique in which every possible solution is examined until an acceptable one is found.

Experimentation: One process by which scientists attempt to understand nature. A phenomenon is observed and/or manipulated so that changes in the phenomenon's state can be seen. The resulting data can be used to derive new models of a process or to confirm an existing model. Experimentation is the complement of theorization.

F

Fuzzy Logic: Traditional Western logic systems assume that things are either in one category or another. Yet in everyday life, we know this is often not precisely so. People aren't just short or tall, they can be fairly short or fairly tall, and besides we differ in the opinions of what height actually corresponds to tall, anyway. The ingredients of a cake aren't just not mixed or mixed, they can be moderately well mixed. Fuzzy logic provides a way of taking our commonsense knowledge that most things are a matter of degree into account when a computer is automatically making a decision. For example, one rice cooker uses fuzzy logic to cook rice perfectly even if the cook put in too little water or too much water.

Feedback: A linking of the output of a system back to the input. Traditionally this can be negative, tending to return the system to a wanted state, or positive tending to diverge from that state. Life employs both methods.

Feedback Network: Network with connections from a layer's output to that layer's input. The feedback connection can be direct or pass through several layers.

Feedforward Neural Network: A topology of an ANN in which neurons in one layer are connected to the neurons in the next layer. The input signals are propagated in a forward direction on a layer-by-layer basis. An example of a feedforward network is a multilayer perceptron.

Firing a Rule: The process of executing a production rule, or more precisely, executing the THEN part of a rule when its IF part is true.

Firing Strength: The degree to which the antecedent part of a fuzzy rule is satisfied. The firing strength may be the result of an AND or an OR operation, and it shapes the output function for the rule, also known as degree of fulfillment.

Fitness: A value assigned to an individual which reflects how well the individual solves the task in hand. A "fitness function" is used to map a chromosome to a fitness value. A "fitness landscape" is the hypersurface obtained by applying the fitness function to every point in the search space.

Fitness Function: A process which evaluates a member of a population and gives a score or fitness. In most cases the goal is to find an individual with the maximum (or minimum) fitness.

Fitness Scaling: Fitness scaling converts the raw fitness scores that are returned by the fitness function to values in a range that is suitable for the selection function. The selection function uses the scaled fitness values to select the parents of the next generation. The selection function assigns a higher probability of selection to individuals with higher scaled values.

Fixed Point: A point in a dynamical system's state space that maps back to itself, i.e., the system will stay at the fixed point if it does not undergo a perturbation.

Flocking: The phenomenon of bird flocking can be explained by simple rules telling an agent to stay a fixed distance from a neighbour. The apparently intelligent behaviour of a flock navigating an obstacle follows directly from the mindless application of these rules.

Frequency: Dependent fitness fitness differences whose intensity changes with changes in the relative frequency of genotypes in the population.

Frequency-Dependent Selection: Selection that involves frequency-dependent fitness; selection of a genotype depending on its frequency in the population.

Frequency-Interdependent Fitness: Fitness that is not dependent upon interactions with other individuals of the same species.

Function: A mapping from one space to another. This is usually understood to be a relationship between numbers. Functions that are computable can be calculated by a universal computer.

Function Approximation: The task of finding an instance from a class of functions that is minimally different from an unknown function. This is a common task for neural networks.

Function Optimization: For a function which takes a set of N input parameters, and returns a single output value, F, function optimization is the task of finding the set(s) of parameters which produce the maximum (or minimum) value of F. Function optimization is a type of value-based problem.

Function Set: The set of operators used in GP. These functions label the internal (non-leaf) points of the parse trees that represent the programs in the population. An example function set might be +, −, *.

Fuzzification: The first step in fuzzy inference; the process of mapping crisp (numerical) inputs into degrees to which these inputs belong to the respective fuzzy sets.

Fuzzy Cluster Analysis: Specifies a membership degree between 0 and 1 for each data sample to each cluster; most fuzzy cluster analysis methods optimize a subjective function that evaluates a given fuzzy assignment of data to clusters; by suitable selection of parameters of the subjective function it is possible to search for clusters of different forms: on the one side solid clusters in form of (hyperdimensional) solid spheres, elliptoids or planes, and on the other side shells of geometrical contures like circles, lines, or hyperboles (shell cluster); latter are especially suitable for image analysis; from the result of a fuzzy cluster analysis a set of fuzzy rules can be obtained to describe the underlying data; these rules can be used to build fuzzy systems like fuzzy classifiers or fuzzy controllers, for example.

Fuzzy C-Means Clustering: A data clustering technique wherein each data point belongs to a cluster to a degree specified by a membership grade.

Fuzzy Inference: The process of reasoning based on fuzzy logic. Fuzzy inference includes four steps: fuzzification of the input variables, rule evaluation, aggregation of the rule outputs, and defuzzification.

Fuzzy Logic: A system of logic developed for representing conditions that cannot be easily described by the binary terms "true" and "false". The concept was introduced by Lotfi Zadeh in 1965. Unlike Boolean logic, fuzzy logic is multi-valued and handles the concept of partial truth (truth values between "completely true" and "completely false").

Fuzzy Operators: AND, OR, and NOT operators. These are also known as logical connectives.

Fuzzy Rule: A conditional statement in the form: IF x is A THEN y is B, where x and y are linguistic variables, and A and B are linguistic values determined by fuzzy sets.

Fuzzy Set: A fuzzy set is any set that allows its members to have different grades of membership (membership function) in the interval [0,1].

Fuzzy Singleton: A fuzzy set with a membership function equal to unity at a single point on the universe of discourse and zero everywhere else.

Fuzzy Variable: A quantity that can take on linguistic values. For example, the fuzzy variable "temperature", might have values such as "hot", "medium", and "cold".

Fuzzy Variables and Fuzzy Logic: Variables that take on multiple values with various levels of certainty and the techniques for reasoning with such variables.

G

Genetic Algorithms: Search algorithms used in machine learning which involve iteratively generating new candidate solutions by combining two high scoring earlier (or parent) solutions in a search for a better solution, so named because of its reliance on ideas drawn from biological evolution.

Gamete: Cells which carry genetic information from their parents for the purpose of sexual reproduction. In animals, male gametes are called sperm, female gametes are called ova. Gametes have a haploid number of chromosomes.

Gaussian: Normally distributed (with a bell-shaped curve) and having a mean at the center of the curve with tail widths proportional to the standard deviation of the data about the mean.

Gene: A basic unit of a chromosome that controls the development of a particular feature of a living organism. In Holland's chromosome, a gene is represented by either 0 or 1.

Gene Pool: The whole set of genes in a breeding population. The metaphor on which the term is based de-emphasizes the undeniable fact that genes actually go about in discrete bodies, and emphasizes the idea of genes flowing about the world like a liquid.

Generalization: The ability of an ANN to produce correct results from data on which it has not been trained.

Generalized Delta Rule: Another name for backpropagation.

Generalized Regression Network: Approximates a continuous function to an arbitrary accuracy, given a sufficient number of hidden neurons.

Generation: An iteration of the measurement of fitness and the creation of a new population by means of reproduction operators.

Generation Equivalent: In a steady state GA, the time taken to create as many new individuals as there is in the population.

Generation Gap: Concept for describing overlapping generations (stationary EA). The generation gap is defined as the ratio of the number of offspring to the size of the parent population.

Generation: One iteration of a genetic algorithm.

Generational GP: Generational genetic programming is the process of producing distinct generations in each iteration of the genetic algorithm.

Genetic Algorithm (GA): Search technique used in computer science to find approximate solutions to optimization and search problems. Genetic algorithms are a particular class of evolutionary algorithms that use techniques inspired by evolutionary biology such as inheritance, mutation, natural selection, and recombination (or crossover).

Genetic Drift: Term used in population genetics to refer to the statistical drift over time of allele frequencies in a finite population due to random sampling effects in the formation of successive generations; in a narrower sense, genetic drift refers to the expected population dynamics of neutral alleles (those defined as having no positive or negative impact on fitness), which are predicted to eventually become fixed at zero or 100% frequency in the absence of other mechanisms affecting allele distributions.

Genetic Fuzzy System: A system combining genetic and fuzzy principles; most frequent approaches: genetic algorithms that adapt and learn the knowledge base of a fuzzy-rule-based system; genetic tuning of fuzzy systems; genetic learning processes in fuzzy systems; hybrid genetic fuzzy systems such as genetic fuzzy clustering or genetic neuro-fuzzy systems.

Genetic Operator: An operator in genetic algorithms or genetic programming, which acts upon the chromosome in order to produce a new individual. Genetic operators include crossover and mutation.

Genetic Program: A program produced by genetic programming.

Genetic Programming (GP): Technique popularized by John Koza, in which computer programs, rather than function parameters, are optimized; GP often uses tree-based internal data structures to represent the computer programs for adaptation instead of the list, or array, structures typical of genetic algorithms.

Genetic Repair: Approach to explain the possible increase of performance by recombination. Accordingly, the task of recombination is to extract the genetic information common to the selected individuals, as this information is likely responsible for fitness increase. A perfect recombination operator should additionally reduce those parts of the genome which are responsible for a decrease in fitness. This is, e.g., statistically realized by ES recombination operators in real-valued search spaces by (partially) averaging out the defective components.

Genotype: The combination of genes that make up an organism. This has no form itself but directs the creation of the phenotype following the interaction of system, dynamics, and environment. Usually regarded as comprising a number of alleles or bits (systems having two states, 0 or 1, off or on).

Genotype/Phenotype Distinction: The distinction between phenotype and genotype is fundamental to the understanding of heredity and development of organisms. The genotype of an organism is the class to which that organism belongs as determined by the description of the actual physical material made up of DNA that was passed to the organism by its parents at the organism's conception. For sexually reproducing organisms that physical material consists of the DNA contributed to the fertilized egg by the sperm and egg of its two parents. For asexually reproducing organisms, for example bacteria, the inherited material is a direct copy of the DNA of its parent. The phenotype of an organism is the class to which that organism belongs as determined by the description of the physical and behavioral characteristics of the organism, for example its size and shape, its metabolic activities, and its pattern of movement.

Global Minimum (Maximum): In a search space, the lowest (or highest) point of the surface, which usually represents the best possible solution in the space with respect to some problem.

Global Optimization: The process by which a search is made for the extremum (or extrema) of a functional which, in evolutionary computation, corresponds to the fitness or error function that is used to assess the performance of any individual.

Goal: A hypothesis that an expert system attempts to prove.

Gradient: A vector of partial derivatives of a function that operates on vectors. Intuitively, the gradient represents the slope of a high-dimensional surface.

Gradient Descent: Process of making changes to weights and biases, where the changes are proportional to the derivatives of network error with respect to those weights and biases. This is done to minimize network error.

H

Heterogeneous Databases: Databases that contain different kinds of data, e.g., text and numerical data.

Haploid: This refers to cell which contains a single chromosome or set of chromosomes, each consisting of a single sequence of genes. An example is a gamete. In EC, it is usual for individuals to be haploid. The solution to GA is a single set of chromosomes (one individual).

Hard Selection: Selection acts on competing individuals. When only the best available individuals are retained for generating future progeny, this is termed "hard selection." In contrast, "soft selection" offers a probabilistic mechanism for maintaining individuals to be parents of future progeny despite possessing relatively poorer objective values.

Hebb's Law: The learning law introduced by Donald Hebb in the late 1940s; it states that if neuron i is near enough to excite neuron j and repeatedly participates in its activation, the synaptic connection between these two neurons is strengthened and neuron j becomes more sensitive to stimuli from neuron i. This law provides the basis for unsupervised learning.

Hedge: A qualifier of a fuzzy set used to modify its shape. Hedges include adverbs such as "very", "somewhat", "quite", "more or less", and "slightly". They perform mathematical operations of concentration by reducing the degree of membership of fuzzy elements (e.g., very tall men), dilation by increasing the degree of membership (e.g., more or less tall men), and intensification by increasing the degree of membership above 0.5 and decreasing those below 0.5 (e.g., indeed tall men).

Heuristic: A strategy that can be applied to complex problems; it usually — but not always — yields a correct solution. Heuristics, which are developed from years of experience, are often used to reduce complex problem solving to more simple operations based on judgment. Heuristics are often expressed as rules of thumb.

Heuristic Search: A search technique that applies heuristics to guide the reasoning, and thus reduce the search space for a solution.

Hidden Layer: A layer of neurons between the input and output layers; called "hidden" because neurons in this layer cannot be observed through the input/output behavior of the neural network. There is no obvious way to know what the desired output of the hidden layer should be.

Hidden Neuron: A neuron in the hidden layer.

Hierarchical: A treelike branching structure where each component has only one owner or higher level component. A 1:N structure.

Hits: The number of hits an individual scores is the number of test cases for which it returns the correct answer (or close enough to it). This may or may not be a component of the fitness function. When an individual gains the maximum number of hits this may terminate the run.

Hopfield Network: A single-layer feedback neural network. In the Hopfield network, the output of each neuron is fed back to the inputs of all other neurons (there is no self-feedback). The Hopfield network usually uses McCulloch and Pitts neurons with the sign activation function. The Hopfield network attempts to emulate characteristics of the associative memory.

Hybrid Systems: Many of Stottler Henke's artificial intelligence software applications use multiple AI techniques in combination. For example, case-based reasoning may be used in combination with model-based reasoning in an automatic diagnostic system. Case-based reasoning, which tends to be less expensive to develop and faster to run, may draw on an historical databases of past equipment failures, the diagnosis of those, and the repairs effected and the outcomes achieved. So CBR may be used to make most failure diagnoses. Model-based reasoning may be used to diagnose less common but expensive failures and also to make fine adjustments to the repair procedures retrieved from similar cases in the case base by CBR.

I

Inference Engine: The part of an expert system responsible for drawing new conclusions from the current data and rules. The inference engine is a portion of the reusable part of an expert system (along with the user interface, a knowledge base editor, and an explanation system) that will work with different sets of case-specific data and knowledge bases.

Implicit Parallelism: The idea that genetic algorithms have an extra built-in form of parallelism that is expressed when a GA searches through a search space. Implicit parallelism depends on the similarities and differences between individuals in the population. The theory posits that GAs process more schemata than there are strings in a population, thus getting something of a free lunch. See also **No Free Lunch**.

Individual: A single member of a population. In EC, each individual contains a chromosome (or, more generally, a genome) which represents a possible solution to the task being tackled, i.e., a single point in the search space. Other information is usually also stored in each individual, e.g., its fitness.

Inference Chain: The sequence of steps that indicates how an expert system applies rules from the rule base to reach a conclusion.

Inference Engine: A basic component of an expert system that carries out reasoning whereby the expert system reaches a solution. It matches the rules provided in the rule base with the facts contained in the database. Also referred to as Interpreter.

Inference Technique: The technique used by the inference engine to direct search and reasoning in an expert system. There are two principal techniques: forward chaining and backward chaining.

Inference: New knowledge inferred from existing facts.

Inheritance: The process by which all characteristics of a class-frame are assumed by the instance-frame. Inheritance is an essential feature of frame-based systems. A common use of inheritance is to impose default features on all instance-frames.

Initialization: The first step of the training algorithm that sets weights and thresholds to their initial values.

Input Layer: The first layer of neurons in an ANN. The input layer accepts input signals from the outside world and redistributes them to neurons in the next layer. The input layer rarely includes computing neurons and does not process input patterns.

Input Neuron: A neuron in the input layer.

Input Space: Range of all possible input vectors.

Input Vector: Vector presented to the network.

Input Weight Vector: Row vector of weights going to a neuron.

Input Weights: Weights connecting network inputs to layers.

Insect Models of Organization: Ethological research on the self-organizing, collective resilience of ant and other insect societies has led to research efforts seeking to emulate their successes.

Instance: A specific object from a class. For example, class "computer" may have instances IBM Aptiva S35 and IBM Aptiva S9C. In frame-based expert systems, all characteristics of a class are inherited by its instances.

Instance-Frame: A **frame** that represents an **instance.**

Intersection: In classical set theory, an intersection between two sets contains elements shared by these sets. For example, the intersection of tall men and fat men contains all men who are tall and fat. In fuzzy set theory, an element may partly belong to both sets, and the intersection is the lowest membership value of the element in both sets.

Inversion: A reordering operator which works by selecting two cut points in a chromosome, and reversing the order of all the genes between those two points.

K

Kohonen Self-Organizing Feature Maps: A special class of ANNs with competitive learning introduced by Teuvo Kohonen in the late 1980s. The Kohonen map consists of a single layer of computation neurons with two types of connections: forward connections from the neurons in the input layer to the neurons in the output layer, and lateral connections between neurons in the output layer. The lateral connections are used to create a competition between neurons. A neuron learns by shifting its weights from inactive connections to active ones. Only the winning neuron and its neighborhood are allowed to learn.

L

LISP: LISP (short for list processing language), a computer language, was invented by John McCarthy, one of the pioneers of artificial intelligence. The language is ideal for representing knowledge (e.g., If a fire alarm is ringing, then there is a fire.) from which inferences are to be drawn.

Lamarckism: A method of heredity that does not apply to genetics but is applicable to social adaptation. Lamarckism posits that acquired traits can be passed from parent to offspring.

Layer: A group of neurons that have a specific function and are processed as a whole. For example, a multilayer perceptron has at least three layers: an input layer, an output layer, and one or more hidden layers.

Learning: The process of acquisition and extinction of modifications in existing knowledge, skills, habits or action tendencies in a motivated organism through experience, practice, or exercise; learning of living organisms is inspiration for machine learning; related topics: supervised learning, unsupervised learning, learning by example, learning from experience, observational learning.

Learning Rate: A positive number less than unity that controls the amount of changes to the weights in the ANN from one iteration to the next. The learning rate directly affects the speed of network training.

Learning Rule: Method of deriving the next changes that might be made in a network or a procedure for modifying the weights and biases of a network.

Linear: Having only a multiplicative factor. If f(x) is a linear function, then f(a+b) = f(a) + f(b) and c f(x) = f(cx) must both be true for all values of a, b, c, and x. Most things in nature are nonlinear.

Linear Activation Function: An activation function that produces an output equal to the net input of a neuron. Neurons with the linear activation function are often used for linear approximation.

Linear Transfer Function: Transfer function that produces its input as its output.

Linearly (In) Separable: Two classes of points are linearly separable if a linear function exists such that one class of points resides on one side of the hyperplane (defined by the linear function), and all points in the other class are on the other side. The XOR mapping defines two sets of points that are linearly inseparable.

Linguistic Value: A language element that can be assumed by a fuzzy variable. For example, the fuzzy variable "income" might assume such linguistic values as "very low", "low", "medium", "high" and "very high". Linguistic values are defined by membership functions.

Linguistic Variable: A variable that can have values that are language elements, such as words and phrases. In fuzzy logic, terms linguistic variable and fuzzy variable are synonyms.

Local Minimum (Maximum): The bottom of a valley or the top of a peak; a point in a search space such that all nearby points are either higher (for a minimum) or lower (for a maximum). In a continuous search space, local minima and maxima have a 0 gradient vector. Note that this particular valley (or peak) may not necessarily be the lowest (or highest) location in the space, which is referred to as the global minimum (maximum).

Local Minimum: The minimum value of a function over a limited range of its input parameters. If a local minimum is encountered during training, the desired behavior of an ANN may never be achieved. The usual method of getting out of a local minimum is to randomize the weights and continue training.

Local Optimum: An easily found optimum in state space, but not guaranteed to be the global optimum.

Local Search: Locating or approximating optimal states with variation operators or search strategies (not necessarily EAs), which explore only a limited part of the search space, the so-called (search) neighborhood. Thus, in general, local optima are found.

M

Mamdani-Type Inference: A type of fuzzy inference in which the fuzzy sets from the consequent of each rule are combined through the aggregation operator and the resulting fuzzy set is defuzzified to yield the output of the system.

Manhattan Distance: The Manhattan distance between two vectors x and y is calculated as $D = \text{sum}(\text{abs}(x\text{-}y))$.

Map: A function that is usually understood to be iterated in discrete time steps.

Mapping: Transforming a input to an output by following a rule or look-up table. It is also referred to as the selective study of 'reality'.

Matrix: A rectangular two-dimensional array of numbers that can be thought of as a linear operator on vectors. Matrix-vector multiplication can be used to describe geometric transformations such as scaling, rotation, reflection, and translation. They can also describe the affine transformation used to construct IFS and MRCM fractals.

McCulloch and Pitts Neuron Model: A neuron model proposed by Warren McCulloch and Walter Pitts in 1943, which is still the basis for most artificial neural networks. The model consists of a linear combiner followed by a hard limiter. The net input is applied to the hard limiter, which produces an output equal to +1 if its input is positive and -1 if it is negative.

Mean: The arithmetical average of a collection of numbers; the center of a Gaussian distribution.

Membership function: A mathematical function that defines a fuzzy set on the universe of discourse. Typical membership functions used in fuzzy expert systems are triangles and trapezoids.

Meta-GA: If a GA is used to set parameters or discover optimal settings for a second GA, the first one is known as a meta-GA.

Micro-GA: A micro-GA is a GA with a small population size (often 5) that has special reinitialization or mutation operators to increase diversity and prevent the natural convergence associated with small population sizes.

Migration: Migration is the exchange of individuals between subpopulations. Migration is used in the regional population model. The spread of information among subpopulations is influenced by the migration topology (i.e., which subpopulations exchange individuals), the migration interval (i.e., how often does an exchange take place), and the migration rate (i.e., number of individuals that are exchanged). These parameters determine whether the subpopulations evolve in a relatively independent way or rather behave like a panmictic population.

Model: In the sciences, a model is an estimate of how something works. A model will usually have inputs and outputs that correspond to its real-world counterpart. An adaptive system also contains an implicit model of its environment that allows it to change its behavior in anticipation of what will happen in the environment.

Multi-Criteria Optimization: Optimization with regard to multiple objective functions aiming at a simultaneous improvement of the objectives. The goals are usually conflicting so that an optimal solution in the conventional sense does not exist. Instead one aims at, e.g., Pareto optimality, i.e., one has to find the Pareto set from which the user can choose a qualified solution.

Multilayer Perceptron (MLP): A type of feedforward neural network that is an extension of the perceptron in that it has at least one hidden layer of neurons. Layers are updated by starting at the inputs and ending with the outputs. Each neuron computes a weighted sum of the incoming signals, to yield a net input, and passes this value through its sigmoidal activation function to yield the neuron's activation value. Unlike the perceptron, an MLP can solve linearly inseparable problems.

Multiobjective: The need to take into account many conflicting variables in order to obtain an optimum fitness. This is a problem due to epistasis.

Mutation: A reproduction operator which forms a new chromosome by making (usually small) alterations to the values of genes in a copy of a single, parent chromosome.

Mutation Probability: A number between zero and one that indicates the probability of mutation occurring in a single gene.

Mutation Rate: Mutation probability of a single gene/object parameter of an individual. With respect to binary representation, the mutation rate is the probability of flipping a single bit position.

Mutation Strength: Usually the standard deviation of the normal distribution with which a single object parameter is mutated. Mutation strength is also a measure for the realized (search) neighborhood size.

N

Natural Selection: The three stage process of variation, selection, reproduction (or persistance) that underlies evolution in all areas (in biology the synthesis of Medelian genetics with natural selection is called neo Darwinism). It is combined within complex systems thinking with self-organization.

Neighborhood: Group of neurons within a specified distance of a particular neuron. The neighborhood is specified by the indices for all the neurons that lie within a radius d of the winning neuron i*
$.N_i(d)=\{j,d_{ij}\leq d\}$

Neo-Darwinism: A synthesis of Darwinism with the mechanisms of genetics; the idea that adaptation equals a combination of variation, heredity, and selection. See also evolution, inheritable, and natural selection.

Net Input: The weighted sum of incoming signals into a neuron plus a neuron's threshold value.

Net Input Vector: Combination, in a layer, of all the layer's weighted input vectors with its bias.

Networks: Connected systems, the properties of which do not entirely depend on the actual units involved but on the dynamics of the interconnections.

Neural Computing: A computational approach to modeling the human brain that relies on connecting a large number of simple processors to produce complex behavior. Neural computing can be implemented on specialized hardware or with software, called artificial neural networks, that simulates the structure and functions of the human brain on a conventional computer.

Neural Network (NN): A network of neurons that are connected through synapses or weights. In this book, the term is used almost exclusively to denote an artificial neural network and not the real thing. Each neuron performs a simple calculation that is a function of the activations of the neurons that are connected to it. Through feedback mechanisms and/or the nonlinear output response of neurons, the network as a whole is capable of performing extremely complicated tasks, including universal computation and universal approximation. Three different classes of neural networks are feedforward, feedback, and recurrent neural networks, which differ in the degree and type of connectivity that they possess.

Neuro-Fuzzy System: Fuzzy system that uses a learning algorithm derived from or inspired by neural network theory to determine its parameters (fuzzy sets and fuzzy rules) by processing data samples; usually represented as special multilayer feedforward neural networks.

Neuron: A simple computational unit that performs a weighted sum on incoming signals, adds a threshold or bias term to this value to yield a net input, and maps this last value through an activation function to compute its own activation. Some neurons, such as those found in feedback or Hopfield networks, will retain a portion of their previous activation.

Niche: In EC, it is often required to maintain diversity in the population. Sometimes a fitness function may be known to be multimodal, and it may be required to locate all the peaks. In this case consider each peak in the fitness function as analogous to a niche. By applying techniques such as fitness sharing, the population can be prevented from converging on a single peak, and instead stable subpopulations form at each peak. This is analogous to different species occupying different niches.

No Free Lunch (NFL): A theorem that states that in the worst case, and averaged over an infinite number of search spaces, all search methods perform equally well. More than being a condemnation of any search method, the NFL theorem actually hints that most naturally occurring search spaces are, in fact, not random.

Node: A point in a graph, usually represented by an ellipse, which represents a specific variable. Some types of nodes are as follows:

Child - A node which has an arrow coming into it, from its parent.

Deterministic - A node with a value completely specified by the values of its parents, with no uncertainty.

Evidence - A node one knows the exact value of, when querying the network for a probability.

Leak - A node used to represent miscellaneous causes, known or otherwise.

Parent - A node which has an arrow leading out of it, to its child.

Query - A node for which one asks the network the probability, given certain evidence.

Nonlinear: A function that is not linear. Most things in nature are nonlinear. This means that in a very real way, the whole is at least different from the sum of the parts.

Non-Terminal : Functions used to link parse tree together. This name may be used to avoid confusion with functions with no parameters which can only act as end points of the parse tree (i.e., leaves) and are part of the terminal set.

Normally Distributed: A random variable is normally distributed if its density function is described as f(x) = 1/sqrt(2*pi*sqr(sigma)) * exp(-0.5*(x-mu)*(x-mu)/sqr(sigma)) where mu is the mean of the random variable x and sigma is the standard deviation.

NP-Complete: A problem type in which any instance of any other NP class problem can be translated to in polynomial time. This means that if a fast algorithm exists for an NP-complete problem, then any problem that is in NP can be solved with the same algorithm.

O

Object: A concept, abstraction or thing that can be individually selected and manipulated, and that has some meaning for the problem at hand. All objects have identity and are clearly distinguishable. Michael Black, Audi 5000 Turbo, IBM Aptiva S35 are examples of objects. In object-oriented programming, an object is a self-contained entity that consists of both data and procedures to manipulate the data.

Object Variables: Parameters that are directly involved in assessing the relative worth of an individual.

Objective Function/Quality Function: Also known as goal function is the function to be optimized, depending on the object parameters (also referred to as search space parameters or phenotype parameters). The objective function constitutes the implementation of the problem to be solved. The input parameters are the object parameters. The output is the objective value representing the evaluation/quality of the individual/phenotype.

Object-Oriented Programming: A programming method that uses objects as a basis for analysis, design, and implementation.

Offspring: An individual that was produced through reproduction, also referred to as a Child.

OPS: A high-level programming language derived from LISP for developing rule-based expert systems.

Optimization: The search for the global optimum, or best overall compromise within a (typically) multivalued system. Where interactions occur many optima are typically present (the fitness landscape is "rugged") and this situation has no analytical solution, generally requiring adaptive solutions.

Order-Based Problem: A problem where the solution must be specified in terms of an arrangement (e.g., a linear ordering) of specific items, e.g. Traveling Salesman Problem, computer process scheduling. Order-based problems are a class of combinatorial optimization problems in which the entities to be combined are already determined.

Ordering phase: Period of training during which neuron weights are expected to order themselves in the input space consistent with the associated neuron positions.

Outer Product: An operation on two vectors that yields a matrix. Given two vectors with the same dimensionality, the outer product is a square symmetric matrix that contains the product of all pairs of elements from the two vectors, i.e., $A[i,j] = x[i]\ y[j]$.

Output Layer: The last **layer** of **neurons** in an ANN. The output layer produces the output pattern of the entire network.

Output Vector: Output of a neural network. Each element of the output vector is the output of a neuron.

Output Weight Vector: Column vector of weights coming from a neuron or input.

Over-Fitting: A state in which an ANN has memorized all the training examples, but cannot generalize. Overfitting may occur if the number of hidden neurons is too big. The practical approach to preventing overfitting is to choose the smallest number of hidden neurons that yields good generalization, also referred to as Over-training.

Over-Training: See **Over-Fitting**.

P

Parallel Processing: A computational technique that carries out multiple tasks simultaneously. The human brain is an example of a parallel information-processing system: It stores and processes information simultaneously throughout the whole biological neural network, rather than at specific locations.

Parallel System: A computer that uses two to thousands of processors at once. Parallel Systems software assigns portions of individual problems to each processor then combines the results. Problems that can be broken into multiple parts, like analyzing large amounts of scientific data, can be solved much faster by Parallel Systems than by single processor systems.

Parallelism: Several agents acting at the same time independently, simultaneous computation similar to that which happens within living systems.

Parent: In a decision tree, a parent node is a node that splits its data between nodes at the next hierarchical level of the tree. The parent node contains a complete data set, while child nodes hold subsets of that set.

Pareto-Optimal: A set of equivalent optimised solutions that all have the same global fitness but embody different compromises or niches between the objectives.

Particle Swarm: A self-organizing system whose global dynamics emerge from local rules.

Particle Swarm Optimization (PSO): Extension of cellular automata; utilizes a population of candidate solutions to evolve an optimal or near-optimal solution to a problem; the degree of optimality is measured by a fitness function defined by the user.

Pattern: A vector.

Pattern Association: Task performed by a network trained to respond with the correct output vector for each input vector presented.

Pattern Classification: A task that neural networks are often trained to do. Given some input pattern, the task is to make an accurate class assignment to the input. For example, classifying many images of letters to one of the twenty-six letters of the alphabet is a pattern classification task.

Pattern Recognition: Identification of visual or audio patterns by computers. Pattern recognition involves converting patterns into digital signals and comparing them with patterns already stored in the memory. Artificial neural networks are successfully applied to pattern recognition, particularly in such areas as voice and character recognition, radar target identification, and robotics.

Perceptron Learning Rule: Learning rule for training single-layer hard-limit networks. It is guaranteed to result in a perfectly functioning network in finite time, given that the network is capable of doing so.

Perceptron: The simplest form of a neural network, suggested by Frank Rosenblatt. The operation of the perceptron is based on the McCulloch and Pitts neuron model. It consists of a single neuron with adjustable synaptic weights and a hard limiter. The perceptron learns a task by making small adjustments in the weights to reduce the difference between the actual and desired outputs. The initial weights are randomly assigned and then updated to obtain an output consistent with the training examples.

Phenotype: The form of the organism. A result of the combined influences of the genotype and the environment on the self-organizing internal processes during development.

Phylogenesis: Refers to a population of organisms. The life span of a population of organisms from pre-historic times until today.

Phylogeny: Evolution of species.

Population: A group of individuals which may interact together, for example by mating, producing offspring, etc. Typical population sizes in EC range from one (for certain evolution strategies) to many thousands (for genetic programming).

Population Size: Number of individuals in a population.

Premature Convergence: A state when a genetic algorithm's population converges to something which is not the solution that is required.

Preprocessing: Transformation of the input or target data before it is presented to the neural network.

Q

Quasi-Newton Algorithm: Class of optimization algorithm based on Newton's method. An approximate Hessian matrix is computed at each iteration of the algorithm based on the gradients.

R

Radial Basis Networks: Neural network that can be designed directly by fitting special response elements where they will do the most good.

Radial Basis Transfer Function: The transfer function for a radial basis neuron is $\text{radbas}(n) = e^{-n^2}$.

Random/Randomness: Without cause; not compressible; obeying the statistics of a fair coin toss.

Reasoning: The process of drawing conclusions or inferences from observations, facts, or assumptions.

Recombination: Recombination is also known as crossover.

Recurrent Neural Network: A network similar to a feedforward neural network except that there may be connections from an output or hidden layer to the inputs. Recurrent neural networks are capable of universal computation.

Recursive: Strictly speaking, a set or function is recursive if it is computable; however, in the usual sense of the word, a function is said to be recursive if its definition makes reference to itself. For example, factorial can be defined as x! = x * (x - 1)! with the base case of 1! equal to 1.

Recursively Enumerable (RE): A potentially infinite <u>set</u> whose members can be enumerated by a universal computer; however, a universal computer may not be able to determine that something is not a member of a recursively enumerable set. The halting set is recursively enumerable but not recursive.

Reductionism: The idea that nature can be understood by dissection. In other words, knowing the lowest-level details of how things work (at, say, the level of subatomic physics) reveals how higher-level phenomena come about. This is a bottom-up way of looking at the universe, and is the exact opposite of holism.

Reproduction Operator: A mechanism which influences the way in which genetic information is passed on from parent(s) to offspring during reproduction. Operators fall into three broad categories: crossover, mutation, and reordering operators.

Reproduction: The creation of a new individual from two parents (sexual reproduction). Asexual reproduction is the creation of a new individual from a single parent.

Roulette Wheel Selection: A method of selecting a particular individual in the population to be a parent with a probability equal to its fitness divided by the total fitness of the population.

Rule Base: The knowledge base that contains a set of production rules.

Rule Evaluation: The second step in fuzzy inference; the process of applying the fuzzy inputs to the antecedents of fuzzy rules, and determining the truth value for the antecedent of each rule. If a given rule has multiple antecedents, the fuzzy operation of intersection or union is carried out to obtain a single number that represents the result of evaluating the antecedent.

Rule-Based Expert System: An expert system whose knowledge base contains a set of production rules.

Rule-Based System: An expert system based on IF-THEN rules for representing knowledge.

S

Schema: A pattern of gene values in a chromosome, which may include "don't care" states. Thus in a binary chromosome, each schema (plural schemata) can be specified by a string of the same length as the chromosome, with each character one of 0, 1, #. A particular chromosome is said to contain a particular schema if it matches the schema (e.g., chromosome 01101 matches schema #1#0#). The order of a schema is the number of non-don't-care positions specified, while the defining length is the distance between the furthest two non-don't-care positions. Thus #1##0# is of order 2 and defining length 3.

Schema Theorem: Theorem devised by Holland to explain the behavior of GAs. In essence, it says that a GA gives exponentially increasing reproductive trials to above average schemata. Because each chromosome contains a great many schemata, the rate of schema processing in the population is very high, leading to a phenomenon known as implicit parallelism. This gives a GA with a population of size N a speedup by a factor of N cubed, compared to a random search.

Schema/Schemata: A similarity template used to analyze genetic algorithms. By using wild-card characters, a schema defines an entire class of strings that may be found in a population.

Search Operators: Processes used to generate new individuals to be evaluated. Search operators in genetic algorithms are typically based on crossover and point mutation. Search operators in evolution strategies and evolutionary programming typically follow from the representation of a solution and often involve Gaussian or lognormal perturbations when applied to real-valued vectors.

Search Space: If the solution to a task can be represented by a set of N real-valued parameters, then the job of finding this solution can be thought of as a search in an N-dimensional space. This is referred to simply as the search space. More generally, if the solution to a task can be represented using a representation scheme, R, then the search space is the set of all possible configurations which may be represented in R.

Selection: The process by which some individuals in a population are chosen for reproduction, typically on the basis of favoring individuals with higher fitness.

Self-Organization: A spontaneously formed higher-level pattern of structure or function that is emergent through the interactions of lower-level objects.

Set: A collection of things, usually numbers. Sets may be infinite in size.

Set Theory: The study of sets or classes of objects. The set is the basic unit in mathematics. Classical set theory does not acknowledge the fuzzy set, whose elements can belong to a number of sets to some degree. Classical set theory is bivalent: the element either does or does not belong to a particular set. That is, classical set theory gives each member of the set the value of 1, and all members that are not within the set a value of 0.

Sigmoid Activation Function: An activation function that transforms the input, which can have any value between plus and minus infinity, into a reasonable value in the range between 0 and 1. Neurons with this function are used in a multilayer perceptron.

Simulate/Simulation: Experimentation in the space of theories, or a combination of experimentation and theorization. Some numerical simulations are programs that represent a model for how nature works. Usually, the outcome of a simulation is as much a surprise as the outcome of a natural event, due to the richness and uncertainty of computation.

Simulated Annealing: Search technique where a single trial solution is modified at random. An energy is defined which represents how good the solution is. The goal is to find the best solution by minimizing the energy. Changes which lead to a lower energy are always accepted; an increase is probabilistically accepted. The probability is given by $\exp(-\Delta E/kT)$, where ΔE is the change in energy, k is a constant, and T is the Temperature. Initially the temperature is high corresponding to a liquid or molten state where large changes are possible and it is progressively reduced using a cooling schedule so allowing smaller changes until the system solidifies at a low energy solution.

Singleton Output Function: An output function that is given by a spike at a single number rather than a continuous curve. In Fuzzy Logic Toolbox, it is only supported as part of a zero-order Sugeno model.

Singleton: See **Fuzzy Singleton**.

Species: In EC the definition of "species" is less clear, since generally it is always possible for a pair of individuals to breed together. It is probably safest to use this term only in the context of algorithms which employ explicit speciation mechanisms.

Stability: Unchanging with time. This can be a static state (nothing changes) or a steady state (resource flows occur). In complex non-equilibrium systems we have multistable states, i.e., many semistable positions possible within a single system.

State: The condition of a system at a particular point or span in time. The concept of states is widely used in computer science and engineering because it allows people to view complicated systems as sets of smaller, simpler units.

Step Activation Function: A hard limit activation function that produces an output equal to +1 if its input is positive and 0 if it is negative.

Stochastic Universal Sampling: The individuals are mapped to contiguous segments of a line, such that each individual's segment is equal in size to its fitness exactly as in roulette-wheel selection. Here equally spaced pointers are placed over the line as many as there are individuals to be selected. Consider NPointer the number of individuals to be selected, then the distance between the pointers are 1/NPointer and the position of the first pointer is given by a randomly generated number in the range [0, 1/NPointer].

Strength: For a classifier system, a classifier's relative ability to win a bidding match for the right to post its message on the message list.

Sub-Population: A population may be sub-divided into groups, known as sub-populations, where individuals may only mate with others in the same group. (This technique might be chosen for parallel processors.) Such sub-divisions may markedly influence the evolutionary dynamics of a population. Sub-populations may be defined by various migration constraints: islands with limited arbitrary migration; stepping-stones with migration to neighboring islands; isolation-by-distance in which each individual mate only with near neighbors.

Subtractive Clustering: A technique for automatically generating fuzzy inference systems by detecting clusters in input-output training data.

Sugeno-Type Inference: A type of fuzzy inference in which the consequent of each rule is a linear combination of the inputs. The output is a weighted linear combination of the consequents.

Sum-Squared Error: Sum of squared differences between the network targets and actual outputs for a given input vector or set of vectors.

Supervised Learning: A type of learning that requires an external teacher, who presents a sequence of training examples to the ANN. Each example contains the input pattern and the desired output pattern to be generated by the network. The network determines its actual output and compares it with the desired output from the training example. If the output from the network differs from the desired output specified in the training example, the network weights are modified. The most popular method of supervised learning is back-propagation.

Survival of the Fittest: The law according to which only individuals with the highest fitness can survive to pass on their genes to the next generation.

Swarm: Disorganized cluster of moving things, usually insects, moving irregularly, chaotically, somehow staying together even while all of them move in apparently random directions; loosely structured collection of interacting agents (elements).

Swarm Intelligence: A field which studies "the emergent collective intelligence of groups of simple agents". In groups of insects, which live in colonies, such as ants and bees, an individual can only do simple tasks on its own, while the colony's cooperative work is the main reason determining the intelligent behavior it shows. Swarm-bot a very complex entity made of many highly sophisticated robot units.

Synapse: A chemically mediated connection between two neurons in a biological neural network, so that the state of the one cell affects the state of the other. Synapses typically occur between an axon and a dendrite, though there are many other arrangements.

Synaptic Weight: See **Weight**.

T

Tan-Sigmoid Transfer Function: Squashing function of the form shown below that maps the input to the interval $(-1,1)$. The function is given by $\frac{1}{1+e^{-n}}$.

Target Vector: Desired output vector for a given input vector.

Terminal Set: A set from which all end (leaf) nodes in the parse trees representing the programs must be drawn. A terminal might be a variable, a constant or a function with no arguments.

Terminals: Terminals are the numeric values, (variables, constants, and zero argument functions) in the parse tree and are always external (leaf) nodes in the tree. The terminals act as arguments for the operator (atom) that is their parent in the tree.

Termination Condition: The conditions which determine the termination of the evolutionary process (examples: number of objective function evaluations, maximum run time, and convergence in the fitness or search space).

Test Set: A data set used for testing the ability of an ANN to generalize. The test data set is strictly independent of the training set, and contains examples that the network has not previously seen. Once training is complete, the network is validated with the test set.

Test Vectors: Set of input vectors (not used directly in training) that is used to test the trained network.

Threshold: A specific value that must be exceeded before the output of a neuron is generated. For example, in the McCulloch and Pitts neuron model, if the net input is less than the threshold, the neuron output is -1. But if the net input is greater than or equal to the threshold, the neuron becomes activated and its output attains a value +1, also referred to as Threshold value.

Threshold Value: see **Threshold**.

Topology: A structure of a neural network that refers to the number of layers in the neural network, the number of neurons in each layer, and connections between neurons, also referred to as Architecture.

Tournament Selection: A mechanism for choosing individuals from a population. A group (typically between 2 and 7 individuals) is selected at random from the population and the best (normally only one, but possibly more) is chosen.

Training: Procedure whereby a network is adjusted to do a particular job. Commonly viewed as an offline job, as opposed to an adjustment made during each time interval, as is done in adaptive training.

Training Set: A data set used for training an ANN.

Training Vector: Input and/or target vector used to train a network.

Trajectory: The path through state space taken by a system. It is the sequence of states or path plotted against time. Two general forms affect fitness, positive-sum and negative-sum.

Traveling Salesman Problem: The traveling salesperson has the task of visiting a number of clients, located in different cities. The problem to solve is: in what order should the cities be visited in order to minimize the total distance traveled (including returning home)? This is a classical example of an order-based problem.

Truncation Selection: Truncation selection is selection with a deterministic choice of the best μ individuals from the λ offspring (parents are not considered), necessary condition: $\lambda > \mu$.

U

Union: In classical set theory, the union of two sets consists of every element that falls into either set. For example, the union of tall men and fat men contains all men who are either tall or fat. In fuzzy set theory, the union is the reverse of the intersection, that is, the union is the largest membership value of the element in either set.

Universal Approximation: Having the ability to approximate any function to an arbitrary degree of accuracy. Neural networks are universal approximators.

Universal Computation: Capable of computing anything that can in principle be computed; being equivalent in computing power to a Turing machine, the lambda calculus, or a post-production system.

Universal Constructor: A machine able to construct any other object (including a copy of itself) given the appropriate instructions.

Universe of Discourse: The Universe of Discourse is the range of all possible values for an input to a fuzzy system.

Unstable: Having a basin of attraction that is 0 in size; being such that the slightest perturbation will forever change the state of a system. A pencil balanced on its point is unstable.

Unsupervised Learning: A type of learning that does not require an external teacher. During learning an ANN receives a number of different input patterns, discovers significant features in these patterns, and learns how to classify input data into appropriate categories, also referred to as *Self-organized learning*.

V

Validation Vectors: Set of input vectors (not used directly in training) that is used to monitor training progress so as to keep the network from overfitting.

Vector: A one-dimensional array of numbers that can be used to represent a point in a multidimensional space.

Vector Optimization: It is typically, an optimization problem wherein multiple objectives must be satisfied. The goals are usually conflicting so that an optimal solution in the conventional sense does not exist. Instead one aims at, e.g., Pareto optimality, i.e., one has to find the Pareto set from which the user can choose a qualified solution.

W

Weight: In a neural network, the strength of a synapse (or connection) between two neurons. Weights may be positive (excitatory) or negative (inhibitory). The thresholds of a neuron are also considered weights, since they undergo adaptation by a learning algorithm.

Weight Function: Weight functions apply weights to an input to get weighted inputs, as specified by a particular function.

Weight Matrix: Matrix containing connection strengths from a layer's inputs to its neurons. The element wij of a weight matrix W refers to the connection strength from input j to neuron i.

Weighted Input Vector: Result of applying a weight to a layer's input, whether it is a network input or the output of another layer.

Widrow-Hoff Learning Rule: Learning rule used to train single-layer linear networks. This rule is the predecessor of the backpropagation rule and is sometimes referred to as the delta rule.

X

XOR: The exclusive-or function; given two Boolean inputs, the output of XOR is 1 if and only if the two inputs are different; otherwise, the output is 0.

Appendix B

List of Abbreviations

A list of abbreviations used in this book is given in this appendix.

ABC	- Ant-Based Control
ACO	- Ant Colony Optimization
AGA	- Adaptive Genetic Algorithms
ANFIS	- Adaptive Neuro-Fuzzy Inference System
ANN	- Artificial Neural Networks
ART	- Adaptive Resonance Theory
ASA	- Adaptive Simulated Annealing
BBF	- Building Block Filtering
BDN	- Binary Decision Units
BP	- Back Propagation
CA	- Cellular Automata
CI	- Computational Intelligence
CPD	- Conditional Probability Distribution
CRG	- Content Representation Graph
CS	- Classifier Systems
DB	- Data Base
DE	- Differential Evolution
DEDP	- Dynamic Economic Dispatch Problem
DFA	- Deterministic Finite-State Automata
DM	- Data Mining
dmEFuNN	- Dynamic Evolving Fuzzy Neural Network
DNA	- Deoxyribonucleic Acid
DP	- Dynamic Programming
DPGA	- Dynamic Parametric Genetic Algorithm
DPSO	- Dissipative Particle Swarm Optimization
DRS	- Discourse Representation Structure
EA	- Evolutionary Algorithms
EC	- Evolutionary Computation
EDA	- Extended Dependency Analysis
EFuNN	- Evolving Fuzzy Neural Network
EP	- Evolutionary Programming
ES	- Evolution Strategies

FACS	- Facial Action Coding System
FATPSO	- Fuzzy Adaptive Turbulence in the Particle Swarm Optimization
FBM	- Fitness-Blind Mutation
FCM	- Fuzzy C-Means
FGCS	- Fifth-Generation Computing Systems
FINEST	- Fuzzy Inference and Neural Network in Fuzzy Inference Software
FIS	- Fuzzy Inference Systems
FL	- Fuzzy Logic
FLC	- Fuzzy Logic Controllers
FUN	- Fuzzy Net
FWR	- Formal Word Representation
GA	- Genetic Algorithm
GAM	- Generalized Additive Models
GAP	- Genetic Algorithm Percentage
GARIC	- Generalized Approximate Reasoning-based Intelligent Control
GCPSO	- Guaranteed Convergence Particle Swarm Optimization
GD	- Gaussian Distribution
GDA	- General Discriminant Function Analysis
GFRBS	- Genetic Fuzzy Rule–Based Systems
GFS	- Genetic Fuzzy Systems
GGA	- Generational Genetic Algorithm
GLM	- General Linear Model
GLM	- Generalized Linear/Nonlinear Models
GP	- Genetic Programming
GPEA	- Geometrical Place Evolutionary Algorithms
GrC	- Granular Computing
GRM	- General Regression Models
GSF	- Global Space Frame
GSO	- Genetic Swarm Optimization
HBGA	- Human-Based Genetic Algorithm
HC	- Hybridization Coefficient
HGA	- Hybrid Genetic Algorithm
HHC	- Higher Harmonic Control
HPA	- Heuristic Path Algorithm
IAGA	- Integrated Adaptive Genetic Algorithm
ILP	- Inductive Logic Programming
INPSO	- Independent Neighborhoods Particle Swarm Optimization
IRL	- Iterative Rule Learning
JPD	- Joint Probability Distribution
LAN	- Local-Area Network
LDS	- Limited Discrepancy Search
LDS	- Linear Dynamical Systems
LISP	- LISt Processing

LP	- Linear Programming
LS	- Local Search
LSD	- Least Squared Deviation
LSE	- Least Square Estimation
LVQ	- Learning Vector Quantizer
MF	- Membership Function
mGA	- messy Genetic Algorithm
MI	- Mutual Information
MIT	- Massachusetts Institute of Technology
MOGA	- Multiobjective Genetic Algorithms
MOP	- Memory Organization Packets
NC	- Neuro-Computing
NEAT	- Neuro Evolution of Augmenting Topologies
NEFCON	- Neuro-Fuzzy Control
NF	- Neuro-Fuzzy
NFC	- Neuro-Fuzzy Controller
NN	- Neural Networks
OCL	- Object Constraint Language
PGA	- Parallel Genetic Algorithm
PLS	- Penalized Least-Squares
PSO	- Particle Swarm Optimization
QAP	- Quadratic Assignment Problem
RA	- Reconstructibility Analysis
RB	- Rule Base
RBF	- Radial Basis Function
RFHN	- Rough-Fuzzy Hopfield Net
SA	- Simulated Annealing
SC	- Soft Computing
SI	- Swarm Intelligence
SOFM	- Self-Organizing Feature Map
SOM	- Self-Organizing Maps
SSE	- Sum-Squared Error
SSGA	- Steady-State Genetic Algorithm
TPSO	- Turbulence in the Particle Swarm Optimization
TSP	- Traveling Salesman Problem
VFSR	- Very Fast Simulated Reannealing

Appendix C

MATLAB Toolboxes Based on CI

A few Computational Intelligence–based MATLAB Toolboxes that are commonly used such as Genetic Algorithm and Direct Search Toolbox, Genetic and Evolutionary Algorithm Toolbox, Genetic Algorithm Toolbox, Genetic Programming Toolbox, Neural Net Toolbox, and Fuzzy Logic Toolbox are discussed in this appendix.

C.1   Genetic Algorithm Toolbox for MATLAB

The Genetic Algorithm Toolbox is a module for use with MATLAB that contains software routines for implementing genetic algorithms (GAs) and other evolutionary computing techniques.

The Genetic Algorithm Toolbox was developed by Andrew Chipperfield, Carlos Fonseca, Peter Fleming, and Hartmut Pohlheim, who are internationally known for their research and applications in this area. The toolbox is a collection of specialized MATLAB functions supporting the development and implementation of genetic and evolutionary algorithms.

Its main features include:

- Support for binary, integer and real-valued representations.

- A wide range of genetic operators.

- High-level entry points to most low-level functions allowing the user greater ease and flexibility in creating GA applications.

- Many variations on the standard GA.

- Support for virtual multiple subpopulations.

Consistent with the open-system approach of other MATLAB toolboxes, the Genetic Algorithm Toolbox is extensible to suit the user's needs. In combination with other MATLAB toolboxes and SIMULINK, the toolbox provides a versatile and powerful environment for exploring and developing genetic algorithms.

The Genetic Algorithm Toolbox is available on an unsupported basis for a modest charge. This includes documentation and allows an unlimited number of users.

Functions of Genetic Algorithm Toolbox

Creating Populations

crtbase	create a base vector
crtbp	create a binary population
crtrp	create a real valued population

Fitness Assignment

ranking	Rank based fitness assignment
scaling	proportional fitness scaling

Selection and Reinsertion

reins	uniform random and fitness based reinsertion
rws	roulette wheel selection
select	High level selection routine
sus	stochastic universal sampling

Mutation Operators

mut	discrete mutation
mutate	High level mutation function
mutbga	Real value mutation

Crossover Operators

recdis	discrete recombination
recint	intermediate recombination
reclin	line recombination
recmut	line recombination with mutation features
recombin	high level recombination function
xovdp	double point crossover
xovdprs	double point reduced surrogate crossover
xovmp	general multi point crossover
xovsh	shuffle crossover
xovshrs	shuffle reduced surrogate crossover
xovsp	single point crossover
xovsprs	single point reduced surrogate crossover

Subpopulation Support

migrate	exchange individuals between subpopulations

Utility Functions

bs2rv	binary string to real value conversion
rep	matrix replication

Demonstration and Other Functions

mpga	multi population genetic algorithm demonstration
objfun1	De Jongs first test function
objharv	harvest function
resplot	result plotting
sga	simple genetic algorithm demonstration

C.2 Fuzzy Logic Toolbox 2.2.7

The Fuzzy Logic Toolbox extends the MATLAB technical computing environment with tools for the design of systems based on fuzzy logic. Graphical user interfaces (GUIs) guides the user through the steps of fuzzy inference system design. Functions are provided for many common fuzzy logic methods, including fuzzy clustering and adaptive neuro-fuzzy learning.

The toolbox lets the user model complex system behaviors using simple logic rules and then implement these rules in a fuzzy inference system. The user can use the toolbox as a stand-alone fuzzy inference engine. Alternatively, fuzzy inference blocks can be used in Simulink to simulate the fuzzy systems within a comprehensive model of the entire dynamic system.

Like all MATLAB toolboxes, the Fuzzy Logic Toolbox can be customized. The user can inspect algorithms, modify source code, and add own membership functions or defuzzification techniques.

Key Features include

- Specialized GUIs for building fuzzy inference systems and viewing and analyzing results

- Membership functions for creating fuzzy inference systems

- Support for AND, OR, and NOT logic in user-defined rules

- Standard Mamdani and Sugeno-type fuzzy inference systems

- Automated membership function shaping through neuro-adaptive and fuzzy clustering learning techniques

- Ability to embed a fuzzy inference system in a Simulink model

- Ability to generate embeddable C code or stand-alone executable fuzzy inference engines

Functions of Fuzzy Logic Toolbox

GUI Tools and Plotting Functions

anfisedit	Open ANFIS Editor GUI
findcluster	Interactive clustering GUI for fuzzy c-means and sub-clustering
fuzzy	Open basic Fuzzy Inference System editor
mfedit	Membership function editor
plotfis	Plot Fuzzy Inference System
plotmf	Plot all membership functions for given variable
ruleedit	Rule editor and parser
ruleview	Rule viewer and fuzzy inference diagram
surfview	Open Output Surface Viewer

Membership Functions

dsigmf	Built-in membership function composed of difference between two sigmoidal membership functions
gauss2mf	Gaussian combination membership function
gaussmf	Gaussian curve built-in membership function
gbellmf	Generalized bell-shaped built-in membership function
pimf	Π−shaped built-in membership function
psigmf	Built-in membership function composed of product of two sigmoidally shaped membership functions
sigmf	Sigmoidally shaped built-in membership function
smf	S-shaped built-in membership function
trapmf	Trapezoidal-shaped built-in membership function
trimf	Triangular-shaped built-in membership function
zmf	Z-shaped built-in membership function

FIS Data Structure Management

addmf	Add membership function to Fuzzy Inference System
addrule	Add rule to Fuzzy Inference System
addvar	Add variable to Fuzzy Inference System
defuzz	Defuzzify membership function
evalfis	Perform fuzzy inference calculations
evalmf	Generic membership function evaluation
gensurf	Generate Fuzzy Inference System output surface
getfis	Fuzzy system properties
mf2mf	Translate parameters between membership functions
newfis	Create new Fuzzy Inference System
parsrule	Parse fuzzy rules
readfis	Load Fuzzy Inference System from file
rmmf	Remove membership function from Fuzzy Inference System
rmvar	Remove variables from Fuzzy Inference System
setfis	Set fuzzy system properties
showfis	Display annotated Fuzzy Inference System
showrule	Display Fuzzy Inference System rules
writefis	Save Fuzzy Inference System to file

Advanced Techniques

anfis	Training routine for Sugeno-type Fuzzy Inference System
fcm	Fuzzy c-means clustering
genfis1	Generate Fuzzy Inference System structure from data using grid partition
genfis2	Generate Fuzzy Inference System structure from data using subtractive clustering
genfis3	Generate Fuzzy Inference System structure from data using FCM clustering
subclust	Find cluster centers with subtractive clustering

Working in Simulink Environment

fuzblock	Simulink fuzzy logic library
sffis	Fuzzy inference S-function for Simulink software

C.3 Neural Network Toolbox 6.0

The Neural Network Toolbox extends MATLAB with tools for designing, implementing, visualizing, and simulating neural networks. Neural networks are invaluable for applications where formal analysis would be difficult or impossible, such as pattern recognition, nonlinear system identification, and control. The Neural Network Toolbox provides comprehensive support for many proven network paradigms, as well as graphical user interfaces (GUIs) that enable the user to design and manage the networks. The modular, open, and extensible design of the toolbox simplifies the creation of customized functions and networks.

Key Features include:

- GUI and quick-start wizard for creating, training, and simulating neural networks

- Support for the most commonly used supervised and unsupervised network architectures

- Comprehensive set of training and learning functions

- Dynamic learning networks, including time delay, nonlinear autoregressive (NARX), layer-recurrent, and custom dynamic

- Simulink blocks for building neural networks and advanced blocks for control systems applications

- Support for automatically generating Simulink blocks from neural network objects

- Modular network representation, enabling an unlimited number of input-setting layers and network interconnections

- Pre- and post-processing functions for improving network training and assessing network performance

- Routines for improving generalization

- Visualization functions for viewing network performance

Functions of Neural Network Toolbox

Analysis Functions	Analyze network properties
Distance Functions	Compute distance between two vectors
Graphical Interface Functions	Open GUIs for building neural networks
Layer Initialization Functions	Initialize layer weights
Learning Functions	Learning algorithms used to adapt networks
Line Search Functions	Line-search algorithms
Net Input Functions	Sum excitations of layer
Network Initialization Function	Initialize network weights
New Networks Functions	Create network architectures
Network Use Functions	High-level functions to manipulate networks
Performance Functions	Measure network performance
Plotting Functions	Plot and analyze networks and network performance
Processing Functions	Preprocess and postprocess data
Simulink Support Function	Generate Simulink block for network simulation
Topology Functions	Arrange neurons of layer according to specific topology
Training Functions	Train networks
Transfer Functions	Transform output of network layer
Utility Functions	Internal utility functions
Vector Functions	Internal functions for network computations
Weight and Bias Initialization Functions	Initialize weights and biases
Weight Functions	Convolution, dot product, scalar product, and distances weight functions

Analysis Functions

errsurf	Error surface of single-input neuron
confusion	Classification confusion matrix
maxlinlr	Maximum learning rate for linear neuron
roc	Receiver operating characteristic

Distance Functions

boxdist	Distance between two position vectors
dist	Euclidean distance weight function
linkdist	Link distance function
mandist	Manhattan distance weight function

Graphical Interface Functions

nctool	Neural network classification tool
nftool	Open Neural Network Fitting Tool
nntool	Open Network/Data Manager
nntraintool	Neural network training tool
nprtool	Neural network pattern recognition tool
view	View a neural network

Layer Initialization Functions

initnw	Nguyen-Widrow layer initialization function
initwb	By-weight-and-bias layer initialization function

Learning Functions

learncon	Conscience bias learning function
learngd	Gradient descent weight/bias learning function
learngdm	Gradient descent with momentum weight/bias learning function
learnh	Hebb weight learning function
learnhd	Hebb with decay weight learning rule
learnis	Instar weight learning function
learnk	Kohonen weight learning function
learnlv1	LVQ1 weight learning function
learnlv2	LVQ2 weight learning function
learnos	Outstar weight learning function
learnp	Perceptron weight and bias learning function
learnpn	Normalized perceptron weight and bias learning function
learnsom	Self-organizing map weight learning function
learnsomb	Batch self-organizing map weight learning function
learnwh	Widrow-Hoff weight and bias learning rule

Line Search Functions

srchbac	1-D minimization using backtracking search
srchbre	1-D interval location using Brent's method
srchcha	1-D minimization using Charalambous' method
srchgol	1-D minimization using golden section search
srchhyb	1-D minimization using hybrid bisection/cubic search

Net Input Functions

netprod	Product net input function
netsum	Sum net input function

Network Initialization Function

initlay	Layer-by-layer network initialization function

Network-Use Functions

adapt	Allow neural network to change weights and biases on inputs
disp	Neural network's properties
display	Name and properties of neural network's variables
init	Initialize neural network
sim	Simulate neural network
train	Train neural network

New Networks Functions

network	Create custom neural network
newc	Create competitive layer
newcf	Create cascade-forward backpropagation network
newdtdnn	Create distributed time delay neural network
newelm	Create Elman backpropagation network
newff	Create feedforward backpropagation network
newfftd	Create feedforward input-delay backpropagation network
newfit	Create a fitting network
newgrnn	Design generalized regression neural network
newhop	Create Hopfield recurrent network
newlin	Create linear layer
newlind	Design linear layer
newlrn	Create layered-recurrent network
newlvq	Create learning vector quantization network
newnarx	Create feedforward backpropagation network with feedback from output to input
newnarxsp	Create NARX network in series-parallel arrangement
newp	Create perceptron
newpnn	Design probabilistic neural network
newpr	Create a pattern recognition network
newrb	Design radial basis network
newrbe	Design exact radial basis network
newsom	Create self-organizing map
sp2narx	Convert series-parallel NARX network to parallel (feedback) form

Performance Functions

mae	Mean absolute error performance function
mse	Mean squared error performance function
msereg	Mean squared error with regularization performance function
mseregec	Mean squared error with regularization and economization performance function
sse	Sum squared error performance function

Plotting Functions

hintonw	Hinton graph of weight matrix
hintonwb	Hinton graph of weight matrix and bias vector
plotbr	Plot network performance for Bayesian regularization training
plotconfusion	Plot classification confusion matrix
plotep	Plot weight and bias position on error surface
plotes	Plot error surface of single-input neuron
plotfit	Plot function fit
plotpc	Plot classification line on perceptron vector plot
plotperf	Plot network performance
plotperform	Plot network performance
plotpv	Plot perceptron input target vectors
plotregression	Plot linear regression
plotroc	Plot receiver operating characteristic
plotsom	Plot self-organizing map
plotsomhits	Plot self-organizing map sample hits
plotsomnc	Plot self-organizing map neighbor connections
plotsomnd	Plot self-organizing map neighbor distances
plotsompos	Plot self-organizing map weight positions
plotsomtop	Plot self-organizing map topology
plottrainstate	Plot training state values
plotv	Plot vectors as lines from origin
plotvec	Plot vectors with different colors
postreg	Postprocess trained network response with linear regression

Processing Functions

fixunknowns	Process data by marking rows with unknown values
mapminmax	Process matrices by mapping row minimum and maximum values to [-1 1]
mapstd	Process matrices by mapping each row's means to 0 and deviations to 1
processpca	Process columns of matrix with principal component analysis
removeconstantrows	Process matrices by removing rows with constant values
removerows	Process matrices by removing rows with specified indices

Simulink Support Function

gensim	Generate Simulink block for neural network simulation

Topology Functions

gridtop	Gridtop layer topology function
hextop	Hexagonal layer topology function
randtop	Random layer topology function

Training Functions

trainb	Batch training with weight and bias learning rules
trainbfg	BFGS quasi-Newton backpropagation
trainbfgc	BFGS quasi-Newton backpropagation for use with NN model reference adaptive controller
trainbr	Bayesian regularization
trainbuwb	Batch unsupervised weight/bias training
trainc	Cyclical order incremental update
traincgb	Powell-Beale conjugate gradient backpropagation
traincgf	Fletcher-Powell conjugate gradient backpropagation
traincgp	Polak-Ribiére conjugate gradient backpropagation
traingd	Gradient descent backpropagation
traingda	Gradient descent with adaptive learning rule backpropagation
traingdm	Gradient descent with momentum backpropagation
traingdx	Gradient descent with momentum and adaptive learning rule backpropagation

trainlm	Levenberg-Marquardt backpropagation
trainoss	One step secant backpropagation
trainr	Random order incremental training with learning functions
trainrp	Resilient backpropagation (Rprop)
trains	Sequential order incremental training with learning functions
trainscg	Scaled conjugate gradient backpropagation

Transfer Functions

compet	C	Competitive transfer function
hardlim		Hard limit transfer function
hardlims		Symmetric hard limit transfer function
logsig		Log-sigmoid transfer function
netinv		Inverse transfer function
poslin		Positive linear transfer function
purelin		Linear transfer function
radbas		Radial basis transfer function
satlin		Saturating linear transfer function
satlins		Symmetric saturating linear transfer function
softmax	S	Softmax transfer function
tansig		Hyperbolic tangent sigmoid transfer function
tribas		Triangular basis transfer function

Utility Functions

calcgx	Calculate weight and bias performance gradient as single vector
calcjejj	Calculate Jacobian performance vector
calcjx	Calculate weight and bias performance Jacobian as single matrix
calcpd	Calculate delayed network inputs
calcperf	Calculate network outputs, signals, and performance
getx	All network weight and bias values as single vector
setx	Set all network weight and bias values with single vector

Vector Functions

combvec	Create all combinations of vectors
con2seq	Convert concurrent vectors to sequential vectors
concur	Create concurrent bias vectors
ind2vec	Convert indices to vectors
minmax	Ranges of matrix rows
normc	Normalize columns of matrix
normr	Normalize rows of matrix
pnormc	Pseudonormalize columns of matrix
quant	Discretize values as multiples of quantity
seq2con	Convert sequential vectors to concurrent vectors
vec2ind	Convert vectors to indices

Weight and Bias Initialization Functions

initcon	Conscience bias initialization function
initsompc	Initialize SOM weights with principal components
initzero	Zero weight and bias initialization function
midpoint	Midpoint weight initialization function
randnc	Normalized column weight initialization function
randnr	Normalized row weight initialization function
rands	Symmetric random weight/bias initialization function
revert	Change network weights and biases to previous initialization values

Weight Functions

convwf	Convolution weight function
dist	Euclidean distance weight function
dotprod	Dot product weight function
mandist	Manhattan distance weight function
negdist	Negative distance weight function
normprod	Normalized dot product weight function
scalprod	Scalar product weight function

C.4   Genetic Algorithm and Direct Search Toolbox 2.2

Genetic Algorithm and Direct Search Toolbox extends the optimization capabilities in MATLAB and Optimization Toolbox with tools for using genetic algorithms, simulated annealing, and direct search. These algorithms are used for problems that are difficult to solve with traditional optimization techniques, including problems that are not well defined or are difficult to model mathematically. The user can also use them when computation of the objective function is discontinuous, highly nonlinear, stochastic, or has unreliable or undefined derivatives.

Genetic Algorithm and Direct Search Toolbox complements other optimization methods to help the user find good starting points. Further the user can then use traditional optimization techniques to refine the solution.

Toolbox functions, accessible through a graphical user interface (GUI) or the MATLAB command line, are written in the open MATLAB language. This means that the user can inspect the algorithms, modify the source code, and create own custom functions.

Key Features include

- Graphical user interfaces and command-line functions for quickly setting up problems, setting algorithm options, and monitoring progress

- Genetic algorithm tools with options for creating initial population, fitness scaling, parent selection, crossover, and mutation

- Direct search tools that implement a pattern search method, with options for defining mesh size, polling technique, and search method

- Simulated annealing and threshold acceptance tools that implement a random search method, with options for defining annealing process, temperature schedule, and acceptance criteria

- Ability to solve optimization problems with nonlinear, linear, and bound constraints

- Functions for integrating Optimization Toolbox and MATLAB routines with the genetic or direct search algorithm

- Support for automatic M-code generation

Functions of Genetic Algorithm and Direct Search Toolbox

Genetic Algorithm

ga	Find minimum of function using genetic algorithm
gamultiobj	Find minima of multiple functions using genetic algorithm
gaoptimget	Obtain values of genetic algorithm options structure
gaoptimset	Create genetic algorithm options structure

Direct Search

patternsearch	Find minimum of function using pattern search
psoptimget	Obtain values of pattern search options structure
psoptimset	Create pattern search options structure

Simulated Annealing

saoptimget	Values of simulated annealing or threshold acceptance algorithm options structure
saoptimset	Create simulated annealing algorithm or threshold acceptance options structure
simulannealbnd	Find unconstrained or bound-constrained minimum of function of several variables using simulated annealing algorithm

C.5 GPLAB - A Genetic Programming Toolbox for MATLAB

GPLAB is a Genetic Programming toolbox for MATLAB. Most of its functions are used as "plug and play" devices, making it a versatile and easily extendable tool, as long as the user has minimum knowledge of the MATLAB programming environment.

Some of the features of GPLAB include:

- 3 modes of tree initialization (Full, Grow, Ramped Half-and-Half) + 3 variations on these

- several pre-made functions and terminals for building trees

- dynamic limits on tree depth or size (optional)

- resource-limited GP (variable size populations) (optional)

- dynamic populations (variable size populations) (optional)

- 4 genetic operators (crossover, mutation, swap mutation, shrink mutation)

- configurable automatic adaptation of operator probabilities (optional)

- steady-state + generational + batch modes, with fuzzy frontiers between them

- 5 sampling methods (Roulette, SUS, Tournament, Lexicographic Parsimony Pressure Tournament, Double Tournament)

- 3 modes of calculating the expected number of offspring (absolute + 2 ranking methods)

- 2 methods for reading input files and for calculating fitness (symbolic regression and parity problems + artificial ant problems)

- runtime cross-validation of the best individual of the run (optional)

- offline cross-validation or prediction of results by any individual (optional)

- 4 levels of elitism

- configurable stop conditions

- saving of results to files (5 frequency modes, optional)

- 3 modes of runtime textual output

- runtime graphical output (4 plots, optional)

- offline graphical output (5 functions, optional)

- runtime measurement of population diversity (2 measures, optional)

- runtime measurement of average tree level, number of nodes, number of introns, tree fill rate (optional)

- 4 demonstration functions (symbolic regression, parity, artificial ant, multiplexer)

No matter how long this list could be, the best feature of GPLAB will always be the "plug and play" philosophy. Any alternative (or collective) function built by the user will be readily accepted by the toolbox, as long as it conforms to the rules pertaining the module in question. Also, there are no incompatibilities between functions and parameters, meaning the user can use any combination of them, even when the user uses their own functions.

GPLAB does not implement:

- multiple subpopulations

- automatically-defined functions

Functions of Genetic Programming Toolbox

The more than one hundred functions provided in the toolbox GPLAB can be divided into several different functional groups. Following is a list of the functions included in each group. The same function may be listed in more than one group.

Demonstration Functions

demo	DEMO runs a symbolic regression problem
demoparity	DEMO runs a parity -3 problem
demoant	DEMO runs an ant

Running the Algorithm and Testing Result

gplab	Runs the GPLAB genetic programming algorithm
testind	Evaluates a GPLAB individual on a different data set

Parameter and State Setting

setparams	Sets the parameter variables for the GPLAB algorithm
resetparams	Resets the parameter variables for the GPLAB algorithm
resetstate	Resets the state variables for the GPLAB algorithm
setoperators	Stores operators info as parameters for the GPLAB algorithm
addoperators	Stores additional operators info as parameters for GPLAB
setfunctions	Stores functions info as parameters for the GPLAB algorithm
addfunctions	Stores additional functions info as parameters for GPLAB
setterminals	Stores terminals info as parameters for the GPLAB algorithm
addterminals	Stores additional terminals info as parameters for GPLAB

Automatic Variable Checking

checkvarsparams	Initializes GPLAB algorithm parameter variables
checkvarsstate	Initializes GPLAB algorithm state variables
checkvarsdata	Fills the dataset variable for the GPLAB algorithm

Description of Parameter and State Variables

availableparams	Describes the GPLAB algorithm parameter variables
availablestate	Describes the GPLAB algorithm state variables

Creation of New Generations

genpop	Creates the initial generation for the GPLAB algorithm
generation	Creates a new generation for the GPLAB algorithm
pickoperator	Draws a genetic operator to apply in the GPLAB algorithm
applyoperator	Apply a genetic operator to create new GPLAB individuals
pickparents	Picks parents from pool for a GPLAB genetic operator
applysurvival	Choose new generation of individuals for GPLAB algorithm
updatestate	Updates the GPLAB algorithm state variables
stopcondition	Checks which stop condition the GPLAB algorithm verifies

Creation of New Individuals

initpop	Creates a new population for the GPLAB algorithm
fullinit	Creates a new GPLAB population with the full method
growinit	Creates a new GPLAB population with the grow method
rampedinit	Creates a new GPLAB population with ramped half-and-half method
newind	Creates a new individual for the GPLAB algorithm
maketree	Creates representation tree for the GPLAB algorithm

Filtering of New Individuals

validateinds	Applies validation procedures to new GPLAB individuals
strictdepth	Applies strict depth filters to a new GPLAB individual
strictnodes	Applies strict size filters to a new GPLAB individual
dyndepth	Applies dynamic depth filters to a new GPLAB individual
dynnodes	Applies dynamic size filters to a new GPLAB individual
heavydyndepth	Applies heavy dynamic depth filters to a new GPLAB individual
heavydynnodes	Applies heavy dynamic size filters to a new GPLAB individual

Protected and Logical Functions

mydivide	MYDIVIDE(X1,X2) returns X1 if X2==0 and X1/X2 otherwise
mylog	MYLOG(X) returns zero if X=0 and LOG(ABS(X)) otherwise
mylog2	MYLOG2(X) returns zero if X=0 and LOG2(ABS(X)) otherwise
mylog10	MYLOG10(X) returns zero if X=0 and LOG10(ABS(X)) otherwise
mysqrt	MYSQRT(X) returns zero if X¡=0 and SQRT(X) otherwise
mypower	MYPOWER(X1,X2) returns 0 if X1 X2 is Inf, or has imaginary part, otherwise returns X1 X2
myif	Calculates the result of an IF-THEN-ELSE statement
kozadivide	KOZADIVIDE(X1,X2) returns 1 if X2==0 and X1/X2 otherwise
kozasqrt	KOZASQRT(X) returns SQRT(ABS(X))
nand	NAND(X1,X2) returns NOT(AND(X1,X2))
nor	NAND(X1,X2) returns NOT(OR(X1,X2))

Artificial Ant Functions

demoant	Demonstration function of the GPLAB toolbox
antmove	Moves the GPLAB artificial ant forward one step
antleft	Turns the GPLAB artificial ant to the left
antright	Turns the GPLAB artificial ant to the right
antprogn2	Executes two actions of the GPLAB artificial ant
antprogn3	Executes three actions of the GPLAB artificial ant
antif	Executes one or other action of the GPLAB artificial ant
antfoodahead	Tests if there is food ahead of the GPLAB artificial ant
antnewpos	Calculates the new location of the GPLAB artificial ant
anteval	Evaluates the tree of a GPLAB artificial ant
antfitness	Measures the fitness of a GPLAB artificial ant
antfitness lib	Measures the fitness of a GPLAB artificial ant, lower is better
anttrail	Interprets a matrix as a food trail for a GPLAB artificial ant
antsim	Simulates the behaviour of the best GPLAB artificial ant
antpath	Stores the path taken by the GPLAB artificial ant

Tree Manipulation

maketree	Creates representation tree for the GPLAB algorithm
treelevel	Counts the number of levels of a GPLAB algorithm tree
nodes	Counts the number of nodes of a GPLAB algorithm tree
intronnodes	Counts the number of intron nodes of a GPLAB tree
tree2str	Translates a GPLAB algorithm tree into a string
findnode	Finds a node in a GPLAB algorithm tree
swapnodes	Swaps nodes (subtrees) between two GPLAB trees
updatenodeids	Updates the node ids of a GPLAB tree

Data Manipulation

xy2inout	Transforms matrices into a GPLAB algorithm data set (input and output)
saveall	Saves all the GPLAB algorithm variables to disk

Expected Number of Children

calcpopexpected	Normalized expected number of offspring for GPLAB
absolute	Calculates expected number of offspring for the GPLAB algorithm
rank85	Calculates expected number of offspring for the GPLAB algorithm
rank89	Calculates expected number of offspring for the GPLAB algorithm

Sampling

sampling	Draws individuals for parenthood in the GPLAB algorithm
roulette	Sampling of GPLAB individuals by the roulette method
sus	Sampling of GPLAB individuals by the SUS method
wheel	Sampling of GPLAB individuals by spinning a wheel
tournament	Sampling of GPLAB individuals by the tournament method
lexictour	Sampling of GPLAB individuals by lexicographic parsimony tournament
doubletour	Sampling of GPLAB individuals by a double tournament method
tourbest	Tournament of GPLAB individuals by size or fitness

Genetic Operators

crossover	Creates new individuals for the GPLAB algorithm by crossover
mutation	Creates a new individual for GPLAB by mutation
shrinkmutation	Creates a new individual for GPLAB by shrink mutation
swapmutation	Creates a new individual for GPLAB by swap mutation

Fitness

calcpopfitness	Calculate fitness values for a GPLAB population
calcfitness	Measures the fitness of a GPLAB individual
regfitness	Measures the fitness of a GPLAB individual
evaluate_tree	Alternative to using 'eval' in 'regfitness' in GPLAB
linearppp	Applies linear parametric parsimony pressure to a GPLAB individual

Survival

applysurvival	Choose new generation of individuals for GPLAB algorithm
fixedpopsize	Chooses fixed number of GPLAB individuals for next generation
resources	Chooses a number of GPLAB individuals for next generation
pivotfixe	Chooses a number of GPLAB individuals for next generation

Limited Resources

low	Applies restrictions to the size of a GPLAB population
steady	Applies restrictions to the size of a GPLAB population
free	Applies no restrictions to the size of a GPLAB population
normal	Decides whether to accept a GPLAB individual into the population
light	Decides whether to accept a GPLAB individual into the population

Dynamic Populations

ajout	Adds a number of GPLAB individuals to the population
suppression	Removes a number of GPLAB individuals from the population

Diversity measures

uniquegen	Calculates genotype-based diversity of a GPLAB population
hamming	Calculates hamming diversity of a GPLAB population

Automatic Operator Probability Adaptation

isoperator	True for GPLAB algorithm operator
setinitialprobs	Sets the initial operator probabilities for the GPLAB algorithm
automaticoperatorprobs	Automatic operator probabilities procedure for GPLAB
moveadaptwindow	Shifts the GPLAB automatic operator probabilities adaptation window

addcredit	Attributes credit to GPLAB individuals in credit list
updateoperatorprobs	Updates GPLAB genetic operator probabilities

Runtime Graphical Output

graphicsinit	Initializes graphics for the GPLAB algorithm
graphicsstart	Initializes first drawing points in the GPLAB graphics
graphicscontinue	Draws past history lines in the GPLAB graphics
graphicsgenerations	Draws data from new generations in the GPLAB graphics

Offline Graphical Output

desired_obtained	Plots desired and obtained functions with GPLAB
accuracy_complexity	Plots accuracy and complexity measures with GPLAB
plotpareto	Plots the pareto front in GPLAB
operator_evolution	Plots operator probabilities evolution with GPLAB
drawtree	Draws the GPLAB trees graphically

Utilitarian Functions

explode	Splits string into pieces
implode	Joins strings with delimiter in between
scale	Maps numbers from one interval to another
normalize	Normalizes vectors
shuffle	Shuffles vectors or matrices
orderby	Orders vectors and matrices according to a pre-defined order
intrand	Generates an integer random number inside an interval
countfind	Counts occurences of numbers
findfirstindex	Finds the first occurences of numbers
isvalid	Validates a value according to a domain
ranking	Ranks the elements of a vector
fixdec	Round towards zero with a specified number of decimals
uniquenosort	Eliminates duplicate rows without altering order
nansum	Calculates mean along any dimension of the N-D array X ignoring NaNs
nullexceeding	Nulls numbers in list so that sum does not exceed limit

Appendix D

Emerging Software Packages

This appendix gives a list of all known commercial emerging software packages related to Computational Intelligence that is available to the users.

D.1 BUGS

BUGS (Better to Use Genetic Systems) is an interactive program for demonstrating the genetic algorithm. The user can evolve lifelike organisms (curves). Playing with BUGS is an easy way to get an understanding of how and why the GA works. In addition to demonstrating the basic genetic operators (selection, crossover, and mutation), it allows users to easily see and understand phenomena such as genetic drift and premature convergence. BUGS is written in C and runs under Suntools and X Windows.

Available at: ftp://www.aic.nrl.navy.mil/pub/galist/src/BUGS.tar.Z

D.2 ComputerAnts

ComputerAnts is a free Windows program that teaches principles of genetic algorithms by breeding a colony of ants on the computer screen. Users create ants, food, poison, and set crossover and mutation rates. Then they watch the colony slowly evolve. Includes extensive on-line help and tutorials on genetic algorithms.

Available at: http://www.bitstar.com

D.3　DGenesis

DGenesis is a distributed implementation of a Parallel GA. It is based on Genesis 5.0. It runs on a network of UNIX workstations. It has been tested with DECstations, microVAXes, Sun Workstations, and PCs running 386BSD 0.1. Each subpopulation is handled by a UNIX process and the communication between them is accomplished using Berkeley sockets.

DGenesis allows the user to set the migration interval, the migration rate, and the topology between the sub-populations. There has not been much work investigating the effect of the topology on the performance of the GA, DGenesis was written specifically to encourage experimentation in this area. It still needs many refinements, but some may find it useful.

Available at: ftp://ftp.aic.nrl.navy.mil/pub/galist/src/ga/dgenesis-1.0.tar.Z

D.4　Ease

Ease - Evolutionary Algorithms Scripting Environment - is an extension to the Tcl scripting language, providing commands to create, modify, and evaluate populations of individuals represented by real number vectors and/or bit strings. With Ease, a standard ES or GA can be written in less than 20 lines of code. Ease is available as source code for Linux and Solaris under the GNU Public License. Tcl version 8.0 or higher is required.

Available at: http://www.sprave.com/Ease/Ease.html

D.5　Evolution Machine

The Evolution Machine (EM) is universally applicable to continuous (real-coded) optimization problems. In the EM we have coded fundamental evolutionary algorithms (genetic algorithms and evolution strategies), and added some of the approaches to evolutionary search.

The EM includes extensive menu techniques with:

- Default parameter setting for un-experienced users.

- Well-defined entries for EM-control by freaks of the EM, who want to leave the standard process control.

- Data processing for repeated runs (with or without change of the strategy parameters).

- Graphical presentation of results: online presentation of the evolution progress, one-, two- and three-dimensional graphic output to analyze the fitness function and the evolution process.

- Integration of calling MS-DOS utilities (Turbo C).

The EM-software is provided in object code, which can be run on PC's with MS-DOS and Turbo C, v2.0, resp. Turbo C++,v1.01. The Manual to the EM is included in the distribution kit.

Available at: ftp://ftp-bionik.fb10.tu-berlin.de/pub/software/Evolution-Machine/

D.6 Evolutionary Objects

EO (Evolutionary Objects) is a C++ library written and designed to allow a variety of evolutionary algorithms to be constructed easily. It is intended to be an "Open source" effort to create the definitive EC library. It has: a mailing list, anon-CVS access, frequent snapshots, and other features.

Available at: http://fast.to/EO

D.7 GAC, GAL

GAC and GAL are packages that have been used for the past few years. GAC is a GA written in C. GAL is my Common Lisp version. They are similar in spirit to John Grefenstette's Genesis, but they don't have all the nice bells and whistles. Both versions currently run on Sun workstations.

Available at: ftp://ftp.aic.nrl.navy.mil/pub/galist/src/ga/GAC.shar.Z

D.8   GAGA

GAGA (GA for General Application) is a self-contained, re-entrant procedure, which is suitable for the minimization of many "difficult" cost functions. Originally written in Pascal by Ian Poole, it was rewritten in C by Jon Crowcroft.

Available at: ftp://ftp://cs.ucl.ac.uk/darpa/gaga.shar

D.9   GAGS

GAGS (Genetic Algorithms from Granada, Spain) is a library and companion programs written and designed to take the heat out of designing a genetic algorithm. It features a class library for genetic algorithm programming, but, from the user point of view, is a genetic algorithm application generator. If the function to be optimized is written, the GAGS surrounds it with enough code to have a genetic algorithm up and running, compiles it, and runs it. GAGS is written in C++, so that it can be compiled in any platform running this GNU utility. It has been tested on various machines.

GAGS includes:

- Steady-state, roulette-wheel, tournament, and elitist selection.

- Fitness evaluation using training files.

- Graphics output through gnuplot.

- Uniform and 2-point crossover, and bit-flip and gene-transposition mutation.

- Variable length chromosomes and related operators.

The application generator gags.pl is written in perl, so this language must also be installed before GAGS.

Available at: http://kal-el.ugr.es/GAGS

D.10 GAlib

GAlib is a C++ library that provides the application programmer with a set of genetic algorithm objects. With GAlib the user can add GA optimization to the program using any data representation and standard or custom selection, crossover, mutation, scaling, and replacement, and termination methods.

GAlib requires a cfront 3.0 compatible C++ compiler. It has been used on the following systems: SGI IRIX 4.0.x (Cfront); SGI IRIX 5.x (DCC 1.0, g++ 2.6.8, 2.7.0); IBM RSAIX 3.2 (g++ 2.6.8, 2.7.0); DEC MIPS ultrix 4.2 (g++ 2.6.8, 2.7.0); SUN SOLARIS 5.3 (g++ 2.6.8, 2.7.0); HP-UX (g++); MacOS (MetroWerks CodeWarrior 5); MacOS (Symantec THINK C++ 7.0); DOS/Windows (Borland Turbo C++ 3.0).

Available at: ftp://lancet.mit.edu/pub/ga/

D.11 GALOPPS

GALOPPS (Genetic Algorithm Optimized for Portability and Parallelism) is a general-purpose parallel genetic algorithm system, written in "C", organized like Goldberg's "Simple Genetic Algorithm". User defines objective function (in template furnished) and any callback functions desired (again, filling in template); can run one or many subpopulations, on one or many PC's, workstations, Mac's, MPP. Runs interactively (GUI or answering questions) or from files, makes file and/or graphical output. Runs easily interrupted and restarted, and a PVM version for Unix networks even moves processes automatically when workstations become busy.

User may choose:

- problem type (permutation or value-type)

- field sizes (arbitrary, possibly unequal, heeded by crossover, mutation)

- among 7 crossover types and 4 mutation types (or define own)

- among 6 selection types, including "automatic" option based on Boltzmann scaling and Shapiro and Pruegel–Bennett statist. Mechanics stuff

- operator probabilities, fitness scaling, amount of output, migration frequency and patterns

- stopping criteria (using "standard" convergence statistics, etc.)

- the GGA (Grouping Genetic Algorithm) reproduction and operators of Falkenauer GALOPPS allows and supports:

  - use of a different representation in each subpopulation, with transformation of migrants

  - inversion on level of subpopulations, with automatic handling of differing field sizes, migrants

  - control over replacement by offspring, including DeJong crowding or random replacement or SGA-like replacement of parents

  - mate selection, using incest reduction

  - migrant selection, using incest reduction, and/or DeJong crowding into receiving subpopulation

  - optional elitism

Available at: http://GARAGe.cps.msu.edu/

D.12 GAMusic

GAMusic 1.0 is a user-friendly interactive demonstration of a simple GA that evolves musical melodies. Here, the user is the fitness function. Melodies from the population can be played and then assigned a fitness. Iteration, recombination frequency, and mutation frequency are all controlled by the user. This program is intended to provide an introduction to GAs and may not be of interest to the experienced GA programmer. GAMusic was programmed with Microsoft Visual Basic 3.0 for Windows 3.1x.

Available at: ftp://wuarchive.wustl.edu/pub/MSDOSUPLOADS/Gen Algs/gamusic.zip

D.13   GANNET

GANNET (Genetic Algorithm/Neural NETwork) is a software package written by Jason Spofford in 1990 which allows one to evolve binary valued neural networks. It offers a variety of configuration options related to rates of the genetic operators. GANNET evolves nets based upon three fitness functions: Input/Output Accuracy, Output "Stability", and Network Size.

The evolved neural network presently has a binary input and binary output format, with neurons that have either 2 or 4 inputs and weights ranging from -3 to $+4$. GANNET allows for up to 250 neurons in a net. Research using GANNET is continuing.

The major enhancement of version 2.0 is the ability to recognize variable length binary strings, such as those that would be generated by a finite automaton. Included is code for calculating the Effective Measure Complexity (EMC) of finite automata as well as code for generating test data.

Available at: http://www.duane.com/ dduane/gannet

D.14   GA Workbench

A mouse-driven interactive GA demonstration program aimed at people wishing to show GA in action on simple function optimizations and to help newcomers understand how GA operates. Features include problem functions drawn on screen using mouse, run-time plots of GA population distribution, peak and average fitness. Useful population statistics displayed numerically, GA configuration (population size, generation gap, etc.) performed interactively with mouse. Requirements include MS-DOS PC, mouse, EGA/VGA display.

Available   at: ftp://wsmr-simtel20.army.mil/pub/msdos/neurlnet/gaw110.zip

D.15   GECO

GECO (Genetic Evolution through Combination of Objects) is an extensible, object-oriented framework for prototyping genetic algorithms in Common Lisp. GECO makes extensive use of CLOS, the Common Lisp Object System, to implement its functionality. The abstractions provided by the classes have been chosen with the intent both of being easily understandable to anyone familiar with the paradigm of genetic algorithms, and of providing the algorithm developer with the ability to customize all aspects of its operation. It comes with extensive documentation, in the form of a PostScript file, and some simple examples are also provided to illustrate its intended use.

Available at: ftp://ftp.aic.nrl.navy.mil/pub/galist/src/ga/GECO-v2.0. README

D.16   Genesis

Genesis is a generational GA system written in C by John Grefenstette. As the first widely available GA program Genesis has been very influential in stimulating the use of GAs, and several other GA packages are based on it. Genesis is available together with OOGA.

Available at: ftp://ftp.aic.nrl.navy.mil/pub/galist/src/genesis.tar.Z

D.17   GENEsYs

GENEsYs is a Genesis-based GA implementation which includes extensions and new features for experimental purposes, such as selection schemes like linear ranking, Boltzmann, (mu, lambda)-selection, and general extinctive selection variants, crossover operators like n-point and uniform crossover as well as discrete and intermediate recombination. Self-adaptation of mutation rates is also possible.

A set of objective functions is provided, including De Jong's functions, complicated continuous functions, a TSP-problem, binary functions, and a fractal function. There are also additional data-monitoring facilities such as recording average, variance and skew of object variables and mutation rates, or creating bitmap-dumps of the population.

Available at: ftp://lumpi.informatik.uni-dortmund.de/pub/GA/src/ GENEsYs-1.0.tar.Z

D.18 GenET

GenET is a "generic" GA package. It is generic in the sense that all problem independent mechanisms have been implemented and can be used regardless of application domain. Using the package forces concentration on the problem: the user has to suggest the best representation, and the best operators for such space that utilize problem-specific knowledge. The package, in addition to allowing for fast implementation of applications and being a natural tool for comparing different models and strategies, is intended to become a depository of representations and operators. Currently, only floating point representation is implemented in the library with few operators.

The algorithm provides a wide selection of models and choices. For example, population models range from generational GA, through steady-state, to (n,m)-EP and (n,n+m)-EP models (for arbitrary problems, not just parameter optimization). (Some are not finished at the moment). Choices include automatic adaptation of operator probabilities and a dynamic ranking mechanism, etc.

Available at: ftp://radom.umsl.edu/var/ftp/GenET.tar.Z

D.19 Genie

Genie is a GA-based modeling/forecasting system that is used for long-term planning. One can construct a model of an environment and then view the forecasts of how that environment will evolve into the future. It is then possible to alter the future picture of the environment so as to construct a picture of a desired future. The GA is then employed to suggest changes to the existing environment so as to cause the desired future to come about.

Available at: ftp://hiplab.newcastle.edu.au/pub/GenieCode.sea.Hqx

D.20 Genitor

Genitor is a modular GA package containing examples for floating-point, integer, and binary representations. Its features include many sequencing operators as well as subpopulation modeling. The Genitor Package has code for several order based crossover operators, as well as example code for doing some small TSPs to optimality.

Available at: ftp://ftp.cs.colostate.edu/pub/GENITOR.tar

D.21   GENlib

GENlib is a library of functions for genetic algorithms. Included are two applications of this library to the field of neural networks. The first one called "cosine" uses a genetic algorithm to train a simple three layer feed-forward network to work as a cosine-function. This task is very difficult to train for a backprop algorithm while the genetic algorithm produces good results. The second one called "vartop" is developing a Neural Network to perform the XOR-function. This is done with two genetic algorithms, the first one develops the topology of the network, the second one adjusts the weights.

Available at: ftp://ftp.neuro.informatik.uni-kassel.de/pub/NeuralNets/GA-and-NN/

D.22   GENOCOP

This is a GA-based optimization package that has been developed by Zbigniew Michalewicz and is described in detail in his book Genetic Algorithms + Data Structures = Evolution Programs. GENOCOP (Genetic Algorithm for Numerical Optimization for COnstrained Problems) optimizes a function with any number of linear constraints (equalities and inequalities).

Available at: ftp://ftp.uncc.edu/coe/evol/genocop2.tar.Z

D.23   GPEIST

The genetic programming environment in Smalltalk (GPEIST) provides a framework for the investigation of Genetic Programming within a ParcPlace VisualWorks 2.0 development system. GPEIST provides program, population, chart and report browsers and can be run on HP/Sun/PC (OS/2 and Windows) machines. It is possible to distribute the experiment across several workstations — with subpopulation exchange at intervals — in this release 4.0a. Experiments, populations and individual genetic programs can be saved to disk for subsequent analysis and experimental statistical measures exchanged with spreadsheets. Postscript printing of charts, programs and animations is supported. An implementation of the Ant Trail problem is provided as an example of the use of the GPEIST environment.

Available at: ftp.cc.utexas.edu:/pub/genetic-programming/code/

D.24   Imogene

Imogene is a Windows 3.1 shareware program, which generates pretty images using genetic programming. The program displays generations of 9 images, each generated using a formula applied to each pixel. (The formulae are initially randomly computed.) The user can then select those images that are preferred. In the next generation, the nine images are generated by combining and mutating the formulae for the most-preferred images in the previous generation. The result is a simulation of natural selection in which images evolve toward the aesthetic preferences.

Imogene supports different color maps, palette animation, saving images to .BMP files, changing the wallpaper to nice images, printing images, and several other features. Imogene works only in 256-color mode and requires a floating point coprocessor and a 386 or better CPU.

Available at: http://www.aracnet.com/ wwir/software.html

D.25   JAG

This Java program implements a simple genetic algorithm where the fitness function takes non-negative values only. It employs elitism. The Java code was derived from the C code in the Appendix of Genetic Algorithms + Data Structures = Evolution Programs, Other ideas and code were drawn from GAC by Bill Spears. Four sample problems are contained in the code: three with bit GENEs and one with double genes.

Available at: ftp://ftp.mcs.drexel.edu/pub/shartley/simpleGA.tar.gz.

D.26   LibGA

LibGA is a library of routines written in C for developing genetic algorithms. It is fairly simple to use, with many knobs to turn. Most GA parameters can be set or changed via a configuration files, with no need to recompile (e.g., operators, pool size and even the data type used in the chromosome can be changed in the configuration file.) Function pointers are used for the genetic operators, so they can easily be manipulated on the fly. Several genetic operators are supplied and it is easy to add more. LibGA runs on many systems/architectures. These include Unix, DOS, NeXT, and Amiga.

Available at: ftp://ftp.aic.nrl.navy.mil/pub/galist/src/ga/libga100.tar.Z

D.27   mGA

mGA is an implementation of a messy GA. Messy GAs overcome the linkage problem of simple genetic algorithms by combining variable-length strings, gene expression, messy operators, and a non-homogeneous phasing of evolutionary processing. Results on a number of difficult deceptive test functions have been encouraging with the messy GA always finding global optima in a polynomial number of function evaluations.

Available at: ftp://gal4.ge.uiuc.edu/pub/src/messyGA/C/

D.28   PGA

PGA is a simple testbed for basic explorations in genetic algorithms. Command line arguments control a range of parameters, there are a number of built-in problems for the GA to solve. The current set includes:

- maximize the number of bits set in a chromosome

- De Jong's functions DJ1, DJ2, DJ3, DJ5

- binary F6, used by Schaffer et al.

- a crude 1-d knapsack problem; when a target and a set of numbers in an external file are specified, GA tries to find a subset that sums as closely as possible to the target

- the "royal road" function(s); a chromosome is regarded as a set of consecutive blocks of size K, and scores K for each block entirely filled with 1s, etc.; a range of parameters

- max contiguous bits

- timetabling, with various smart mutation options; capable of solving a good many real-world timetabling problems

Lots of GA options: rank, roulette, tournament, marriage-tournament, spatially-structured selection; one-point, two-point, uniform or no crossover; fixed or adaptive mutation; one child or two; etc. Default output is curses-based, with optional output to file; can be run non-interactively too for batched series of experiments. Chromosomes are represented as character arrays. PGA has been used for teaching for a couple of years now, and has been used as a starting point by a fair number of people for their own projects. So it's reasonably reliable.

Available at: ftp://ftp.dai.cd.ac.uk/pub/pga/pga-3.1.tar.gz

D.29 PGAPack

PGAPack is a general-purpose, data-structure-neutral parallel genetic algorithm library. It is intended to provide most capabilities desired in a genetic algorithm library, in an integrated, seamless, and portable manner.

Features include:

- Callable from Fortran or C

- Runs on uniprocessors, parallel computers, and workstation networks

- Binary-, integer-, real-, and character-valued native data types

- Full extensibility to support custom operators and new data types

- Easy-to-use interface for novice and application users.

- Multiple levels of access for expert users

- Extensive debugging facilities

- Large set of example problems

- Detailed users guide

- Parameterized population replacement

- Multiple choices for selection, crossover, and mutation operators

- Easy integration of hill-climbing heuristics

Available at: ftp://info.mcs.anl.gov/pub/pgapack/pgapack.tar.Z

D.30   SGA-C, SGA-Cube

SGA-C is a C-language translation and extension of the original Pascal SGA code presented in Goldberg's book. It has some additional features, but its operation is essentially the same as that of the Pascal version. SGA-Cube is a C-language translation of Goldberg's SGA code with modifications to allow execution on the nCUBE 2 Hypercube Parallel Computer. When run on the nCUBE 2, SGA-Cube can take advantage of the hypercube architecture, and is scalable to any hypercube dimension. The hypercube implementation is modular, so that the algorithm for exploiting parallel processors can be easily modified.

In addition to its parallel capabilities, SGA-Cube can be compiled on various serial computers via compile-time options. In fact, when compiled on a serial computer, SGA-Cube is essentially identical to SGA-C. Each of these programs is distributed in the form of a Unix file.

Available at: ftp://ftp.aic.nrl.navy.mil/pub/galist/src/ga/sga-c.tar.Z

D.31   Splicer

Splicer is a genetic algorithm tool created by the Software Technology Branch (STB) of the Information Systems Directorate at NASA/Johnson Space Center with support from the MITRE Corporation. Splicer has well-defined interfaces between a GA kernel, representation libraries, fitness modules, and user interface libraries.

The representation libraries contain functions for defining, creating, and decoding genetic strings, as well as multiple crossover and mutation operators. Libraries supporting binary strings and permutations are provided, others can be created by the user.

Fitness modules are typically written by the user, although some sample applications are provided. The modules may contain a fitness function, initial values for various control parameters, and a function, which graphically displays the best solutions.

Splicer provides event-driven graphic user interface libraries for the Macintosh and the X11 window system (using the HP widget set); a menu-driven ASCII interface is also available though not fully supported. The extensive documentation includes a reference manual and a user's manual; an architecture manual and the advanced programmer's manual are currently being written.

Available if mailed to: ¡bayer@galileo.jsc.nasa.gov¿

D.32   Trans-Dimensional Learning

This is a Windows 3.1 artificial neural network and GA program (shareware). TDL allows users to perform pattern recognition by utilizing software that allows for fast, automatic construction of Neural Networks, mostly alleviating the need for parameter tuning. Evolutionary processes combined with semi-weighted networks (hybrid cross between standard weighted neurons and weightless n-level threshold units) generally yield very compact networks (i.e., reduced connections and hidden units). By supporting multi-shot learning over standard one-shot learning, multiple data sets (characterized by varying input and output dimensions) can be learned incrementally, resulting in a single coherent network. This can also lead to significant improvements in predictive accuracy (Trans-dimensional generalization). Graphical support and several data files are also provided.

Available at: http://pages.prodigy.com/upso

D.33   WOLF

This is a simulator for the G/SPLINES (genetic spline models) algorithm, which builds spline-based functional models of experimental data, using crossover and mutation to evolve a population toward a better fit. It is derived from Friedman's MARS models. The original work was presented at ICGA-4, and further results including additional basis function types such as B-splines have been presented at the NIPS-91 meeting.

Available at: ftp://riacs.edu/pub/wolf-4.0.tar.Z

D.34   XGenetic

XGenetic is an ActiveX control for the implementation of a genetic algorithm in any language that accepts ActiveX interfaces. Such languages include, but are not limited to: Visual Basic, Visual C++, Delphi, etc. Written in Visual Basic 6.0, XGenetic is flexible in implementation to allow the user to easily define the parameters for their particular scenario, be it forecasting, scheduling, or the myriad of other uses for the genetic algorithm.

Features:

- Data Types: Bit, Integer, Real

- Selection Operators: Roulette, Tournament, Stochastic Universal Sampling, Truncation, Random

- Crossover Operators: N-Point (1 point, 2 point, 3 point, etc.,), Uniform, Arithmetic

- Mutation Operators: Uniform, Boundary

There are two versions of the software available. The shareware version of the product is available freely off the net (address below). It includes the program file (xgen.ocx) and documentation (including a sample program) in three formats.

Available at: http://www.winsite.com/info/pc/win95/demo/xgen-sw.zip

D.35   XFUZZY: A Simulation Environment for Fuzzy Logic Control Systems

XFUZZY is an X-Windows based simulation tool for fuzzy controllers that runs on Sun workstations. The kernel of XFUZZY consists of the software implementation of a fuzzy inference engine. The modules defining the behavior of the controller (Fuzzifier, Rules Set, and Defuzzifier) and the system being controlled (model) interact with this inference engine. The environment is completed with a user interface, which, through a group of menus, eases the tasks of editing the membership functions, rules, and the system model, as well as the graphic presentation of the simulation results.

The closing of the feedback loop of the complete system (controller + system under control) is accomplished by means of a program in C, provided by the user according to the directions contained in a template file supplied by the environment. This module is dynamically linked to the inference engine, allowing the user to perform the whole tuning of the controller without exiting the environment.

Available at: ftp.cnm.us.es:/pub/Xfuzzy11.tar.Z

D.36 ART*Enterprise

ART*Enterprise is the latest of the family of rule-based development environments originating with ART in the mid-1980s. It is a development environment for enterprise-wide applications, incorporating rules, a full object system which includes features currently not present in C++ or Smalltalk, and a large collection of object classes for UI development across platforms (from Windows to OS/2 to Unix), access to databases (SQL-based and ODBC-based), and multi-person development. The ART*Enterprise environment provides a forward chaining engine where backward chaining can be implemented, though it is not supported directly. ART*Enterprise also provides a CBR kernel for those who are interested in incorporating it into their applications.

Available at: http://www.brightware.com/

D.37 COMDALE/C, COMDALE/X, and ProcessVision

COMDALE/C is a real-time expert system designed for industrial process monitoring and control. COMDALE/C allows requests for justification of recommendations, conclusions, and control actions without interrupting the decision making process. It can deal with uncertainty in knowledge and data, and has an open architecture and time-based reasoning. Other features include:

- full object-oriented configuration

- full networking capabilities

- alarm processing

- an interrupt driven controller

- trending and historical data collection

- time-scheduled events

- a realtime database, and interfaces with DCSs, PLCs and other I/O devices.

COMDALE/X is an off-line consultative expert system which queries the user for information required to make its decisions. COMDALE/X is included with COMDALE/C as the development tool for real-time expert systems. COMDALE/X has the capability to incorporate hypertext documents with the reasoning abilities of the expert system to produce expert hyper manuals which provide information and generate advice through an easy to use interface.

ProcessVision is a real-time process monitoring and control software package. Based on an open and modular architecture, ProcessVision provides a graphical operator interface; intuitive object-oriented display configuration, smart alarming, sensor validation, hot standby, and unlimited connectivity to all the process instrumentation in one global environment.

Available at: http://www.comdale.com/

Appendix E

Research Projects

A brief description of research projects from various IEEE journals, Research centers, and Universities are given in this appendix.

E.1 An Evolutionary Algorithm for Global Optimization Based on Level-Set Evolution and Latin Squares

In this project, the level-set evolution is exploited in the design of a novel evolutionary algorithm (EA) for global optimization. An application of Latin squares leads to a new and effective crossover operator. This crossover operator can generate a set of uniformly scattered offspring around their parents, has the ability to search locally, and can explore the search space efficiently. To compute a globally optimal solution, the level set of the objective function is successively evolved by crossover and mutation operators so that it gradually approaches the globally optimal solution set. As a result, the level set can be efficiently improved. Based on these skills, a new EA is developed to solve a global optimization problem by successively evolving the level set of the objective function such that it becomes smaller and smaller until all of its points are optimal solutions. Furthermore, it can be proved that the proposed algorithm converges to a global optimizer with probability one. Numerical simulations are conducted for 20 standard test functions. The performance of the proposed algorithm is compared with that of eight EAs that have been published recently and the Monte Carlo implementation of the mean-value-level-set method. The results indicate that the proposed algorithm is effective and efficient.

E.2 Evolving Problems to Learn about Particle Swarm Optimizers and Other Search Algorithms

Evolutionary computation (EC) is used to automatically find problems, which demonstrate the strength and weaknesses of modern search heuristics. In particular, particle swarm optimization (PSO), differential evolution (DE), and covariance matrix adaptation-evolution strategy (CMA-ES) are analyzed. Each evolutionary algorithm is contrasted with the others and with a robust non-stochastic gradient follower (i.e., a hill climber) based on Newton-Raphson. The evolved benchmark problems yield insights into the operation of PSOs, illustrate benefits and drawbacks of different population sizes, velocity limits, and constriction (friction) coefficients. The fitness landscapes made by genetic programming reveal new swarm phenomena, such as deception, thereby explaining how they work and allowing to devise better extended particle swarm systems. The method could be applied to any type of optimizer.

E.3 Analog Genetic Encoding for the Evolution of Circuits and Networks

This project describes a new kind of genetic representation called analog genetic encoding (AGE). The representation is aimed at the evolutionary synthesis and reverse engineering of circuits and networks such as analog electronic circuits, neural networks, and genetic regulatory networks. AGE permits the simultaneous evolution of the topology and sizing of the networks. The establishment of the links between the devices that form the network is based on an implicit definition of the interaction between different parts of the genome. This reduces the amount of information that must be carried by the genome, relatively to a direct encoding of the links. The application of AGE is illustrated with examples of analog electronic circuit and neural network synthesis. The performance of the representation and the quality of the results obtained with AGE are compared with those produced by genetic programming.

E.4 A Runtime Analysis of Evolutionary Algorithms for Constrained Optimization Problems

Although there are many evolutionary algorithms (EAs) for solving constrained optimization problems, there are few rigorous theoretical analyses. This project presents a time complexity analysis of EAs for solving constrained optimization. It is shown when the penalty coefficient is chosen properly, direct comparison between pairs of solutions using penalty fitness function is equivalent to that using the criteria "superiority of feasible point" or "superiority of objective function value." This project analyzes the role of penalty coefficients in EAs in terms of time complexity. The results show that in some examples, EAs benefit greatly from higher penalty coefficients, while in other examples, EAs benefit from lower penalty coefficients. This project also investigates the runtime of EAs for solving the 0-1 knapsack problem and the results indicate that the mean first hitting times ranges from a polynomial-time to an exponential time when different penalty coefficients are used.

E.5 Solving the Register Allocation Problem for Embedded Systems Using a Hybrid Evolutionary Algorithm

Embedded systems are unique in the challenges they present to application programmers, such as power and memory space constraints. These characteristics make it imperative to design customized compiler passes. One of the important factors that shape runtime performance of a given embedded code is the register allocation phase of compilation. It is crucial to provide aggressive and sophisticated register allocators for embedded devices, where the excessive compilation time can be tolerated due to high demand on code quality. Failing to do a good job on allocating variables to registers (i.e., determining the set of variables to be stored in the limited number of registers) can have serious power, performance, and code size consequences. This project explores the possibility of employing a hybrid evolutionary algorithm for register allocation problem in embedded systems. The proposed solution combines genetic algorithms with a local search technique. The algorithm exploits a novel, highly specialized crossover operator that takes into account domain-specific information. The results from the implementation based on synthetic benchmarks and routines that are extracted from well-known benchmark suites clearly show that the proposed approach is very successful in allocating registers to variables. In addition, the experimental evaluation also indicates that it outperforms a state-of-the-art register allocation heuristic based on graph coloring for most of the cases experimented.

E.6   Semantic Understanding of General Linguistic Items by Means of Fuzzy Set Theory

Modern statistical techniques used in the field of natural language processing are limited in their applications by the fact they suffer from the loss of most of the semantic information contained in text documents. Fuzzy techniques have been proposed as a way to correct this problem through the modeling of the relationships between words while accommodating the ambiguities of natural languages. However, these techniques are currently either restricted to modeling the effects of simple words or are specialized in a single domain. In this project, a novel statistical-fuzzy methodology is proposed to represent the actions described in a variety of text documents by modeling the relationships between subject-verb-object triplets. The research will focus in the first place on the technique used to accurately extract the triplets from the text, on the necessary equations to compute the statistics of the subject-verb and verb-object pairs, and on the formulas needed to interpolate the fuzzy membership functions from these statistics and on those needed to defuzzify the membership value of unseen triplets. Taken together, these sets of equations constitute a comprehensive system that allows the quantification and evaluation of the meaning of text documents, while being general enough to be applied to any domain. In the second phase, this project proceeds to experimentally demonstrate the validity of the new methodology by applying it to the implementation of a fuzzy classifier conceived especially for this research. This classifier is trained using a section of the Brown Corpus, and its efficiency is tested with a corpus of 20 unseen documents drawn from three different domains. The positive results obtained from these experimental tests confirm the soundness of the new approach and show that it is a promising avenue of research.

E.7 Fuzzy Evaluation of Heart Rate Signals for Mental Stress Assessment

Mental stress is accompanied by dynamic changes in autonomic nervous system (ANS) activity. Heart rate variability (HRV) analysis is a popular tool for assessing the activities of autonomic nervous system. This project presents a novel method of HRV analysis for mental stress assessment using fuzzy clustering and robust identification techniques. The approach consists of 1) online monitoring of heart rate signals, 2) signal processing (e.g., using the continuous wavelet transform to extract the local features of HRV in time-frequency domain), 3) exploiting fuzzy clustering and fuzzy identification techniques to render robustness in HRV analysis against uncertainties due to individual variations, and 4) monitoring the functioning of autonomic nervous system under different stress conditions. The experiments involved 38 physically fit subjects (26 male, 12 female, aged 18-29 years) in air traffic control task simulations. The subjective rating scores of mental workload were assessed using NASA Task Load Index. Fuzzy clustering methods have been used to model the experimental data. Further, a robust fuzzy identification technique has been used to handle the uncertainties due to individual variations for the assessment of mental stress.

E.8 An Ant Colony Optimization Approach to the Probabilistic Traveling Salesman Problem

The Probabilistic Traveling Salesman Problem (PTSP) is a TSP problem where each customer has a given probability of requiring a visit. The goal is to find an *a priori* tour of minimal expected length over all customers, with the strategy of visiting a random subset of customers in the same order as they appear in the *a priori* tour. The question of whether and in which context an a priori tour found by a TSP heuristic can also be a good solution for the PTSP is addressed. This question is answered by testing the relative performance of two ant colony optimization algorithms, Ant Colony System (ACS) introduced by Dorigo and Gambardella for the TSP, and a variant of it probabilistic ACS (pACS) which aims to minimize the PTSP objective function. It is found that the probability configuration of customers pACS and ACS are promising algorithms for the PTSP.

E.9   Neural Processing of Symbolic Data

Connectionist systems constitute powerful adaptive machine learning tools which are particularly suited for large scale and noisy learning problems. One major problem, however, is given by the fact that typical neural system are restricted to flat vector representations such that their applicability in domains with additional symbolic or structured knowledge is limited. In various concrete projects, new models as well as theoretical background have been developed which investigate neural methods for non-Euclidian, symbolic, and structured data.

References

[1] Ackley, D. H., Hinton, G. E., & Sejnowski, T. J. (1985). A learning algorithm for Boltzmann machines. Cognitive Science, 9(1), 147-169.

[2] Ahalt, S. C., Krishnamurthy, A. K., Chen, P., & Melton, D. (1990). Competititve learning algorithms for vector quantization. Neural Networks, 3, 277-290.

[3] Almasi, G. S., & Gottlieb, A. (1989). Highly Parallel Computing. The Benjamin/Cummings Publishing Company, Inc.

[4] Almeida, L. B. (1987). A learning rule for asynchronous perceptrons with feedback in a combinatorial environment. In Proceedings of the First International Conference on Neural Networks (Vol. 2, pp. 609 - 618). IEEE.

[5] Amit, D. J., Gutfreund, H., & Sompolinsky, H. (1986). Spin-glass models of neural networks. Physical Review A, 32(2), 1007-1018.

[6] Anderson, J. A. (1977). Neural models with cognitive implications. In D. LaBerge & S. J. Samuels (Eds.), Basic Processes in Reading Perception and Comprehension Models (p. 27-90). Hillsdale, NJ: Erlbaum.

[7] Anderson, J. A., & Rosenfeld, E. (1988). Neurocomputing: Foundations of Research. Cambridge, MA: The MIT Press.

[8] Barto, A. G., & Anandan, P. (1985). Pattern-recognizing stochastic learning automata. IEEE Transactions on Systems, Man and Cybernetics, 15, 360-375.

[9] Barto, A. G., Sutton, R. S., & Anderson, C. W. (1983). Neuronlike adaptive elements that can solve difficult learning problems. IEEE Transactions on Systems, Man and Cybernetics, 13, 834-846.

[10] Barto, A. G., Sutton, R. S., & Watkins, C. (1990). Sequential decision problems and neural networks. In D. Touretsky (Ed.), Advances in Neural Information Processing II.

[11] Blelloch, G., & Rosenberg, C. R. (1987). Network learning on the Connection Machine. In Proceedings of the Tenth International Joint Conference on Artificial Intelligence, pp.323-326.

[12] Breiman, L., Friedman, J. H., Olshen, R. A., & Stone, C. J. (1984). Classification and Regression Trees. Wadsworth and Brooks/Cole.

[13] Carpenter, G. A., & Grossberg, S. (1987a). A massively parallel architecture for a self-organizing neural pattern recognition machine. Computer Vision, Graphics, and Image Processing.

[14] Carpenter, G. A., & Grossberg, S. (1987b). ART 2: Self-organization of stable category recognition codes for analog input patterns. Applied Optics, 26(23), 4919-4930.

[15] Cottrell iG.W., Munro, P., & Zipser, D. (1987). Image compression by back-propagation: A demonstration of extensional programming (Tech. Rep. No. TR 8702). USCD, Institute of Cognitive Sciences.

[16] Cybenko, G. (1989). Approximation by superpositions of a sigmoidal function. Mathematics of Control, Signals, and Systems, 2(4), 303-314.

[17] Eckmiller, R., Hartmann, G., & Hauske, G. (1990). Parallel Processing in Neural Systems and Computers. Elsevier Science Publishers B.V.

[18] Fahrat, N. H., Psaltis, D., Prata, A., & Paek, E. (1985). Optical implementation of the Hopfield model. Applied Optics, 24, 1469-1475.

[19] Feldman, J. A., & Ballard, D. H. (1982). Connectionist models and their properties. Cognitive Science, 6, 205-254.

[20] Feynman, R. P., Leighton, R. B., & Sands, M. (1983). The Feynman Lectures on Physics. Reading (MA), Menlo Park (CA), London, Sidney, Manila: Addison-Wesley Publishing Company.

[21] Flynn, M. J. (1972). Some computer organizations and their effectiveness. IEEE Transactions on Computers, C-21, 948-960.

[22] Friedman, J. H. (1991). Multivariate adaptive regression splines. Annals of Statistics, 19, 1-141.

[23] Fritzke, B. (1991). Let it grow: Self-organizing feature maps with problem dependent cell structure. In T. Kohonen, O. Simula, & J. Kangas (Eds.), Proceedings of the 1991 International Conference on Artificial Neural Networks (pp. 403-408). North-Holland/Elsevier Science Publishers.

[24] Fukushima, K. (1975). Cognitron: A self-organizing multilayered neural network. Biological Cybernetics, 20, 121-136.

[25] Fukushima, K. (1988). Neocognitron: A hierarchical neural network capable of visual pattern recognition. Neural Networks, 1, 119-130.

[26] Gielen, C., Krommenhoek, K., & Gisbergen, J. van. (1991). A procedure for self-organized sensor-fusion in topologically ordered maps. In T. Kanade, F. C. A. Groen, & L. O. Hertzberger (Eds.), Proceedings of the Second International Conference on Autonomous Systems (pp. 417-423). Elsevier Science Publishers.

[27] Gorman, R. P., & Sejnowski, T. J. (1988). Analysis of hidden units in a layered network trained to classify sonar targets. Neural Networks, 1(1), 75-89.

[28] Grossberg, S. (1976). Adaptive pattern classification and universal recoding I & II. Biological Cybernetics, 23, 121-134, 187-202.

[29] Group, O. U. (1987). Parallel Programming of Transputer Based Machines: Proceedings of the 7th Occam User Group Technical Meeting. Grenoble, France: IOS Press.

[30] Gullapalli, V. (1990). A stochastic reinforcement-learning algorithm for learning real-valued functions. Neural Networks, 3, 671-692.

[31] Hartigan, J. A. (1975). Clustering Algorithms. New York: John Wiley & Sons.

[32] Hartman, E. J., Keeler, J. D., & Kowalski, J. M. (1990). Layered neural networks with Gaussian hidden units as universal approximations. Neural Computation, 2(2), 210-215.

[33] Hebb, D. O. (1949). The Organization of Behaviour. New York: Wiley.

[34] Hecht-Nielsen, R. (1988). Counterpropagation networks. Neural Networks, 1, 131-139.

[35] Hertz, J., Krogh, A., & Palmer, R. G. (1991). Introduction to the Theory of Neural Computation. Addison Wesley.

[36] Hesselroth, T., Sarkar, K., Smagt, P. van der, & Schulten, K. (1994). Neural network control of a pneumatic robot arm. IEEE Transactions on Systems, Man, and Cybernetics, 24(1), 28-38.

[37] Hestenes, M. R., & Stiefel, E. (1952). Methods of conjugate gradients for solving linear systems. Nat. Bur. Standards J. Res., 49, 409-436.

[38] Hillis, W. D. (1985). The Connection Machine. The MIT Press.

[39] Hopfield, J. J. (1982). Neural networks and physical systems with emergent collective computational abilities. Proceedings of the National Academy of Sciences, 79, 2554-2558.

[40] Hopfield, J. J. (1984). Neurons with graded response have collective computational properties like those of two-state neurons. Proceedings of the National Academy of Sciences, 81, 3088-3092.

[41] Hopfield, J. J., & Tank, D. W. (1985). "Neural" computation of decisions in optimization problems. Biological Cybernetics, 52, 141-152.

[42] Kohonen, T. (1977). Associative Memory: A System-Theoretical Approach. Springer-Verlag.

[43] Kohonen.T. (1982). Self-organized formation of topologically correct feature maps. Biological Cybernetics, 43, 59-69.

[44] Lippmann, R. P. (1987). An introduction to computing with neural nets. IEEE Transactions on Acoustics, Speech, and Signal Processing, 2(4), 4-22.

[45] Lippmann, R. P. (1989). Review of neural networks for speech recognition. Neural Computation, 1, 1-38.

[46] Minsky, M., & Papert, S. (1969). Perceptrons: An Introduction to Computational Geometry. The MIT Press.

[47] Pearlmutter, B. A. (1989). Learning state space trajectories in recurrent neural networks. Neural Computation, 1(2), 263-269.

[48] Pearlmutter, B. A. (1990). Dynamic Recurrent Neural Networks (Tech. Rep. Nos. CMU-CS: 90-196). Pittsburgh, PA 15213: School of Computer Science, Carnegie Mellon University.

[49] Pineda, F. (1987). Generalization of back-propagation to recurrent neural networks. Physical Review Letters, 19(59), 2229-2232.

[50] Polak, E. (1971). Computational Methods in Optimization. New York: Academic Press.

[51] Pomerleau, D. A., Gusciora, G. L., Touretzky, D. S., & Kung, H. T. (1988). Neural network simulation at warp speed: How we got 17 million connections per second. In IEEE Second International Conference on Neural Networks (Vol. II, pp. 143-150).

[52] Press, W . H., Flannery, B. P., Teukolsky, S. A., & Vetterling, W. T. (1986). Numerical Recipes: The Art of Scientific Computing. Cambridge: Cambridge University Press.

[53] Rumelhart, D. E., Hinton, G. E., & Williams, R. J. (1986). Learning representations by backpropagating errors. Nature, 323, 533-536.

[54] Sanger, T. D. (1989). Optimal unsupervised learning in a single-layer linear feedforward neural network. Neural Networks, 2, 459-473.

[55] Singer, A. (1990). Implementations of Artificial Neural Networks on the Connection Machine (Tech. Rep. Nos. RL90-2). Cambridge, MA: Thinking Machines Corporation.

[56] Sofge, D., & White, D. (1992). Applied learning: Optimal control for manufacturing. In D. Sofge & D. White (Eds.), Handbook of Intelligent Control, Neural, Fuzzy, and Adaptive Approaches. Van Nostrand Reinhold, New York.

[57] Tomlinson, M. S., Jr., & Walker, D. J. (1990). DNNA: A digital neural network architecture. In Proceedings of the International Conference on Neural Networks (Vol. II, pp. 589-592).

[58] Werbos, P. J. (1974). Beyond Regression: New Tools for Prediction and Analysis in the Behavioral Sciences. Unpublished doctoral dissertation, Harvard University.

[59] Werbos, P. W. (1990). A menu for designs of reinforcement learning over time. In W. T. M. III, R. S. Sutton, & P. J. Werbos (Eds.), Neural Networks for Control. MIT Press/Bradford.

[60] Widrow, B., Winter, R. G., & Baxter, R. A. (1988). Layered neural nets for pattern recognition. IEEE Transactions on Acoustics, Speech, and Signal Processing, 36(7), 1109-1117.

[61] Bezdek, J.C. Pattern Recognition with Fuzzy Objective Function Algorithms. Plenum Press, New York, 1981.

[62] Brown, M. & Harris, C.J. Neurofuzzy Adaptive Modelling and Control. Prentice Hall, Hemel-Hempstead, UK, 1994.

[63] Brown, C. & Harris, C.J. A nonlinear adaptive controller: A comparison between fuzzy logic control and neurocontrol. IMA J. Math. Control and Info., 8(3):239-265, 1991.

[64] Cox, E. The Fuzzy Systems Handbook. AP Professional, 1994.

[65] Driankov, D., Hellendoorn, H., & Reinfrank, M. An Introduction to Fuzzy Control. Springer-Verlag, Berlin, 1993.

[66] Friedman, J. H. Multivariate adaptive regression splines. The Annals of Statistics, 19(1):1-141, 1991.

[67] Friedman, J. H. & Stuetzle, W. Projection pursuit regression. J. American Statistical Association, 76(376):817-823, 1981.

[68] Harris, C.J. Moore, C. G., & Brown, M. Intelligent Control: Some Aspects of Fuzzy Logic and Neural Networks. World Scientific Press, London & Singapore, 1993.

[69] Hunt, K. J., Haas, R., & Brown, M. On the functional equivalence of fuzzy inference systems and spline-based networks. Int. J. Neural Systems, 6(2):171-184, 1995.

[70] Hwang, J. N., Lay, S. R., Maechler, Martin, M. R., & Schimert, J. Regression modeling in back-propagation and projection pursuit learning. IEEE Trans. on Neural Networks, 5(3):342-353, 1994.

[71] Jang, J. S. R., & Sun, C. T. Neuro-fuzzy modeling and control. Proc. IEEE, 83(3):378-406, 1995.

[72] Kosko, B. Neural Networks and Fuzzy Systems. Prentice Hall, Englewood Cliffs, NJ, 1992.

[73] Kosko, B. Fuzzy Thinking: The New Science of Fuzzy Logic. Harper and Collins, 1994.

[74] Lane, S. H., Handelman, D. A., & Gelfand, J. J. Theory and development of higher order CMAC neural networks. IEEE Control Systems Mag., pages 23-30, April 1992.

[75] Laviolette, M & Seaman, J. W. The efficacy of fuzzy representations of uncertainty. IEEE Trans. on Fuzzy Systems, 2(1):4-15, 1994.

[76] McNeill, D., & Freiberger, P. Fuzzy Logic. Simon & Schuster, New York, NY, 1993.

[77] Mills, D, J., Brown, M., & Harris, C. J. Training neurofuzzy systems. In Proc. IFAC Symposium on AI in Real-Time Control, pp. 213-218, Valencia, 1994.

[78] Pedrycz, W. Fuzzy Control and Fuzzy Systems. Research Studies Press, John Wiley and Sons, Taunton, 2nd edition, 1993.

[79] Sutton, R., & Jess, I. M. A design study of a self-organizing fuzzy autopilot for ship control. IMechE, Proc. Instn. Mech. Engrs., 205:35-47, 1991.

[80] Wang, L. X. Adaptive Fuzzy Systems and Control: Design and Stability Analysis. Prentice Hall, Englewood Cliffs, NJ, 1994.

[81] Werntges, H. W. Partitions of unity improve neural function approximation. In Proc. IEEE Int. Conf. Neural Networks, pages 914-918, San Francisco, CA, 1993. Vol. 2.

[82] Zadeh, L. A. Fuzzy sets. Information and Control, 8:338-353, 1965.

[83] Zadeh, L. A. Outline of a new approach to the analysis of complex systems and decision processes. IEEE Trans. on System Man and Cybernetics, 3(1):28-44, 1973.

[84] Zimmermann, H. J. Fuzzy Set Theory & its Applications. Kluwer Academic Press, Boston, MA, 2nd edition, 1993.

[85] Zadeh, L. A. "Fuzzy algorithms," Info. & Ctl., Vol. 12, 1968, pp. 94-102.

[86] Zadeh, L. A. "Making computers think like people," IEEE. Spectrum, 8/1984, pp. 26-32.

[87] Korner, S. "Laws of thought," Encyclopedia of Philosophy, Vol. 4, MacMillan, NY: 1967, pp. 414-417.

[88] Baldwin, J. F. "Fuzzy logic and fuzzy reasoning," in Fuzzy Reasoning and Its Applications, E.H. Mamdani and B.R. Gaines (eds.), London: Academic Press, 1981.

[89] Esragh, F., & Mamdani, E. H. "A general approach to linguistic approximation," in Fuzzy Reasoning and Its Applications, Academic Press, 1981.

[90] Fox, J. "Towards a reconciliation of fuzzy logic and standard logic," Int. Jrnl. of Man-Mach. Stud., Vol. 15, 1981, pp. 213-220.

[91] Haack, S. "Do we need fuzzy logic?" Int. Jrnl. of Man-Mach. Stud., Vol. 11, 1979, pp. 437-445.

[92] Radecki, T. "An evaluation of the fuzzy set theory approach to information retrieval," in R. Trappl, N.V. Findler, & W. Horn, Progress in Cybernetics and System Research, Vol. 11: Proceedings of a Symposium Organized by the Austrian Society for Cybernetic Studies, Hemisphere Publ. Co., NY: 1982.

[93] Kruse, R., Gebhardt, J., & Klawon, F. "Foundations of Fuzzy Systems", Wiley, Chichester 1994.

[94] Zimmermann, H. J. Fuzzy Sets, Decision Making and Expert Systems, Boston, Kluwer 1987.

[95] Jang, R. Neuro-Fuzzy Modeling: Architectures, Analyses and Applications, PhD Thesis, University of California, Berkeley, July 1992

[96] Juang Chia Feng, Lin Chin Teng, An Online Self Constructing
      Neural Fuzzy Inference Network and its Applications, IEEE Trans-
      actions on Fuzzy Systems, Vol 6, No.1, pp. 12-32, 1998.

[97] Sulzberger, S. M., Tschicholg-Gurman, N. N., & Vestli, S. J., FUN:
      Optimization of Fuzzy Rule Based Systems Using Neural Networks,
      In Proceedings of IEEE Conference on Neural Networks, San Fran-
      cisco, pp. 312-316, March 1993.

[98] Tano, S., Oyama, T., & Arnould, T., Deep combination of Fuzzy
      Inference and Neural Network in Fuzzy Inference, Fuzzy Sets and
      Systems, 82(2) pp. 151-160, 1996.

[99] Kasabov, N., & Qun Song. Dynamic Evolving Fuzzy Neural Net-
      works with "m-out-of-n" Activation Nodes for On-line Adaptive
      Systems, Technical Report TR99/04, Department of information
      science, University of Otago, 1999. .

[100] Nauck, D., & Kruse, R. Neuro-Fuzzy Systems for Function Approx-
      imation, 4th International Workshop Fuzzy-Neuro Systems, 1997.

[101] Lin, C. T., & Lee, C. S. G. Neural Network based Fuzzy Logic Con-
      trol and Decision System, IEEE Transactions on Comput. (40(12):
      pp. 1320-1336, 1991.

[102] Bherenji, H. R., & Khedkar, P. Learning and Tuning Fuzzy Logic
      Controllers through Reinforcements, IEEE Transactions on Neural
      Networks, Vol (3), pp. 724-740, 1992.

[103] Abraham, A., & Nath, B. Evolutionary Design of Neuro-Fuzzy Sys-
      tems - A Generic Framework, In Proceedings of the 4th Japan -
      Australia joint Workshop on Intelligent and Evolutionary Systems,
      Japan, November 2000.

[104] Abraham, A., & Nath, B. Designing Optimal Neuro-Fuzzy Sys-
      tems for Intelligent Control, In proceedings of the Sixth Interna-
      tional Conference on Control Automation Robotics Computer Vi-
      sion, (ICARCV 2000), Singapore, December 2000.

[105] Klir, G. J., & Yuan, B. Fuzzy sets and fuzzy logic. Theory and
      applications. Prentice Hall, 1997.

[106] Hush, D. R., & Horne, B. G. Progress in Supervised Neural Net-
      works. What's new since Lippmann? IEEE Signal Processing Mag-
      azine, January 1993, Vol.10, No. 1, pp. 32-43.

[107] Grabusts, P. A study of clustering algorithm application in RBF
      neural networks, Scientific Proceedings of Riga Technical Univer-
      sity, 2001, pp. 50-57.

[108] Kosko, B. "Hidden patterns in combined and adaptive knowledge networks," Int. J. Approx. Reas., vol. 2, pp. 377-393, 1988.

[109] Omlin, C. W., & Lee Giles, C. "Rule revision with recurrent neural networks," IEEE Trans. Knowl. Data Eng., vol. 8, pp. 183-188, 1996

[110] Opitz, D. W., & Shavlik, J. W., "Connectionist theory refinement: Genetically searching the space of network topologies," J. Artif. Intell. Res., Vol. 6, pp. 177-209, 1997.

[111] Kasabov, N., & Woodford, B. "Rule insertion and rule extraction from evolving fuzzy neural networks: Algorithms and applications for building adaptive, intelligent expert systems," in Proc. IEEE Conf. Fuzzy Syst. FUZZ-IEEE'99, Seoul, Korea, Aug. 1999, pp. III-1406-III-1411.

[112] Mitra, S., & Pal, S. K., "Fuzzy multilayer perceptron, inferencing and rule generation," IEEE Trans. Neural Networks, vol. 6, pp. 51-63, Jan. 1995.

[113] Mitra, S., De, S. K., & Pal, S. K. "Knowledge-based fuzzy MLP for classification and rule generation," IEEE Trans. Neural Networks, vol. 8, pp. 1338-1350, 1997.

[114] Banerjee, M., Mitra, S., & Pal, S. K., "Rough fuzzy MLP: Knowledge encoding and classification," IEEE Trans. Neural Networks, Vol. 9, pp. 1203-1216, 1998.

[115] Taha, I. A., & Ghosh, J. "Symbolic interpretation of artificial neural networks," IEEE Trans. Knowl. Data Eng., vol. 11, pp. 448-463, 1999.

[116] Nauck, D., Klawonn, F., & Kruse, R. Foundations of Neuro-Fuzzy Systems. Chichester, U.K.: Wiley, 1997.

[117] Kosko, B. Neural Networks and Fuzzy Systems. Englewood Cliffs, NJ: Prentice-Hall, 1991.

[118] Takagi, H. "Fusion technology of fuzzy theory and neural network-Survey and future directions," in Proc. Int. Conf. Fuzzy Logic Neural Networks, Iizuka, Japan, 1990, pp. 13-26. .

[119] Buckley, J. J., & Hayashi, Y. "Numerical relationship between neural networks, continuous functions and fuzzy systems," Fuzzy Sets Syst., Vol. 60, no. 1, pp. 1-8, 1993.

[120] Mamdani, E. H., & Assilian, S. "An experiment in linguistic synthesis with a fuzzy logic controller," Int. J. Man-Mach. Stud., Vol. 7, pp. 1-13, 1975.

[121] Takagi, T., & Sugeno, M. "Fuzzy identification of systems and its application to modeling and control," IEEE Trans. Syst., Man, Cybern., Vol. SMC-15, pp. 116-132, 1985.

[122] Buckley, J. J., & Feuring, T. Fuzzy and Neural: Interactions and Applications, ser. Studies in Fuzziness and Soft Computing. Heidelberg, Germany: Physica-Verlag, 1999.

[123] Bramer, M.A. "Automatic induction of classification rules from examples using NPRISM", Research and Development in Intelligent Systems XVI, Springer-Verlag, pp. 99-121, 2000.

[124] Tsumoto, S. "Modelling medical diagnostic rules based on rough sets", Rough Sets and Current Trends in Computing, Lecture Notes in Artificial Intelligence, 1424, Springer-Verlag, Berlin, pp. 475-482, 1998.

[125] Zadeh, L. A. "Towards a theory of fuzzy information granulation and its centrality in human reasoning and fuzzy logic", Fuzzy Sets and Systems, 19, 111-127, 1997.

[126] Knoblock, C.A. Generating Abstraction Hierarchies: An Automated Approach to Reducing Search in Planning, Kluwer Academic Publishers, Boston, 1993.

[127] Pawlak, Z. Rough Sets, Theoretical Aspects of Reasoning about Data, Kluwer Academic Publishers, Dordrecht, 1991.

[128] Salthe, S. N. Evolving Hierarchical Systems, Their Structure and Representation, Columbia University Press, 1985.

[129] Whyte, L. L., Wilson, A. G., & Wilson, D. (Eds.) Hierarchical Structures, American Elsevier Publishing Company, Inc., New York, 1969.

[130] Zadeh, L. A. Some reflections on soft computing, granular computing and their roles in the conception, design and utilization of information/intelligent systems, Soft Computing, 23-25, 1998.

[131] Shaw, E. "The schooling of fishes," Sci. Am., Vol. 206, pp. 128-138, 1962.

[132] Bonabeau, E., Dorigo, M., & Theraulaz, G. Swarm Intelligence: From Natural to Artificial Systems. NY: Oxford Univ. Press, 1999.

[133] Arkin, R. Behavior-Based Robotics. Cambridge, MA: MIT Press, 1998.

[134] Beni, G., & Wang, J. "Swarm intelligence in cellular robotics systems," in Proceeding of NATO Advanced Workshop on Robots and Biological System, 1989.

[135] Reynolds, C. "Flocks, herds, and schools: A distributed behavioral model," Comp. Graph, Vol. 21, No. 4, pp. 25-34, 1987.

[136] Terzopoulos, D., Tu, X., & Grzeszczuk, R. "Artificial fishes with autonomous locomotion, perception, behavior, and learning in a simulated physical world," in Artificial Life I, p. 327, MIT Press, 1994.

[137] Millonas, M. "Swarms, phase transitions, and collective intelligence," in Artificial Life III, Addison-Wesley, 1994.

[138] Sims, K. "Evolving 3d morphology and behavior by competition," in Artificial Life I, p. 353, MIT Press, 1994.

[139] Dudek, G., and et al., "A taxonomy for swarm robots," in IEEE/RSJ Int. Conf. on Intelligent Robots and Systems, (Yokohama, Japan), July 1993.

[140] Hackwood, S., & Beni, S. "Self-organization of sensors for swarm intelligence," in IEEE Int. Conf. on Robotics and Automation, (Nice, France), pp. 819-829, May 1992.

[141] Brooks, R., "Intelligence without reason," tech. rep., Artificial Intelligence Memo. No. 1293, 1991.

[142] Ueyama, T., Fukuda, T., & Arai, F. "Configuration of communication structure for distributed intelligent robot system," in Proc. IEEE Int. Conf. on Robotics and Automation, pp. 807-812, 1992.

[143] Mataric, M. "Minimizing complexity in controlling a mobile robot population," in IEEE Int. Conf. on Robotics and Automation, (Nice, France), May 1992.

[144] Ueyama, T., Fukuda, T., & Sugiura, T. "Self-organization and swarm intelligence in the society of robot being," in Proceedings of the 2nd International Symposium on Measurement and Control in Robotics, 1992.

[145] Fukuda, T., Funato, D., Sekiyam, K., & Arai, F. "Evaluation on flexibility of swarm intelligent system," in Proceedings of the 1998 IEEE International Conference on Robotics and Automation, pp. 3210-3215, 1998.

[146] Mogilner, A., & Edelstein-Keshet, L. "A non-local model for a swarm," Journal of Mathematical Biology, vol. 38, pp. 534-570, 1999.

[147] Jin, K., Liang, P., & Beni, G. "Stability of synchronized distributed control of discrete swarm structures," in IEEE International Conference on Robotics and Automation, pp. 1033-1038, 1994.

[148] Beni, G. & Liang, P. "Pattern reconfiguration in swarms — Convergence of a distributed asynchronous and bounded iterative algorithm," IEEE Trans. on Robotics and Automation, Vol. 12, June 1996.

[149] Passino, K., & Burgess, K. Stability Analysis of Discrete Event Systems. New York: John Wiley and Sons Pub. 1998.

[150] Koza, J. R. Genetic Programming. The MIT Press, Cambridge, Massachusetts, 1992.

[151] Reynolds, C. W. Flocks, herds, & schools: A distributed behavioral model. Computer Graphics, 21(4):25-34, July 1987.

[152] Bourgine, P. Autonomy, Abduction, Adaptation, in: (Magnenat Thalmann N, Thalmann D, EDS), Proc. Computer Animation '94, IEEE Computer Society Press, 1994.

[153] De Castro L & Von Zuben F (2001), Learning and Optimization Using the Clonal Selection Principle. IEEE Transactions on Evolutionary Computation, Special Issue on Artificial Immune Systems, 6(3), pp. 239-251.

[154] Chao, D. L. & Forrest, S. "Information immune systems," Genetic Programming and Evolvable Machines, vol. 4, no. 4, pp. 311-331, 2003.

[155] Atmar, W. (1992) The philosophical errors that plague both evolutionary theory and simulated evolutionary programming. Proceedings of the First Annual Conference on Evolutionary Programming, 27-34. San Diego, CA: Evolutionary Programming Society.

[156] Back, T., Hoffmeister, F., & Schwefel, H.-P. (1991) A survey of evolution strategies. Proceedings of the Fourth International Conference on Genetic Algorithms, 2- 9.

[157] Belew, R. K., & Booker, L. B. (eds.) (1991) Proceedings of the Fourth International Conference on Genetic Algorithms.

[158] La Jolla, C. A., Morgan Kaufmann. Booker, L. B. (1992) Recombination distributions for genetic algorithms. Proceedings of the Foundations of Genetic Algorithms Workshop.

[159] Vail, C. O. Morgan Kaufmann. Box, G. E. P. (1957) Evolutionary operation: A method of increasing industrial productivity. Applied Statistics, Vol. 6, 81-101.

[160] Davis, L. (1989) Adapting operator probabilities in genetic algorithms. Proceedings of the Third International Conference on Genetic Algorithms, 60-69.

[161] La Jolla, CA: Morgan Kaufmann. de Garis, H. (1990) Genetic programming: modular evolution for darwin machines. Proceedings of the 1990 International Joint Conference on Neural Networks, 194-197.

[162] Lawrence Erlbaum. De Jong, K. A. (1975) An analysis of the behavior of a class of genetic adaptive systems. Doctoral Thesis, Department of Computer and Communication Sciences. University of Michigan, Ann Arbor.

[163] De Jong, K., & Spears, W. (1991) Learning concept classification rules using genetic algorithms. Proceedings of the Twelfth International Joint Conference on Artificial Intelligence, 651-656.

[164] Morgan Kaufmann. De Jong, K. A. (1992) Are genetic algorithms function optimizers? Proceedings of the Second International Conference on Parallel Problem Solving from Nature.

[165] Eshelman, L. J., & Schaffer, J. D. (1991) Preventing premature convergence in genetic algorithms by preventing incest. Proceedings of the Fourth International Conference on Genetic Algorithms, 115-122.

[166] Fogel, D. B. (1992) An analysis of evolutionary programming. Proceedings of the First Annual Conference on Evolutionary Programming, 43-51.

[167] Fogel, D. B., & Atmar, J. W. (EDS.) (1992) Proceedings of the First Annual Conference on Evolutionary Programming.

[168] La Jolla, CA: Evolutionary Programming Society Fraser, A. S. (1957) Simulation of genetic systems by automatic digital computers. Australian Journal of Biological Science, 10, 484-491.

[169] Fujiko, C., & Dickinson, J. (1987) Using the genetic algorithm to generate LISP source code to solve the prisoner's dilemma. Proceedings of the Second International Conference on Genetic Algorithms, 236-240. Cambridge, MA.

[170] Lawrence Erlbaum. Goldberg, D. E. (1989a) Sizing populations for serial and parallel genetic algorithms. Proceedings of the Third International Conference on Genetic Algorithms, 70- 79. Fairfax, VA

[171] Goldberg, D. E. (1989b) Genetic Algorithms in Search, Optimization & Machine Learning. Reading, MA: Addison-Wesley.

[172] Grefenstette, J. (1989) A system for learning control strategies with genetic algorithms. Proceedings of the Third International Conference on Genetic Algorithms, 183-190.

[173] Harp, S. A., Samad, T., & Guha, A. (1991) Towards the genetic synthesis of neural networks. Proceedings of the Fourth International Conference on Genetic Algorithms, 360-369.

[174] Fogel, D. B. "Evolutionary Computing", IEEE Press 2002.

[175] Whitely, D. A Genetic Algorithm Tutorial, Computer Science Department, Colorado University.

[176] Buckles, B. P., & Petry, F. E. "Genetic Algorithms" IEEE Press 1992.

[177] Back, T., Hoffmcister, F., Schwefel, H. P. "A survey of Evolution Strategies",University of Dortmund, Department of Computer Science XI

[178] Rudilph, G. "Convergence of Evolutionary Algorithms in General Search Spaces", JCD Informatik Centrum Dortmund.

[179] Back, T. "Self-Adaption in Genetic Algorithms", in Proceedings of the 1st European Conference on Aritificial Life, F.J. Varela & P. Bourgine, Eds. Cambridge, MA: MIT Press, pp. 263-271, 1992.

[180] Back, T., & Schwefel, H.P. "Evolutionary computation: An overview", Evolutionary Computation, 1996., Proceedings of IEEE Internation Conference on, pp. 20-29, 1996.

[181] Bremermann, H.J. "The evolution of intelligence. The nervous system as a model of its environment", Technical report, no.1, contract no. 477(17), Dept. Mathematics, Univ. Washington, Seattle, July, 1958.

[182] Sumathi, S., Hamsapriya, T., & Surekha, P., "Evolutionary Intelligence: An Introduction to theory and applications using MAT-LAB", Springer-Verlag, 2008.

[183] Fogel, D.B., Ghozeil, A. "Schema processing under proportional selection in the presence of random effects". IEEE Transactions on Evolutationary Computation, Vol. 1, Iss. 4, pp 290-293, 1997.

[184] Gehlhaar, D. K., & Fogel, D. B. "Tuning evolutionary programming for conformationally flexible molecular docking," in Proc. 5th

Annu. Conf on Evolutionary Programming. Cambridge, MA: MIT Press, 1996, pp. 419-429.

[185] Greene, W.A., "Dynamic load-balancing via a genetic algorithm". Tools with Artificial Intelligence, Proceedings of the 13th International Conference, pp. 121 -128, 2001.

[186] Greenwood, G.W., Gupta, A., & McSweeney, K. "Scheduling tasks in multiprocessor systems using evolutionary strategies", Evolutionary Computation, 1994. IEEE World Congress on Computational Intelligence. Proceedings of the First IEEE Conference on, Vol.1, pp. 345 -349, 1994.

[187] Grefenstette, J. J. "Deception considered harmful". In L. D. Whitley, ED., Foundations of Genetic Algorithms 2. Morgan Kaufmann, 1993.

[188] Goldberg. D. E. Genetic Algorithms in Search, Optimization, and Machine Learning. Addison-Wesley, 1989.

[189] Minglun, G., & Yee-Hong, Y. "Multi-resolution stereo matching using genetic algorithm", Stereo and Multi-Baseline Vision, 2001. (SMBV 2001). Proceedings. IEEE Workshop on, pp 21 -29, 2001.

[190] Hansen, N., & Ostermoior, A. "Adapting arbitrary normal mutation distributions in evolution strategies: The covariance matrix adaptation". Evolutionary Computation, 1996, Proceedings of IEEE International Conference, pp. 312-317, 1996.

[191] Hajela, P., & Lin, C.Y. "Genetic search strategies in multicriterion optimal design", in Structural Optimization, vol 4, pp. 99-107, 1992.

[192] Hatanaka, T., Uosaki, K., Tanaka, H., & Yamada, Y. "System parameter estimation by evolutionary strategy", SICE '96. Proceedings of the 35th SICE Annual Conference. International Session Papers, pp. 1045 -1048, 1996.

[193] Horn, J., Nafpliotis, N. Multiobjective optimization using the niched pareto genetic algorithm, IlliGAL Report 93005, Illinois Genetic Algorithms Lab., Univ. Illiniois, Urbana-Champagn, July 1993.

[194] Knowles, J., & Corne, D. "The Pareto archived evolution strategy: A new baseline algorithm for Pareto multiobjective optimization", Evolutionary Computation, 1999. CEC 99. Proceedings of the 1999 Congress on, Vol 1, pp.98-105, 1999.

[195] Kotani, M., Ochi, M., Ozawa, S., & Akazawa, K. "Evolutionary discriminant functions using genetic algorithms with variable-length chromosome". Neural Networks, 2001. Proceedings. IJCNN '01. Inernational Joint Conference on, Vol. 1, pp. 761-766, 2001.

[196] Louchet, J. "Stereo analysis using individual evolution strategy", Pattern Recognition, 2000. Proceedings. 15th International Conference on, Vol. 1, pp. 908 -911, 2000.

[197] Madureira, A. Ramos, C. do Carmo Silva, S. "A Coordination Mechanism for Real World Scheduling Problems using Genetic Algorithms", Evolutionary Computation, 2002. CEC '02. Proceedings of the 2002 Congress on, Vol. 1, pp. 175 -180, 2002.

[198] Nix, E.A., Vose, M.D. "Modelling genetic algorithms with Markov Chains", Ann. Math Artif. Intell, Vol. 5, pp. 79-88, 1992.

[199] Ostertag, M., Nock, E., & Kiencke, U. "Optimization of airbag release algorithms using evolutionary strategies", Control Applications, 1995. Proceedings of the 4th IEEE Conference on, pp. 275-280, 1995.

[200] Pettit, E., & Swigger, K. M. "An analysis of genetic-based pattern tracking" in Pric. National Conf. on AI, AAAO '83, pp. 327-332, 1983.

[201] Poli, R. "Why the schema theorem is correct also in the presence of stochastic effects". Evolutionary Computation, 2000. Proceedings of the 2000 Congress on, Vol. 1, 2000, pp. 487-492.

[202] Radcliffe, N. J. "Schema Processing". In: Handbook of Evolutionary Computation (T. Back, D.B. Fogel & Z. Michalewicz, EDS.) pp. B.2.5-1.10, Oxford University Press, 1997.

[203] Rogers, A. & Prugel-Bennett, A. Modelling the Dynamics of a Steady State Genetic Algorithm. Foundations of Genetic Algorithms - 5 pp. 57-68, 1999.

[204] Rudolph, G. "Convergence of evolutionary algorithms in general search spaces", Evolutionary Computation, 1996., Proceedings of IEEE International Conference on, pp 50-54, 1996.

[205] Prugel-Bennett, A., & Shaprio, J.L. 1994. "An analysis of genetic algorithms using statistical mechanics". Physical Review Letters 72, no. 9: 1305-1309.

[206] Schaffer, J.D. "Multiple-Objective optimization using genetic algorithm". Proc. Of the First Int. Conf. on Genetic Algorithms, pp. 93- 100, 1985.

[207] 207. Schwefel, H. P. Numerical Optimization of Computer Models. Wiley, Chichester, 1981.

[208] Smith, J., & Fogarty, T.C. "Self adaptation of mutation rates in a steady state genetic algorithm", in Proc. 3rd IEEE Conf. on Evolutionary Computation. Piscataway, NJ: IEEE Press, pp 318-323, 1996.

[209] Whitley, D., "A Genetic Algorithm Tutorial", Computer Science Department, Colorado State University, whiteky@cs.colostate.edu

[210] Zitzler, E., & Thiele, L. "Multiobjective evolutionary algorithms: a comparartive case study and the strength Pareto approach", Evolutionary Computation, IEEE Transactions on, Vol. 3, Issue 4, pp 257-271, 1999.

[211] 211. Bishop, C. M. Neural Networks for Pattern Recognition. Oxford University Press, 1995.

[212] MacKay, J. C. D. Information Theory, Inference and Learning Algorithms. Cambridge University Press, 2003.

[213] Abraham, A. Intelligent Systems: Architectures and Perspectives, Recent Advances in Intelligent Paradigms and Applications, Abraham A., Jain L., & Kacprzyk J. (EDS.), Studies in Fuzziness and Soft Computing, Springer Verlag Germany, Chap. 1, pp. 1-35, 2002.

[214] Abraham, A. Meta-Learning Evolutionary Artificial Neural Networks, Neurocomputing Journal, Elsevier Science, Netherlands, 2003 (in press).

[215] Abraham, A., & Nath B. Evolutionary Design of Fuzzy Control Systems - An Hybrid Approach, The Sixth International Conference on Control, Automation, Robotics and Vision, (ICARCV 2000), CD-ROM Proceeding, Wang J.L. (Ed.), ISBN 9810434456, Singapore, 2000.

[216] Abraham, A., & Nath B. Evolutionary Design of Neuro-Fuzzy Systems - A Generic Framework, In Proceedings of The 4-th Japan-Australia JointWorkshop on Intelligent and Evolutionary Systems, Namatame, A. et al (EDS.), Japan, pp. 106-113, 2000.

[217] Bezdek, J. C. Pattern Recognition with Fuzzy Objective Function Algorithms, New York: Plenum Press, 1981.

[218] Cordon, O., Herrera F., Hoffmann F., & Magdalena L. Genetic Fuzzy Systems: Evolutionary Tuning and Learning of Fuzzy Knowledge Bases, World Scientific Publishing Company, Singapore, 2001.

[219] Edwards, R., Abraham A., & Petrovic-Lazarevic, S. Expert Behaviour Modeling Using EvoNF Approach, The International Conference on Computational Science (ICCS 2003), Springer-Verlag, Lecture Notes in Computer Science- Volume 2660, Sloot P.M.A. et al (EDS.), pp. 169-178, 2003.

[220] Hall, L. O., Ozyurt, I. B., & Bezdek J. C., Clustering with a Genetically Optimized Approach, IEEE Transactions on Evolutionary Computation, Vol. 3, No. 2, pp. 103-112, 1999.

[221] Jang J. S. R. ANFIS: Adaptive-Network-Based Fuzzy Inference System, IEEE Transactions in Systems Man and Cybernetics, Vol. 23, No. 3, pp. 665-685, 1993.

[222] Jayalakshmi, G. A., Sathiamoorthy, S., & Rajaram, A Hybrid Genetic Algorithm — A New Approach to Solve Traveling Salesman Problem, International Journal of Computational Engineering Science, Vol. 2, No. 2, pp. 339-355, 2001.

[223] Lotfi, A. Learning Fuzzy Inference Systems, PhD Thesis, Department of Electrical and Computer Engineering, University of Queensland, Australia, 1995.

[224] Pedrycz, W. (ED.), Fuzzy Evolutionary Computation, Kluwer Academic Publishers, U. S. A., 1997.

[225] Sanchez, E., Shibata, T., & Zadeh, L. A. (EDS.), Genetic Algorithms and Fuzzy Logic Systems: Soft Computing Perspectives, World Scientific Publishing Company, Singapore, 1997.

[226] Stepniewski, S. W., & Keane, A. J., Pruning Back-propagation Neural Networks Using Modern Stochastic Optimization Techniques, Neural Computing & Applications, Vol. 5, pp. 76-98, 1997.

[227] Wang, L. X., & Mendel, J. M., Backpropagation Fuzzy System as Nonlinear Dynamic System Identifiers, In Proceedings of the First IEEE International conference on Fuzzy Systems, San Diego, U. S. A., pp. 1409-1418, 1992.

[228] Wang, L. X., & Mendel, J. M., Generating Fuzzy Rules by Learning from Examples, IEEE Transactions in Systems Man and Cybernetics, Vol. 22, pp. 1414 - 1427, 1992.

[229] Wang, L. X., Adaptive Fuzzy Systems and Control, Prentice Hall Inc, U. S. A., 1994.

[230] Wang, X., Abraham A. & Smith K. A, Soft Computing Paradigms for Web Access Pattern Analysis, Proceedings of the 1st Interna-

tional Conference on Fuzzy Systems and Knowledge Discovery, pp. 631-635, 2002.

[231] Zadeh, L. A., Roles of Soft Computing and Fuzzy Logic in the Conception, Design and Deployment of Information/Intelligent Systems, Computational Intelligence: Soft Computing and Fuzzy-Neuro Integration with Applications, Kaynak O. et al (Eds.), pp. 1-9, 1998.

[232] Honavar, V. Toward Learning Systems That Use Multiple Strategies and Representa-tions. In: Artificial Intelligence and Neural Networks: Steps Toward Principled Integration. Honavar, V. and Uhr, L. (Ed.) New York: Academic Press, 1994. - pp. 615-644.

[233] Gavrilov, A.V., & Novickaja J.V. The Toolkit for development of Hybrid Expert Systems. - 5-th Int. Symp. "KORUS-2001", Tomsk: TPU, 2001. - Proceedings. - Vol.1. - pp. 73-75.

[234] Gavrilov, A. V. The model of associative memory of intelligent system. - The 6-th Russian-Korean International Symposium on Science and Technology. Proceedings. - No-vosibirsk, 2002. - Vol. 1. - pp. 174-177.

[235] Carpenter, G. A., and Grossberg, S. Pattern Recognition by Self-Organizing Neural Net-works, Cambridge, MA, MIT Press, 1991.

[236] Bonissone, P. P., Badami, V., Chiang, K. H., Khedkar, P. S., Marcelle, K., & Schutten, M. J. "Industrial applications of fuzzy logic at General Electric," Proc. IEEE, Vol. 83, pp. 450-465, Mar. 1995.

[237] Chen, Y.T., & Bonissone, P. P. "Industrial applications of neural networks at General Electric," General Electric CRD, Schenectady, NY, Tech. Inform. Series 98CRD79, Oct. 1998.

[238] Zadeh, L. A. "Fuzzy logic and soft computing: Issues, contentions and perspectives," in Proc. IIZUKA'94: 3rd Int. Conf Fuzzy Logic, Neural Nets and Soft Computing, Iizuka, Japan, 1994, pp. 1-2.

[239] Dubois, D., & Prade, H. "Soft computing, fuzzy logic, and Artificial Intelligence," Soft Comput. Fusion of Foundations, Methodologies Applicat., Vol. 2, no. 1, pp. 7-11, 1998.

[240] Bouchon-Meunier, B., Yager, R., & Zadeh, L. A. Fuzzy Logic and Soft Computing. Singapore: World Scientific, 1995.

[241] Bonissone, P. P. "Soft computing: The convergence of emerging reasoning technologies," Soft Comput. Fusion of Foundations, Methodologies Applicat., Vol. 1, No. 1, pp. 6-18, 1997.

[242] Black, M. "Vaguenes: An exercise in logical analysis," Phil. Sci., vol. 4., pp. 427-455, 1937.

[243] Ruspini, E. H., Bonissone, P. P., & Pedycz, W. Handbook of Fuzzy Computation. Bristol, U.K.: Inst. Phys., 1998.

[244] Pok, Y.M., & Xu, J.X. "Why is fuzzy control robust," in Proc. 3rd IEEE Int. Conf. Fuzzy Systems (FUZZ-IEEE'94), Orlando, FL, 1994, pp. 1018-1022.

[245] Shafer, G. A Mathematical Theory of Evidence. Princeton, NJ: Princeton Univ. Press, 1976.

[246] Hornick, K., Stinchcombe, M., & White, H. "Multilayer feedforward networks are universal approximators," Neural Networks, Vol. 2, pp. 359-366, 1989.

[247] Moody, J., & Darken, C. "Fast learning in networks of locally tuned processing units," Neural Computation, Vol. 1, No. 2, pp. 281-294, 1989.

[248] Fraser, A. S. "Simulation of genetic systems by automatic digital computers, I: Introduction," Australian J. Bio. Sci., Vol. 10, pp. 484-491, 1957.

[249] Holland, J. Adaptation in Natural and Artificial Systems. Cambridge, MA: MIT Press, 1975.

[250] Koza, J. Genetic Programming: On the Programming of Computers by Means of Natural Selection. Cambridge, MA: MIT Press, 1992.

[251] Fogel, D. B. Evolutionary Computation. New York: IEEE Press, 1995.

[252] Back, T., Fogel, D. B., & Michalewicz, Z. Handbook of Evolutionary Computation. New York: Oxford, 1997.

[253] Takagi, T., & Sugeno, M. "Fuzzy identification of systems and its applications to modeling and control," IEEE Trans. Syst., Man, Cybern., Vol. 15, pp. 116-132, Jan.-Feb. 1985.

[254] Jang, J. S. R. "ANFIS: Adaptive-network-based-fuzzy-inference-system," IEEE Trans. Syst., Man, Cybern., Vol. 23, pp. 665-685, May-June 1993.

[255] Karr, C. L. "Design of an adaptive fuzzy logic controller using genetic algorithms," in Proc. Int. Conf. Genetic Algorithms (ICGA'91), San Diego, CA, 1991, pp. 450-456.

[256] Abraham, A. (2004) Meta-Learning Evolutionary Artificial Neural Networks, Neurocomputing Journal, Elsevier Science, Netherlands, Vol. 56c, pp. 1-38.

[257] Ahuja, R. K., Orlin, J. B., Tiwari, A. A greedy genetic algorithm for the quadratic assignment problem. Computers and Operations Research, 27, 917-934, 2000.

[258] Clerc, M., & Kennedy, J. The particle swarm: Explosion stability and convergence in a multi-dimensional complex space, IEEE Transaction on Evolutionary Computation 6(1), pp. 58-73, 2002.

[259] Dozier G, Bowen J, Homaifar A. Solving Constraint Satisfaction Problems Using Hybrid Evolutionary Search, IEEE Transactions on Evolutionary Computation, 2(1), pp. 23-33, 1998.

[260] Eberhart, R. C., & Kennedy, J. A new optimizer using particle swarm theory. In Proceedings of 6th Internationl Symposium on Micro-Machine and Human Science, Nagoya, Japan, IEEE Service Center, Piscataway, NJ, pp. 39-43, 1995.

[261] Martnez, C. H., Pedrajas, G. C. Hybridization of Evolutionary Algorithms and Local Search by Means of a Clustering Method, IEEE Transactions on Systems, Man and Cybernetics, Part B, 36(3), pp. 534-545, 2006.

[262] Fatourechi, M., Bashashati, A., Ward, R. K., & Birch, G. A Hybrid Genetic Algorithm Approach for Improving the Performance of the LF-ASD Brain Computer Interface, In Proccedings of the International Conference on Acoustics, Speech, and Signal Processing (ICASSP05), Philadelphia, pp. 345-348, 2005.

[263] Fleurent. C., & Ferland, J. Genetic hybrids for the quadratic assignment problems. In Quadratic Assignment and Related Problems, P.M. Pardalos & H. Wolkowicz (Eds.), DIMACS Series in Discrete Mathematics and Theoretical Computer Science, Vol. 16, AMS: Providence, Rhode Island, 1994, pp. 190-206, 1994.

[264] Fogel, D. B., An introduction to simulated evolutionary optimization. IEEE Transaction on Neural Networks, 5(1), pp. 3-14, 1994.

[265] Ganesh, K., & Punniyamoorthy, M. Optimization of continuous-time production planning using hybrid genetic algorithms-simulated annealing, International Journal of Advanced Manufacturing Technology, 26(1), pp. 148-154, 2004.

[266] Grefenstette, J. J. Incorporating problem specific knowledge into genetic algorithms. In: Davis, L. (Ed.), Genetic Algorithms & Sim-

ulated Annealing. Morgan Kaufmann, Los Altos, CA, pp. 42-60, 1987.

[267] Grimaldi, E. A., Grimacia, F., Mussetta, M., Pirinoli, P., & Zich, R. E. A new hybrid genetical swarm algorithm for electromagnetic optimization, In Proceedings of International Conference on Computational Electromagnetics and its Applications, Beijing - China, pp. 157-160, 2004.

[268] Harik, G. R., & Goldberg, D. E. Linkage learning through probabilistic expression. Computer Methods in Applied Mechanics and Engineering 186 (2-4), pp. 295-310, 2000.

[269] Herrera, F, & Lozano, M. Adaptation of Genetic Algorithm Parameters based on Fuzzy Logic Controllers. Genetic Algorithms and Soft Computing, F. Herrera, J. L. Verdegay (Eds.), pp. 95-125, 1996.

[270] Jeong, S. J., Lim, S. J., & Kim, K. S. Hybrid approach to production scheduling using genetic algorithm and simulation, International Journal of Advanced Manufacturing Technology, 28(1), pp. 129-136, 2005.

[271] Keedwell, E., & Khu, S. T. A hybrid genetic algorithm for the design of water distribution networks, Engineering Applications of Artificial Intelligence, 18(4), pp. 461-472, 2005.

[272] Kennedy, J, & Eberhart, R. C. Particle swarm optimization, In Proceedings of IEEE International Conference on Neural Networks, Perth, Australia, pp. 1942-1948, 1995.

[273] Kennedy, J. The Particle Swarm: Social adaptation of knowledge, In Proceedings of IEEE International Conference on Evolutionary Computation, Indianapolis, IN, 1997, pp. 303-308, 1997.

[274] Kim, J. H., Myung, H., Jeon, J.Y. Hybrid evolutionary programming with fast convergence for constrained optimization problems, IEEE Conference on Intelligent Systems for the 21 Century, Vancouver, Canada, pp. 3047-3052, 1995.

[275] Koza, J. R. Genetic Programming II: Automatic Discovery of Reusable Programs, MIT Press, Cambridge, MA, 1994.

[276] Koza, J. R., Bennett, F. H., Andre, D., & Keane, M. A. Genetic Programming III: Darwinian Invention and Problem Solving, Morgan Kaufmann, 1999.

[277] Koza, J. R., Keane, M. A., Streeter, M. J., Mydlowec, W., Yu, J., & Lanza, G. Genetic Programming IV: Routine Human-Competitive Machine Intelligence, Kluwer Academic Publishers, 2003.

[278] Lee, M. A., & Takagi, H. Dynamic control of genetic algorithms using fuzzy logic techniques. In: S. Forrest, ED., Proc. of the Fifth Int. Conf. on Genetic Algorithms pp. 76-83, 1993.

[279] Liaw, C. F. A hybrid genetic algorithm for the open shop scheduling problem. European Journal of Operational Research 124 (1), pp. 28-42, 2000.

[280] Lin, Y. C., Hwang, K. S., & Wang, F. S. A mixed-coding scheme of evolutionary algorithms to solve mixed-integer nonlinear programming problems, Computers and Mathematics with Applications, 47(8-9), pp. 1295-1307, 2004.

[281] Lo, C. C., & Chang, W. H. A Multiobjective Hybrid Genetic Algorithm for the Capacitated Multipoint Network Design Problem, IEEE Transactions on Systems, Man and Cybernetics - Part B, 30(3), pp. 461-470, 2000.

[282] Liu, H., Abraham, A., & Zhang, W. A Fuzzy Adaptive Turbulent Particle Swarm Optimization, International Journal of Innovative Computing and Applications, Vol. 1, Issue 1, pp. 39-47, 2007.

[283] Maa, C. Y., & Shanblatt, M. A. A two-phase optimization neural network, IEEE Transaction on Neural Networks, 3(6), pp. 1003-1009, 1992.

[284] Magyar. G., Johnsson. M., & Nevalainen, O. An Adaptive Hybrid Genetic Algorithm for the Three-Matching Problem, IEEE Transactions on Evolutionary Computation, 4(2), pp. 135-146, 2000.

[285] Menon, P. P., Bates, D. G., & Postlethwaite, I. Hybrid Evolutionary Optimisation Methods for the Clearance of Nonlinear Flight Control Laws, In Proceedings of the 44th IEEE Conference on Decision and Control, and the European Control Conference, Seville, Spain, pp. 4053-4058, 2005.

[286] Misevicius, A. Genetic algorithm hybridized with ruin and recreate procedure: Application to the quadratic assignment problem, Knowledge-Based Systems, 16, pp. 261-268, 2003.

[287] Myung, H., & Kim, J. H. Hybrid evolutionary programming for heavily constrained problems Biosystems, 38(1), pp. 29-43, 1996.

[288] Neppalli, V. R., Chen, C. L., & Gupta, J. N. D. Genetic algorithms for the two stage bicriteria owshop problem. European Journal of Operational Research 95, pp. 356-373, 1996.

[289] Von Neumann, J. In: Burks, A. (Ed.), Theory of Self-Reproducing Automata. University of Illinois Press, Champaign, IL, 1966.

[290] Oman, S., Cunningham, P. Using case retrieval to seedgenetic algorithms. International Journal of Computational Intelligence and Applications 1 (1), pp. 71-82, 2001.

[291] Schlottmann, F., & Seese, D. A hybrid heuristic approach to discrete multiobjective optimization of credit portfolios, Computational Statistics and Data Analysis, 47(2), pp. 373-399, 2004.

[292] Shi, X. H., Liang, Y. C., Lee, H, P., Lu, C., & Wang, L. M. An improved GA and a novel PSO-GA-based hybrid algorithm, Information Processing Letters, 93(5), pp. 255-261, 2005.

[293] Sinha, A., & Goldberg, D. E. A Survey of Hybrid Genetic and Evolutionary Algorithms, Technical Report, 2003.

[294] Storn, R., & Price, K. Differential evolution — A simple and efficient adaptive scheme for global optimization over continuous spaces. Journal of Global Optimization, 11(4), pp. 341-359, 1997.

[295] Wang, L., Zhang, L., & Zheng, D. Z. An effective hybrid genetic algorithm for flow shop scheduling with limited buffers, Computers and Operations Research, 33(10), pp. 2960-2971, 2006.

[296] Grosan, C., Abraham, A., & Nicoara, M. Search Optimization Using Hybrid Particle Sub-Swarms and Evolutionary Algorithms, International Journal of Simulation Systems, Science and Technology, UK, Volume 6, Nos. 10 and 11, pp. 60-79, 2005.

[297] Andrews, R., Diederich, J., & Tickle, A. B. A survey and critique of techniques for extracting rules from trained artificial networks. Technical report, Queensland University of Technology, 1995.

[298] Andrews, R. & Geva, S. Rules and local function networks. In Proceedings of the Rule Extraction From Trained Artificial Neural Networks Workshop, Artificial Intelligence and Simulation of Behaviour, Brighton UK, 1996.

[299] Barnden, J. A., & Holyoak, K. J. Advances in connectionist and neural computation theory, volume 3. Ablex Publishing Corporation, 1994.

[300] Benitz, J., Castro, J., & Requena, J. I. Are artificial neural networks black boxes? IEEE Transactions on Neural Networks, 8(5):1156-1164, 1997.

[301] Bogacz, R., & Giraud-Carrier, C. A novel modular neural architecture for rulebased and similarity-based reasoning. In Hybrid Neural Systems (this volume). Springer-Verlag, 2000.

[302] Bologna, G. Symbolic rule extraction from the DIMLP neural network. In Hybrid Neural Systems (this volume). Springer-Verlag, 2000.

[303] Churchland, P. S., & Sejnowski, T. J. The Computational Brain. MIT Press, Cambridge, MA, 1992.

[304] Fu, L.M. Neural Networks in Computer Intelligence. McGraw-Hill, Inc., New York, NY, 1994.

[305] Gallant, S. I. Neural Network Learning and Expert Systems. MIT Press, Cambridge, MA, 1993.

[306] Holldobler, S., Kalinke, Y., & Wunderlich, J. A recursive neural network for reexive reasoning. In Hybrid Neural Systems (this volume). Springer-Verlag, 2000. 307. Honkela, S. Self-organizing maps in symbol processing. In Hybrid Neural Systems (this volume). Springer-Verlag, 2000.

[307] Alba, E., & Tomassini, M. "Parallelism & evolutionary algorithms," IEEE Transactions on Evolutionary Computation, vol. 6, pp. 443-462, 2002.

[308] Bertoluzza, C., Gil, M. A., & Ralescu, D.A. Statistical Modeling, Analysis & Management of Fuzzy Data. Springer-Verlag., 2003.

[309] Coello, C.A., Van Veldhuizen, D.A., & Lamont, G.B. Evolutionary Algorithms for Solving Multi-Objective Problems. Kluwer Academic Publishers, 2002.

[310] Cordon, O., M.J. del Jess, Herrera, F., & Lozano, M. "MOGUL: A Methodology to Obtain Genetic fuzzy rule-based systems Under the iterative rule Learning approach," International Journal of Intelligent Systems, Vol. 14, pp. 1123-1153, 1999.

[311] Cordon, O., Gomide, F., Herrera, F., Hoffmann, F., & Magdalena, L. "Ten years of genetic fuzzy systems: Current framework & new trends," Fuzzy Sets & Systems, Vol. 41, pp. 5-31, 2004.

[312] Cordon, F., Herrera, F., Hoffmann, F., & Magdalena, L.Genetic Fuzzy Systems. Evolutionary Tuning & Learning of Fuzzy Knowledge Bases. World Scientific, 2001.

[313] Deb, K. Multi-Objective Optimization using Evolutionary Algorithms. John Wiley & Sons, 2001

[314] De Jong, K.A., Spears, W.M., Gordon, D.F. "Using genetic algorithms for concept learning," Machine Learning, vol. 13, pp. 161-188, 1993.

[315] Diettereich, T. "Approximate statistical tests for comparing supervised classification learning algorithms," Neural Computation, vol. 10, pp. 1895- 1924, 1998.

[316] Dubois, D., Prade, H., & Sudamp, T., "On the representation, measurement, & discovery of fuzzy associations," IEEE Trans. on Fuzzy Systems, Vol. 13, pp. 250-262, 2005.

[317] Goldberg, D. E. Genetic Algorithms in Search, Optimization, & Machine Learning. Addison-Wesley, 1989.

[318] Herrera, F., Lozano, M., & Verdegay, J.L. "A Learning Process for Fuzzy Control Rules using Genetic Algorithms," Fuzzy Sets & Systems, Vol. 100, 143- 151, 1998.

[319] Konar, A. Computational Intelligence: Principles, Techniques & Applications. Springer-Verlag, 2005.

[320] Kweku-Muata, Osey-Bryson, "Evaluation of decision trees: a multicriteria approach," Computers & Operations Research, Vol. 31, pp. 1933-1945, 2004.

[321] Yager, R.R., & Filev, D.P. Essentials of Fuzzy Modeling & Control, John Wiley & Sons, 1994.

[322] Wang, H., Kwong, S., Jin, Y., Wei, W., & Man, K.F. "Multiobjective hierarchical genetic algorithm for interpretable fuzzy rule-based knowledge extraction," Fuzzy Sets & Systems, vol. 149, pp. 149-186, 2005.

[323] Wong, M.L., & Leung, K.S. Data Mining Using Grammar-Based Genetic Programming & Applications. Kluwer Academics Publishers, 2000.

[324] Bhatt, R. B. & Gopal, M. "On fuzzy-rough sets approach to feature selection," Pattern Recognition Letters, Vol. 26, No. 7, pp. 965-975, 2005.

[325] Bian, H., & Mazlack, L. "Fuzzy-Rough Nearest-Neighbor Classification Approach," Proceeding of the 22nd International Conference of the North American Fuzzy Information Processing Society (NAFIPS), pp. 500-505, 2003.

[326] Boixader, D., Jacas, J., & Recasens, J. "Upper & lower approximations of fuzzy sets," International Journal of General Systems, Vol. 29, No. 4, pp. 555-568, 2000.

[327] Chen, W.C., Chang, N.B., & Chen, J.C. "Rough set-based hybrid fuzzy-neural controller design for industrial wastewater treatment," Water Research, Vol. 37, No. 1, pp. 95-107, 2003.

[328] Chen, D., Zhang, W.X., Yeung, D., & Tsang, E.C.C. "Rough approximations on a complete completely distributive lattice with applications to generalized rough sets," Information Sciences, Vol. 176, No. 13, pp. 1829-1848, 2006.

[329] Hirano, S., & Tsumoto, S. "Rough Clustering & Its Application to Medicine," Journal of Information Science, Vol. 124, pp. 125-137, 2000.

[330] Hoppner, F., Kruse, R., Klawonn, F., & Runkler, T. Fuzzy Cluster Analysis: Methods for Classification, Data Analysis & Image Recognition, John Wiley & Sons, 2000.

[331] Jensen, R., & Shen, Q. "Fuzzy-Rough Data Reduction with Ant Colony Optimization," Fuzzy Sets & Systems, Vol. 149, No. 1, pp. 5-20, 2005.

[332] Jensen, R., & Shen, Q. "Fuzzy-Rough Feature Significance for Fuzzy Decision Trees," Proc. 2005 UK Workshop on Computational Intelligence, pp. 89-96, 2005.

[333] Kuncheva, L.I. "Fuzzy rough sets: application to feature selection," Fuzzy Sets & Systems, Vol. 51, No. 2, pp. 147-153, 1992.

[334] Lingras, P. "Unsupervised Learning Using Rough Kohonen Neural Network Classifiers," Proc. Symposium on Modelling, Analysis & Simulation, CESA'96 IMACS Multiconference, pp. 753-757, 1996.

[335] Lingras, P., & Davies, C. "Applications of Rough Genetic Algorithms," Computational Intelligence, Vol. 17, No. 3, pp. 435-445, 2001.

[336] Pawlak, Z. "Rough Sets," International Journal of Information & Computer Sciences Vol. 11, pp. 145-172, 1982.

[337] Pawlak, Z. Rough Sets: Theoretical Aspects of Reasoning about Data. Kluwer Academic Publishing, 1991.

[338] Qina, K., & Pei, Z. "On the topological properties of fuzzy rough sets," Fuzzy Sets & Systems, Vol. 151, No. 3, pp. 601-613, 2005.

[339] Radzikowska, A.M., & Kerre, E.E. "A comparative study of fuzzy rough sets," Fuzzy Sets & Systems, Vol. 126, No. 2, pp. 137-155, 2002.

[340] Wang, X.Z., Ha, Y., & Chen, D. "On the reduction of fuzzy rough sets," Proc. 2005 International Conference on Machine Learning & Cybernetics, Vol. 5, pp. 3174-3178, 2005.

[341] Sumathi, S., Deepa, S.N., & Sivanandham, S.N. "Introduction to Neural Networks Using MATLAB 6.0", Tata McGraw-Hill Company Ltd, New Delhi, June 2005.

[342] Sumathi, S., Deepa, S.N., & Sivanandham, S.N. "Introduction to Fuzzy Logic Using MATLAB", Springer-Verlag, Germany, 2006.

[343] Kosko, B. Neural Networks & fuzzy Systems: A Dynamical System Approach to Machine Intelligence, Prentice Hall, 1991

[344] Goldberg, D.E. "Messy genetic algorithms: motivation, analysis, & first results", Complex Systems, 3, 493-530, 1989.

[345] Chowdhury, M., & Li, Y. "Messy genetic algorithm–based new learning method for structurally optimized neuro-fuzzy controllers", Proc. IEEE Int. Conf. Industrial Technology, Shanghai, China, 274-279, 1996

[346] Bonarini, A. "Delayed reinforcement, fuzzy Q-learning & fuzzy logic controllers", in Genetic Algorithms & Soft Computing, Herrera, F. & Verdegay, J. L. (EDS.), Physica-Verlag., 1996

[347] Li, Y. Neural & Evolutionary Computing, Fourth-Year Course Notes (2RPX), Department of Electronics & Electrical Engineering, University of Glasgow, 1995

[348] Koza, J.R. 1994. Genetic Programming II: Automatic Discovery of Reusable Programs. The MIT Press.

[349] Koza, J.R., Bennett, H. B., Andre, D., & Keane, M. A. 1999. Genetic Programming III: Darwinian Invention & Problem Solving. Morgan Kaufmann Publishers.

[350] Alpigini, James J., "Rough Sets & Current Trends in Computing: Third International", Springer, 2002

[351] Alba, E., & Cotta, C. Online Tutorial on Evolutionary Computation, Advances in Soft Computing, Roy R., Furuhashi T., Chawdhry P.K. (eds.), pp. 603-604, Springer-Verlag, London, 1999

[352] Porwal, Alok, Carranza E. J. M., & Hale, M. "A Hybrid Neuro-Fuzzy Model for Mineral Potential Mapping", Springer Netherlands, 2005.

[353] Thierens, Dirk, & Goldberg, David E. "Convergence Models of Genetic Algorithm Selection Schemes", Springer-Verlag, London, 1994.

[354] Lewis, Roger J. "An Introduction to Classification and Regression Tree (CART) Analysis", Department of Emergency Medicine Harbor-UCLA Medical Center Torrance, California.

[355] Timofeev, Roman. Thesis on "Classification and Regression Trees (CART) Theory and Applications" CASE - Center of Applied Statistics and Economics Humboldt University, Berlin.

[356] Haskell, Richard E. "Neuro-Fuzzy Classification and Regression Trees", Computer Science and Engineering Department, Oakland University, Rochester, Michigan, U. S. A.

[357] Lynch, Stephen, "Dynamical Systems with Applications Using Mathematica", Birkhuser, Boston (2007). ISBN 0-8176-4482-2.

[358] Lynch, Stephen. "MATLAB Code on Hopfield and BPN networks", Birkhuser, Boston (2007). ISBN 0-8176-4482-2.

[359] Weng, Juyang. "Regression Trees", Department of Computer Science and Engineering Michigan State University East Lansing, MI.

[360] Menczer, Filippo, Street, W. Nick, & Degeratu, Melania "Evolving Heterogeneous Neural Agents by Local Selection", MIT Press.

Internet Sources

References

[1] http://upetd.up.ac.za/thesis/available/etd-06102005-095510/unrestricted/02chapter2.pdf

[2] http://www.bmf.hu/conferences/sisy2006/32_Saletic.pdf

[3] http://cogprints.org/5358/1/06-CIdef.pdf

[4] http://www.cs.bham.ac.uk/∼ jxb/PUBS/IR.pdf

[5] http://www.iau.dtu.dk/∼jj/pubs/logic.pdf

[6] http://icosym.cvut.cz/dynlabmodules/ihtml/dynlabmodules/syscontrol/main.html

[7] http://members.aol.com/btluke/fuzzy01.htm

[8] http://media.wiley.com/product_data/excerpt/73/07803349/0780334973.pdf

[9] http://www.comp.nus.edu.sg/~ pris/FuzzyLogic/ModelDetailed1.html

[10] http://egweb.mines.edu/faculty/msimoes/tutorials/Introduction_fuzzy_logic/Intro_Fuzzy_Logic.pdf

[11] http://citeseer.ist.psu.edu/cache/papers/cs2/121/

[12] http:zSzzSzfalklands.globat.comzSz

[13] http://arxiv.org/ftp/cs/papers/0405/0405011.pdf

[14] http://www.fmi.uni-sofia.bg/fmi/statist/education/textbook/ENG/stcart.html

[15] http://www.iau.dtu.dk/~ jj/pubs/nfmod.pdf

[16] http://www.cs.rtu.lv/dssg/download/publications/2002/Grabusts-RA-2002.pdf

[17] http://ieeexplore.ieee.org/iel5/72/18371/00846746.pdf

[18] http://www.geocities.com/francorbusetti/saweb.pdf

[19] http://www.rzg.mpg.de/~rfs/comas/Helsinki/helsinki04/CompScience/csep/csep1.phy.ornl.gov/mo/node32.html

[20] http://www.idi.ntnu.no/emner/it3708/lectures/evolalgs.pdf

[21] http://www.faqs.org/faqs/ai-faq/genetic/part1/preamble.html

[22] http://members.aol.com/btluke/codschm.htm

[23] http://openmap.bbn.com/~ thussain/publications/1998_cito98tutorial.pdf

[24] http://www.cs.cmu.edu/Groups/AI/html/faqs/ai/genetic/part2/faq-doc-3.html

[25] http://web.cecs.pdx.edu/~mm/cs410-510-winter-2005/ga-tutorial.pdf

[26] http://www.ceng.metu.edu.tr/~ucoluk/darwin/node10.html

[27] http://www.lcc.uma.es/~ccottap/semEC/cap02/cap_2.html

[28] www.cs.bgu.ac.il/~sipper/ga.html

[29] http://www.spaceandmotion.com/Charles-Darwin-Theory-Evolution.htm

[30] http://www.complexity.org.au/ci/vol04/thornton/building.htm

[31] http://library.thinkquest.org/18242/ga_math.shtml

[32] http://www.genetic-programming.com/gppreparatory.html

[33] http://www.generation5.org/content/2001/ga_math.asp

[34] http://www.cimms.ou.edu/~lakshman/Papers/ga/node8.html

[35] http://evonet.lri.fr/CIRCUS2/node.php?node=161

[36] http://www.massey.ac.nz/~mgwalker/gp/intro/introduction_to_gp.pdf

[37] http://www.ece.osu.edu/~passino/swarms.pdf

[38] http://www.mathworks.com/products/fuzzylogic/description1.html

[39] http://gplab.sourceforge.net/

[40] http://www.nd.com/products/genetic/mutation.htm

[41] http://www.scholarpedia.org/article/Swarm_Intelligence

[42] http://evonet.lri.fr/evoweb/resources/software/record.php?id=15

[43] http://www.mathworks.com/products/fuzzylogic/description1.html

[44] http://www.nd.com/products/genetic/mutation.htm

[45] http://www.mathworks.de/matlabcentral

[46] http://www.ece.osu.edu/~passino/ICbook/ic_code.html

[47] http://www.softcomputing.net/publication.html

[48] http://www.psychology.mcmaster.ca/3W03/sigproc.html

[49] https://www.dacs.dtic.mil/techs/neural/neural11.php

[50] http://www.nd.com/products/genetic/mutation.htm

[51] www.mathworks.de/matlabcentral

[52] http://www.ece.osu.edu/~passino/ICbook/ic_code.html

[53] www.geneticprogramming.com/Tutorial/

[54] http://www.waset.org/ijcs/v1/v1-1-3.pdf

[55] http://ww2.lafayette.edu/~reiterc/mvp/fuzz_auto/fuzzlife_4d_pp.pdf

[56] ieeexplore.ieee.org/book/0780334817.excerpt.pdf

[57] Mustafa UCAK, "MATLAB Code on Coin Detection", www.math works.com

[58] www.cs.umt.edu/CS/COURSES/CS555/lect9_3.pdf

[59] http://www.edc.ncl.ac.uk/highlight/rhjanuary2007g02.php/

[60] http://www.talkorigins.org/faqs/genalg/genalg.html

[61] Rakesh Patel, "GUI based Fuzzy Washing Machine", www.math works.com

[62] Leo Chen, "A Fuzzy Controller in MATLAB", www.mathworks. com

[63] Kyriakos Tsourapas, "Ant Algorithm for the Quadratic Assignment Problem", www.mathworks.com

[64] http://en.wikipedia.org/wiki/Holland's_schema_theorem

[65] http://neo.lcc.uma.es/TutorialEA/semEC/cap06/cap_6.html

[66] http://www.genetic-programming.com/gppreparatory.html

[67] http://www.geniqmodel.com/Koza_GPs.html

[68] http://www.cis.syr.edu/m̃ohan/pso/

[69] Emre Akbas, "Simplified Fuzzy Art Map", www.mathworks.com

[70] staff.washington.edu/paymana/papers/globecom01_2.pdf

[71] http://www.aaai.org/Papers/AAAI/2008/AAAI08-004.pdf

[72] http://news.bbc.co.uk/1/hi/technology/7549059.stm

[73] Guanglei Xiong, "Comparative Case Study on Fuzzy c-means clustering", www.mathworks.com

[74] http://www.itl.nist.gov/div898/handbook/

[75] http://math.fullerton.edu/mathews/n2003/GradientSearchMod.html

[76] Aravind Seshadri, "Multi-objective optimization using Evolutionary Algorithm", www.mathworks.com

[77] http://www.dtreg.com/crossvalidation.htm

[78] www.esr.ruhr-uni-bochum.de/rt1/syscontrol/node119.html

[79] http://www.nd.com/products/genetic/mutation.htm

[80] http://people.msoe.edu/b̃lessing/cs470/Fuzzification.pdf

[81] Saloman Danaraj, "Solving economic dispatch problem using Genetic Algorithm", www.mathworks.com

[82] Wesam Elshamy, "Simulation of the movement of swarm to minimize the objective functions", www.mathworks.com

[83] Sahba Jahromi "Ant Colony Optimization to determine the shortest path", www.mathworks.com

[84] http://www.ece.osu.edu/ passino/

[85] Brian Birge, "Behavior of Particle Swarm Optimization", www.mathworks.com

[86] Nikolaus Hansen, "Evolutionary Strategy for Non linear function minimization", http://www.lri.fr/hansen/cmaes_inmatlab.html

[87] Heikki Koivo, "NeuroFuzzy System Using Simulink", http://profile.iiita.ac.in/rkkarne_b03/SPEECH%20RECOGNITION %20BY%20B2003107/papers/FVAnfis2.pdf

[88] Kenneth Price & Rainer Storn, "Differential Evolutionary Optimizer", http://www.icsi.berkeley.edu/storn/code.html

[89] Janos Madar, "System Identification", University of Veszprem, 2005, http://www.fmt.vein.hu/softcomp/publications.html

[90] Sara, "Symbolic Regression problem using Genetic Programming Toolbox", prdownloads.sourceforge.net/gplab/gplab.manual.1.1.pdfdownload.

Index

Printed and bound by CPI Group (UK) Ltd, Croydon, CR0 4YY

23/10/2024

01777671-0020